Statistical Physics:
Equilibrium and Non-Equilibrium Aspects

Statistical Physics :
Equilibrium and Non-Equilibrium Aspects

J.K. Bhattacharjee

ALLIED PUBLISHERS LIMITED
NEW DELHI MUMBAI KOLKATA CHENNAI NAGPUR
AHMEDABAD BANGALORE HYDERABAD LUCKNOW

ALLIED PUBLISHERS PRIVATE LIMITED

1/13-14 Asaf Ali Road, **New Delhi**–110002
Ph.: 011-23239001 • E-mail: delhi.books@alliedpublishers.com

87/4, Chander Nagar, Alambagh, **Lucknow** - 226005
Ph.: 0522-4012850 • E-mail: appltdlko9@gmail.com

17 Chittaranjan Avenue, **Kolkata**–700072
Ph.: 033-22129618 • E-mail: cal.books@alliedpublishers.com

15 J.N. Heredia Marg, Ballard Estate, **Mumbai**–400001
Ph.: 022-42126969 • E-mail: mumbai.books@alliedpublishers.com

60 Shiv Sunder Apartments (Ground Floor), Central Bazar Road,
Bajaj Nagar, **Nagpur**–440010
Ph.: 0712-2234210 • E-mail: ngp.books@alliedpublishers.com

F-1 Sun House (First Floor), C.G. Road, Navrangpura,
Ellisbridge P.O., **Ahmedabad**–380006
Ph.: 079-26465916 • E-mail: ahmbd.books@alliedpublishers.com

751 Anna Salai, **Chennai**–600002
Ph.: 044-28523938 • E-mail: chennai.books@alliedpublishers.com

The Hebbar Sreevaishnava Sabha, Sudarshan Complex - 2
No. 22, Seshadri Road, **Bangalore** - 560009
Ph.: 080-22262081 • E-Mail: bngl.books@alliedpublishers.com

3-2-844/6 & 7 Kachiguda Station Road, **Hyderabad**–500027
Ph.: 040-24619079 • E-mail: hyd.books@alliedpublishers.com

Website: www.alliedpublishers.com

Edition, 2016

ISBN 81-7764-036-4

SPONSORING EDITOR	:	JYOTI MEHROTRA
COPYEDITOR	:	D.S. SHASTRI
DTP	:	S.J.I. SERVICES
COVER	:	KAM STUDIO

Published by Sunil Sachdev and printed by Ravi Sachdev at Allied Publishers Limited, Printing Division, A-104 Mayapuri, Phase-II, New Delhi - 110 064

TO SHAYAK

Acknowledgements

The financial support for the preparation of the manuscript was provided by IIT, Kanpur. The satisfaction of teaching the undergraduates at the IIT, Kanpur provided the drive to write. Support in the form of arranging summer and semester visits where I could be relieved of teaching and administrative duties have been provided by Peter G.J. Lucas of the University of Manchester, England and Richard A. Ferrell of the University of Maryland, USA. Some of the chapters had been read very carefully by Sashikant Upadhyay when he was a final year undergraduate. I am grateful for his comments and corrections. Finally my sincere thanks to Mr. Suresh Gopal and the editorial team at Allied Publishers who have been very patient and most helpful all along.

Preface

This book developed primarily from the statistical mechanics courses that I have taught at the IIT, Kanpur over the past few years. The first part deals with equilibrium statistical mechanics and is tailored to meet the needs of the standard statistical mechanics courses offered at the Master's level in Indian universities. In this context, the contents of the second chapter which recalls some developments in nonlinear dynamics in the last two decades and their possible relevance to the basic tenets of statistical mechanics may seem too short and sketchy. However, a fair number of excellent texts and monographs on the subject exist and the instructor will not have any difficulty in expanding on the material of Chapter Two. We have left out discussion of the renormalization group from Part I of the book. Mean field theory of critical phenomena has been done quite extensively for both the Van der Waal's gas and the Ising model. If there is time to discuss the role of fluctuations, then chapters 20-23 (Part-III) provide a fairly detailed introduction.

In Part II of the book, the approach to equilibrium has been discussed. Along with a selection of topics from Part III thus can form the basis for an advanced course in statistical mechanics. In trying to decide what to include in Part III, we never really evolved a satisfactory criterion. Turbulence has been included but discussion of KPZ and related models have been left out. Polymers have been excluded. The excellent text 'Scaling Concepts in Polymer Physics' makes any attempt to write a short self contained chapter almost redundant.

Roughening and Crumpling too have been virtually left out. Some elementary discussion of random systems has been included. The decision to exclude any discussion of the Hubbard model, statistical mechanics of anyon etc. is probably unwise but at some point the line had to be drawn. Correlated electron systems having been left out, and no discussion of high temperature superconductivity has been possible.

Contents

Chapter One

Recalling Thermodynamics

1.1 Introduction

In this chapter we will briefly recall the formulation of equilibrium thermodynamics. Our main purpose will be to introduce definitions and terminologies for the macroscopic characterization of a system in equilibrium (that is a system not changing with time) and not dwell at length on the science of thermodynamics. Any macroscopic system (number of molecules in such a system is of the order of the Avogadro number) will be termed a thermodynamic system. The measurable parameters of such a system (pressure, volume, temperature, magnetization, etc.) are called thermodynamic parameters and the thermodynamic state is specified by specifying the values of the required number of thermodynamic parameters. Not all parameters are independent and the functional relationship among thermodynamic parameters in the equilibrium state is called the equation of state. For a gas in thermal and mechanical equilibrium, the macroscopic parameters are the pressure P, the volume V, the temperature T and the number of molecules N. The equation of state is a functional relation between P, T and V/N. We note that the pressure and temperature are intensive quantities in that if two systems with equal values of these parameters are put in contact, the final pressure and temperature remain the same. On the other hand, the volume V and the number N are extensive quantities in that if two systems are put in contact these quantities simply add.

The ratio V/N which is the volume per molecule is, quite clearly, an intensive quantity. Intensive quantities are independent of the amount of material, while extensive quantities are directly proportional to it.

A thermodynamic transformation is a change in the thermodynamic state of a system and can be effected by varying the external conditions. If the variation occurs slowly the process is quasistatic and if the transformation reverses itself when the external conditions are reversed, then the process is reversible. A reversible process occurs so slowly that equilibrium is maintained at every stage of development.

The concept of work is identical to that in mechanics. If a gas at pressure P undergoes a change dV in its volume, then the work done by it is PdV. Heat is what the system absorbs (or emits) if its temperature changes without any work being done. A system is thermally isolated if it cannot exchange heat with its surroundings. Processes occurring in a system in thermal isolation are called adiabatic. A heat reservoir is a source (or sink) of heat at constant temperature whose temperature cannot change on withdrawal or addition of heat.

1.2 First Law of Thermodynamics

If an amount of heat ΔQ is added to a system which performs in the process an amount of work ΔW, then the first law of thermodynamics asserts that

$$\Delta E = \Delta Q - \Delta W \qquad \text{... (1.2.1)}$$

depends only on the initial and final states of the system. This defines an internal energy E for the system to within an additive constant since Eq. (1.2.1) does not specify the absolute value of E. The additive constant corresponds to the value of E for a chosen reference state. The internal energy is an extensive quantity which follows from the saturation property of the molecular forces.

In an infinitesimal process, we have

$$dE = dQ - dW, \qquad \text{... (1.2.2)}$$

and from the definition of E, dE is a perfect differential, a property not shared by dQ and dW. If the system under consideration is a gas characterized by its pressure P, volume V and temperature T, then the internal energy will be a function of any two of the above quantities since the equation of state provides a relation between P, V and T. Choosing any two of P, V, and T at a time, we have the three possibilties

$$dE = \left(\frac{\partial E}{\partial T}\right)_V dT + \left(\frac{\partial E}{\partial V}\right)_T dV \qquad \text{... (1.2.3a)}$$

$$dE = \left(\frac{\partial E}{\partial T}\right)_P dT + \left(\frac{\partial E}{\partial P}\right)_T dP \qquad \text{... (1.2.3b)}$$

$$dE = \left(\frac{\partial E}{\partial V}\right)_P dV + \left(\frac{\partial E}{\partial P}\right)_V dP \qquad \text{... (1.2.3c)}$$

The specific heat is defined as $C = \underset{\Delta T \to 0}{\text{Lt}} \frac{\Delta Q}{\Delta T}$ and clearly has to be determined by the nature of the process (constant volume or constant pressure). For the constant volume specific heat

$$C_v = \left(\frac{\partial Q}{\partial T}\right)_V \qquad \text{... (1.2.4)}$$

Using the fact that $dW = PdV$, we note

$$dQ = \left(\frac{\partial E}{\partial T}\right)_V dT + \left(P + \frac{\partial E}{\partial V}\right)_T dV \qquad \text{... (1.2.5)}$$

and hence

$$C_v = \left(\frac{\partial E}{\partial T}\right)_V \qquad \text{... (1.2.6)}$$

To tackle the constant pressure specific heat we note

$$dQ = \left(\frac{\partial E}{\partial T}\right)_P dT + \left(\frac{\partial E}{\partial P}\right)_T dP + PdV$$

$$= \left[\left(\frac{\partial E}{\partial T}\right)_P + P\left(\frac{\partial V}{\partial T}\right)_P\right] dT + \left[\left(\frac{\partial E}{\partial P}\right)_T + P\left(\frac{\partial V}{\partial P}\right)_T\right] dP$$

and hence

$$C_P = \left(\frac{\partial Q}{\partial T}\right)_P = \left(\frac{\partial E}{\partial T}\right)_P + P\left(\frac{\partial V}{\partial T}\right)_P$$

$$= \frac{\partial}{\partial T}(E ++PV)\Big)_P = \left(\frac{\partial H}{\partial T}\right)_P \quad \text{... (1.2.7)}$$

where $H = E + PV$ is the enthalpy of the system.

At this point we introduce the concept of an ideal gas in thermodynamics. An ideal gas is defined to have the equation of state

$$PV = NkT \quad \text{... (1.2.8)}$$

From our considerations in Chapter 11, we will see that for the molecules in an ideal gas the only interaction is an elastic collision between two molecules. For the internal energy of an ideal gas, one has to consider the experimental observation on the free expansion of an ideal gas. A free expansion is the escape of an ideal gas from a volume V into vacuum. There is no pressure opposing the expansion and hence the work done is zero. The experimental observation is that the temperature of the gas does not change in the process. This implies that $\Delta Q = 0$ in the process since any ΔQ if present would eit cause work to be done or change the temperature. Both being a $\Delta Q = 0$. With $\Delta Q = \Delta W = 0$, the first law implies that $\Delta E = 0$. Since e volume of the gas has changed, but not its temperature, we can conclude that the internal energy E is not a function of the volume V. We can now integrate Eq. (1.2.6) to write for an ideal gas

$$E = C_v T + \text{constant} \quad \text{... (1.2.9)}$$

From Eq. (1.2.7), we obtain

$$C_P = \frac{\partial}{\partial T}(E + PV)\,|_P = C_v + P\left(\frac{\partial V}{\partial T}\right)_P$$

$$= C_v + N_k \quad \text{... (1.2.10)}$$

For one mole of the gas $kN = R$ and thus the difference of molar specific heats at constant pressure and constant volume is given by

$$C_P - C_v = R \quad \text{... (1.2.11)}$$

1.3 Second Law of Thermodynamics

There are two equivalent formulations for the second law of thermodynamics. The one due to Kelvin asserts that a process whose only final result is to extract heat from a reservoir and convert it completely into work is impossible. The formulation of Clausius states that a process whose sole outcome is to extract heat from a reservoir at a temperature T_1 and deliver it to a reservoir at a temperature $T_2 > T_1$ is impossible. To show that they are equivalent statements we need to show that if one is violated, so is the other. To begin with, let us assume that the Kelvin statement is false and it is possible to extract heat from a reservoir at temperature T_1 and convert it completely into work. It is then possible to convert this work completely into heat and deliver it to a heat reservoir at any temperature T_2. In particular if $T_2 > T_1$, we have engineered a situation which contradicts the Clausius formulation.

To show that a violation of the Clausius formulation leads to a negation of the Kelvin statement we define an engine as a device which operates in a cycle (initial and final states are identical) in the following steps:

i) extracts heat Q_1 from a reservoir at temperature T_1
ii) deposits heat Q_2 to a reservoir at temperature $T_2 < T_1$
iii) performs an amount of work $W = Q_1 - Q_2$.

The engine is said to operate between temperatures T_1 and T_2.

We now assume that the Clausius statement is violated and we can extract an amount of heat Q_1 from a reservoir at temperature T_2 and transfer it completely to a reservoir at temperature $T_1 > T_2$. Now operate an engine between T_1 and T_2 with the steps (i), (ii) and (iii) described above. The result is that an amount of heat $Q_1 - Q_2$ is extracted from the reservoir at temperature T_2 and converted completely into work—a contradiction of Kelvin's statement.

At this point we are ready to introduce the concept of a Carnot engine. It is designed to perform the functions listed in (i)–(iii) above in a reversible manner. The Carnot engine consists of four steps.

a) An isothermal expansion when the system absorbs an amount Q_1 of heat from a reservoir at temperature T_1.
b) An adiabatic expansion where work is done by the system.
c) An isothermal contraction where the system rejects an amount of heat Q_2 to a reservoir at a temperature $T_2 < T_1$.
d) An adiabatic contraction where work is done on the system to bring the system back to the starting point.

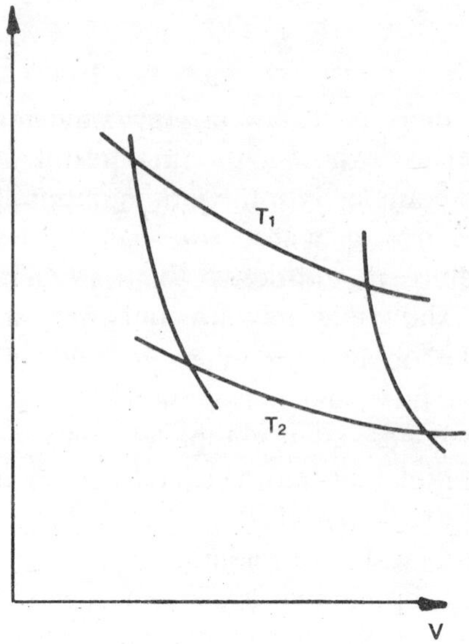

Fig. 1.1: Carnot cycle

The schematic arrangement is shown in Fig. 1.1.

In a cyclic process there is no change in internal energy and hence $Q_1 - Q_2 = W$, where W is the work done by the system. We now show that if $W > 0$ then Q_2 and Q_1 must be positive. Clearly, we must have $Q_2 \neq 0$ as otherwise an immediate violation of the Kelvin statement occurs. Now assume $Q_2 < 0$. The system then absorbs an amount of heat Q_1 from the reservoir at temperature T_1 and an amount Q_2 from the reservoir at temperature T_2. The work done is $Q_1 - Q_2$ and this work can be completely converted into heat and delivered to the reservoir at temperature T_1. Since Q_1 had been drawn from this reservoir, the net effect is a delivering of an amount $- Q_2$ of heat to this reservoir. The process that we have just described leads to this: An amount of heat $- Q_2$ is absorbed from the reservoir at temperature T_2 and delivered to the reservoir at temperature T_1. But $T_1 > T_2$ and hence for $Q_2 < 0$ this violates the Clausius form of the second law. If W and $Q_2 > 0$, then $Q_1 = Q_2 + W > 0$. Similarly, if $W < 0$, then $Q_2 < 0$ and $Q_1 < 0$ and what we have is a refrigerator. The efficiency of an engine is defined as $\eta = W/Q_1 = 1 - Q_2/Q_1$.

What makes the Carnot engine important is the fact that it is the most efficient engine operating between two given temperatures T_1 and T_2. The proof is as follows:

The Carnot engine C absorbs an amount of heat Q_1 from the reservoir at T_1 and delivers Q_2 to the reservoir at T_2 performing an amount of work $W = Q_1 - Q_2$. An arbitrary engine X absorbs Q'_1 from T_1, delivers Q'_2 to T_2 and performs an amount of work $W' = Q'_1 - Q'_2$. We now perform a process where X operates between T_1 and T_2 and C operates in reverse. The process consists of N_2 operations of X and N_1 reverse operations of C, such that

$$\frac{Q_1}{Q'_1} = \frac{N_2}{N_1} \qquad \text{... (1.3.1)}$$

which can always be done if N_1 and N_2 are large enough. We now have a process where the amount of heat extracted from the reservoir T_2 is

$$Q = N_1 Q_2 - N_2 Q'_2 \qquad \text{... (1.3.2)}$$

The total work done is

$$\begin{aligned} W &= N_2 \left(Q'_1 - Q'_2\right) - N_1 \left(Q_1 - Q_2\right) \\ &= N_1 Q_2 - N_2 Q'_2 = Q \end{aligned} \qquad \text{... (1.3.3)}$$

Thus we would have heat extracted from a reservoir and completely converted into work violating Kelvin's form of the second law if $W > 0$. Hence we must have $W \leq 0$. This leads to

$$N_1 Q_2 \leq N_2 Q'_2$$

$$\frac{Q_2}{Q'_2} \leq \frac{N_2}{N_1} = \frac{Q_1}{Q'_1}$$

$$\frac{Q_2}{Q_1} \leq \frac{Q'_2}{Q'_1}$$

i.e.

$$\eta_C = 1 - \frac{Q_2}{Q_1} \geq 1 - \frac{Q_2'}{Q_1'} = \eta_X \qquad \text{... (1.3.4)}$$

Hence the efficiency of X cannot exceed that of C. If X is another Carnot engine, the equality holds and we have the result that all Carnot engines have the same efficiency when operating between the same temperatures.

This property of the Carnot engine can be used to define an absolute temperature scale. It is done by defining the absolute temperatures of the two reservoirs as θ_1 and θ_2 (where $\theta_1 > \theta_2$) such that

$$\frac{\theta_2}{\theta_1} = 1 - \eta \qquad \text{... (1.3.5)}$$

where η is the efficiency of the Carnot engine. We now imagine a series of Carnot engines, performing the same amount of work and operating such that the heat rejected by one is the heat absorbed by the next. For the n th engine the heat absorbed is Q_n and the heat rejected is Q_{n+1}, so that

$$W = Q_n - Q_{n+1} \qquad \text{... (1.3.6)}$$

The $(n + 1)$ th engine absorbs Q_{n+1}, and discards Q_{n+2}, so that

$$W = Q_{n+2} - Q_{n+1} = Q_n - Q_{n-1} = \ldots \qquad \text{... (1.3.7)}$$

For the n th Carnot engine, if θ_n and θ_{n+1} are the temperatures of the hot and cold reservoirs respectively, then the efficiency is $1 - \frac{\theta_{n+1}}{\theta_n}$ and we get

$$1 - \frac{\theta_{n+1}}{\theta_n} = \frac{W}{Q_n} = 1 - \frac{Q_{n+1}}{Q_n}$$

This leads to

$$\frac{\theta_n}{Q_n} = \frac{\theta_{n+1}}{Q_{n+1}} = x \text{ (Independent of } n) \qquad \text{... (1.3.8)}$$

and hence

$$\theta_n - \theta_{n+1} = x\left(Q_n - Q_{n+1}\right) = x\,W, \qquad \text{... (1.3.9)}$$

which is independent of n and thus provides the equal intervals for a temperature scale with the choice $x\,W = 1$. It should be noted that the scale is indepedent of the specific material used in the cycle. Further, we see from Eq. (1.3.8) that θ cannot be equal to or smaller than 0. If we use an ideal gas as the operating substance, then it can be shown that the ideal gas temperature T is the same as the absolute temperature θ above. This is seen as follows:

For the isothermal expansion process at temperature T_1, the amount of heat absorbed Q_1 is clearly ($dU = 0$ as $dT = 0$)

$$Q_1 = \int_{V_1}^{V_2} P dV = \int_{V_1}^{V_2} \frac{RT_1}{V}\, dV = RT_1 \ln \frac{V_2}{V_1} \qquad \text{... (1.3.10)}$$

(we have taken one mole of the gas).
For the isothermal contraction at temperature T_2, the heat lost Q_2 is

$$Q_2 = \int_{V_4}^{V_3} PdV = RT_2 \ln \frac{V_3}{V_4} \qquad \text{.. (1.3.11)}$$

The adiabatic expansion from V_2 to V_3 gives

$$0 = \int_{T_1}^{T_2} C_v \frac{dT}{T} + R \int_{V_2}^{V_3} \frac{dV}{V} \qquad \text{... (1.3.12)}$$

leading to

$$R \ln \frac{V_3}{V_2} = C_v \ln \frac{T_1}{T_2} \qquad \text{... (1.3.13)}$$

The adiabatic contraction from V_4 to V_1 leads to

$$0 = \int_{T_2}^{T_1} C_v \frac{dT}{T} + R \int_{V_4}^{V_1} \frac{dV}{V}$$

yielding

$$R \ln \frac{V_4}{V_1} = C_v \ln \frac{T_1}{T_2} \qquad \text{... (1.3.14)}$$

Comparison of Eqs. (1.3.13) and (1.3.14) shows

$$\frac{V_4}{V_1} = \frac{V_3}{V_2} \text{ or } \frac{V_2}{V_1} = \frac{V_3}{V_4} \qquad \text{... (1.3.14a)}$$

yielding from Eqs. (1.3.10) and (1.3.11)

$$\frac{Q_1}{Q_2} = \frac{T_1}{T_2} \qquad \text{... (1.3.15)}$$

Hence the ideal gas temperature scale and the absolute scale introduced above are identical.

1.4 Entropy

In this section we explore the possibility of associating a perfect differential with the infinitesimal amount of heat transfer dQ. This is done by proving the following theorem:

In a cyclic process O where the temperature is always defined,

$$\oint \frac{dQ}{T} \leq 0$$

To prove it, we break up the cycle O into N infinitesimal parts and associate a temperature T_i with the i th part. If Q_i is the heat absorbed by the i th part, then it suffices to show $\sum_{i=1}^{N} \frac{Q_i}{T_i} \leq 0$. The theorem follows as $N \longrightarrow \infty$.

We imagine N Carnot engines operating between each of the N parts of the system and a reservoir at temperature T_0 ($\geq T_i$ for all i). The i th Carnot engine operating between T_0 and T_i draws an amount $Q_0^{(i)}$ of heat from T_0 and deposits an amount Q_i to T_i. From Eq. (1.3.8), we have

$$Q_0^{(i)} = Q_i \, T_0/T_i \qquad \text{... (1.4.1)}$$

We now consider an operation consisting of

i) A cycle of each of the N Carnot engines

ii) A cycle of the process O

The net result of the combined operation of (i) and (ii) is to draw an amount of heat $\sum_i Q_0^{(i)} = T_0 \sum (Q_i/T_i)$ from the reservoir at temperature T_0 and convert it entirely into work. This is impossible unless $\sum_i Q_0^{(i)} \leq 0$ and hence $\sum_i Q_i/T_i \leq 0$.

If the cycle O is a reversible cycle, we can run all the operations in the reverse and get $\sum Q_i/T_i \geq 0$. Hence for a reversible cycle both $\sum_i Q_i/T_i \leq 0$ and $\sum_i Q_i/T_i \geq 0$ needs to be satisfied and thus $\sum Q_i/T_i = 0$ for a reversible cycle. As a result, we have the corollary that for a reversible cycle $\int dQ/T = 0$ and hence for a reversible transformation $\int dQ/T$ is independent of path. This allows one to define a state function S by the relation

$$S(A) = \int_{\alpha}^{A} \frac{dQ}{T} \qquad \text{... (1.4.2)}$$

where α is a reference state and A an arbitrary state. The definition, clearly, assumes that the state A can be reached from α through successive reversible transformations. The state function S (A) defined above is the entropy.

We note immediately that S(A) has the following properties:

i) If A and B are two states, then

$$\int_{A}^{B} \frac{dQ}{T} \leq S(B) - S(A) \qquad \text{... (1.4.3)}$$

We connect A and B by two paths, one reversible and the other irreversible. Consider the closed circuit which takes one from A to B by the irreversible path and from B to A by the reversible one. For the closed path (using the theorem in the beginning of this section)

$$0 \geq \oint \frac{dQ}{T} = \int_{A}^{B} \frac{dQ}{T} + S(A) - S(B)$$

which establishes Eq. (1.4.3).

ii) The entropy of a thermally isolated system never decreases.

For the thermally isolated system $dQ = 0$ and hence from Eq. (1.4.3), $S(B) \geq S(A)$. The equality holds for a reversible transformation.

It must be emphasized that in using the above principle the question of thermal isolation must be carefully implemented. As an illustration consider the reversible isothermal compression of an ideal gas from volume V_1 to V_2. Since in an isothermal process, the temperature and hence the internal energy of an ideal gas is unchanged, we have

$$(\Delta S)_{\text{gas}} = \frac{\Delta Q}{T} = R \int_{V_1}^{V_2} \frac{dV}{V} = R \ln \frac{V_2}{V_1} \qquad \text{... (1.4.4)}$$

But $V_1 > V_2$ and hence ΔS is negative.

The entropy of the gas decreases but then it is not a thermally isolated system. The constant temperature is maintained by connection to a heat bath at temperature T which supplies the heat required for the compression and thus loses entropy of the amount $\Delta\, Q/T$. Consequently $(\Delta\, S)_{gas} + (\Delta\, S)_{bath} = 0$ and the thermally isolated system (gas + bath) undergoes no entropy change in the reversible process.

For a reversible transformation $dQ = T\,dS$ and the first law assumes the form

$$T\,dS = dE + dW = dE + P\,dV,$$

$$= \left(\frac{\partial E}{\partial T}\right)_V dT + \left[P + \left(\frac{\partial E}{\partial V}\right)_T\right] dV \qquad \text{... (1.4.5)}$$

Entropy S is a state function and hence will be determined by any two independent quantities like V and T, V and P, P and T, V and E and so on. Expressing S as a function of E and V, we find from the first step of Eq. (1.4.5)

$$\left(\frac{\partial S}{\partial E}\right)_V = \frac{1}{T} \qquad \text{... (1.4.6)}$$

Expressing S as a function of V and T, it is clear from Eq. (1.4.5) that

$$C_v = \left(\frac{\partial E}{\partial T}\right)_T = T\left(\frac{\partial S}{\partial T}\right)_V \qquad \text{... (1.4.7a)}$$

and

$$\left(\frac{\partial S}{\partial V}\right)_T = P + \left(\frac{\partial E}{\partial V}\right)_T \qquad \text{... (1.4.7b)}$$

If we consider S to be a function of P and T, it is straightforward to show

$$C_P = T\left(\frac{\partial S}{\partial T}\right)_P \qquad \text{... (1.4.8)}$$

For the difference of specific heats we get

$$C_P - C_v = T\left[\left(\frac{\partial S}{\partial T}\right)_P - \left(\frac{\partial S}{\partial T}\right)_V\right]$$

$$= T\left(\frac{\partial S}{\partial V}\right)_T \left(\frac{\partial V}{\partial T}\right)_P \qquad \text{... (1.4.9)}$$

To obtain the second line of Eq. (1.4.9), write $dS = \left(\frac{\partial S}{\partial V}\right)_T dV + \left(\frac{\partial S}{\partial T}\right)_V dT$ and then form $\left(\frac{\partial S}{\partial T}\right)_P$. The expression on the right hand side of Eq. (1.4.9) will be further simplified in the next section.

1.5 Thermodynamic Potentials

In this section we introduce two quantities with dimensions of energy which are responsible for determining equilibrium conditions. The first is the Helmholtz free energy (F)

$$F = E - TS, \qquad \text{... (1.5.1)}$$

with

$$\begin{aligned} dF &= dE - T\,dS - S\,dT \\ &= -P\,dV - S\,dT \end{aligned} \qquad \text{... (1.5.2)}$$

For a system at constant volume, thermal equilibrium (T = constant) leads to $dF = 0$ and hence Helmholtz free energy has to be a minimum.

The other quantity is the Gibbs free energy G, defined as

$$G = F + PV = E - TS + PV \qquad \text{... (1.5.3)}$$

with

$$\begin{aligned} dG &= dE - T\,dS - S\,dT + P\,dV + V\,dP \\ &= V\,dP - S\,dT \end{aligned} \qquad \text{... (1.5.4)}$$

since $dE - T\,dS + P\,dV = 0$ [see Eq. (1.4.5)]

For a system at constant pressure, thermal equilibrium ($dT = 0$) leads to $dG = 0$, i.e. the Gibbs free energy has to be a minimum.

From Eqs. (1.5.2) and (1.5.4), we can easily see

$$\left(\frac{\partial P}{\partial T}\right)_V = \left(\frac{\partial S}{\partial V}\right)_T \qquad \text{... (1.5.5)}$$

and

$$\left(\frac{\partial V}{\partial T}\right)_P = -\left(\frac{\partial S}{\partial P}\right)_T \qquad \text{... (1.5.6)}$$

by imposing the conditions that dF and dG are perfect differentials. Using the above relations in Eq. (1.4.7b) and Eq. (1.4.9), we obtain

$$\left(\frac{\partial E}{\partial V}\right)_T = T\left(\frac{\partial P}{\partial T}\right)_V - P \qquad \text{... (1.5.7)}$$

and

$$C_p - C_v = T\left(\frac{\partial P}{\partial T}\right)_V \left(\frac{\partial V}{\partial T}\right)_P$$

$$= T\left(\frac{\partial P}{\partial V}\right)_T \left[\left(\frac{\partial V}{\partial T}\right)_p\right]^2$$

$$= -TV\left[V\left(\frac{\partial P}{\partial V}\right)_T\right]\left[\frac{1}{V}\frac{\partial V}{\partial T}\right]^2$$

$$= VT\, x_T\, \alpha^2 \qquad \text{... (1.5.8)}$$

where $\alpha = \frac{1}{V}\left(\frac{\partial V}{\partial T}\right)$ is the expansion coefficient and $x_T = -V\left(\frac{\partial P}{\partial V}\right)_T$ is the isothermal bulk modulus. Relations of the form (1.5.5) and (1.5.6) are known as Maxwell's relations and two more can be obtained by considering the perfect differentials dE and dH.

1.6 Work Done in Coming to Equilibrium

In this section, we consider a couple of situations in which a system not in thermodynamic equilibrium to begin with comes to equilibrium as a result of interactions.

In this first case we consider a thermally isolated system consisting of several bodies not in thermal equilibrium with each other. While equilibrium is being established, the system is capable of doing work on external objects. Depending upon how the system comes to equilibrium, the final equilibrium state (energy, entropy, etc.) will be different and also the amount of work done will accordingly differ. What we are interested in finding out is under what condition will the amount of work done be a maximum.

In discussing the amount of work, we are not concerned with the work that could be done in equilibrium, e.g. the work done in a general expansion of the system. Consequently, we can assume that the total volume of the system is unchanged, i.e. the initial and final volumes are the same, although it could vary during the process.

Let the original energy of the system be E_0. The final state energy will be a function of the entropy S of the final state and we will denote it by $E(S)$. Since no heat is exchanged, the work done W by the system must equal the change in energy, i.e.

$$W = E_0 - E(S) \qquad \text{... (1.6.1)}$$

Differentiating with respect to S, at constant volume

$$\frac{\partial W}{\partial S} = -\left(\frac{\partial E}{\partial S}\right)_V = -T \qquad \text{... (1.6.2)}$$

by Eq. (1.4.6). Thus the derivative is always negative and W can only decrease with increasing S. The entropy of a thermally isolated system can never decrease [see Eq. (1.4.3)] and hence for W to be a maximum, one should have the entropy constant in the process. In other words, the work done will be maximized if the change occurs in a reversible manner.

As an example of the above assertion, we will give an alternative derivation of the efficiency of a reversible Carnot engine working between two temperatures T_1 and T_2. To do so, we consider the problem of transferring an amount of energy δE from a body at temperature T_1 to a body at temperature T_2 ($T_1 > T_2$). If the transfer is by contact, then no work is done and the process is irreversible. The entropy change of body at T_1 is $-\delta E/T_1$ and that for the body at T_2 is $\delta E/T_2$. The total change of entropy is $\delta E\,(1/T_2 - 1/T_1) > 0$ signifing that the process is irreversible.

To maximize the work, a reversible transfer of energy has to occur and some other body (a working medium) has to be brought in as a part of the system. The working medium must be capable of performing a reversible cyclic process. The process must be carried out in such a manner that the bodies between which direct transfer of energy occurs are at the same temperature. So, the working medium at temperature T_1 is brought into contact with the hotter body and a transfer of energy of amount δE_1 occurs isothermally. The medium now expands adiabatically so that it cools to the temperature T_2. It now transfers an amount of energy δE_2 to the cooler body and then by an adiabatic compression the working medium is brought back to the temperature T_1. The change in entropy of the hotter body is $-\delta E_1/T_1$ and that of the cooler body is $\delta E_2/T_2$. The total change of entropy is $\delta E_1/T_1 - \delta E_2/T_2$ (the working medium is back to its original state and hence there is no change in entropy) and for a reversible process $\delta E_1/T_1 = \delta E_2/T_2$. The change in energy is $\delta E_2 - \delta E_1 = -\left(1 - \frac{T_2}{T_1}\right)\delta E_1$ and thus from Eq. (1.6.1), the work done (the maximum work since the process is reversible) is

$$W = \left(1 - \frac{T_2}{T_1}\right)\delta E_1 \qquad \text{... (1.6.3)}$$

The efficiency is the work done divided by the expended energy (δE_1) and hence

$$\eta = 1 - \frac{T_2}{T_1}$$

which is the efficiency of the Carnot engine derived in Eq. (1.3.5).

The next situation that we envisage is a body at temperature T and pressure P in an external medium whose temperature is T_0 and pressure is P_0. The body can do work on some object which is thermally isolated from the body and the medium. The medium, together with the object and the body, constitutes a closed system. The medium is considered large enough that the process of energy and heat exchange cannot affect its temperature and pressure. As a result, T_0 and P_0 are regarded as constants. The work done by the body on a thermally isolated object would, in the absence of the medium, be simply equal to the difference in energy between its final and initial states. In the presence of the medium, this is no longer true and the question arises as to what is the maximum amount of work that the body can do.

If work is done by the body in going from state 1 to state 2, then if the body goes from state 2 to state 1, some external source has to do work on the body. The latter work will be a minimum, when the work done by the body is a maximum. So one can as well calculate the minimum work done by a thermally isolated source of work in taking the body from state 2 to state 1.

The total change in energy E of the body gets contribution from three sources:

i) the work W done by the medium

ii) the heat exchanged with the medium—this can be written as $-T_0 \Delta S_0$, where ΔS_0 is the change in entropy of the medium

iii) the work, $P_0 \Delta V_0$, done by the medium due to the change in volume ΔV_0 of the medium

Thus

$$\Delta E = W + P_0 \Delta V_0 - T_0 \Delta S_0 \quad \text{... (1.6.4)}$$

Since the total volume of the body and the medium remains constant, $\Delta V_0 = -\Delta V$ and the law of increase of entropy implies $\Delta S + \Delta S_0 \geq 0$ (the entropy of the thermally isolated source of work does not vary). We now obtain

$$\begin{aligned} W &= \Delta E + T_0 \Delta S_0 - P_0 \Delta V_0 \\ &= \Delta E + P_0 \Delta V + T_0 \Delta S_0 \\ &\geq \Delta E - T_0 \Delta S + P_0 \Delta V \end{aligned} \quad \text{... (1.6.5)}$$

The equality holds for the reversible process and hence it is the reversible process for which the minimum work is done on the body. The value of the minimum work is

$$\begin{aligned} W_{min} &= \Delta E - T_0 \Delta S + P_0 \Delta V \\ &= \Delta \left(E - T_0 \Delta S + P_0 V\right) \end{aligned} \quad \text{... (1.6.6)}$$

If the change occurs reversibly and the body is in an equilibrium state at every instant during the process, we can write for the differential of W_{min} as

$$\begin{aligned} d W_{min} &= dE - T_0 \, dS + P_0 \, dV \\ &= T \, dS - P \, dV - T_0 \, dS + P_0 \, dV \\ &= \left(T - T_0\right) dS + \left(P_0 - P\right) dV \end{aligned} \quad \text{... (1.6.7)}$$

We note from Eq. (1.6.6.), that if the volume and temperature of the body remain constant in the process, with the temperature equal to that of the medium (i.e. $T = T_0$), then

$$W_{\min} = \Delta F \qquad \text{... (1.6.8)}$$

and that if the temperature and pressure are held constant at the values corresponding to those of the medium, then

$$W_{\min} = \Delta G \qquad \text{... (1.6.9)}$$

Now, suppose that the body in an external medium is left to itself and no work is done on it. Spontaneous irreversible processes will bring it to equilibrium and Eq. (1.6.5) with $W = 0$ shows

$$\Delta\left(E - T_0 S + P_0 V\right) \le 0 \qquad \text{... (1.6.10)}$$

Thus the quantitiy $E - T_0 S + P_0 V$ keeps on decreasing reaching a minimum at equilibrium. For a process at constant temperature and volume it is the Helmholtz free energy which decreases and for a process at constant pressure and temperature it is the Gibbs free energy which decreases in accordance with the results of Sec. 1.5.

1.7 Third Law of Thermodynamics

The second law of thermodynamics defines the entropy to within an additive constant. The third law provides the constant by stating that the entropy at the absolute zero (T = 0) is an universal number and can be taken to be equal to zero (Nernst Heat Theorem).

It immediately follows that all specific heats vanish as $T \to 0$. If $C(T)$ is the specific heat along any reversible path, then

$$S(T) = \int_0^T \frac{dQ}{T} = \int_0^T \frac{CdT}{T} \qquad \text{... (1.7.1)}$$

If $S(T) \to 0$ as $T \to 0$, then it is clear that $C(T)$ must go to zero as $T \to 0$.

The expansion coefficient also vanishes as $T \to 0$. This is seen as follows:

$$\left(\frac{\partial C_P}{\partial P}\right)_T = \frac{\partial}{\partial P}\left[T\left(\frac{\partial S}{\partial T}\right)_P\right]_T$$

$$= T \frac{\partial}{\partial T}\left[\left(\frac{\partial S}{\partial P}\right)_T\right]_P$$

$$= - T\left(\frac{\partial^2 V}{\partial T^2}\right)_P \qquad \text{... (1.7.2)}$$

by using Eq. (1.5.6). We now write

$$V\alpha = \left(\frac{\partial V}{\partial T}\right)_P = -\left(\frac{\partial S}{\partial P}\right)_T = -\frac{\partial}{\partial P}\int_0^T \frac{C_P\,dT}{T}$$

$$= -\int_0^T \frac{1}{T}\frac{\partial C_P}{\partial P}\,dT$$

$$= \int_0^T \left(\frac{\partial^2 V}{\partial T^2}\right)_P dT \quad \text{(integration along path of constant pressure)}$$

$$= \left(\frac{\partial V}{\partial T}\right)_P - \left[\left(\frac{\partial V}{\partial T}\right)_P\right]_{T=0} \qquad \text{... (1.7.3)}$$

The right hand side of Eq. (1.7.3) tends to zero as $T \to 0$ and hence $\alpha \to 0$.

In a similar fashion it is possible to show that $\left(\frac{\partial P}{\partial T}\right)_V \to 0$ as $T \to 0$. From the first line of Eq. (1.7.3) it is clear that $\alpha V/C_p \to$ a finite constant as $T \to 0$. [This is seen by writing $C_P = a(P)\,T^x$ as $T \to 0$ and then it is clear that as $T \to 0$, $\alpha V/C_P \to a'(P)/x$.).] Now consider adiabatic cooling by releasing the pressure.

The governing equation is

$$T\,dS = C_P\,dT + T\left(\frac{\partial S}{\partial P}\right)_T dP = C_P\,dT - T\left(\frac{\partial V}{\partial T}\right)_P dP,$$

leading to

$$\frac{dT}{dP} = \frac{\alpha T V}{C_P} \text{ for } dS = 0 \qquad \text{... (1.7.4)}$$

Since $\alpha V/C_p$ is finite for $T \to 0$, we note that, as $T \to 0$, to produce a finite dT, infinitely large dP will be required. This leads to the alternative formulation of the third law in the form of the statement that the absolute zero of temperature cannot be attained.

We have briefly recalled the salient features of equilibrium thermodynamics which is based on the three phenomenological laws that we have considered and provides relationships between macroscopic quantities. It is, however, incapable of providing absolute values for the macroscopic quantities. As an example, the laws of thermodynamics provide a relation between C_p and C_v (Eq. 1.5.8) but cannot provide values for C_V and C_P in terms of microscopic quantities. This is where equilibrium statistical mechanics comes in. The passage from the microscopic dynamics of the individual atoms and molecules to the macroscopic thermodynamic description where the knowledge of the microscopic effects are lost will be the topic of our discussion in the next chapter.

Problems

1.1 The representation of a cyclic process ABCA on any gas is as shown in the Fig. 1.2. Calculate the efficiency of the process and draw your conclusions from the answer you get.

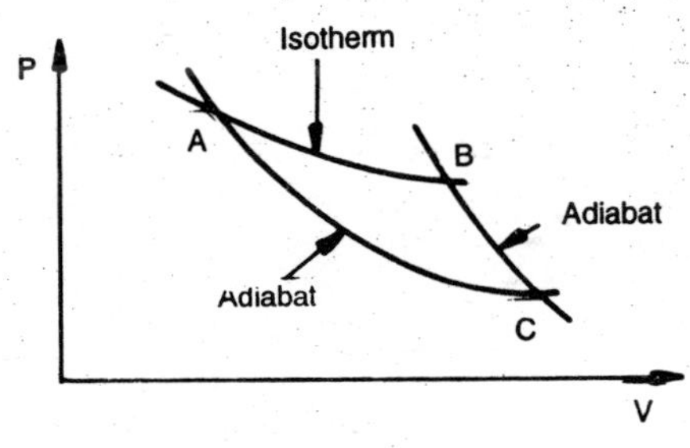

Fig. 1.2

1.2 A process where application of pressure brings about a change of temperature is called throttling. It takes place at constant enthalpy $H(H = E + PV)$. Find an expansion for the throttling coefficient $J = \left(\frac{\partial T}{\partial P}\right)_H$ in terms of C_P, V, T and α (expansion coefficient). Evaluate J for an ideal gas. If the equation of state of a real gas is $PV = NkT + \frac{N^2}{V} \cdot f(T)$, then find J. Discuss the possibility of cooling the gas by this isenthalpic increase of pressure.

1.3 An adiabatic process is one in which the system is thermally insulated from its surrounding i.e. no heat exchange can occur. Show that for an ideal gas PV^{γ} = constant for such a process where $\gamma = C_p/C_v$.

1.4 An acceptable equation of state for one mole of a real gas is $\left(P + \frac{a}{V^2}\right)(V - b) = RT$, where a and b are constants (Van der Waals equation). The following problems are based on the above equation of state:

a) Find the internal energy E, the entropy S and the free energy F per unit mole of the above gas.

b) Consider a free expansion of a real gas. This is done by taking a thermally insulated rigid container and dividing it into two parts. One part is vacuum and the other containing the gas and the separating walls has a valve. The valve is opened and the gas expands into the empty chamber. This is a free expansion. Since it is an expansion into vaccum, no work is done. If the initial volume and temperature are V_1 and T_1 and the final volume is V_2, find the final temperature.

c) A gas at pressure P_1 is steadily transferred to a vessel where the pressure is P_2. During the process, P_1 and P_2 do not change (e.g. transfer of gas through a porous partition where the pressure on each side of the partition is maintained by suitable pistons). The gas is thermally isolated from an external medium. If the initial temperature is T_1, find the final temperature T_2.

1.5 A vertical cylinder contains N moles of an ideal gas and is closed off by a piston of mass M and area A. The specific heat at constant volume is independent of temperature. The whole system is thermally isolated. Initially the piston is clamped in position so that the gas has volume V_0 and temperature T_0. The piston is now released and after some oscillation comes to rest in a final equilibrium situation corresponding to a larger volume of gas. Find the final temperature in terms of T_0, V_0 and the other parameters.

1.6 A gas in a vertical cylindrical container supports a freely moving piston of mass m. The common cross-sectional area is A. Atmospheric pressure is P_0 and in equilibrium, the volume of gas is V_0. The piston is now given a small displacement and found to oscillate with a frequency ω. The frequency is low enough for internal equilibrium but high enough for the gas not to be able to exchange heat with its surroundings. Find ω in terms of m, g, A, P_0, V_0 and γ.

1.7 Two vessels contain two identical ideal gases at the same temperature T and with equal number of particles N but at different pressure P_1 and P_2. The vessels are then connected. Find the change in entropy.

1.8 Two identical ideal gases at the same pressure P and containing the same number of particles N but at different temperatures T_1 and T_2 are in vessels with volumes V_1 and V_2. The vessels are then connected. Find the change in entropy.

1.9 Find the work done on an ideal gas and the quantity of heat it gains on going through a cyclic process, consisting of two isochoric and two isobaric processes. The gas goes from a state with pressure and volume P_1, V_1 to states P_1, P_2; P_2, V_1 and P_1, V_1 again.

1.10 Repeat Problem 1.9 with two isobaric and two isothermal processes.

1.11 Repeat Problem 1.9 with two isobaric and two adiabatic processes.

1.12 Determine the maximum work that can be obtained by connecting vessels containing two identical gases at the same temperature T_0 and with equal numbers of particles but having different volumes V_1 and V_2.

1.13 Find the work that must be done on an ideal gas in order to compress it from pressure P_1 and P_2 at a constant temperature equal to that of the surrounding medium.

1.14 Gas at temperature T_0 flows from a large thermally isolated reservoir into an empty thermally isolated vessel, the gas pressure in the reservoir remaining constant. Find the change in gas temperature.

1.15 Vapour pressure is the pressure at which a gas can coexist in equilibrium with an infinite volume of its own liquid. This is denoted by $P_\infty(T)$. The pressure $P_r(T)$ at which the gas coexists with a droplet of radius r is higher than $P_\infty(T)$. The purpose of the present problem is to find $P_r(T)$. Suppose a liquid droplet is placed in an external medium which exerts a pressure P on the droplet. The work done on expansion is

$$dW = P\,dV - S\,dA$$

where S is surface tension and dA is the increase in surface area. In this case, the First law takes the form

$$dE = dQ - P\,dV + S\,dA$$

Integrate to find

$$E = \frac{4}{3}\pi r^3 E_\infty + 4\pi S r^2$$

where E_∞ is the energy of an infinite droplet. Similarly the Gibbs potential is

$$G = \frac{4}{3}\pi r^3 G_\infty + 4\pi S r^2$$

Equilibrium at a given temperature and pressure requires that G be minimized. Write down the total G as

$$G_{tot} = M_2 g_2 + M_1 g_1 + 4\pi S r^2$$

where M_1 and M_2 are the mass of the droplet and the vapour respectively and g_1 and g_2 are the respective Gibbs potential per unit mass for the infinite volume limit. For equilibrium $\delta G = 0$. If an amount of liquid ∂M_1 evaporates ($\delta M_1 = -\delta M_2$), obtain the equilibrium condition as (ρ is the liquid density)

$$g_2 - g_1 = \frac{2S}{\rho r}$$

Differentiate with respect to pressure at constant temperature and using the ideal gas law for the vapour phase whose density is $\rho' << \rho$, obtain

$$P_r(T) = P_\infty(T)\, e^{2Sm/\rho kTr}$$

where m is the mass of a gas atom.

1.16 For equilibrium at temperature T_0 and pressure P_0 the function

$$G = E - T_0 S + P_0 V$$

is a minimum. This implies that a variation ΔG for variation of parameters about equilibrium must be ≥ 0. Using S and V as the independent variables, show that for $\Delta G \geq 0$, we must have

$$C_v \geq 0 \text{ and } \left(\frac{\partial P}{\partial V}\right)_T \leq 0.$$

Chapter Two
Basis of Statistical Mechanics

2.1 Introduction

In the previous chapter we learnt that macroscopic systems are characterized by a few variables like pressure, volume, temperature, etc. At a microscopic level, however, this description is not viable; the description must be in terms of properties of the microscopic parts of the macroscopic system. For a mass of gas in a container, the thermodynamic state is specified by the pressure, temperature and number density but the description in terms of the microscopic parts entails the specification of the state of each individual molecule. The number of molecules is of order N, where $N \sim 10^{23}$ (Avogadro number) and each individual molecule requires three coordinates of position and three coordinates of momentum for the specification of its state. Thus, the total number of variables that are required for the description of the microscopic state is $6N$. The bridge between the microscopic and macroscopic descriptions is provided by statistical mechanics.

The first thing to note is that if we specify the total energy of the N-particle system, then there exist an extremely large number of microscopic states that are available to the system corresponding to this energy. Specifying the total energy specifies the macroscopic state of an isolated system but allows for a large number of microscopic states corresponding to this macroscopic state. To specify the microscopic state of the system it is convenient to work in terms

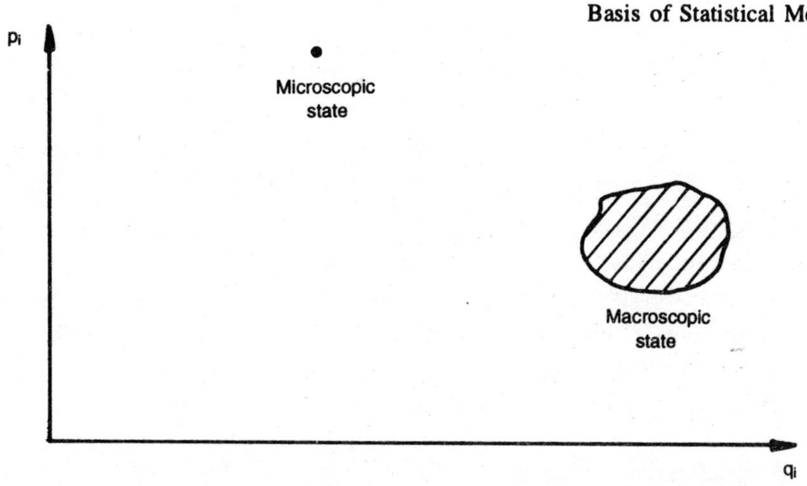

Fig. 2.1: Phase space for N-particles. The axes are the 3N coordinates and the 3N momenta.

of a $6N$-dimensional space of coordinates and momenta of the N particles. This $6N$-dimensional space is called the 'phase space' of the system and a microscopic state is simply a point in this phase space. A macroscopic state on the other hand corresponds to a large number of microstates, (microscopic state) and hence corresponds to a volume in phase space (Fig. 2.1). If we consider a particular microscopic state of the system and watch how it evolves in time, we will be obtaining a phase trajectory. A statistical description ensues if instead of accurately specifying the trajectory (which is virtually impossible for a large system), we specify the probability dw of its being in a volume $d\Gamma$ of the phase space Γ. This probability can be obtained by noting the time dt the system spends in the volume $d\Gamma$ and forming the ratio $dw = \underset{T\to\infty}{\text{Lt}}\, dt/T$ where T is the time over which the trajectory is tracked. If we consider any quantity $f(t)$ measured along the trajectory, then its average would be $\bar{f} = \underset{T\to\infty}{\text{Lt}} \frac{1}{T}\int_0^T f(t)dt$.

An alternative statistical description of the situation could be obtained if one considered the various points in the phase space volume as representing different copies of the same system and called the collection of these copies an ensemble. If $\bar{N}$ be the total number of copies corresponding to Γ and $d\bar{N}$ be the number of copies corresponding to an elementary volume $d\Gamma$, then the probability of the system lying in the state $d\Gamma$ could be written as $d\bar{w} = \underset{N\to\infty}{\text{Lt}}\, d\bar{N}/\bar{N} = \rho(\{q_i\}, \{p_i\})\, d\Gamma$, $\{q_i\}$ is the set of the $3N$ position coordinates and $\{p_i\}$ is the set of the $3N$ momentum coordinates and $\rho(\{q_i\}, \{p_i\})$ (normalized to unity) is a density in the ensemble space. The average value of any function f of the set $\{q_i\}$, $\{p_i\}$ would then be obtained as $\langle f \rangle = \int f \rho d\Gamma$. The quantity ρ is the distribution function, and in forming the averages of arbitrary functions f with it, we are assuming that ρ corresponds to an equilibrium situation (no time dependence)—the achieving of equilibrium obviously being determined by the longest time scale τ_L set by the interactions in the system. This is what we will repeatedly come across in the later chapters when we start from a non-equilibrium situation, the interactions (e.g. the collision between gas molecules) take the system to equilibrium and if we make our time of observation $T >> \tau_L$, then we are permitted to talk about an equilibrium ρ independent of time.

The success of statistical mechanics now depends on three crucial assumptions:

i) The two descriptions described above are completely equivalent and for any functions $f(\{q_i\}, \{p_i\})$, the two averages $\bar{f}$ and $\langle f \rangle$ are equal. This is the ergodic hypothesis.
ii) If one calculates the volume Γ filled by a trajectory in phase sapce—the trajectory being controlled by appropriate constants of motion (this point could be tricky as constants of motion may be determined by the competition between equilibration time and the time of observation)—then the entropy S is given by $S = k \ln \Gamma$. Here k is the universal constant, known as Boltzmann's constant, which we have encountered before. This is Boltzmann's hypothesis.
iii) The large number of microstates which are compatible with the single macrostate are all equally probable. This is the postulate of equal a priori probabilities.

As is apparent, the rule (iii) is a method of calculation, rule (ii) provides the bridge between the microscopic description and a thermodynamic one, while rule (i) ensures that the situation is internally consistent.

The fundamental question that one asks is whether there is a unifying theme behind the three rules stated above and the point we would like to make in the remainder of this chapter is that the microscopic dynamics will generate this unifying theme. The microscopic dynamics will more often than not be chaotic—this means that motion will be sensitive to initial conditions, a small volume of initial points will be stretched and spread out all over phase space—there will be ergodicity, mixing and loss of correlation and the three rules will appear quite justified.

To end this section, we note that thermodynamic systems need not be isolated, but can exchange energy with the surrounding and yet one can characterize the system by its energy. In the statistical mechanical description, the energy is now no longer a constant but an ensemble average of the energy can be constructed. What transpires is that the statistical fluctuations about the mean energy must be infinitesimal in the thermodynamic limit and hence the thermodynamic description of characterizing a system by its energy makes sense. In the next chapter, when we discuss the canonical and grand canonical ensembles, this point will become abundantly clear.

2.2 Ergodic Hypothesis

To understand the content of the ergodic hypothesis we anticipate the discussion of Langevin equations in Chapter 16. For a variable $q(t)$, the Langevin equation is

$$\dot{q} = \frac{\partial S}{\partial q} + f(t) \qquad \text{... (2.2.1)}$$

where $f(t)$ is a noise term with the correlations as prescribed in Eqs. (16.4.5a) –(16.4.6b). The main point now is that $f(t)$ is random and hence $q(t)$ will be random, which is an obvious statement. If we consider an arbitrary function $g(q)$, averages can be calculated in two different ways:

i) we evaluate $q(t)$ at every instant of time and then form the average

$$\overline{g}(q) = \underset{T \to \infty}{\text{Lt}} \frac{1}{T} \int_0^T g(q(t))\, dt \qquad \text{... (2.2.2)}$$

ii) we recognize the fact that as $t \to \infty$, an equilibrium distribution $P_{eq}(q)$ for q is obtained and hence

$$< g(q) > = \int_{-\infty}^{\infty} g(q)\, P_{eq}(q)\, dq \qquad \text{... (2.2.3)}$$

By virtue of the properties of the noise $f(t)$, the two averages are equal, i.e. $\overline{g}(q) = < g(q) >$. The lesson to be learnt is that, given a dynamics and a statistical description [induced by the noise $f(t)$], there are two different ways of forming averages and the two averages are indeed equal—the requirement of the ergodic hypothesis.

The major difference between the above dynamics and the dynamics of the N-particle system of Sec. 2.1 is that the dynamics of the N-particle system is deterministic, guided by the Hamiltonian equations

$$\dot{q}_i = \frac{\partial H}{\partial p_i} \qquad \text{... (2.2.4)}$$

$$p_i = -\frac{\partial H}{\partial q_i} \qquad \text{... (2.2.5)}$$

The phase space is the $6N$-dimensional space of the variables $\{q_i\}$ and $\{p_i\}$ and at time $t = t_0$, a state is a point in this space. As time goes on, the state evolves producing a trajectory and in principle Eqs. (2.2.4) and (2.2.5) can be integrated providing $\{q_i(t)\}$ and $\{p_i(t)\}$ at every instant of time. If we consider a function $g(\{q_i\}, \{p_i\})$, then it can be averaged over a trajectory and the average defined as

$$< \overline{g(\{q_i\},\{p_i\})} > = \underset{T\to\infty}{Lt} \frac{1}{T} \int_{t_0}^{t_0+T} g(\{q_i(t)\}, \{p_i(t)\})dt \qquad \text{... (2.2.6)}$$

The alternative description is in terms of the ensemble of points in phase space (each point corresponding to the same macroscopic state) characterized by an equilibrium density of points $\rho(\{q_i\}, \{p_i\})$. The point to be stressed is that if one waits long enough, an equilibrium distribution will be reached. An obvious example is to consider a box of volume V separated into two halves by a partition. We consider n moles of an ideal gas trapped in one half with the other empty. The box is in contact with a heat reservoir at temperature T. If at time $t = 0$, the partition is removed, the mechanical equilibirium will be disturbed—the gas will spread out in the total volume V and eventually an equilibrium situation will be reached with the pressure halved.

An average of $g(\{q_i\}, \{p_i\})$ can be formed over this equilibrium ensemble and this is defined as

$$< g(\{q_i\}, \{p_i\}) > = \int \rho(\{q_i\}, \{p_i\})\, g(\{q_i\}\ \{p_i\})\, d^{3N} q_i \,.\, d^{3N} p_i \qquad \text{... (2.2.7)}$$

where $d^{3N} q_i\, d^{3N} p_i$ is the elementary volume in phase space. The assertion of the ergodic hypothesis is

$$< g > = \bar{g} \qquad \text{... (2.2.8)}$$

To clarify the implications of a time-independent $\rho(\{q_i\}, \{p_i\})$, we need to follow the time development of $\rho(\{q_i\}, \{p_i\}, t)$. The relevant principle for determining the dynamics of ρ is a conservation law—conservation of the number of ensembles. In a given elementary volume $d^{3N} q\, d^{3N} p$, the number of ensembles is $\rho\, d^{3N} q\, d^{3N} p$ and this number can change only if in the course of the evolution some ensemble points enter or leave the volume across the surface which bounds it. The result is a local conservation law

$$\frac{\partial \rho}{\partial t} + \vec{\nabla} . (\rho \vec{v}) = 0 \qquad \text{... (2.2.9)}$$

where v is the velocity $(\{\dot{q}_i\}, \{\dot{p}_i\})$ in the $6N$-dimensional space. Writing out Eqn. (2.2.9) explicitly.

$$\frac{\partial \rho}{\partial t} + \sum_i \frac{\partial}{\partial q_i} (\rho \dot{q}_i) + \sum_i \frac{\partial}{\partial p_i} (\rho \dot{p}_i) = 0$$

or

$$\frac{\partial \rho}{\partial t} + \sum_i \left(\frac{\partial \rho}{\partial q_i} \dot{q}_i + \frac{\partial \rho}{\partial p_i} \dot{p}_i \right) + \sum \rho \left(\frac{\partial \dot{q}_i}{\partial q_i} + \frac{\partial \dot{p}_i}{\partial p_i} \right) = 0 \qquad \text{... (2.2.10)}$$

But, $\dfrac{\partial \dot{q}_i}{\partial q_i} + \dfrac{\partial \dot{p}_i}{\partial p_i} = \dfrac{\partial}{\partial q_i} \left(\dfrac{\partial H}{\partial p_i} \right) + \dfrac{\partial}{\partial p_i} \left(- \dfrac{\partial H}{\partial q_i} \right) = 0$

where we have used Eqs. (2.2.4) and (2.2.5) and thus Eq. (2.2.10) reduces to

$$\frac{d\rho}{dt} = \frac{\partial \rho}{dt} + \sum_i \left(\frac{\partial \rho}{\partial q_i} \dot{q}_i + \frac{\partial \rho}{\partial p_i} \dot{p}_i \right) = 0 \qquad \text{... (2.2.11)}$$

An alternative proof uses the fact that the volume element $d^{3N} p\, d^{3N} q$ is preserved by Hamilton's equations and hence number conservation implies $\dfrac{d\rho}{dt} = 0$ (see any classical mechanics text).

To have an equilibrium distribution, we require an even stronger condition on ρ, namely $\dfrac{\partial \rho}{\partial t} = 0$. This implies that

$$\frac{\partial \rho}{\partial q_i}\dot{q}_i + \frac{\partial \rho}{\partial p_i}\dot{p}_i = 0$$

or

$$\frac{\partial \rho}{\partial q_i}\frac{\partial H}{\partial p_i} - \frac{\partial \rho}{\partial p_i}\frac{\partial H}{\partial q_i} = 0 \qquad \text{... (2.2.12)}$$

a condition which can be easily satisfied if $\rho = \rho\,(H(\{q_i\}, \{p_i\}))$ since then

$$\frac{\partial \rho}{\partial q_i} = \rho'\frac{\partial H}{\partial q_i} \qquad \text{... (2.2.13a)}$$

and

$$\frac{\partial \rho}{\partial p_i} = \rho'\frac{\partial H}{\partial p_i} \qquad \text{... (2.2.13b)}$$

and Eq. (2.2.12) is clearly satisfied. In the above the prime denotes differentiation with respect to H. Hence the equilibrium distribution can be chosen to have the from

$$\rho(\{q_i\}, \{p_i\}) = \rho\,(H(\{q_i\}, \{p_i\})) \qquad \text{... (2.2.14)}$$

Thus there is a constraint on the equilibrium distribution. We now note that the ergodicity statement [Eq. (2.2.8)] provides the bridge between a statistical description and microscopic dynamics. This statement was a natural consequence of stochastic dynamics when we considered the noise-driven process of Eq. (2.2.1). In deterministic dynamics, the above ergodicity statement is less accessible but points to the fact that if there is some characteristic of randomness in the deterministic dynamics, then Eq. (2.2.8) would certainly be understandable. Our point will be that deterministic chaos provides the underlying randomness that justifies Eq. (2.2.8).

2.3 Boltzmann's Hypothesis

To begin with let us use the laws of thermodynamics to find the entropy of a perfect gas. The starting point is

$$T\,dS = dE + P\,dV \qquad \text{... (2.3.1)}$$

Now, for a perfect monatomic gas at a temperature T, the energy is $E = 3/2\ NkT$ (we are considering N molecules) and the equation of state is $PV = NkT$, so that Eq. (2.3.1) becomes

$$T\,dS = \frac{3}{2}\,Nk\,dT + \frac{NkT}{V}\,dV$$

or

$$dS = \frac{3}{2}\,Nk\,\frac{dT}{T} + \frac{Nk}{V}\,dV \qquad \text{... (2.3.2)}$$

Integrating,

$$S = \frac{3}{2} N\,k\,\ln T + N\,k\,\ln V + S_0$$

$$= k\,\ln\,(VT^{3/2})^N + S_0$$

$$= k\,\ln\left(V\left(\frac{2E}{3Nk}\right)^{3/2}\right)^N + S_0 \qquad \text{... (2.3.3)}$$

We now need to look at the volume occupied by the trajectories in phase space. We consider an extremely small spread Δ in the total energy E, i.e. the total energy lies between E and $E+\Delta$, $\Delta << E$ and in fact as $N \to \infty$, $\Delta/E \to 0$. The volume in phase space is defined by the restriction

$$E < \sum_{i=1}^{3N} \frac{p_i^2}{2m} < E + \Delta \qquad \text{... (2.3.4)}$$

To calculate the volume, we proceed to find the volume in phase space occupied by all trajectories corresponding to energies $\leq E$. This volume $\Gamma(E)$ is given by

$$\Gamma(E) = \frac{1}{h^{3N/2}} \int_V d^{3N}\,q_i \int_{0<\frac{p^2}{2m}<E} d^{3N}\,p_i \qquad \text{... (2.3.5)}$$

where $p^2 = \sum_{i=1}^{N} p_i^2$. To appreciate why h enters a classical formula, see Problem (2.2) Clearly,

$$\Gamma(E) = (2\pi\hbar)^{-3N/2} V^N \times (\text{volume of } 3N\text{-dimensional sphere of radius } (2mE)^{1/2}), \qquad \ldots \textbf{(2.3.6)}$$

where V is the volume of the container. Now recall (Problem 2.3) that the volume of a sphere of radius R in a D-dimensional space is $2\pi^{D/2}R^D/\overline{\Gamma}(D/2)$ and hence (note that $\overline{\Gamma}$ is the usual gamma function)

$$\Gamma(E) = V^N \, 2\pi^{3N/2} \, (2mE)^{3N/2}/3N\,\overline{\Gamma}\left(\frac{3N}{2}\right)(2\pi\hbar)^{3N/2} \qquad \ldots \textbf{(2.3.7)}$$

The volume $\Gamma(E, \Delta)$ which is the volume of the annulus lying between spheres of radii E and $E + \Delta$ is thus obtained as

$$\Gamma(E, \Delta) \approx \frac{\partial\Gamma(E)}{\partial E}\,\Delta = \frac{V^N\,(2\pi)^{3N/2}\,(2mE)^{3N/2}\,\Delta}{(2\pi\hbar)^{3N/2}\,\overline{\Gamma}(3N/2)\;2E} \qquad \ldots \textbf{(2.3.8)}$$

Thus,

$$\begin{aligned}
\ln\Gamma(E, \Delta) &= N\ln V + \frac{3N}{2}\ln 2mE + \frac{3N}{2}\,.\,\ln 2\pi \\
&\quad + \ln\frac{\Delta}{2E} - \frac{3N}{2}\left(\ln\frac{3N}{2} - 1\right) - \frac{3N}{2}\ln(2\pi\hbar) \\
&= N\ln V + \frac{3N}{2}\ln\frac{mE}{3N/2} + \frac{3N}{2}(1 - \ln h) + \ln\frac{\Delta}{E} \\
&\approx N\ln V + \frac{3N}{2}\ln\frac{E}{N} + \frac{3N}{2}\left(\ln\frac{m}{h} + 1\right)
\end{aligned} \qquad \ldots \textbf{(2.3.9)}$$

where we have made the approximation of N being very large and dropped terms of O(1). Comparing with Eq. (2.3.3), we note that the Boltzmann hypothesis $S = k\ln\Gamma(E, \Delta)$ is indeed satisfied if S_0, the undetermined constant in thermodynamics, is fixed from Eq. (2.3.9).

At this point, we make a small digression to point out a difficulty with Eq. (2.3.3). Let us imagine a container of volume $2V$ divided into equal volumes carrying N molecules each of the same gas. The entropy of the total system is

$$S_{tot} = 2N\ln V + 3N\ln E/N + 2S_0 \qquad \ldots \textbf{(2.3.10)}$$

If the partition is removed and the gases allowed to mix, then the final entropy is that of $2N$ molecules occupying a volume $2V$ and thus

$$S_{\text{final}} = 2N \ln 2V + 3N \ln \frac{E}{N} + 2S_0 \qquad \text{... (2.3.11)}$$

leading to an entropy change $S = S_{\text{final}} - S_{\text{tot}} = 2N \ln 2$. This cannot be correct as the mixing process described is clearly reversible and there can be no change in entropy. This is known as Gibbs paradox. The error lies in a wrong enumeration of the number of available states for N totally indistinguishable particles. In this case one has to recognize the fact that $N!$ permutations of the particles among themselves cannot lead to new states and hence the number of available states has to be changed from Eq. (2.3.5) to

$$\Gamma(E) = \frac{1}{N!}\left(\frac{1}{2\pi\hbar}\right)^{3N/2} \int_V d^{3N} q_i \int_{0<p_i^2/2m<E} d^{3N} p_i \qquad \text{...(2.3.12)}$$

with the corresponding change in $\Gamma(E, \Delta)$. With this change in counting, Eq. (2.3.3) is altered to

$$S = Nk \ln \frac{V}{N} + \frac{3}{2} Nk \ln \frac{E}{N} + S_0 \qquad \text{... (2.3.13)}$$

and Gibbs paradox is removed as ΔS now does turn out to be zero.

After this digression, we return to a consideration of the Boltzmann relation $S = k \ln \Gamma(E)$ (Note that for large N, it does not matter whether we consider $\ln \Gamma(E)$ or $\ln \Gamma(E, \Delta)$). The thing to note is that entropy being extensive is $O(N)$ for a N-particle system and thus the region of phase space $\Gamma(E)$ covered by a trajectory is $O(e^N)$, which is the same as $O(V^N)$. This fact is significant because it means that any particular trajectory is totally uncorrelated with any other trajectory and can wander over the allowed phase space independent of the other trajectories. Once again this loss of correlation (in different trajectories independent of each other) is a characteristic of chaotic dynamics, so that the justification behind the Boltzmann hypothesis lies in the chaotic nature of the microscopic dynamics.

Finally, we note that ergodicity (ability of a trajectory to cover all of phase space) does not imply loss of correlation. This can be seen most easily by considering the mapping $\theta_{n+1} = \theta_n + \Omega \pmod{2\pi}$, where Ω is irrational.

Clearly the successive iterates of an arbitrary initial θ_0 cover the entire circumference and thus assures ergodicity. A moment's thought, however, convinces one that there can be no loss of correlation and thus we arrive at the conclusion that ergodicity does not necessarily lead to loss of correlation. However, if we have a loss of correlation, ergodicity will be assured.

2.4 Equal A Priori Probability

We now consider an isolated system where the energy lies between E and $E + \Delta$, with $\Delta << E$ and enumerate the number of microstates satisfying the total energy constraint. The equal a priori probability hypothesis asserts: All microstates compatible with the given macrostate (energy between E and $E + \Delta$) are equally probable.

An ensemble satisfying this energy constraint is a microcanonical ensemble.

We first verifv that if two microcanonical ensembles are put in contact, the entropies add. System '1' has energy lying between E_1 and $E_1 + \Delta$ and System '2' has energy between E_2 and $E_2 + \Delta$.

When put in contact let the final ensemble have an energy lying between E and $E + 2\Delta$. If $\Gamma(E_1, \Delta)$ is the number of available states for '1' and $\Gamma(E_2, \Delta)$ the number of states for '2', then the number of states where the energy lies between $E_1 + E_2$ and $E_1 + E_2 + 2\Delta$ is $\Gamma(E_1, \Delta)\cdot \Gamma(E_2, \Delta)$. The number of states $\Gamma(E, \Delta)$ available to the final ensemble is then given by

$$\Gamma(E, \Delta) = \sum_{E_1 + E_2 < E < E_1 + E_2 + 2\Delta} \Gamma(E_1, \Delta)\, \Gamma(E_2, \Delta) \qquad \text{... (2.4.1)}$$

Let $E_1 = \bar{E}_1$ and $E_2 = \bar{E}_2$ be the values of E_1 and E_2 for which the product $\Gamma(E_1, \Delta)\,\Gamma(E_2, \Delta)$ has its maximum value. Then

$$\Gamma(E, \Delta) \geq \Gamma(\bar{E}_1, \Delta)\, \Gamma(\bar{E}_2, \Delta) \qquad \text{... (2.4.2)}$$

The number of terms in the sum of the right hand side of Eq. (2.4.1) is of the order of E/Δ. This follows from the fact that E_1 (or E_2) must be bounded from below for stability and if we call this lower bound zero, then the range of E_1 is from 0 to E, approximately in steps of Δ and hence the foregoing estimate of the number of terms. Clearly,

$$\Gamma(E, \Delta) \leq \frac{E}{\Delta}\ \Gamma(\bar{E}_1,\ \Delta)\ (\bar{E}_2,\ \Delta) \qquad \text{... (2.4.3)}$$

and thus

$$\Gamma(\bar{E}_1,\ \Delta)\ \Gamma(\bar{E}_2,\ \Delta) \leq \Gamma(E,\ \Delta) \leq \frac{E}{\Delta}\ \Gamma(\bar{E}_1, \Delta)\ \Gamma(\bar{E}_2,\ \Delta)$$

or

$$\ln\ \Gamma(\bar{E}_1,\ \Delta) + \ln\ \Gamma(\bar{E}_2,\ \Delta) \leq \ln\ \Gamma(E,\ \Delta) \leq \ln\ \Gamma(\bar{E}_1,\ \Delta) + \ln\ \Gamma(\bar{E}_2,\ \Delta) + \ln\ \frac{E}{\Delta} \qquad \text{... (2.4.4)}$$

Now if system '1' has N_1 particles and system '2' has N_2, then $\ln\ \Gamma(\bar{E}_1,\ \Delta)$ is $O(N_1)$ and $\ln\ \Gamma(\bar{E}_2,\ \Delta)$ is $O(N_2)$ while $\ln\ \Gamma(E,\ \Delta)$ is $O\ (N_1 + N_2)$. In the thermodynamic limit, $\ln\ \frac{E}{\Delta}$ is consequently negligible as it can at best be of $O(\ln N)$. Thus the equation (2.4.4) becomes

$$k\ \ln\ \Gamma(E,\ \Delta) = k\ \ln\ \Gamma\ (\bar{E}_1,\ \Delta) + k\ \ln\ \Gamma(\bar{E}_2,\ \Delta)$$

or

$$S_{\text{combined}} = S_1 + S_2 \qquad \text{... (2.4.5)}$$

where S_1 and S_2 are the entropies of the individual systems '1' and '2'.

The above exercise gives more information than exhibiting the additive property of entropy. It shows that when the two systems are put in contact, the energies of the systems in equilibrium are given by $\bar{E}_1$ and $\bar{E}_2$, where $\bar{E}_1$ and $\bar{E}_2$ are such that the product $\Gamma(E_1)\ \Gamma(E_2)$ is maximized, i.e. for $E_1 = \bar{E}_1$ and $E_2 = \bar{E}_2$,

$$\delta\ [\Gamma(E_1)\ \Gamma(E_2)] = 0 \qquad \text{... (2.4.6)}$$

subject to

$$\delta\ E_1 + \delta\ E_2 = 0 \qquad \text{... (2.4.7)}$$

since the total energy $E_1 + E_2$ does not change apart from fluctuations of $O(\Delta)$ which are negligible on the macroscopic scale. From Eq. (2.4.6)

$$\frac{1}{\Gamma(E_1)} \frac{\partial \Gamma(E_1)}{\partial E_1}\bigg|_{\bar{E}_1} \delta E_1 + \frac{1}{\Gamma(E_2)} \frac{\partial \Gamma(E_2)}{\partial E_2}\bigg|_{\bar{E}_2} \delta E_2 = 0$$

or

$$\frac{\partial \ln \Gamma(E_1)}{\partial E_1}\bigg|_{\bar{E}_1} \delta E_1 + \frac{\partial \ln \Gamma(E_2)}{\partial E_2}\bigg|_{\bar{E}_2} \partial E_2 = 0 \qquad \text{... (2.4.8)}$$

Now using Eq. (2.4.7) with a Lagrange multiplier λ, we have

$$\left(\frac{\partial \ln \Gamma(E_1)}{\partial E_1}\bigg|_{E_1} + \lambda\right) \delta E_1 + \left(\frac{\partial \ln \Gamma(E_2)}{\partial E_2}\bigg|_{\bar{E}_2} + \lambda\right) \delta E_2 = 0 \qquad \text{... (2.4.9)}$$

leading to

$$\frac{\partial}{\partial E_1} \ln \Gamma(E_1)\bigg|_{\bar{E}_1} = -\lambda = \left|\frac{\partial \ln \Gamma(E_2)}{\partial E_2}\right|_{\bar{E}_2} \qquad \text{... (2.4.10)}$$

Using Boltzmann's hypothesis

$$\left(\frac{\partial S_1}{\partial E_1}\right)_{\bar{E}_1} = \left(\frac{\partial S_2}{\partial E_2}\right)_{\bar{E}_2} \qquad \text{... (2.4.11)}$$

which, on using the thermodynamic relation $T^{-1} = (\partial S/\partial E)_{V,N}$, leads to $T_1 = T_2$ and thus the important result that if two systems are put in contact and come to equilibrium, then the equilibrium condition implies that the temperature of the two systems are equal, which is the expected result.

It should be borne in mind that this equal a priori probability for microcanonical ensemble is all that one requires for obtaining the probability distribution for the other ensembles. The next chapter will establish this. To end this section we once more raise the question that we claim is the unifying theme behind the statistical mechanics assumption namely, does chaotic dynamics justify all the three tenets discussed in Sec. 2.1. As we saw at the end of the previous section, loss of correlation between trajectories would justify Boltzmann's hypothesis and ergodicity, but would it also justify the equal a priori probabilities. We will discuss the problem in the next section.

2.5 Dynamics, Dynamical Systems and Chaos

As we have repeatedly emphasized, the ultimate justification of the postulates of statistical mechanics lies in the microscopic dynamics exhibiting chaotic trajectories. This leads to loss of correlation due to the sensitive dependence on initial conditions. This particular feature of dynamics is what we intend to explore in this section. To do so, we do not need to worry about an N-particle system—only a few degrees of freedom are necessary to establish these concepts. Consequently, in this section we will not talk about large N, rather we will discuss simple systems like oscillators, forced oscillators or two coupled oscillators. We first need to establish certain terminologies. The equations of motion in each case will obviously be Newton's law which has the general form

$$\ddot{q} = f(q, p, t) \qquad \text{... (2.5.1)}$$

where q is the position coordinate and p is the momentum (we take the mass to be unity for convenience) and the force on the right hand side of Eq. (2.5.1) is the most general possible including, if need be, an explicit dependence on time. It is clear that this second order differential equation can be written as two first order differential equations by writing

$$\dot{q} = p \qquad \text{... (2.5.2a)}$$

$$\dot{p} = f(q, p, t) \qquad \text{... (2.5.2b)}$$

We consider a few examples:

Harmonic oscillator

$$\ddot{X} + \omega^2 X = 0 \qquad \text{... (2.5.3)}$$

This can be written as

$$\dot{X} = Y \qquad \text{... (2.5.3a)}$$

$$\dot{Y} = -\omega^2 X \qquad \text{... (2.5.3b)}$$

Damped harmonic oscillator

$$\ddot{X} + 2k\,\dot{X} + \omega^2 X = 0 \qquad \text{... (2.5.4)}$$

This can be written as

$$\dot{X} = Y \quad \text{... (2.5.4a)}$$

$$\dot{Y} = -\,2kY - \omega^2 X \quad \text{... (2.5.4b)}$$

Non-linear damped oscillator

$$\ddot{X} + 2k\,\dot{X} + \omega^2 X + \lambda X^3 = 0$$

This can be cast as

$$\dot{X} = Y \quad \text{... (2.5.5a)}$$

$$\dot{Y} = -\,2kY - \omega^2 X - \lambda X^3 \quad \text{... (2.5.5b)}$$

Forced damped harmonic oscillator

$$\ddot{X} + 2k\,\dot{X} + \omega^2 X = f_0 \cos \Omega t \quad \text{... (2.5.6)}$$

This can be recast as

$$\dot{X} = Y \quad \text{... (2.5.6a)}$$

$$\dot{Y} = -\,2kY - \omega^2 X + f_0 \cos \Omega t \quad \text{... (2.5.6b)}$$

Yet another way of rewriting it yields

$$\dot{X} = Y \quad \text{... (2.5.7a)}$$

$$\dot{Y} = -\,2kY - \omega^2 X + f_0 Z \quad \text{... (2.5.7b)}$$

$$\dot{Z} = -\,\Omega\sqrt{1 - Z^2} \quad \text{... (2.5.7c)}$$

Forced damped non-linear oscillator

$$\ddot{X} + 2k\,\dot{X} + \omega^2 X + \lambda X^3 = f_0 \cos \Omega t \quad \text{... (2.5.8)}$$

which can be put in the form

$$\dot{X} = Y \quad \text{... (2.5.8a)}$$

$$\dot{Y} = -\,2kY - \omega^2 X - \lambda X^3 + f_0 \cos \Omega t \quad \text{... (2.5.8b)}$$

From the above it is clear that the equation of motion can always be written as a set of first order differential equations. A set of first-order differential equations is called a dynamical system. The dynamical system can be linear [Eqs. (2.5.3a) and (2.5.3b), Eqs. (2.5.4a) and (2.5.4b) and Eqs. (2.5.6a) and (2.5.6b)] or non-linear [Eqs. (2.5.5a) and (2.5.5b) and the set of Eqs. (2.5.8a) and (2.5.8b)]. It can also have a right hand side which has no explicit time dependence [Eqs. (2.5.5a) and (2.5.5b)] in which case we have an autonomous system, or a right hand side which has an explicit time dependence [Eqs. (2.5.6a) and (2.5.6b)] in which case the system is called non-autonomous. The number of first order equations in the system constitutes the dimension of the dynamical system. It should be noted that by re-defining variables a non-autonomous system can be made into an autonomous one but the dimensionality changes [Eqs (2.5.6a) and (2.5.6b); and Eqs (2.5.7a), (2.5.7b) and (2.5.7c)]. Thus an n-dimensional autonomous dynamical system is characterized by n variables $x_1, x_2, \ldots\ldots, x_n$ satisfying

$$\begin{aligned} \dot{x}_1 &= f_1\ (x_1,\ x_2, \ldots\ldots\ldots, x_n) \\ \dot{x}_2 &= f_2\ (x_1,\ x_2, \ldots\ldots\ldots, x_n) \\ &\vdots \\ &\vdots \\ \dot{x}_n &= f_n\ (x_1,\ x_2, \ldots\ldots\ldots, x_n) \end{aligned} \quad \ldots \textbf{(2.5.9)}$$

Initial conditions on $x_1, x_2, \ldots\ldots\ldots, x_n$ have to be specified and then integrating the above system yields a trajectory which is the set $\{x_i(t)\}$ for all values of t. The space spanned by $\{x_i\}$ is the phase space of the dynamical system and the phase trajectory is a path traced out in this space. Trajectories may be closed, e.g. those for the system represented by Eqs. (2.5.3a) and (2.5.3b) or open but heading for a definite point in phase space, e.g. those for the system represented by Eqs. (2.5.4a) and (2.5.4b).

A few representative trajectories are shown in Figs. 2.2a and 2.2b.

The different trajectories correspond to different initial conditions and it should be noted that a common feature of the above trajectories is that two neighbouring trajectories triggered by two neighbouring initial conditions remain close (Fig. 2. 1a) or come closer together (Fig. 2.1b). This is what is meant by trajectories being correlated—if initial conditions are close the trajectories remain close as well—the 'memory' of initial conditions is always retained. Closed trajectories are also obtained from Eqs. (2.5.6a) and (2.5.6b) but these are different from those shown in Fig. (2.2a). The closed trajectories

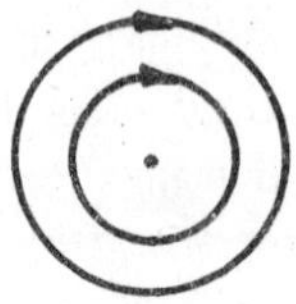

Fig. 2.2a: Trajectories, initially close, always remain so.

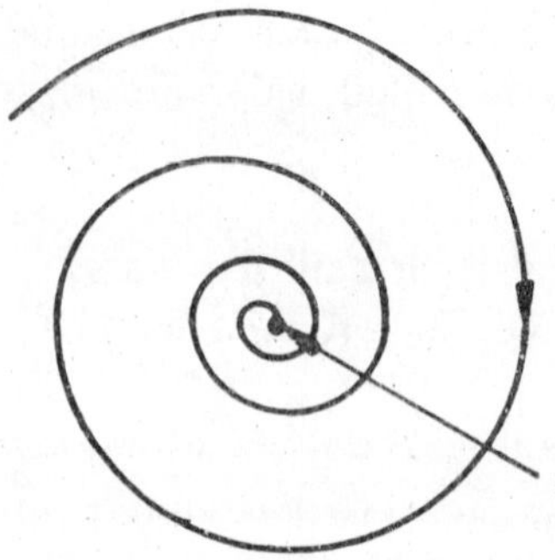

Fig. 2.2b: Trajectories end on the fixed point and hence come closer together.

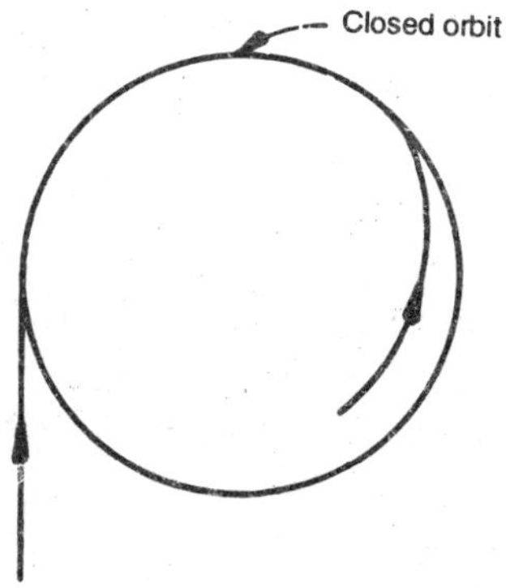

Fig. 2.3: Trajectories end up on the fixed closed orbit.

(Fig. (2.2a)) for the system of Eqs. (2.5.3a) and (2.5.3b) are obtained because of a conservation law which leads to $\omega^2 x^2 + y^2 =$ const., but the closed trajectory of the system of Eqs. (2.5.6a) and (2.5.6b) is obtained because of the balance between dissipative forces and the periodic external forces. Once again neighbouring trajectories remain close to each other since independent of the initial conditions the trajectories fall on the same closed orbit (Fig. 2.3).

Qualitatively different solutions are obtained for the system of Eqs. (2.5.8a) and (2.5.8b) if the various parameters (k, f_0 and Ω) are properly adjusted. Since an analytical solution is not obtainable in this case, we have to consider even simpler systems to explore these qualitatively different solutions. This cannot be achieved by differential equations but by maps which are obtainable from differential equations through a technique due to Poincare: For an n-dimensional dynamical system, this map (Poincare map) is constructed by considering an $(n - 1)$ dimensional hypersurface and noting the successive intersections of an evolving trajectory with this hypersurface. The j-th intersection is an $(n-1)$ dimensional vector $\{x_i^{(j)}\}$, where i ranges from 1 to $n-1$. The map is constructed as the recurrence relation

$$\{x_i^{i+1}\} = F_i\left(\{x_i^{(j)}\}\right) \qquad \text{... (2.5.10)}$$

For a system with dissipation (e.g. damped oscillator) the simplest such map can be one-dimensional, as in

$$x^{(j+1)} = f\ \{x^{(j)}\} \qquad \text{... (2.5.11)}$$

whereas for a conservative system, the simplest such map must be two-dimensional. The conservation law of the dynamical system must lead to a conserved quantity in the map and the quantity which can be preserved in two dimensions under a mapping is the area. So for Hamiltonian systems, the simplest mapping is a two-dimensional area-preserving one.

At first we discuss the dissipative system, where a simple one-dimensional mapping of the sort shown in Eq. (2.5.11) can result. The simplest of such mapping exhibiting complicated behaviour is the Bernoulli shift, defined by

$$x_{n+1} = 2\,x_n \pmod 1 \qquad \text{... (2.5.11)}$$

i.e.

$$x_{n+1} = \begin{cases} 2\,x_n & \text{if } x_n < \dfrac{1}{2} \\ 2\,x_n - 1 & \text{if } x_n > \dfrac{1}{2} \end{cases} \qquad \text{... (2.5.12)}$$

To appreciate the action of this map let us write any number between 0 and 1 in the binary representation as the infinite series:

$$a_0\left(\frac{1}{2}\right)^0 + a_1\left(\frac{1}{2}\right)^1 + a_2\left(\frac{1}{2}\right)^2 + \ldots$$

where the a's can be 0 or 1 and $a_0 = 0$. If the number has to be greater than half, then $a_1 = 1$ while for a number less than half $a_1 = 0$. An arbitrary number, thus can be represented by the string $(a_1, a_2, a_3, \ldots)$. The action of the map on this string is clearly to shift the string one step to the left, i.e. $(a_1, a_2, a_3, \ldots)$ goes to $(a_2, a_3, a_4, \ldots)$. The extreme sensitivity to initial conditions of this operation can now become apparent. Let us assume that we have two starting points very close to each other, i.e. the strings have the first few a's identical. Specifically, let us suppose that the strings differ at the 100th place, i.e. we have two starting points x_0 and $\overline{x}_0$ such that

$$x_0 = (a_1, a_2, a_3, \ldots\ldots\ldots\ldots a_{100} \ldots\ldots\ldots) \quad \ldots (2.5.13a)$$

$$\overline{x}_0 = (a_1, a_2, a_3, \ldots\ldots\ldots\ldots \overline{a}_{100} \ldots\ldots\ldots) \quad \ldots (2.5.13b)$$

Clearly x_0 and $\overline{x}_0$ differ by $1/2^{100}$. After 100 iterations by the Bernoulli shift operation

$$x_{100} = (a_{100}, a_{101}, a_{102}, \ldots) \quad \ldots (2.5.14a)$$

$$\overline{x}_{100} = (\overline{a}_{100}, \overline{a}_{101}, \overline{a}_{102}, \ldots) \quad \ldots (2.5.14b)$$

and these numbers differ by O(1). This is the extreme sensitivity to intial conditions that for certain parameter values can be an intrinsic feature of a dynamical system (the system has got to be non-linear and at least three-dimensional if it is autonomous and at least two-dimensional if non-autonomous). The extreme sensitivity to initial conditions can be formalized by an index called the Lyapunov index λ. For the one dimensional map $x_{n+1} = f(x_n)$, we define it as

$$\lambda = \underset{n \to \infty}{\mathrm{Lt}} \frac{1}{N} \ln \left| \frac{f^N(x_0 + \varepsilon) - f^N(x_0)}{\varepsilon} \right|$$

$$= \underset{N \to \infty}{\mathrm{Lt}} \frac{1}{N} \ln \left| \frac{df^N(x_0)}{dx} \right|$$

$$= \underset{N \to \infty}{\mathrm{Lt}} \frac{1}{N} \ln \left| \prod_{i=0}^{N-1} \left. \frac{df}{dx} \right|_{x_i} \right|$$

$$= \underset{n \to \infty}{\mathrm{Lt}} \frac{1}{N} \sum_{i=0}^{n-1} \left| \ln \left. \frac{df}{dx} \right|_{x_i} \right| \quad \ldots (2.5.15)$$

In the above, $f^N(x)$ is the N times iterated function $[f^2(x) = f\{f(x)\}]$ and so on) and the successive iterates of x_0 are $x_1, x_2, \ldots$ Clearly from the first line of Eq. (2.5.15), $f^N(x_0 + \varepsilon)$ will differ by orders of magnitude from

$f^N(x_0)$ if $\lambda > 0$. Thus if the mapping shows sensitivity to initial conditions, the Lyapunov index $\lambda > 0$. If $\lambda \leq 0$, then we have the iterates maintaining their separation or coming closer. For the Bernoulli shift discussed above it should be clear that $\lambda = \ln 2 > 0$, which is consistent with the sensitivity to initial conditions that we have demonstrated.

We now demonstrate that the Bernoulli shift generates a random sequence. To do so we consider a well known random sequence—the coin toss. We will show that given any sequence of heads and tails in a coin toss experiment, the Bernoulli shift can reproduce it. To do so we make a convention: if the coin toss produces heads, then a number greater than half will be generated and if a coin toss produces tails, then a number less than half will be generated. Now let an arbitrary sequence be

H T T H H H T T H T H T H H T T T ...

Accordingly, we start with the initial number for generating a Bernoulli sequence as

1 0 0 1 1 1 0 0 1 0 1 0 1 1 0 0 0 ...

This number is greater than half and corresponds to H. Now a shift moves things to the left by a step and we generate a T since the new number is less than half. As we go on generating the new numbers, the random toss sequence is generated.

The dynamics described above is called chaotic—the main characteristic is the existence of a positive Lyapunov index—this is what leads to neighbouring initial conditions being stretched apart—the points do not go away to infinity since the map folds (the modulo one (Eq. 2.5.11) in the Bernoulli shift is responsible for the folding). They remain confined in a finite interval but an infinitesimal separation becomes a finite separation after sufficiently many time steps. The chaotic dynamics leads to loss of correlation (e.g. the effect of the randomness in the Bernoulli shift) and to the iterates being spread out over an interval often with equal probability.

The systems that one deals with in statistical mechanics are mainly Hamiltonian systems and hence the mapping if, it exists, would exhibit some invariance properties. As stated before, the simplest such mapping has got to be two-dimensional. We now illustrate the above points with such a two-dimensional area-preserving map:

$$\left.\begin{aligned} x_{n+1} &= 2x_n \\ y_{n+1} &= y_n/2 \end{aligned}\right\} \quad 0 \le x_n \le \frac{1}{2},\ 0 \le y_n \le 1 \qquad \text{... (2.5.16)}$$

$$\left.\begin{aligned} x_{n+1} &= 2x_n - 1 \\ y_{n+1} &= \frac{1}{2}(1 + y_n) \end{aligned}\right\} \quad \frac{1}{2} \le x_n \le \frac{1}{2},\ 0 \le y_n \le 1$$

It is obvious that the map is area preserving. It is equally plain that there is stretching in the x-direction identical to the Bernoulli shift and then a folding back. Initial points separated in the x-direction consequently rapidly diverge with a positive Lyapunov index ln 2. However, for separations in the y-direction alone the Lyapunov index is $-$ ln 2. The two-dimensional system has two Lyapunov indices, but it suffices that only one of them be positive for the dynamics to be chaotic. In general, for an n-dimensional system, there are n Lyapunov indices. Necessary and sufficient condition for chaotic behaviour is to have the largest Lyapunov index positive.

It would be interesting to find out if under the above chaos-exhibiting mapping, something resembling the postulate of equal a priori probability emerges. Consequently we ask whether an arbitrary probability distribution $f(x_0, y_0, n = 0)$ defined on the unit square with the normalization $\int f(x_0, y_0, n)\, dx_0\, dy_0 = 1$ will redistribute itself. Because of Liouville's theorem, it is clear that $f(x_0, y_0, n = 0)$ cannot change, i.e. $f(x_n, y_n, n) = f(x_{n-1}, y_{n-1}, n-1)$. So, we choose to work with a derived probability distribution, defined as

$$W\left(x_n, n\right) = \int_0^1 f\left(x_n, y_n, n\right) dy_n \qquad \text{... (2.5.17)}$$

The recursion relation for W can be found as follows:

$$\begin{aligned} W\left(x_n, n\right) &= \int_0^{1/2} f\left(x_n, y_n, n\right) dy_n + \int_{1/2}^1 f\left(x_n, y_n, n\right) dy_n \\ &= \int_0^{1/2} f\left(x_{n-1}, y_{n-1}, n-1\right) dy_n + \int_{1/2}^1 f\left(x_{n-1}, y_{n-1}, n-1\right) dy_n \\ &= \int_0^{1/2} f\left(\frac{x_n}{2}, 2y_n, n-1\right) dy_n + \int_{1/2}^1 f\left(\frac{1+x_n}{2}, y_n, n-1\right) dy_n \end{aligned}$$

$$= \left[W\left(\frac{x_n}{2}, n-1\right) + W\left(\frac{1+x_n}{2}, \ n-1\right)\right] \qquad \text{... (2.5.18)}$$

As $n \rightarrow \infty$, one obtains $W(x_n, n) \rightarrow 1$. This is an example of equal a priori probability in a restricted sense.

The main postulates of statistical mechanics seem to depend on the microscopic dynamics being chaotic. So it is important to ask, how natural is it for a Hamiltonian system to be chaotic?

2.6 The Prevalence of Chaos

We first note that for an integrable system, the motion in phase space cannot be ergodic. For a 2N-dimensional system, integrability means setting up a canonical transformation to the action angle variables with the N action variables as constants of motion. The trajectory thus lies on an N- dimensional torus and certainly cannot cover the 2N–1 dimensional energy hypersuface. Thus for a Hamiltonian $H(\{q\}, \{p\})$, which is integrable, we seek a canonical transformation to a set $\{J\}, \{\theta\}$, such that $H = H_0(\{J\})$. If this can be done,

$$\dot{J} = \frac{\partial H_0}{\partial \theta} = 0$$

$$\dot{\theta} = \frac{\partial H_0}{\partial J} = \omega\ (\{J\})$$

The resulting integrals are

$$J = \text{constant} \qquad \text{... (2.6.2)}$$

and

$$\theta = \omega(J)\ t + \text{constant}$$

The canonical transformation is generated by $S(q, J)$ which satisfies

$$p = \frac{\partial S}{\partial q}, \quad \theta = \frac{\partial S}{\partial J} \qquad \text{... (2.6.3)}$$

As an example consider the harmonic oscillator,

$$H = \frac{1}{2}\left(p^2 + \omega^2 q^2\right)$$

We use the generator S to write,

$$\frac{1}{2}\left[\left(\frac{\partial S}{\partial q}\right) + \omega^2 q^2\right] = H_0 (J)$$

so that

$$\left(\frac{\partial S}{\partial q}\right)^2 = H_0 - \omega^2 q^2$$

and J is determined by

$$J = \frac{1}{2\pi} \oint \frac{\partial S}{\partial q} dq = \frac{H_0}{\omega}$$

leading to

$$H_0 = \omega J$$

and the equations of motion

$$J = \text{constant}$$

$$\theta = \omega t + \text{constant} \qquad \dots (2.6.4)$$

For a $2N$-dimensional integrable system, there would be effectively N constants $J_1, J_2, \dots J_N$ with N frequencies $\omega_1, \omega_2, \dots \omega_N$ which, in general, would depend on the $\{J\}$ and thus the motion would take place on an N-dimensional torus. For $N = 2$, the corresponding torus is shown in Fig. 2.4. It is obvious that trajectories are confined on an N-dimensional torus.

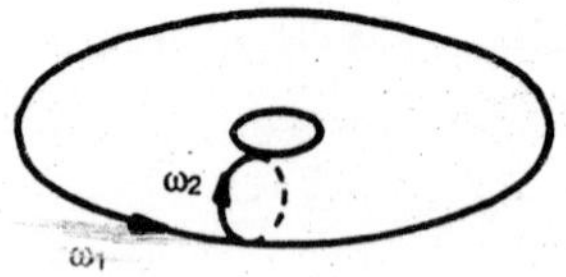

Fig. 2.4: A 2-torus.

We now ask what happens if this unperturbed system with trajectories on an N-torus is subject to a perturbation, i.e. to $H_0(\vec{J})$ we add $\varepsilon H_1(\vec{J}, \vec{\theta})$. If the system is still integrable, then we will be able to define a new set of action variables $\vec{J'}$, and a new generating function $S'(\vec{J'}, \vec{\theta})$, such that $J_i' = \dfrac{\partial S'}{\partial \theta_i}$, $\theta_i = \dfrac{\partial S'}{\partial J_i'}$ and

$$H\left(\frac{\partial S'}{\partial \theta_i}, \theta_i\right) = \bar{H}_o(J'_i) \quad \text{... (2.6.5)}$$

We anticipate the expansion,

$$S'(\vec{J'}, \vec{\theta}) = \vec{\theta}.\vec{J'} + \in S_1(\vec{J'}, \theta) \quad \text{... (2.6.6)}$$

Inserting this expansion in Eq. (2.6.5) and working to O ($\in$), we get

$$H = H_0(\vec{J}) + \varepsilon H_1(\vec{J}, \vec{\theta}) = H_0\left(\frac{\partial S'}{\partial \vec{\theta}} - \varepsilon \frac{\partial S_i}{\partial \vec{\theta}}\right) + \varepsilon H_1(\vec{J}, \theta)$$

$$= H_0(\vec{J}) - \varepsilon \frac{\partial H_0}{\partial \vec{J}} \cdot \frac{\partial S_1}{\partial \vec{\theta}} + \varepsilon H_1(\vec{J}, \theta) \quad \text{... (2.6.7)}$$

For integrability, the right hand side has to be independent of θ, and hence the coefficient of ε must vanish. Thus to O (ε),

$$\vec{\omega} \cdot \frac{\partial S_1}{\partial \theta} = H_1(\vec{J}, \theta) \quad \text{... (2.6.8)}$$

where $\vec{\omega} = \partial H_0 / \partial \vec{J}$ are the frequencies of the unperturbed system. Expanding S_1 and H_1 in Fourier series as

$$S_1(\vec{J}, \theta') = \sum_{\vec{K}} S_{1,\vec{K}}(\vec{J}) e^{2\pi i \vec{K} \cdot \vec{\theta'}}$$

$$H_1(\vec{J}, \theta') = \sum_{\vec{K}} H_{1,\vec{K}}(\vec{J}) e^{2\pi i \vec{K} \cdot \vec{\theta'}} \quad \text{... (2.6.9)}$$

where the components of the vector $\vec{K}$ are integers, we obtain

$$S(\vec{J}, \vec{\theta}) = \vec{\theta} \cdot \vec{J} + i\varepsilon \sum_{\vec{K}} \frac{H_{1,\vec{K}}(\vec{J})}{2\pi \vec{K} \cdot \vec{\omega}(\vec{J})} + 0(\varepsilon^2) \quad \text{... (2.6.10)}$$

Clearly Eq. (2.6.10) does not work if $\vec{K}\cdot\vec{\omega} = 0$ for some values of $\vec{K}$. The values of $\vec{J}'$ so obtained define the resonant tori. These are typically the tori which are destroyed for any small $\varepsilon > 0$. The resonant tori are dense in the phase space of the unperturbed Hamiltonian. There are however, a very large number of strongly non-resonant tori. These satisfy

$$|\vec{K}\cdot\vec{\omega}| > C/|m|^{N+1} \qquad \text{... (2.6.11)}$$

where $|m| = |m_1| + |m_2| + \dots |m_N|$ and C is independent of m. For $\vec{\omega}$ satisfying Eq. (2.6.11), the series for S_1 converges if H_1 is analytic in θ. Under these conditions, similar series for S_2, S_3 ... etc. also converge. One would still have to argue the convergence of the sum of S_1, S_2, ... etc. to the desired action S'. This requires a more sophisticated perturbation theory than we have given. A superconvergent perturbation theory leads to the KAM (Kolmogorov-Arnold-Moser) theorem that for small ε most of the tori of the unpertubed integrable Hamiltonian survive. This would violate the ergodic hypothesis, but we note that since the resonant tori are dense, arbitrarily near the surviving tori there will be regions of phase space where the trajectories are not on the surviving tori. It is here that the trajectories are chaotic and the success of statistical mechanics lies on the existence of these dense regions containing chaotic orbits. For larger ε, of course, the surviving tori are destroyed and ergodicity etc. are obvious.

Problems

2.1 Langevin Dynamics: Consider a variable satisfying the Langevin equation $\dot{x} = -\Gamma x + f(t)$, where Γ is a constant and $f(t)$ is a random noise satisfying the correlation function $\langle f(t)\, f(t')\rangle = 2\Gamma\,\delta(t-t')$. Solve explicity for $x(t)$ and find the average $\underset{T\to\infty}{\text{Lt}}\ \frac{1}{T}\int_0^T \langle x^2(t)\rangle\, dt$. Now, consider the probability distribution associated with the random variable x. It will be shown later that as $t \to \infty$, this distribution acquires the time-independent form $p^*(x) = N e^{-x^2/2}$, where N is a normalization constant. Find the expectation value $\overline{x^2}$ of x^2 with this $p(x)$ and compare it with the average obtained before.

2.2 States in phase space: Consider a particle confined in a large cube of side L. The allowed energies of the particle are characterized by three integers n_1, n_2, n_3 and expressed as $E = \frac{\pi^2 \hbar^2}{2mL^2}\left(n_1^2 + n_2^2 + n_3^2\right)$. A particular set of values of n_1, n_2 and n_3 corresponds to a microstate. You are required to find the number of states accessible to the particle if the energy lies between E and $E + \Delta E$ in the limit of $L \to \infty$. (In this limit the number reduced to the volume of a spherical annulus. Your answer should reduce to $V \,.\, d^3 p/\hbar^3$).

2.3 D-dimensional sphere: Find the surface area of a unit sphere in D-dimensions. (Consider a D-dimensional space and evaluate the integral $\int_{-\infty}^{\infty} dx_1 \int_{-\infty}^{\infty} dx_2 \ldots \int_{-\infty}^{\infty} dx_D \, e^{-(x_1^2 + x_2^2 + \ldots x_D^2)}$ in two different ways.)

2.4 Differential equations and maps: Consider a bar of moment of inertial I and length l. It is fastened at one end to a frictionless pivot. The other end is subjected to a vertical periodic impulsive force of strength K/l applied at time $t = 0, \tau, 2\tau, \ldots$. Neglect gravity. Write down the equations of motion for the angle θ and the canonically conjugate momentum p_θ. Let p_n and θ_n be the values of p_θ and θ at the time $t = n\tau + O^+$, where O^+ is a positive infinitesimal. Setting $\tau/I = 1$, show that

$$p_{n+1} = p_n + k \sin \theta_{n+1}$$

$$\theta_{n+1} = (\theta_n + p_n) \text{ modulo } 2\pi$$

Find the fixed points that lie in the strip $\pi > p > -\pi$. Determine the stability as a function of K.

2.5 Consider the $2x$ modulo 1 map of Eq. (2.5.11) with noise, i.e. $y_{n+1} = 2y_n$ modulo 1 + noise. Assume that the noise is such that it changes randomly all the digits a_j in the binary representation of y for $j \geq 50$. Assume that you are given exact observations $y_0, y_1, y_2 \ldots y_T$ of the noisy orbit for a time T ($T >> 50$). Show that there is an initial condition x_0 such that the exact true orbit $x_0, x_1 \ldots x_T$ followed by the noiseless map, shadows the noisy orbit, i.e. in particular, show that x_0 can be so chosen that $|y_n - x_n| \leq 2^{-49}$ for all n from $n = 0$ to $n = T$.

Chapter Three

Canonical and Grand Canonical Ensembles

3.1 Canonical Ensemble

In the last chapter, we asserted that the postulate of equal a priori probability and the concept of microcanonical ensembles are all that we require to find the probability distribution in situations where the system under consideration did exchange energy and particles with its environment. Expoundir g this theme and constructing the connection between this statistical description and the thermodynamic one will be the purpose of this chapter. We begin with the situation where our system has a fixed number of particles, but can exchange energy with its environment. The question to be asked is what is the probability of the system having an energy E, when in equilibrium with its surrounding.

The technique to be used must employ a microcanonical ensemble according to our contention. So we consider our system (say S) together with the environment as an isolated system having a total energy lying between E_{tot} and $E_{tot} + \Delta$ and with all microstates corresponding to this macrostate equally probable. The probability $\rho(E)$ of our system S having an energy E is consequently proportional to the number of states $\Gamma(E)$ available to it, which, in turn, is equal to the number of states $\Gamma_e(E_{tot} - E)$ available to the environment, i.e.

$$\rho(E) \propto \Gamma_s(E) = \Gamma_e(E_{tot} - E) \qquad \text{... (3.1.1)}$$

The environment is much bigger than the system and hence the environment energy $E_e \cong E_{tot}$ and $E << E_{tot}$ or E_e. Consequently the E-dependence can be expanded in a Taylor series with only the lowest term retained. With this in mind, we now recast Eq. (3.1.1) as

$$\rho(E) \propto \Gamma_e(E_{tot} - E) \cong \Gamma_e(E_e - E)$$

$$= \Gamma_e(E_e) - E\frac{\partial \Gamma_e}{\partial E_e} + O(E^2)$$

$$\cong \Gamma_e(E_e)\left\{1 - E\frac{\partial \ln \Gamma_e}{\partial E_e}\right\} \qquad \text{... (3.1.2)}$$

This leads to,

$$\ln \rho(E) = \ln \Gamma_e(E_e) + \ln\left\{1 - E\frac{\partial \ln \Gamma_e}{\partial E_e}\right\}$$

$$= \ln \Gamma_e(E_e) - E\frac{\partial \ln \Gamma_e}{\partial E_e} + O(E^2)$$

$$\cong \ln \Gamma_e(E) - E\frac{\partial \ln \Gamma_e}{\partial E_e} \qquad \text{... (3.1.3)}$$

We now recognize that the entropy S_e of the environment is k in Γ_e and hence Eq. (3.1.3) can be written as

$$\ln \rho(E) = \ln \Gamma_e(E_e) - \frac{E}{k}\frac{\partial S_e}{\partial E_e} \qquad \text{... (3.1.4)}$$

Making use of the fact that $\partial S_e/\partial E_e = 1/T_e$ specifies the temperature of the environment and that in equilibrium the system temperature $T = T_e$, yields

$$\ln \rho(E) = \ln \Gamma_e(E_e) - \frac{E}{kT}$$

or

$$\rho(E) = e^{\Gamma_e(E_e)}\, e^{-E/kT}$$

$$= A\, e^{-E/kT} \qquad \text{... (3.1.5)}$$

where A is a constant and the last line follows since $\Gamma_e(E_e)$ is a constant so far as S is concerned. We thus find that if the system S is in thermal equilibrium with its environment and is characterized by a temperature T, then the probability of its having an energy E is given by Eq. (3.1.5). The distribution should be normalized to unity, which implies that

$$A = \left(\sum_{\text{all microstates}} e^{-E/kT} \right)^{-1}$$

$$= Z^{-1}$$

where

$$Z = \sum_{\text{microstates}} e^{-E/kT} \qquad \text{... (3.1.7)}$$

is called the canonical partition function of the system. In future, we will use the notation $\sum_E$ as the sum over all microstates.

We now have to rationalize the fact that the thermodynamic state is characterized by a definite energy, while the statistical description above envisages a distribution of energy. If one has a distribution, then a definite number can be obtained by considering the mean of the distribution. The mean will be obtained repeatedly if the fluctuation about the mean is small. That is where the contact with the macroscopic description must lie. The above distribution should give a very small fluctuation about the mean. This is what we now establish in the following.

The mean energy is obviously

$$\langle E \rangle = \frac{1}{Z} \sum_E E\, e^{-E/kT}$$

A derivative gives

$$\frac{\partial \langle E \rangle}{\partial T} = \frac{1}{Z} \sum \frac{E^2}{kT^2} e^{-E/kT}$$

$$- \frac{1}{Z^2} \frac{\partial Z}{\partial T} \sum E\, e^{-E/kT}$$

$$= \frac{1}{kT^2}\left[\frac{1}{Z}\sum E^2 \, e^{-E/kT}\right]$$

$$- \frac{1}{Z^2}\left(\frac{1}{kT^2}\sum E \, e^{-E/kT}\right)\sum E \, e^{-E/kT}$$

$$= \frac{1}{kT^2}\left[< E^2 > - < E^2 >\right]$$

$$= <(\Delta E)^2 > /kT^2 \qquad \dots (3.1.9)$$

where $<(\Delta E)^2>$ is the fluctuation about the mean, defined as

$$<(\Delta E)^2> = <(E - <E>)^2> = <E^2> - <E>^2 \qquad \dots (3.1.10)$$

From Eq. (3.1.9),

$$<(\Delta E)^2> = kT^2 \frac{\partial}{\partial T} <E> = kT^2 C \qquad \dots (3.1.11)$$

where C is the specific heat, which is $O(N)$, *where* N is the number or particles in the system.

The relative fluctuation is

$$\frac{<(\Delta E)^2>^{1/2}}{<E>} = \frac{k^{1/2} T C^{1/2}}{<E>} = O\left(\frac{1}{N^{1/2}}\right) \qquad \dots (3.1.12)$$

since $C^{1/2}$ is $O(N^{1/2})$ and $<E>$ is $O(N)$.

In the thermodynamic limit, $N \rightarrow \infty$ and hence the relative fluctuation decreases enormously. This is the reason why the statistical description is also an adequate description of the macroscopic thermodynamics.

We make the above point and the connection between statistical mechanics and thermodynamics even more transparent by rewriting the canonical distribution through the following considerations:

$$Z = \sum_E e^{-E/kT}$$

$$= \int \Gamma(E) \, e^{-E/kT} \, dE$$

$$= \int e^{\frac{S(E)}{k} - \frac{E}{kT}} dE$$

$$= \int e^{-1/k\,(E - TS)}\, dE \qquad \text{... (3.1.13)}$$

In the above $\Gamma(E)$ is the number of microstates with energy between E and $E + dE$ and we have made use of Boltzmann's hypothesis in the form $S = k \ln \Gamma(E)$.

We now examine the integrand of Eq. (3.1.3) and note that $E - TS$ is of O (N) and hence for very large N, the integrand will be dominated by the minimum of $E - TS$ (provided of course $E - TS$ has a minimum). The minimum of $E - TS$ is obtained as $E = \overline{E}$, where

$$1 = T\left(\frac{\partial S}{\partial E}\right)_{\overline{E}}$$

or

$$\left(\frac{\partial S}{\partial E}\right) = \frac{1}{T} \qquad \text{... (3.1.14)}$$

But this is the thermodynamic relation between entropy and temperature and hence $\overline{E}$ is the thermodynamic energy that one is familiar with. We expand $E - TS$ about $\overline{E}$ to write,

$$E - TS = \overline{E} + E - \overline{E} - TS(\overline{E}) - T\left(\frac{\partial S}{\partial E}\right)_{\overline{E}} \Delta E$$

$$- \frac{T}{2}\left(\frac{\partial^2 S}{\partial E^2}\right)_{\overline{E}} (\Delta E)^2 + O((\Delta E)^3)$$

$$= \overline{E} - TS(\overline{E}) - \frac{1}{2} T \left(\frac{\partial^2 S}{\partial E^2}\right)_{\overline{E}} (\Delta E)^2$$

$$= \overline{F} - \frac{T}{2}\frac{\partial}{\partial E}\left(\frac{1}{T}\right)(\Delta E)^2$$

$$= \overline{F} + \frac{T}{2}\frac{1}{T^2}\frac{1}{\partial E/\partial T}(\Delta E)^2$$

$$= \bar{F} + \frac{(\Delta E)^2}{2TC} \qquad \text{... (3.1.15)}$$

where $\bar{F} = \bar{E} - TS(\bar{E})$ is the thermodynamic free energy (Helmholtz free energy) and is the minimum value of $E - TS$ provided the correction term in Eq. (3.1.15) is positive which will happen for all $C > 0$, i.e. for positive specific heat. With the above relation, Eq. (3.1.14) can be written as

$$Z = e^{-\bar{F}/kT} \int e^{-(\Delta E)^2/2kT^2C}\, dE \qquad \text{... (3.1.16)}$$

Thus, the canonical distribution corresponds to a Gaussian fluctuation about the mean (thermodynamic) energy. The integral in Eq. (3.1.16) is clearly $O(C^{1/2})$ which is $O(N^{1/2})$ and hence

$$\ln Z = -\bar{F}/kT + \ln\{O(N^{1/2})\} \qquad \text{... (3.1.17)}$$

The first term on the right hand side is $O(N)$ and hence dominates in the thermodynamic limit leading to (we drop the bar now)

$$F \text{ (thermodynamic free energy)} = -kT \ln Z \qquad \text{... (3.1.18)}$$

the desired connection between a statistical mechanical quantity Z and the thermodynamic quantity F.

The connection between statistical mechanics and thermodynamics also ensues from using Boltzmann's relation in the form $\langle S \rangle = k \langle \ln \Gamma(E) \rangle$ by noting $\rho(E)$ is inversely proportional to $\Gamma(E)$ and hence the thermodynamic entropy

$$S = \langle S \rangle = -k \langle \ln (e^{-E/kT}/Z) \rangle$$

$$= k \ln Z + \langle E \rangle / T$$

$$= k \ln Z + E/T$$

or $F = -kT \ln Z$

3.2 Alternative Derivation

In this section, we take up the study of the canonical distribution from a different point of view. The emphasis now is on the canonical ensemble aspect and the fact that the distribution that we are seeking is an equilibrium distribution is established by appealing to the volumes occupied by different distributions in phase space. We assume our given system S can reside in various energy states characterized by the discrete energies $\varepsilon_1, \varepsilon_2, \cdots \varepsilon_n, \cdots$ This occurrence of discrete energies is usual in quantum mechanics—it should be borne in mind, however, that one is not dealing with quantum statistics as yet—and no quantum mechanical question about the state of the system is being asked. We simply assume the existence of discrete energy levels which is, however, a consequence of quantum mechanics.

Consider N mental copies of our system S and assume that n_1 of them have energy ε_1, n_2 have energy ε_2 and so on.

Clearly,

$$\sum n_i = N \qquad \text{... (3.2.1a)}$$

$$\sum \varepsilon_i n_i = E \qquad \text{... (3.2.1b)}$$

A simple example would be a collection of non-interacting gas molecules in thermal equilibrium. In this case, a single molecule would constitute our system S and the whole collection (all other molecules are mental copies of our chosen one) would be the entire ensemble.

A given configuration (n_i systems having energy ε_i) can be achieved in $P(\{n_i\})$ ways where $P(\{n_i\})$ is clearly given by

$$P(\{n_i\}) = \frac{N!}{\prod_i n_i!} \qquad \text{... (3.2.2)}$$

How does the set $\{n_i\}$ distribute itself in equilibrium? This is where the criterion for an equilibrium distribution comes in. We will see in our discussions of the non-equilibrium situations that if one waits long enough, an equilibrium distribution is set up—a system prepared in a non-equilibrium state has an extremely large probability of ending up in the equilibrium state. Now, every distribution occupies a volume in phase space. If the the natural tendency for an arbitrary distribution is to approach the equilibrium distribution, then it is reasonable to assume that the equilibrium distribution occupies the maximum volume in phase space. Thus the desired distribution is obtained by maximizing

P of Eq. (3.2.2) subject to the constraints of Eqs. (3.2.1a) and (3.2.1b). The maximization is clearly archieved by the set $\{\bar{n}_i\}$, where

$$\bar{n}_i = \bar{A}\, e^{-\beta \varepsilon_i} \qquad \text{... (3.2.3)}$$

((we maximize $\ln P = \ln N! - \sum_i \ln n_i!$

$$= N \ln N - N - \sum n_i \ln n_i + N = N \ln N - \sum_i n_i \ln n_i \, .$$

The differential $\delta \ln P = 0$ leads, after some elementary algebraical manipulation, to $\sum_i (\ln n_i)\, \delta n_i = 0$. The constraints $\sum_i \delta n_i = \sum_i \varepsilon_i\, \delta n_i = 0$ are taken care of by the Lagrange multipliers α and β and we have written $\bar{A} = e^{-\alpha}$ in Eq. (3.2.3).)

Consequently, the probability p_i of finding the system in an energy state ε_i is given by

$$p_i = A\, e^{-\beta\varepsilon_i} = \left(\sum_i e^{-\beta\varepsilon_i}\right)^{-1} e^{-\beta\varepsilon_i}$$

$$= \frac{1}{Z}\, e^{-\beta\varepsilon_i} \qquad \text{... (3.2.4)}$$

where Z, the canonical partition function, is $\sum_i e^{-\beta\varepsilon_i}$. To fix the Lagrange multiplier β, we may apply the result to the ideal gas in a volume V and calculate the pressure. This fixes the constant β as $(kT)^{-1}$. The distribution of Eq. (3.2.4) now agrees completely with that of Eq. (3.1.5).

We now want to demonstrate that the set $\{\bar{n}_i\}$ of Eq. (3.2.3) indeed maximizes $P(\{n_i\})$. The probability of seeing a given set $\{n_i\}$ among all possible sets is

$$\Omega(\{n_i\}) = \frac{P(\{n_i\})}{\sum_{\{n_j\}} P(\{n_j\})} \qquad \text{... (3.2.5)}$$

In principle, we would like to demonstrate that for $\{n_i\} \rightarrow \{\bar{n}_i\}$, $\Omega\,\{n_i\} \rightarrow 1$, but that being difficult, we intend to show that for $\{n_i\} \rightarrow \{\bar{n}_i\}$, the fluctuations in $\{n_i\}$ tend to zero. One begins by generalizing Eq. (3.2.2) to

$$P(\{n_i\}) = \frac{N!}{\prod_i n_i\,!}\, g_i^{n_i}, \qquad \text{... (3.2.6)}$$

where all the g_i are to be set equal to unity at the end of the calculation. We first calculate the average number $<n_i>$ of ensemble systems with energy ε_i as

$$<n_i> = \sum_{\{n_j\}} n_i\ \Omega\,(\{n_j\})$$

$$= g_i \frac{\partial}{\partial g_i}\left[\ln \sum_{\{n_j\}} P(\{n_j\})\right] \qquad \text{... (3.2.7)}$$

We now note the following identity

$$g_i \frac{\partial}{\partial g_i} <n_i> = g_i \frac{\partial}{\partial g_i} \sum_{\{n_j\}} \frac{n_i\ P(\{n_j\})}{\sum_{\{n_k\}} P(\{n_k\})}$$

$$= \sum_{\{n_i\}} \frac{n_1^2\ P(\{n_j\})}{\sum_{\{n_k\}} P(\{n_k\})} - \left[\sum_{\{n_k\}} \frac{n_i\ P(\{n_k\})}{\sum_{\{n_j\}} P(\{n_j\})}\right]^2$$

$$= <n_i^2> - <n_i>^2$$

$$= <(\Delta\, n_i)^2> \qquad \text{... (3.2.8)}$$

Hence the ensemble fluctuation in n_i is given by

$$<(\Delta n_i)^2>^{1/2} = \underset{\{g_i\} \rightarrow 1}{\text{Lt}} \left[g_i \frac{\partial}{\partial g_i} <n_i>\right]^{1/2} \qquad \text{... (3.2.9)}$$

It should be emphasized that these averages and fluctuations refer to a space of ensembles—averages are taken over various kinds of ensembles that can exist and the averages have nothing to do with the ensemble averages discussed in the previous section which were obtained by working within one ensemble, namely the canonical ensemble.

With the P changed to the form given by Eq. (3.2.6), the method of maximizing P leads to

$$\bar{n}_i = \bar{A}_i \, g_i \, e^{-\beta\varepsilon_i} = N \frac{g_i e^{-\beta\varepsilon_i}}{\sum g_j \, e^{-\beta\varepsilon_j}} \qquad \text{... (3.2.10)}$$

If we consider all the g's equal to 1, except g_i, which we take close to unity, them $\bar{n}_i \propto g_i$ and we can make the identification $\bar{n}_i \cong \langle n_i \rangle$ (see Eq. (3.2.7)). This leads to

$$g_i \frac{\partial}{\partial g_i} \langle n_i \rangle \cong g_i \frac{\partial}{\partial g_i} \bar{n}_i \cong \langle \bar{n}_i \rangle \cong \langle n_i \rangle \qquad \text{.... (3.2.11)}$$

Thus,

$$\frac{\{(\Delta n_i)^2\}^{1/2}}{\langle n_i \rangle} = \left\{ \frac{1}{\langle n_i \rangle} \right\}^{1/2}, \qquad \text{... (3.2.12)}$$

leading to the result that if $\langle n_i \rangle = \bar{n}_i$, then the fluctuation in $\langle n_i \rangle$ can be made arbitrarily small by making $\langle n_i \rangle$ large. Hence in the space of all distributions, almost all distributions are of the canonical form and if one starts from an arbitrary distribution, then there is an overwhelming probability that the canonical distribution will be obtained.

It is worthwhile pointing out exactly what is ignored in Eq. (3.2.11). To do so we differentiate directly Eq. (3.2.7) and obtain,

$$g_i \frac{\partial}{\partial g_i} \langle n_i \rangle = g_i \frac{\partial}{\partial g_i} \ln \sum_{\{n_j\}} P(\{n_j\}) + g_i \frac{\partial^2}{\partial g_i^2} \ln \sum_{\{n_i\}} P(\{n_j\})$$

$$= \sum_{\{n_j\}} n_j \frac{P(\{n_j\}}{\sum_{\{n_k\}} P\{n_k\})} + g_i \frac{\partial^2}{\partial g_i^2} \ln \sum_{\{n_j\}} P(\{n_j\})$$

$$= <n_i> + g_i \frac{\partial^2}{\partial g_i^2} \ln \sum_{\{n_j\}} P\ (\{n_j\}) \qquad \text{... (3.2.13)}$$

It is the second term on the right hand side of Eq. (3.2.13) that is dropped when we obtain $<(\Delta n_i)^2> \equiv <n_i>$.

To conclude this section, we note that the canonical partition function $Z = \sum_i e^{-\varepsilon_i/kT}$ need not always converge. A collection of non-ineracting hydrogen atoms is the classic example, where one is left with a clearly divergent sum of the form $\sum_{n=1}^{\infty} n^2\ e^{c/n^2}$, where n is an interger and c is a constant.

3.3 Varying Number of Particles

As a prelude to discussing distributions for systems where the particle number may vary, we first study the macroscopic properties of such systems. As stated briefly in Chapter 1, whenever there is an extra variable that can change (i.e. apart from temperature and volume), one has to think of the forces which can cause such changes and the resultant work done. When the number of particles changes, the relevant force is the chemical potential μ and the second law of thermodynamics takes the form

$$T\,dS = dE + p\,dV - \mu\ dN \qquad \text{(3.3.1)}$$

The sign of the last term can be appreciated from the fact that if a certain amount of heat is supplied to the system (positive $T\,dS$), there will be a propensity for the particles to leave, i.e. a negative dN, provided of course $\mu > 0$. Clearly from the above form of the second law

$$\left(\frac{\partial S}{\partial N}\right)_{E,V} = -\frac{\mu}{T} \qquad \text{... (3.3.2.)}$$

This is a useful relation for obtaining the chemical potential. Using Boltzmann's hypothesis we have obtained the entropy of an ideal gas in Sec. 2.3 as

$$S = k\,N\left[\ln \frac{V}{N} + \ln \left(\frac{E}{N}\right)^{3/2} + \frac{3}{2} \ln \frac{2\,\pi\, m}{h^2} + \frac{5}{2}\right], \qquad \text{... (3.3.3)}$$

leading to

$$-\frac{\mu}{T} = \left(\frac{\partial S}{\partial N}\right)_{E,V} = k\left[\ln\frac{V}{N} + \ln\left(\frac{E}{N}\right)^{3/2} + \frac{3}{2}\ln\frac{2\pi m}{h^2}\right]$$

$$= k\ln\left[\left(\frac{2\pi mkT}{h^2}\right)^{3/2}\frac{V}{N}\right]$$

or

$$e^{-\mu/kT} = \frac{V}{N}\cdot\left(\frac{2\pi mkT}{h^2}\right)^{3/2} \quad \text{... (3.3.4)}$$

We can now generate the different thermodynamic potentials. Corresponding to the Helmholtz free energy F, we define a variation $\overline{F}$ as

$$\overline{F} = E - TS - \mu N \quad \text{... (3.3.5)}$$

with

$$\overline{dF} = dE - T\,dS - S\,dT - \mu\,dN - N\,d\mu$$

$$= -S\,dT - p\,dV - N\,d\mu \quad \text{... (3.3.6)}$$

This leads to

$$N = -\left(\frac{\partial\overline{F}}{\partial\mu}\right)_{T,V} \quad \text{... (3.3.7)}$$

and also shows that for a thermodynamic system at constant T, V and μ, $\overline{F}$ is the quantity which has to be minimized at equilibrium. The Gibbs free energy G is defined in the usual manner, but its differential form changes to account for Eq. (3.3.1)

$$G = E - TS + PV$$

$$dG = dE - T\,dS + P\,dV + V\,dP$$

$$= -S\,dT + V\,dP + \mu\,dN \quad \text{... (3.3.8)}$$

leading to

$$\left(\frac{\partial G}{\partial T}\right)_{P,N} = -S, \left(\frac{\partial G}{\partial P}\right)_{T,N} = V, \left(\frac{\partial G}{\partial N}\right)_{T,P} = \mu \quad \text{... (3.3.9)}$$

We readily see that a thermodynamic system at constant T, P and N is characterized by the minimum of the Gibbs free energy.

We now set up an interesting connection between G and μ. The free energy can be expressed as function of any three independent variables and we choose P, T and N for our purpose. Clearly, the energy is proportional to N and so must have the form (Note that changing N, does not change P or T)

$$G = N\,\Phi\,(P,T) \quad \text{... (3.3.10)}$$

The differential reads

$$dG = \left(\frac{\partial G}{\partial P}\right)_{N,T} dP + \left(\frac{\partial G}{\partial T}\right)_{N,P} dT + \Phi\,dN$$

$$= V\,dP - S\,dT + \frac{G}{N}\,dN \quad \text{... (3.3.11)}$$

Comparing with Eq. (3.3.8), we see

$$G = \mu\,N \quad \text{... (3.3.12)}$$

the chemical potential is simply the Gibbs free energy per particle. The free energy $\overline{F}$, defined above, now becomes

$$\overline{F} = E - TS - \mu\,N$$

$$= E - TS - G$$

$$= E - TS - (E - TS + PV)$$

$$= -\,PV \quad \text{... (3.3.13)}$$

From Eq. (3.3.7), we now obtain

$$N = \frac{\partial}{\partial \mu} \,|\,(PV)\,|_{T,V} \quad \text{... (3.3.14)}$$

This relation is extremely useful for calculating the chemical potential of a gas as will become evident later.

3.4 Grand Canonical Ensemble

We will now discuss the probability distribution of a system which has variable energy and a variable number of particles. The thermodynamic state of such a system is characterized by a definite temperature and a definite chemical potential—this is made possible by the system being in equilibrium with a surrounding with which it can exchange energy and particles. Our derivation will be almost identical to the derivation of ρ for the canonical ensemble in Sec. 3.1. We will focus on our system S which is in contact with the environment and the combination of S and its enviornment constitutes an isolated system characterized by E_{tot}, N_{tot} and V_{tot} for energy, particle number and volume. At any instant, our system S has energy E, particle number N and volume V (fixed) where $E, N, V << E_{tot}, N_{tot}, V_{tot}$ respectively. The probability that S has energy E and particle number N, is proportional to the number of microstates for which this occurs which is exactly equal to the number of microstates for which the environment has energy $E_{tot} - E$ and particle number $N_{tot} - N$. Thus

$$\rho\ (E, n) \propto \Gamma\,(E,N) \propto \Gamma_e\left(E_{tot} - E,\ N_{tot} - N\right)$$
$$\cong \Gamma_e\left(E_e - E,\ N_e - N\right) \qquad \text{... (3.4.1)}$$

In the last step we have assumed that the environment energy and particle number are very nearly equal to the total. Working to O (E/E_e) and (N/N_e), we get

$$\rho\ (E,N) \propto \Gamma_e\left(E_e, N_e\right) - E\left(\frac{\partial \Gamma_e}{\partial E_e}\right)_N - N\left(\frac{\partial \Gamma_e}{\partial N_e}\right)_E$$

$$= \Gamma_e\left(E_e, N_E\right)\left[1 - E\left(\frac{\partial \ln \Gamma_e}{\partial E_e}\right)_N - N\left(\frac{\partial \ln \Gamma_e}{\partial N_e}\right)_E\right]$$

$$= \Gamma_e\left(E_e, N_e\right)\left[1 - \frac{E}{k}\left(\frac{\partial S}{\partial E_e}\right)_N - \frac{N}{k}\left(\frac{\partial S}{\partial N}\right)_E\right]$$

$$= \Gamma_e\left(E_e, N_e\right)\left[1 - \frac{T}{kT} + \frac{\mu N}{kT}\right] \qquad \text{... (3.4.2)}$$

where the Boltzmann relation and the definition of T and μ have been made use of in lines three and four of Eq. (3.4.2).

Once again working to linear order in E and N

$$\ln \rho\ (E, N) \cong \ln \Gamma_e \left(E_e\,, N_e\right) - \frac{(E - \mu N)}{kT},$$

leading to

$$\rho\ (E, N) = A\ e^{\frac{-(E - \mu N)}{kT}} \qquad \text{... (3.4.3)}$$

where A is a constant as far as S is concerned, the system having nothing to do with E_e and N_e. Thus, the grand canonical distribution function is given by Eq. (3.4.3) and the grand partition function G is obtained by summing over all the microstates, so that

$$Q = \sum_{E,N} e^{-(E - \mu N)/kT}\ . \qquad \text{... (3.4.4)}$$

where, as before, $\sum\limits_{E,N}$ stands for sum over all microstates.

We can now write the distribution function as

$$\rho\ (E,N) = \frac{1}{Q}\ e^{-(E - \mu N)/kT} \qquad \text{... (3.4.5)}$$

Once again, the contact with thermodynamics depends crucially on the fact that the distribution is very sharply centred around $\overline{E}$ and $\overline{N}$ in the limit of $\overline{N} \to \infty$ and hence we must demonstrate that the above distribution does lead to very small fluctuations in <E> and <N>.

We begin with the fluctuation in the average particle number, by noting that

$$\langle N \rangle = \frac{1}{Q} \sum_{E,N} e^{-\beta(E - \mu N)} \qquad \text{... (3.4.6)}$$

$$\left(\frac{\partial \langle N \rangle}{\partial \mu}\right)_{V,T} = \frac{\beta}{Q} \sum N^2\ e^{-\beta(E-\mu N)} - \frac{\beta}{Q^2}\left(\sum N\, e^{-\beta(E - \mu N)}\right)^2$$

$$= \beta \left[<N^2> - <N>^2 \right]$$

$$= \beta < (\Delta N)^2> \qquad \dots (3.4.7)$$

where $\Delta N = N - <N>$. To proceed further, we relate the derivative $\partial <N>/\partial \mu$ to the susceptibility. Our aim of showing vanishingly small fluctuation has of course already been satisfied by Eq. (3.4.7) since $\partial <N>/\partial \mu$ is $O(<N>)$ and hence $<(\Delta N)^2>^{1/2} / <N>$ is $O(<N>^{-1/2})$ and becomes vanishingly small as $<N> \to \infty$. In an identical manner the smallness of $<(\Delta E)>^{1/2} / <E>$ can be established as we will explicitly demonstrate later. Now,

$$\left| \frac{\partial <N>}{\partial \mu} \right|_{V,T} = \left| \frac{\partial <N>}{\partial P} \right|_{V,T} \Big/ \left| \frac{\partial \mu}{\partial P} \right|_{V,T}$$

$$= \frac{N}{V} \left| \frac{\partial <N>}{\partial P} \right|_{V,T} \qquad \dots (3.4.8)$$

where we have made use of Eq. (3.3.14). Now P is a function of N/V and T and hence differentiating with respect to N and differentiating with respect to V are going to be simply related as

$$\left| \frac{\partial P}{\partial V} \right|_{N,T} = - \frac{N}{V} \left(\frac{\partial P}{\partial N} \right)_{V,T} \qquad \dots (3.4.9)$$

Consequently Eq. (3.4.8) becomes

$$\left(\frac{\partial N}{\partial \mu} \right)_{V,T} = \frac{N}{V} \left| \frac{\partial N}{\partial P} \right|_{V,T} = - \frac{N}{V} \cdot \frac{N}{V} \cdot \left(\frac{\partial V}{\partial P} \right)_{N,T}$$

$$= - n^2 N \frac{\partial}{\partial P} \left| (V/N) \right|_{N,T}$$

$$= n^2 N \chi_T , \qquad \dots (3.4.10)$$

where $n = N/V$, the number density and $\chi_T = - \partial/\partial P \, (1/n) \, |_{N,T}$ is the isothermal susceptibility, Thus,

$$<(\Delta N)^2> = kT \cdot n^2 N \chi_T \qquad \dots (3.4.11)$$

and as expected $<(\Delta N)^2>^{1/2} / N$ does go to zero for $N \to \infty$, unless χ_T blows up which it can near a second-order phase transition.

To find the energy fluctuation, we proceed by finding

$$\frac{\partial <E>}{\partial \beta} = \frac{\partial}{\partial \beta} \frac{1}{Q} \sum E\, e^{-\beta(E - \mu N)}$$

$$= -\frac{1}{Q} \sum E(E - \mu N)\, e^{-\beta(E - \mu N)} + \frac{1}{Q^2} \sum E\, e^{-\beta(E - \mu N)} \times \sum e^{-\beta(E - \mu N)}$$

$$= -<(\Delta E)^2> + [<EN> - <E><N>] \qquad \text{... (3.4.12)}$$

and

$$\frac{\partial <E>}{\partial \mu} = \frac{\beta}{Q} \sum E N e^{-\beta(E - \mu N)} - \frac{\beta}{Q^2} \sum E e^{-\beta(E - \mu N)} \times \sum N e^{-\beta(E - \mu N)}$$

$$= \beta [<EN> - <E><N>]$$

$$= \beta <(\Delta E)(\Delta N)> \qquad \text{... (3.4.13)}$$

Combining Eqs. (3.4.12) and (3.4.13), we find

$$<(\Delta E)^2> = kT^2 \frac{\partial <E>}{\partial T} + \mu kT \frac{\partial <E>}{\partial \mu} \qquad \text{... (3.4.14)}$$

Once again $<(\Delta E)^2>$ is O (N) and hence $<(\Delta E)^2>^{1/2} / <E>$ is O ($N^{-1/2}$) which is vanishingly small as $N \to \infty$. Thus the validity of the thermodynamic distribution is established for the macroscopic system and we now establish the passage from the grand partition function to a thermodynamic variable. Before doing so it is worth pointing out that Eq. (3.4.13) shows that the fluctuations in E and N are not independent of each other and gives the strength of the correlation of the fluctuation.

Boltzmann's formula once more provides the quantitative connection with thermodynamics:

$$S = -k < \ln \frac{1}{Q} e^{-\beta(E - \mu N)} >$$

$$= \frac{1}{T} <E - \mu N> + k \ln Q$$

or

$$TS = E - \mu N + kT \ln Q$$

or

$$-kT \ln Q = E - TS - \mu N = \overline{F} = -PV,$$

leading to

$$PV = kT \ln Q \qquad \text{... (3.4.15)}$$

We have made use of Eq. (3.3.13) in the final step. Combining Eqs. (3.4.15) and (3.3.14), we can obtain the equation of state of a gas.

We finally show that the grand canonical distribution is equivalent to a Gaussian distribution in fluctuation in energy and particle number. To do so, we write the grand canonical partition function as

$$Q = \int e^{-\beta(E - \mu N)} \Gamma(E,N)\, dE\, dN$$

$$= \int e^{-(E - TS - \mu N)/kT}\, dE\, dN \qquad \text{... (3.4.16)}$$

Now the argument of the exponential in the above integrand is O(N) and in the limit of $N \to \infty$ will be dominated by the minimum of $E - TS - \mu N$, provided the expression does have a minimum. The conditions for extremization are

$$1 = T \left(\frac{\partial S}{\partial E} \right)_{V,N} \qquad \text{... (3.4.17a)}$$

and

$$-\mu = T\left(\frac{\partial S}{\partial N}\right)_{V,E} \qquad \text{... (3.4.17b)}$$

The solutions of the above equations give $E = \bar{E}$ and $N = \bar{N}$. Now, we write

$$E - TS - \mu N = \bar{E} - \mu\bar{N} - TS(\bar{E}, \bar{N})$$

$$-\frac{T}{2}\left[\left(\frac{\partial^2 S}{\partial E^2}\right)_{N,V}(\Delta E)^2 + 2\frac{\partial^2 S}{\partial E\,\partial N}\Delta E\ \Delta N + \left(\frac{\partial^2 S}{\partial N^2}\right)_{E,V}(\Delta N)^2\right]$$

$$= -PV + \frac{T}{2}\left[\alpha(\Delta E)^2 + 2\beta(\Delta E)(\Delta N) + \gamma(\Delta N)^2\right] \qquad \text{... (3.4.18)}$$

where we have recognized that $\bar{E}$ and $\bar{N}$ are the thermodynamic values of $\bar{E}$ and $\bar{N}$ [see Eqs. (3.4.17a) and (3.4.17 b)] and set

$$\alpha = -\frac{\partial^2 S}{\partial E^2}, \quad \beta = -\frac{\partial^2 S}{\partial E \partial N}, \quad \gamma = -\frac{\partial^2 S}{\partial N^2} \qquad \text{... (3.4.19)}$$

For $E - TS - \mu N$ to have a minimum, the quadratic form in the square brackets in Eq. (3.4.18) must be positive definite, which can be achieved if $\alpha > 0, \gamma > 0$ and $\alpha\,\gamma > \beta^2$. We can now write

$$Q = e^{PV/kT}\int d(\Delta E)\, d(\Delta N)\, e^{-\frac{1}{2K}\left[\alpha(\Delta E)^2 + 2\beta(\Delta E)(\Delta N) + \gamma(\Delta N)^2\right]} \qquad \text{... (3.4.20)}$$

The integral is $O(N)$ and hence taking the logarithm in Eq. (3.4.20),

$$kT \ln Q = PV + O(\ln N) \qquad \text{... (3.4.21)}$$

As $N \to \infty$, Eq. (13.4.15) is regained as PV is $O(N)$ and completely dominates. We can use Eq. (3.4.20) to obtain the mean value of the fluctuation $(\Delta E)^2$, $(\Delta N)^2$ and $(\Delta E)(\Delta N)$ and, as promised, Eq. (3.4.20) shows that the grand canonical distributions can be viewed as quadratic distribution of fluctuations in E and N.

Problems

3.1 Consider a body placed in an external field, such that its energy $E(p,q)$ can be written as

$$E(p,q) = E_0(p,q) + V(p,q)$$

where $V(p, q)$ is a "potential" representing the contribution of the external field to the energy. Treating V as a perturbation show that the free energy is given by (to second order in V)

$$F = F_0 + \overline{V} - \frac{1}{2T} \langle (V - \overline{V})^2 \rangle$$

where F_0 is the free energy in the absence of V.

3.2 Find the canonical distribution function for a body rotating with angular velocity Ω.

3.3 Find the mean square fluctuation of the energy using V and T as the independent variables.

3.4 With variables V and T, find

(i) $\langle \Delta T\, \Delta P \rangle$, (ii) $\langle \Delta V\, \Delta P \rangle$, (iii) $\langle \Delta S\, \Delta V \rangle$, (iv) $\langle \Delta S\, \Delta T \rangle$

3.5 Consider a system distributed over its accessible states r in accordance with an arbitrary probability distribution p_r and let its entropy be defined as $S = -k \Sigma P_r \ln P_r$. We have the normalization $\Sigma P_r = 1$. For a canonical distribution $P_r^{(0)} = e^{-\beta E_r} / \Sigma_r e^{-\beta E_r}$. We constrain the mean energies to be the same, i.e.

$$\sum P_r^{(0)} E_r = \sum P_r E_r = E$$

The entropy for the canonical distribution is $S_0 = -k \Sigma P_r^{(0)} \ln P^{(0)}$. Show that $S - S_0 = k \sum_r P_r \ln P_r^{(0)}/P_r$ and that $S_0 \geq S$, with the equality holding for $P_r = P_r^{(0)}$. Hence for a given mean energy, the canonical distribution maximizes the entropy.

Chapter Four

Statistical Mechanics of Non-interacting Systems

4.1 Ideal Gas

In this chapter we will illustrate the use of the ensembles, general statistical mechanical concepts and their connection with thermodynamics by taking up examples of simple non-interacting classical systems. The best known of such systems is the ideal gas—a collection of N identical molecules confined in a volume V at a temperature T. The general problem is to calculate the pressure P as a function of N, V and T. This relation is called the equation of state. The energy of the N-particle system is

$$E = \sum_{i=1}^{N} \frac{p_i^2}{2m} \qquad \text{... (4.1.1)}$$

where it is to be understood that each p_i^2 is $p_i^2 = p_{ix}^2 + p_{iy}^2 + p_{iz}^2$, when we are dealing with three-dimensional space. Fixing N, we use the canonical distribution to write the partition function as

$$Z = \int e^{-\sum \frac{p_i^2}{2mkT}} \, d\Gamma$$

$$= \int e^{-\sum \frac{p_i^2}{2mkT}} \frac{d^{3N}p\, d^{3N}q}{h^{3N}} \cdot \frac{1}{N!}$$

$$= \frac{V^N}{N!} \frac{1}{h^{3N}} \left[\int e^{-\frac{p^2}{2mkT}} d^3 p \right]^N$$

$$= \frac{V^N}{N!} \left(\frac{2m\pi kT}{h^2} \right)^{\frac{3N}{2}} \qquad \text{... (4.1.2)}$$

Note that in the second step the $N!$ has entered the denominator for correct counting when the particles are indistinguishable. It should be pointed out that a single gas molecule could be taken to constitute one of the systems in a canonical ensemble with the rest of the molecules providing the environment and in that case the single molecule partition function Z_s would be

$$Z_s = \int \frac{d^3 q\, d^3 p}{h^3} e^{-p^2/2mkT}$$

$$= \left(\frac{2\pi mkT}{h^2} \right)^{\frac{3}{2}} V \qquad \text{... (4.1.3)}$$

Since we are dealing with an N-particle system with no interaction, the full partition function would be expected to be

$$Z = Z(N) = Z_s^N \qquad \text{(4.1.4)}$$

and indeed would be true for distinguishable particles. But for identical (classical) particles, we note that Eqs. (4.1.3) and (4.1.4) do not reproduce Eq. (4.1.2) and in this case Eq. (4.1.4) has to be modified to

$$Z = Z(N) = Z_s^N / N! \text{ (identical classical particles)} \qquad \text{... (4.1.5)}$$

We are now in a position to make contact with thermodynamics. This comes from Eq. (3.1.18), whence we write the free energy as

$$F = -kT \ln Z = -kT \left[N \ln V/N + \frac{3N}{2} \ln \frac{2\pi m kT}{h^2} + N \right] \qquad \text{... (4.1.6)}$$

Thermodynamic quantities follow by differentiation

Equation of state: $P = -\left(\frac{\partial F}{\partial V}\right)_T = \frac{NkT}{V}$ ... **(4.1.7)**

The well known relation for ideal gases is reproduced.

Mean energy: The thermodynamic energy E is found as

$$E = < E > = \frac{1}{Z}\sum_i \int \frac{p_i^2}{2m} e^{-\Sigma \frac{p_i^2}{2mkT}} d\Gamma$$

$$= 3N\, kT\; \Gamma(3/2)/\Gamma\left(\frac{1}{2}\right) = \frac{3}{2} NkT \quad \text{... (4.1.8)}$$

Alternatively, from Eq. (4.1.6)

$$S = -\left(\frac{\partial F}{\partial T}\right)_V = k\left[N \ln\frac{V}{N} + \frac{3N}{2} \ln \frac{2\pi\, mkT}{h^2} + \frac{5}{2}N\right] \quad \text{... (4.1.9)}$$

Now

$$E = F + TS = \frac{3}{2} NkT, \text{ in agreement with Eq. (4.1.8).}$$

Specific heat: This follows from differentiation of the energy as

$$C = \frac{\partial E}{\partial T} = \frac{3}{2} Nk \quad \text{... (4.1.10)}$$

It should be noted that a constant volume is implicit in the above differentiation and what we have obtained is the specific heat at constant volume.

Fluctuation in <E>:

$$< (\Delta E)^2 > = kT^2 \frac{\partial E}{\partial T} = \frac{3}{2} N (kT)^2, \quad \text{... (4.1.11)}$$

Consequently,

$$\frac{<(\Delta E)^2>^{1/2}}{<E>} = \left(\frac{3}{2}N\right)^{1/2} \cdot \frac{kT}{\frac{3}{2}NkT} = \left(\frac{2}{3}\right)^{1/2} \cdot \frac{1}{\sqrt{N}} \quad \text{... (4.1.12)}$$

As expected, the fractional fluctuation vanishes for $N \rightarrow \infty$, and <E> is indeed the thermodynamic energy.

4.2 Ideal Gas in the Grand Canonical Ensemble

The ideal gas is a sufficiently important example to warrant yet another treatment. We now consider the collection of N-particles in the box of volume V to be in contact with a surrounding with which it can exchange both energy and particles. The grand canonical partition function is then

$$G = \sum_N \frac{1}{N!} \int e^{-\sum_i \frac{p_i^2}{2mkT}} e^{\frac{\mu N}{kT}} \frac{d^{3N}p\, d^{3N}q}{h^{3N}}$$

$$= \sum \frac{V^N}{N!} \left(\frac{2\pi mkT}{h^2}\right)^{3N/2} e^{\mu N/kT}$$

$$= \sum_N \frac{z^N}{N!} = e^z \quad \text{... (4.2.1)}$$

where $z = V\left(\frac{2\pi mkT}{h^2}\right)^{3/2} \cdot e^{\mu/kT}$. The thermodynamic connection comes from

$$PV = kT \ln Q = kT z, \quad \text{... (4.2.2)}$$

while we use Eq. (3.3.14) to obtain the chemical potential as

$$N = \frac{\partial}{\partial \mu}(PV) = kT \frac{\partial z}{\partial \mu} = z \quad \text{... (4.2.3)}$$

Thus,

$$e^{\mu/kT} = \frac{N}{V}\left(\frac{h^2}{2\pi m kT}\right)^{3/2} \quad \text{... (4.2.4)}$$

in agreement with Eq. (3.3.4) obtained from the microcanonical picture. Elimination of z between Eqs. (4.2.2) and (4.2.3) yields the equation of state as $PV = NKT$.

We end this section by establishing that the fluctuations in N and E are negligible in the grand canonical ensemble. Using Eq. (3.4.10), Eq. (4.2.3) and the definition of z

$$< (\Delta N)^2 > = kT \frac{\partial N}{\partial \mu} = kT \frac{\partial}{\partial \mu} (z) = kT \frac{z}{kT} = z = N$$

or

$$\frac{< (\Delta N)^2 >^{1/2}}{N} = \frac{1}{N^{1/2}} \qquad \text{... (4.2.5)}$$

For the fluctuation in E, Eq. (3.4.17) yields the relevant information. First, we calculate the expectation value of E as

$$<E> = \frac{1}{Q} \sum_N \frac{1}{N!} \int \frac{d^{3N} q \, d^{3N} p}{h^{3N}} \left(\sum \frac{p_i^2}{2m} \right) e^{-\Sigma \frac{p_i^2}{2mkT}} e^{\frac{\mu N}{kT}}$$

$$= \frac{3}{2} \frac{kT}{Q} \sum_N \frac{1}{(N-1)!} z^N = \frac{3}{2} \frac{kTz}{Q} e^z$$

$$= \frac{3}{2} kTz = \frac{3}{2} NkT,$$

in agreement with Eq. (4.1.4). Now, we find

$$<(\Delta E)^2 > = kT^2 \frac{\partial <E>}{\partial T} + \mu kT \frac{\partial <E>}{\partial \mu}$$

$$= kT^2 \left[\frac{3}{2} Nk + \frac{3}{2} kT \frac{\partial N}{\partial T} \right] + \frac{3\mu}{2} (kT)^2 \frac{\partial N}{\partial \mu}$$

$$= \frac{15}{4} N (kT)^2$$

after straightforward algebraic manipulation. The fractional fluctuation is

$\frac{[<(\Delta E)^2>]^{1/2}}{<E>} = \frac{\sqrt{15}}{2}\,\frac{2}{3}\,\frac{1}{\sqrt{N}} = \sqrt{\frac{5}{3N}}$ which vanishes in the thermodynamic limit.

4.3 Space-dependent Distribution

4.3.1 In a Gravitational Potential

We consider a collection of non-interacting molecules in a given volume V where the gravitational force acts in the negative z-direction. The whole mass of gas is at a constant temperature T so that equilibrium concepts apply. The energy of a single molecule is

$$E = \frac{p^2}{2m} + mgz \qquad \text{... (4.3.1)}$$

and the probability of finding the molecule with energy between E and $E + dE$ is

$$P\,(E) = \frac{1}{Z}\, e^{\frac{p^2}{2mkT}}\, e^{-\frac{mgz}{kT}}$$

The number of molecules with z-coordinate between z and $z + dz$ is found by integrating over all the other coordinates, i.e. p_x, p_y, p_z, x and y. Thus

$$P\,(z)\;dz = \frac{1}{Z}\int dx\;dy \int \frac{d^3p}{h^3}\, e^{-\frac{p^2}{2mkT}}\, e^{-\frac{mgz}{kT}}\;dz$$

$$= P\,(0)\; e^{-mgz/kT}\;dz, \qquad \text{... (4.3.2)}$$

where $P(0)$ is a constant which correctly normalizes the distribution. In the true atmosphere this distribution does not obtain as the temperature does not remain constant throughout.

4.3.2 In a Centrifuge

We consider a collection of macromolecules (large organic molecules) rotated with an angular velocity in an ultra centrifuge. In the rotating frame of reference, the molecules feel a centrifugal force $m\,\omega^2 r$ radially outward. This force can be obtained from a potential

$$\vec{F} = m\omega^2 r\,\hat{r} = -\frac{\partial V}{\partial r}\,\hat{r}$$

($\hat{r}$ is the unit vector in the radial direction)
or

$$V = -\frac{1}{2}\,m\,\omega^2 r^2 \qquad \dots (4.3.3)$$

so that the total energy of a molecule in the rotating frame is

$$E = \frac{p^2}{2m} - \frac{1}{2}\,m\omega^2 r^2 \qquad \dots (4.3.4)$$

with the probability distribution

$$P(E) = \frac{1}{Z}\,e^{-\frac{p^2}{2mkT}}\,e^{\frac{m\omega^2 r^2}{2kT}} \qquad \dots (4.3.5)$$

Arguments identical to those in Subsection 4.3.1 lead to

$$P(r)\,dr = P(0)\,e^{\frac{m\omega^2 r^2}{2kT}}\,dr \qquad \dots (4.3.6)$$

for the radial distribution of the molecules.

As expected, the molecules crowd towards larger radii. We have taken the rotation speed to be very high and consequently ignored the effect of the Coriolis force. Also note that the number distribution must be the same in rotating and non-rotating frames.

The above distribution leads to a practical method of obtaining the molecular weight m. By determining the concentrations $\rho(r_1)$ and $\rho(r_2)$ at two radii r_1 and r_2, we find from Eq. (4.3.6)

$$\frac{\rho(r_1)}{\rho(r_2)} = e^{\frac{1}{2}m\omega^2\left(r_1^2 - r_2^2\right)/kT}$$

or

$$m = \frac{2kT}{\omega^2\left(r_1^2 - r_2^2\right)} \ln \left(\frac{\rho(r_1)}{\rho(r_2)}\right) \qquad \text{... (4.3.7)}$$

4.4 Oscillators in Contact with Heat Bath

4.4.1 Classical Oscillators

We now consider a set of N oscillators in three dimensional space in equilibrium with a heat bath characterized by a temperature T. The energy of each oscillator is

$$E = \frac{p^2}{2m} + \frac{1}{2} k_s r^2, \qquad \text{... (4.4.1)}$$

where k_s is the spring constant. The partition function is

$$Z = Z_s^{3N} = \left(\int \frac{dpdx}{h} e^{-\frac{p^2}{2mkT}} e^{-\frac{1}{2}\frac{k_s x^2}{kT}}\right)^{3N}, \qquad \text{... (4.4.2)}$$

where Z_s is the single oscillator partition function in one dimension and the average energy

$$E = N \cdot E_s = 3N \cdot \frac{1}{Z_s} \int \frac{dpdx}{h} \left(\frac{p^2}{2m} + \frac{1}{2} k_s x^2\right) \times e^{-\frac{p^2}{2mkT}} e^{-\frac{1}{2}\frac{k_s x^2}{kT}}$$

$$= 3\,NkT \qquad \text{... (4.4.3)}$$

Note that all our results so far—the ideal gas and the oscillators—give a total energy which corresponds to ½ kT for each degree of freedom (i.e. each independent quadratic term in the energy expression). This is a general result known as the classical equipartition theorem. Note that the quadratic form for the energy is essential. The reader should verify that the equipartition theorem is not valid for the quartic oscillator, e.g. where the energy expression (in one dimension) is $E = \frac{p^2}{2m} + \frac{1}{2} \lambda x^4$.

4.4.2 Quantum Oscillators

We now consider a set of one-dimensional oscillators in thermal equilibrium but with the energy levels given by quantum mechanics. The energy levels are

$$E_n = \left(n + \frac{1}{2}\right)\hbar\,\omega \qquad \text{... (4.4.4)}$$

The single oscillator partition function is

$$Z_s = \sum e^{-\left(n+\frac{1}{2}\right)\frac{\hbar\,\omega}{kT}}$$

$$= e^{-\hbar\,\omega/2kT}\left(1 - e^{-\frac{\hbar\,\omega}{kT}}\right)^{-1} \qquad \text{... (4.4.5)}$$

The average energy E_s is given by

$$E_s = \frac{\hbar\,\omega}{Z}\sum\left(n + \frac{1}{2}\right)e^{-\left(n+\frac{1}{2}\right)\frac{\hbar\,\omega}{kT}}$$

$$= \hbar\,\omega\left[\frac{1}{e^{\hbar\,\omega/kT} - 1} + \frac{1}{2}\right] \qquad \text{... (4.4.6)}$$

For N oscillators in three dimensions

$$E = 3N\,E = 3N\,\hbar\,\omega\left[\frac{1}{e^{\hbar\,\omega/kT} - 1} + \frac{1}{2}\right] \qquad \text{... (4.4.7)}$$

The point to note is that for $kT >> \hbar\,\omega$, the first term within the brackets dominates and we can expand the exponential to find the leading terms giving

$$E = 3NkT \qquad \text{... (4.4.8)}$$

in agreement with the classical result of Eq. (4.4.3). On the other hand, for $T \to 0$, it is the second term within the brackets in Eq. (4.4.7) that dominates and provides the average energy for $T = 0$. It corresponds to the quantum fluctuations and is the biggest departure from the classical result.

4.5 Specific Heat of Solids

The foregoing discussion of the statistical mechanics of oscillators has a direct application in the evaluation of specific heat of solids. The model of the solids that we assume considers atoms (nucleus together with the tightly bound core electrons constituting an atom) arranged on a regular lattice with the effect of loosely bound electrons ignored in the present picture. The atoms vibrate with small amplitude about their mean positions on the lattice and the energy of the solid comes from the energy associated with these simple harmonic vibrations. Assuming that the oscillators are governed by classical mechanics and considering the molar specific heat ($Nk = R$), we see from Eq. (4.4.3)

$$C = \frac{\partial}{\partial T} E = 3R \qquad \text{... (4.5.1)}$$

independent of temperature and material. This universal result is correct at high temperatures and is known as Dulong and Petit's law. At low temperatures substantial deviation from this result occurs and as $T \rightarrow 0$, the specific heat tends to zero in accordance with Nernst's heat theorem.

To treat the low temperature situation, we consider the vibrations to be governed by quantum mechanics and further assume that each oscillator is characterized by the same frequency. This is the Einstein model of solids. The molar specific heat now follows from a differentiation of Eq. (4.4.7) to be

$$C = 3R \left(\frac{\hbar\,\omega}{kT}\right)^2 \frac{e^{\hbar\,\omega/kT}}{(e^{\hbar\,\omega/kT} - 1)^2} \qquad \text{... (4.5.2)}$$

For $kT >> \hbar\,\omega$, we recover $C = 3R$ as expected, but for $T \rightarrow 0$, we find (with $T_E = \hbar\,\omega/k$)

$$C \cong 3R \left(\frac{\hbar\,\omega}{kT}\right)^2 e^{-\hbar\,\omega/kT} = 3R \left(\frac{T_E}{T}\right)^2 e^{-T_E/T} \qquad \text{... (4.5.3)}$$

making the specific heat vanish exponentially at low temperatures. This removes the difficulty with Nernst's heat theorem but is not borne out experimentally. For $T << T_E = \hbar\,\omega/k$, the decrease of C is not exponential as Eq. (4.5.3) would suggest. Instead C vanishes as an algebraic function of T and the reason for the discrepancy lies in the drastic assumption of all oscillators having the same frequency.

We begin by considering the situation at low frequencies, where the vibrational modes must correspond to the normal modes of oscillation of an elastic medium of the same size and the same elastic constants as the crystal we are considering. At low temperatures, the high frequency modes are frozen out and hence the low frequency modes will give an adequate description. Hence if we correctly enumerate the low frequency vibrational modes, the low temperature behaviour will be correctly obtained.

We consider the solid in the shape of a cube with side L. The low frequency vibrations that we are considering will form standing waves inside the solid with the possible wavelengths in the x, y and z directions as

$$\lambda_i = L/n_i, \; i = x, y \text{ and } z \text{ and } n_i \text{ is an integer} \qquad \text{... (4.5.4)}$$

If the propagating velocity is c, then the possible frequencies are

$$\nu_i = c \cdot n_i/L$$

or

$$\nu^2 = \frac{c^2}{L^2}\left(n_x^2 + n_y^2 + n_z^2\right) \qquad \text{... (4.5.5)}$$

If we now ask how many modes of vibration exist between frequencies ν and $\nu + d\nu$, then the issue is one of finding the number of n_x, n_y and n_z that satisfy Eq. (4.5.5). The different points in n_x, n_y and n_z space satisfying $n_x^2 + n_y^2 + n_z^2 \leq \nu^2 L^2/c^2$, can be taken in the limit of very large L to form a continuum and occupy the entire sphere of radius $\nu L/c$. The volume of the annulus between radii corresponding to ν and $\nu + d\nu$, then gives the number of available points (n_x, n_y, n_z), and thus the number of available modes. This is clearly $4\pi\, n^2\, dn = 4\pi\, (\nu^2/c^3)\, L^3\, d\nu$. This number corresponds to a given value of c. But there are three possible waves—two transverse with velocity c_T and one longtitudinal with velocity c_L. So the total number of modes of vibration $g(\nu)\, d\nu$ between a frequency ν and $\nu + d\nu$ is

$$g(\nu)\, d\nu = 4\pi\, \nu^2\, V\left[\frac{2}{c_T^3} + \frac{1}{c_L^3}\right] d\nu$$

$$= \frac{12 V \pi \nu^2}{c^3} \cdot d\nu \qquad \text{... (4.5.6)}$$

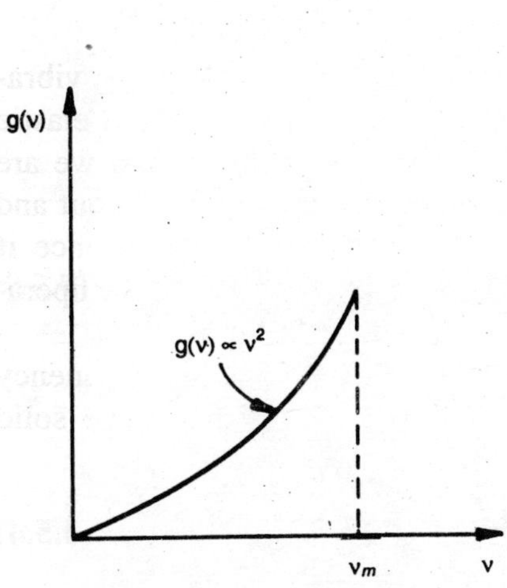

Fig. 4.1: The density of states

where

$$\frac{2}{c_T^3} + \frac{1}{c_L^3} = \frac{3}{c^3}\,.$$

The total number of modes cannot exceed $3N$ for N three-dimensional oscillators and hence there must be an upper cut-off ν_m on the available frequencies such that (see Fig. 4.1)

$$3N = \int_0^{\nu_m} g(\nu)\, d\nu = \frac{12\pi V}{c^3} \times$$

$$\times \int_0^{\nu_m} \nu^2\, d\nu = \frac{4\pi V}{c^3}\, \nu_m^3 \qquad \text{... (4.5.7)}$$

This upper limit corresponds to an upper limit on the wave-number and hence a lower limit on the wavelength beyond which the elastic continuum which we have assumed loses meaning. This lower limit on the wavelength corresponds approximately to the lattice spacing.

We can now evaluate the mean energy (we drop the zero-point term since it cannot give rise to specific heat) and find

$$E = \int g(\nu)\, \frac{h\nu}{e^{h\nu/kT} - 1}\, d\nu$$

$$= \frac{12\pi V}{c^3} \int_0^{\nu_m} h\nu\, \frac{\nu^2\, d\nu}{e^{h\nu/kT} - 1}$$

$$= \frac{12\pi V}{c^3}\, \frac{(kT)^4}{h^3} \int_0^{h\nu_m/kT} \frac{x^3\, dx}{e^x - 1} \qquad \text{... (4.5.8)}$$

Now, if we consider very low temperatures ($kT << \hbar\ \nu_m$), then the upper limit on the integration can be raised to infinity and we obtain

$$E = \frac{12\pi V}{c^3} \cdot \frac{3N c^3}{4\pi V \nu_m^3} \frac{(kT)^4}{h^3} \int_0^\infty \frac{x^3\, dx}{e^x - 1}$$

$$= 9NkT \left(\frac{kT}{h\,\nu_m}\right)^3 I, \qquad \text{... (4.5.9)}$$

where

$$I = \int_0^\infty x^3 \left(e^x - 1\right)^{-1} dx$$

$$= \int_0^\infty x^3\, e^{-x} \left[\sum_{n=0}^{\infty} e^{-nx}\right] dx \text{ (expanding } (1 - e^{-x})^{-1} \text{ in a geometric series)}$$

$$= \sum_{n=0}^{\infty} \frac{1}{(n+1)^4} \int_0^\infty y^3\, e^{-y}\, dy$$

$$= \Gamma(4) \sum_{m=1}^{\infty} \frac{1}{m^4} = \frac{\pi^4}{15} \qquad \text{... (4.5.10)}$$

(Using the definition of the gamma function $\Gamma(x) = \int_0^\infty t^{x-1} e^{-t}\, dt$. Note that $\Gamma(x + 1) = x!$ if x is an integer)

Thus for $T << T_D = \hbar\, \nu_m/k$, we have

$$E = N \cdot \frac{3}{5} \pi^4\, kT \left(\frac{T}{T_D}\right)^3 \qquad \text{... (4.5.11)}$$

and hence the specific heat comes out as

$$C = \frac{\partial E}{\partial T} = N \cdot \frac{12}{5} \pi^4 \cdot k \cdot \left(\frac{T}{T_D}\right)^3 \qquad \text{... (4.5.12)}$$

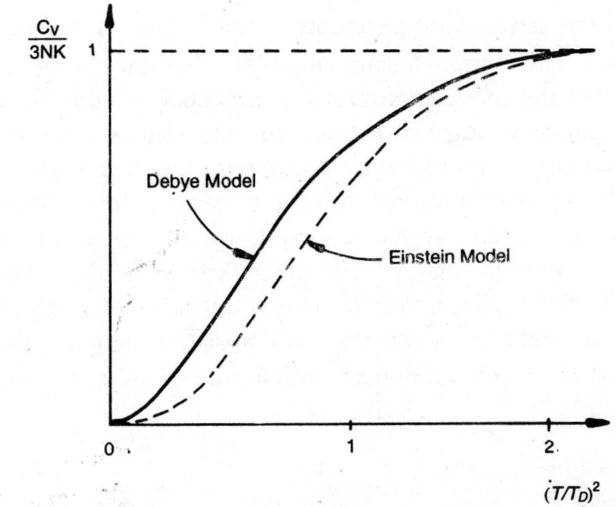

Fig. 4.2: Specific heat in the Debye and Einstein models

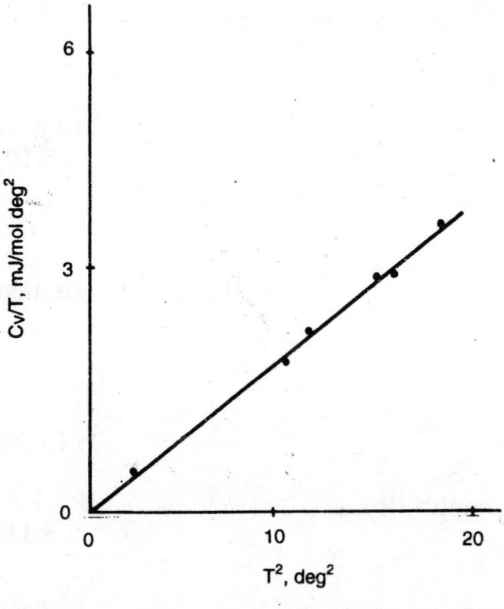

Fig. 4.3: Comparison of experimental data with Debye theory for KCl. It is conventional to plot C_v/T, so that the electronic contribution (Chapter 7) if present can show up as a non-zero intercept.

Thus the zero in the density of states at low frequencies (see Eq. (4.5.6) mutes the strong temperature dependence in the specific heat that arises in the Einstein model (Fig 4.2). This model of the specific heat of solids is due to Debye and the T^3-law of Eq. (4.5.12) is borne out very well experimentally (Fig. 4.3). Needless to say, at the higher frequencies this model breaks down as enumeration of the density of states becomes more complicated.

4.6 Model for Paramagnetism

In one of the problems at the end of this chapter we will show that classical physics cannot exhibit paramagnetism or diamagnetism. Yet in this section we will be discussing paramagnetism in the context of classical statistical mechanics. This somewhat paradoxical situation can be understood when we point out that the model assumes the existence of individual magnetic moments (the origin of which lies in quantum mechanics) and what we intend to discuss is how a collection of such magnetic moments (i.e. atoms with magnetic moments) will orient themselves when put in an external magnetic field at a given temperature T. The simplest model, which we begin working with assumes that the individual magnetic moments can only align with or against the direction of the field. If we call the magnetic moment μ, then in the external field H, there are only two energy values possible, namely $E_1 = -\mu H$ ($\vec{\mu}$ and $\vec{H}$ parallel) and $E_2 = \mu H$ ($\vec{\mu}$ and $\vec{H}$ antiparallel). The single moment partition function is clearly

$$Z_s = e^{\mu H/kT} + e^{-\mu H/kT} = 2\cosh\frac{\mu H}{kT} \qquad \text{... (4.6.1)}$$

For the N-particle system, the free energy is

$$F = -kT \ln Z = -NkT \ln Z_s$$

$$= -NkT \ln 2\cosh\frac{\mu H}{kT} \qquad \text{... (4.6.2)}$$

The magnetization follows as:

$$M = -\frac{\partial F}{\partial H} = \mu N \tan h\frac{\mu H}{kT} \qquad \text{... (4.6.3)}$$

For $kT >> \mu H$, we have

$$M = \mu^2 NH/kT \qquad \text{... (4.6.4)}$$

which leads to the high temperature susceptibility as

$$\chi = \left.\frac{\partial M}{\partial H}\right|_{H\to 0} = \frac{N\mu^2}{kT} \qquad \text{(4.6.5)}$$

the usual Curie law for temperature dependence of susceptibilities. At very low temperatures, $M \to \mu N$, and the approach of M to μN as $T \to 0$ is given by

$$M \cong \mu N \left[1 - e^{-\frac{2\mu H}{kT}}\right] \qquad \dots (4.6.6)$$

[An alternative way of arriving at Eq. (4.6.3) simply requires the evaluation of $<\mu>$ using the canonical distribution and this immediately leads to

$$M = N < \mu > = N \frac{\mu\, e^{\mu H/kT} - \mu e^{-\mu H/kT}}{e^{\mu H/kT} + e^{-\mu H/kT}} = N \mu \tanh \frac{\mu H}{kT}$$

where $e^{\mu H/kT}$ is the probability for $+\mu$ and $e^{-\mu H/kT}$ is the probability for $-\mu$.]

This model is important enough to be explored from the point of view of density of states and Boltzmann's hypothesis. If the total number of moments be N and the total energy E, then we need to evaluate $\Gamma(E,N)$—the total number of configurations under these conditions. If n_1 be the number of spins pointing parallel to the field and n_2, the number pointing antiparallel, then the total energy is given by,

$$E = \left(n_2 - n_1\right)\mu H \qquad \dots (4.6.7)$$

Together with $N = n_1 + n_2$, this yields

$$n_2 = \frac{1}{2}\left(\frac{E}{\mu H} + N\right) \qquad \dots (4.6.8a)$$

$$n_1 = \frac{1}{2}\left(N - \frac{E}{\mu H}\right) \qquad \dots (4.6.8b)$$

The number of configurations is clearly

$$\Gamma(N,E) = \frac{N!}{n_1!\, n_2!} = \frac{N!}{\left\{\frac{1}{2}\left(N + \frac{E}{\mu H}\right)\right\}!\left\{\frac{1}{2}\left(N - \frac{E}{\mu H}\right)\right\}!} \qquad \dots (4.6.9)$$

Now Boltzmann's relation leads to ($N \gg 1$).

$$S = k \ln \Gamma = k\left[\ln N! - \ln \frac{1}{2}\left(N + \frac{E}{\mu H}\right)! - \ln\left\{\frac{1}{2}\left(N - \frac{E}{\mu H}\right)\right\}!\right]$$

$$= k\left[N \ln N - N - \frac{1}{2}\left(N + \frac{E}{\mu H}\right)\ln\frac{1}{2}\left(N + \frac{E}{\mu H}\right)\right.$$

$$\left. - \left[\frac{1}{2}\left(N - \frac{E}{\mu H}\right)\ln\frac{1}{2}\left(N - \frac{E}{\mu H}\right) + N\right]\right.$$

$$= k\left[N \ln N - \frac{1}{2}\left(N + \frac{E}{\mu H}\right)\ln\left(N + \frac{E}{\mu H}\right)\right.$$

$$\left. - \left[\frac{1}{2}\left(N - \frac{E}{\mu H}\right)\ln\left(N - \frac{E}{\mu H}\right) + N \ln 2\right]\right. \quad \text{... (4.6.10)}$$

The condition for thermal equilibrium now yields

$$\frac{1}{T} = \left(\frac{\partial S}{\partial E}\right)_N = -\frac{k}{2\mu H}\left[\ln\left(N + \frac{E}{\mu H}\right) - \ln\left(N - \frac{E}{\mu H}\right)\right]$$

$$= \frac{k}{2\mu H} \ln \frac{N - E/\mu H}{N + E/\mu M}$$

or

$$\frac{N - \dfrac{E}{\mu H}}{N + \dfrac{E}{\mu H}} = e^{\frac{2\mu H}{kT}}$$

or $E = -N\mu H \tanh \dfrac{\mu H}{kT}$ ... **(4.6.11)**

This energy is to be understood as $-MH$, where M is the magnetization and thus

$$M = \mu N \tanh \frac{\mu H}{kT} \quad \text{... (4.6.12)}$$

in exact agreement with Eq. (4.6.3).

All the answers are as expected. At very high temperatures the parallel and antiparallel configurations have equal probability and hence the magnetization goes to zero as $T \to \infty$. On the other hand as $T \to 0$, the antiparallel configuration which requires greater energy is frozen out and all the moments are lined up with the field leading to $M = \mu N$.

Having discussed this simple model which has all the essential ingredients, we briefly discuss the more realistic situation where the atom is characterized by a toal angular momentum **J** and the magnetic moment $\vec{\mu}$ points in the direction of **J** with the relation

$$\vec{\mu} = g_J \mu_B \mathbf{J}, \qquad \text{... (4.6.13)}$$

where μ_B is the Bohr magneton and g_J is the Land e g-factor. In a uniform external field, H, in the z-direction, the possible energies of the atom are given by

$$E = -\vec{\mu} \cdot \mathbf{H} = -g_J \mu_B H J_z \qquad \text{... (4.6.14)}$$

where we know from quantum mechanics that J_z has $2J + 1$ allowed values running from $-J$ to $+J$ in steps of unity. The single atom partition function is clearly

$$Z_s = e^{\beta g_J \mu_B JH} + e^{\beta g_J \mu_B (J-1)H} + \dots$$

$$+ \dots e^{-\beta g_j \mu_\beta JH}$$

$$= \frac{\sinh\left(g_J \mu_B H\left(J + \frac{1}{2}\right)\right)/kT}{\sinh\left(\frac{g_j \mu_B H}{2kT}\right)} \qquad \text{... (4.6.15)}$$

The free energy for N atoms is $F = -NkT \ln Z_s$ and differentiation with respect to H, leads to the magnetization

$$M = \partial F / \partial H = \frac{N g_j \mu_B}{2} \times B_J\left(\frac{\mu_B g_j H}{2kT}\right) \qquad \text{... (4.6.16)}$$

where $B_J(x)$ is the Brillouin function

$$B_J(x) = (2J + 1) \coth\{(2J + 1)x\} - \coth x \qquad \text{... (4.6.17)}$$

At high temperatures $x = \dfrac{\mu_B\, g_J\, H}{2kT} << 1$ and $B_J(x) \sim \dfrac{4}{3}\, x\, J(J+1)$, so that the Curie susceptibility is given by

$$\chi = \frac{N J (J+1)\, g_J^2\, \mu_B^2}{3kT} \qquad \text{... (4.6.18)}$$

For very low temperatures, $B_J(x) \to 2J$ and

$$M \cong N g_J\, \mu_B\, J \qquad \text{... (4.6.19)}$$

i.e. all atoms are oriented parallel to the field.

4.7 Electronic Energy in a Monatomic Gas

When we dealt with the ideal gas in Sec. 4.1, we took the energy as purely translational. However, the electronic motion within each atom leads to definite energies and the total energy of an atom moving with momentum p should be taken to be

$$E = \frac{p^2}{2m} + \varepsilon_n \qquad \text{... (4.7.1)}$$

where ε_n is the nth eigenvalue of the electronic energy. The total partition function (single atom) is clearly

$$Z_s = \sum_n \int d\Gamma\, e^{-\frac{p^2}{2mkT}}\, e^{-\varepsilon_n/kT}$$

$$= \left(\frac{2\pi m kT}{h^2}\right)^{3/2} V \cdot \overline{Z} = Z_{\text{trans}} \cdot \overline{Z} \qquad \text{... (4.7.2)}$$

where $\overline{Z} = \Sigma\, e^{-\varepsilon_n/kT}$ and the remainder Z_{trans} is simply the relevant partition function of the ideal gas obtained in Sec. 4.1. Now if the temperature T is much higher than the ionization temperature, then there are no bound electrons and the sum $\overline{Z}$ is meaningless. Thus $\overline{Z}$ is a significant quantity only if $T \le T_{\text{ion}}$. Now the ionization energy is of the order of $\varepsilon_1 - \varepsilon_0$, and hence if

$T << T_{ion}$, it is safe to assume that all electrons are in the ground state ε_0. Arbitrarily choosing $\varepsilon_0 = 0$, we have $\bar{Z}$ = degeneracy of the ground state. If the ground state is $L = 0$ (no orbital angular momentum), then the degeneracy is $(2S + 1)$, where S is the total spin angular momentum.

For the $L = 0$ ground state there is no fine structure splitting ($\mathbf{L} \cdot \mathbf{S} = 0$), but if $L \neq 0$ then the fine structure splittings do exist and the splitting energies ε_J may be comparable to kT. So one finds the sum $\bar{Z}$ to be

$$\bar{Z} = \sum (2J + 1) e^{-\varepsilon_J/kT} \qquad \text{... (4.7.3)}$$

This is because in the case of fine structure splitting, the degeneracy of a given level is $2J + 1$ and for a given L and S, the J values can range from $L + S$ to $|L - S|$ in steps of unity. Now if T is high, i.e. $KT >> \varepsilon_J$, then all $e^{-\varepsilon_J/kT} \sim 1$ and

$$\bar{Z} = \sum_{|L-S|}^{|L+S|} 2J + 1 = (2L + 1)\ (2S + 1) \qquad \text{... (4.7.4)}$$

If, on the other hand, $T << \varepsilon_J$, then again all the atoms settle for the lowest ε_J and $\bar{Z} = (2J_m + 1)$, where the J_m is the J–value for which ε_J is lowest.

4.8 Diatomic Molecules

4.8.1 Heteronuclear Molecules

In the previous section we saw that the electronic energy for the monatomic gas is important only if the temperature is much lower than the ionization temperature. Similarly, for a molecule, the binding effects are important only if the temperature is much lower than the dissociation temperature T_{diss} (A typical estimate would be a molecular binding energy between 1 and 2 eV, so that $kT_{diss} \cong 1.5$ eV or $T_{diss} \sim 1.7 \times 10^4$ K). As before, the gas molecule will be assumed to be in its lowest electronic state. For a diatomic molecule which we are interested in, the total internal energy (the effect of the translational energy has already been taken into account in dealing with the perfect gas; the mass appearing there now stands for the molecular weight of the gas) can be written as

$$E_{int} = E_{el} + E_{rot} + E_{vib} \qquad \text{... (4.8.1)}$$

In Eq. (4.8.1) E_{el} is the energy associated with all Coulombic interaction and electronic motion about nuclei. For the electronic ground state $E_{el} = \varepsilon_0$. The energy associated with the rotation E_{rot} is the energy due to the orbital motion of the two nuclei about the centre of mass and can be modelled by a rigid rotator and E_{vib} is the energy associated with the vibrational motion of the nuclei about their equilibrium positions along the line joining the nuclei. The latter can be modelled by a one-dimensional oscillator. Consequently,

$$E_{int} = \varepsilon_0 + \frac{\hbar^2}{2I} K(K+1) + \hbar\,\omega \left(n + \frac{1}{2}\right) \qquad \text{... (4.8.2)}$$

where I is the moment of inertia given by $I = \dfrac{m_1 m_2}{m_1 + m_2} \cdot r_0^2$ where m_1 and m_2 are the masses of the two nuclei and r_0 is the equilibrium separation; K is the angular momentum quantum number and ω is the frequency of oscillation of the two-nuclear system along the line joining them. As in Eq. (4.7.2), the total partition function (including translation) can be written as

$$Z = Z_{trans} \cdot \bar{Z} = Z_{trans}\, Z_{el} \cdot Z_{rot} \cdot Z_{vib} \qquad \text{... (4.8.3)}$$

where

$$Z_{el} = e^{-\varepsilon_0/kT} \qquad \text{... (4.8.4a)}$$

$$Z_{rot} = \sum_{K=0}^{\infty} (2K+1)\, e^{\frac{-\hbar^2 K(K+1)}{2IkT}} \qquad \text{... (4.8.4b)}$$

and

$$Z_{vib} = \sum_{n=0}^{\infty} e^{-\hbar\,\omega \left(n + \frac{1}{2}\right)\frac{1}{kT}} \qquad \text{... (4.8.4c)}$$

The free energies, entropies and specific heats calculated from the separate parts add to give the total amounts in each case.

Evaluating Z_{rot} is difficult in general, but we note that for high temperatures, i.e. $kT >> \hbar^2/2I$, the high values of K play an important part in the sum and we can approximate Z_{rot} by

$$Z_{rot} \cong \int_0^\infty 2K\, e^{-\hbar^2 K^2/2IkT}\, dK = \frac{2IkT}{\hbar^2} \qquad \text{... (4.8.5)}$$

The contribution to the specific heat in this range of temperature is clearly Nk. For very low temperatures, on the other hand, the first two terms in the sum give a good approximation and

$$Z_{rot} \cong 1 + 3\, e^{-\hbar^2/IkT} \qquad \text{... (4.8.6)}$$

with the contribution to the rotational specific heat being $C_{rot} \sim 3\, Nk\, (\hbar^2/ITk)^2\, e^{-\hbar^2/IkT}$. As for Z_{vib} it is exactly identical to the situation in Sec. 4.4 and the expression for Z_{vib} is given by Eq. (4.4.5). At high temperatures ($kT >> \hbar\,\omega$), the contribution to the specific heat is Nk, while at low temperatures it falls exponentially. We finally note that, at very high temperatures, the molar specific heat of the diatomic molecule is

$$C_{diatomic} = C_{trans} + C_{rot} + C_{vib} = \left(\frac{3}{2} + 1 + 1\right) R = \frac{7}{2} R \qquad \text{... (4.8.7)}$$

4.8.2 Homonuclear Molecules

We consider the specific case of the hydrogen molecule (H_2). The new complication is that at low temperatures where quantum effects are important, the exchange of two hydrogen nuclei (protons) should lead to the wavefunction changing sign (for our purpose here the electronic degrees of freedom are considered frozen). The total wavefunction of the two nuclei can be written as

$$\Psi\left(\mathbf{r}_1, \mathbf{r}_2\right) = \Psi_{orb}\left(\mathbf{r}_1, \mathbf{r}_2\right) \times \Psi_{spin} \qquad \text{... (4.8.8)}$$

The parity of the orbital part is given by $(-)^K$ and is even for $K = 0, 2, 4,$ etc. when interchanging the two nuclei does not change the orbital wavefunction, while for $K = 1,3,5,$ ... etc., interchanging the two nuclei changes the sign of the wavefunction. As for the Ψ_{spin}, the wavefunction is symmetric (even under interchange) if $S = 1$ and antisymmetric if $S = 0$. Thus for $S = 1$, the possible K values are the odd integers while for $S = 0$, the allowed K values are the even integers. The degeneracy of the spin wavefunction is $2S + 1$, and hence to one state of $S = 0$, there are three states of $S = 1$ and hence the rotational partition function must now be written as

$$Z_{rot} = \frac{1}{4}\left[\sum_{K_{even}} (2K + 1)\, e^{-\frac{\hbar^2 K(K+1)}{2IkT}} + 3\sum_{K_{odd}} (2K + 1)\, e^{-\frac{\hbar^2 K(K+1)}{2IkT}}\right] \quad \text{... (4.8.9)}$$

The factor of 1/4 is for normalization. The $S = 1$ state with the larger statistical weight is called the *ortho* hydrogen and $S = 0$ is called *para* hydrogen. The above formula for Z_{rot} assumes that the *ortho* and *para* hydrogen are in thermal equilibrium with each other with the ratio x of the number of *ortho* molecules to the number or *para* molecules being given by

$$x = \frac{3\sum_{K_{odd}} (2K + 1)\, e^{-\frac{(K+1)\hbar^2}{2IkT}}}{\sum_{K_{even}} (2K + 1)\, e^{-\frac{K(K+1)\hbar^2}{2IkT}}} \quad \text{... (4.8.10)}$$

In practice, this equilibrium is very difficult to set up. As discussed on several occasions, the setting up of equilibrium entails collisions which keep changing the states. But collisions between molecules do not change the nuclear spin and thus the ratio x is fixed by observation, rather than by assumption of thermal equilibrium. For all practical purposes x is a constant of motion, having nothing to do with equilibrium. The *ortho* and *para* molecules can be taken to be separately in equilibrium at temperature T with the ratio x fixed by an experimental determination. At room temperatures, $x \cong 3$. Instead of the partition function having the additive form of Eq. (4.8.9), it is the free energies which add and

$$F_{rot} = F_{rot}^{(para)} + F_{rot}^{(ortho)}$$

$$= -kT\left[\ln Z_{rot}^{para} + x \ln Z_{ror}^{ortho}\right] \quad \text{... (4.8.11)}$$

where

$$Z_{\text{para}} = \sum_{K_{\text{even}}} (2K + 1)\, e^{-\frac{\hbar^2}{2IkT} K(K+1)} \qquad \text{... (4.8.12a)}$$

and

$$Z_{\text{ortho}} = \sum_{K_{\text{odd}}} (2K + 1)\, e^{-\frac{\hbar^2}{2IkT} K(K+1)} \qquad \text{... (4.8.12b)}$$

The specific heat calculated from the above F_{rot} is in very good agreement with the experiments while the specific heat obtained from Eq. (4.8.9) is not.

4.9 Solid-Vapour Equilibrium

We consider an application of the principle of equality of chemical potential to describe the equilibrium between two systems which can exchange particles. The two systems are a solid and its vapour in equilibrium. The equilibrium is dynamic in the sense that evaporation of atoms continually occurs from the solid surface but there is an equal number of atoms which are condensing on the surface and thus an equilibrium is maintained. Applying the condition of equality of chemical potential for the gas phase and the solid phase we obtain the vapour pressure, p, on the solid surface.

The calculation proceeds by evaluating the chemical potential for the vapour and the solid phases. We consider for simplicity a vapour that is monatomic, i.e. equilibrium between a noble gas and its solid form. The chemical potential can now be written down as [see Eq. (4.2.4)]

$$\mu_{\text{vap}} = kT \left[\ln n - \frac{3}{2} \ln \frac{2\pi mkT}{h^2} \right] \qquad \text{... (4.9.1)}$$

where $n = N/V$ in the number density.

Next we need to consider the chemical potential of the solid. To do this we note if we write $\mu_{\text{solid}} = G_s/N$, where G_s is the Gibbs potential, then a calculation of G would be necessary, Now writing $G_s = PV_s + E - TS$, we note that solids are characterized by a very high density compared to the vapour and hence the term PV_s may be dropped as a first approximation; then $G_s \cong F_s = E - TS$ and we can calculate F_s within the canonical ensemble. For this, we require a model—we shall take the Einstein model already discussed in connection with the specific heat. Now the free energy calculated on the basis of this model will not include the cohesive energy E_0 per molecule. The meaning of E_0 is that the potential energy of a solid with all its N atoms at rest at their equilibrium positions is $-NE_0$ compared to the situation where the N atoms are all widely separated from each other. Thus $F_s = -NE_0 +$ contribution from the Einstein model. The contribution of the Einstein model is

$$F_s^{(E)} = -NkT \ln Z = -TN\left[\ln e^{-T_E/2T} - \ln\left(1 - e^{-T_E/T}\right)\right] \qquad \text{... (4.9.2)}$$

where the partition function expression has been taken from Eq. (4.4.5) and T_E is defined in Eq. (4.5.3). We will be interested in temperature $T >> T_E$ and in this limit, we have

$$F_s^{(E)} = \frac{3}{2} N kT_E + 3NkT \ln \frac{T_E}{T} \qquad \text{... (4.9.3)}$$

so that

$$\mu_{solid} \cong F_s/N = -E_0 + \frac{3}{2} kT_E + 3kT \ln \frac{T_E}{T} \qquad \text{... (4.9.4)}$$

The equality $\mu_{vap} = \mu_{solid}$ leads to

$$\ln\left[\left(\frac{T_E}{T}\right)^3 \left(\frac{2\pi mkT}{h^2}\right)^{3/2} \frac{1}{n}\right] = \frac{E_0 - \frac{3}{2} kT_E}{kT}$$

or

$$n\left(\frac{T}{T_E}\right)^3 \left(\frac{h^2}{2\pi mkT}\right)^{3/2} = e^{-\frac{(E_0 - \frac{3}{2} kT_E)}{kT}} \qquad \text{... (4.9.5)}$$

whence using $p = nkT$,

$$p = kT\left(\frac{T_E}{T}\right)^3 \left(\frac{2\pi mkT}{h^2}\right)^{3/2} e^{-\frac{\left(E_0 - \frac{3}{2} kT_E\right)}{kT}}$$

$$= \frac{(2\pi m)^{3/2}}{(kT)^{1/2}} \nu^3 e^{-\frac{(E_0 - \frac{3}{2} kT_E)}{kT}}, \qquad \dots \textbf{(4.9.6)}$$

where $h\nu = k T_E$. The temperature dependence of the vapour pressure expressed in Eq. (4.9.6) has been experimentally confirmed.

Problems

4.1 The nuclei of atoms in a certain crystalline solid have spin 1. According to quantum theory it can exist in three possible quantum states characterized by the quantum number $m = 1, 0, -1$. Since the charge distribution in a nucleus is not spherically symmetric, the energy depends on the angle between the spin orientation and the local electric field. Accordingly, the energy for the states $m = \pm 1$ is ε, and that for $m = 0$ is 0. Find the nuclear contribution to the internal energy, entropy and specific heat. By directly calculating the number of accessible states, find the entropy at very high and very low temperatures. Compare with your previous answer.

4.2 A wire of radius r_0 coincides with the axis of a metal cylinder of radius R and length L ($L >> R$). The wire is maintained at a positive potential V with respect to the cylinder. The system is at a fairly high temperature T. The electrons emitted from the hot metal form a dilute gas filling the cylinder and in equilibrium with it. The density is low enough to neglect the electrostatic interaction between the electrons. In thermal equilibrium, find the dependence of the electric charge density on the radial distance r.

4.3 Two atoms interact with the mutual potential energy of interaction of the form

$$U = U_0 \left[\left(\frac{a}{x}\right)^{12} - 2\left(\frac{a}{x}\right)^{6} \right]$$

where x is the separation of the particles. The system is in equilibrium at a temperature T, such that $kT << U_0$, but high enough for the applicability of classical statistical mechanics. Calculate the mean separation $\bar{x}(T)$ and the expansion coefficient

$$\alpha = \frac{1}{\bar{x}} \frac{\partial \bar{x}}{\partial T}$$

(assuming the temperature to the fairly low, set up a perturbative calculation and work to the first non-vanishing order).

4.4 Consider a collection of N_0 non-interacting magnetic moments in a volume V and at a temperature T. The situation is to be described classically. Consider an external field H applied in the z-direction. Find the mean magnetic moment $\bar{M}$ in the direction of the field.

4.5 A thin-walled vessel of volume V, kept at constant temperature, contains a gas which slowly leaks out through a small hole of area A. The leakage back into the vessel is negligible. Find the time required for the pressure in the vessel to decrease to $1/e$ of its initial value. The answer should be in terms of A, V and the mean molecular speed.

4.6 Helium atoms can be absorbed on the surface of a metal, an amount of work φ being then necessary to remove a helium atom from the metal surface to infinity. The helium atoms are completely free to move on the two-dimensional metal surface. If such a metal surface is in contact with helium gas at a pressure P, and the whole system is in equilibrium at temperature T, what is the mean number of atoms absorbed per unit area of the metal surface?

4.7 a) Find the free energy of a single quantum oscillator of frequency ω.
b) For a crystal, the above expression needs to be summed over all the frequencies of lattice vibrations. Call this contribution to the free energy $F_0(T)$. If there is a fractional change Δ in the volume of the crystal, the total free energy is

$$F(T, \Delta) = \frac{1}{2} \beta\Delta^2 + F_0(\tau),$$

where β is the bulk modulus. If the change in freguency $\delta\omega$ of a mode due to the volume change is written as $\delta\omega = -r\omega\Delta$, where r is a constant independent of the mode, then show that minimization of F with respect to Δ, leads to

$$\Delta = r\,E(\tau)/\beta\,,$$

where $E(\tau)$ is the energy density.

4.8 Consider a classical system of N non-interacting diatomic molecules enclosed in a box of volume V at temperature T. The energy for a single molecule is taken to be

$$E\left(\bar{p}_1, \bar{p}_2, \bar{r}_1, \bar{r}_2\right) = \frac{1}{2m}\left(p_1^2 + p_2^2\right) + \varepsilon\left|r_{12} - r_0\right|$$

where ε and r_0 are given positive constants and $r_{12} = |\bar{r}_{12} - \bar{r}_0|$. Find the specific heat as a function of temperature.

Chapter Five
Interacting Classical Systems

5.1 The Non-ideal Gas

We now consider N molecules enclosed in a volume V interacting via a pairwise potential $v(r_{ij})$, where $r_{ij} = |\mathbf{r} - \mathbf{r}_j|$, the magnitude of the separation vector between particles i and j. The potential corresponds to an attractive force at large distances and a very strong repulsion at short distances. A typical $v(r)$ is shown in Fig. (5.1).

For later purposes, we define an auxiliary function

$$f(r) = e^{-\beta v(r)} - 1, \qquad \text{... (5.1.1)}$$

shown in Fig. (5.2).

The potential $v(r)$ has a zero at $r = \sigma$, which corresponds roughly to the molecular diameter. The function $f(r)$ is zero wherever $v(r)$ is so.

The energy of the system of N particles is given by

$$E = \sum \frac{p_i^2}{2m} + \sum_{\text{pairs}} v(r_{ij}) \qquad \text{... (5.1.2)}$$

and the canonical partition function Z for the N identical molecules is given by

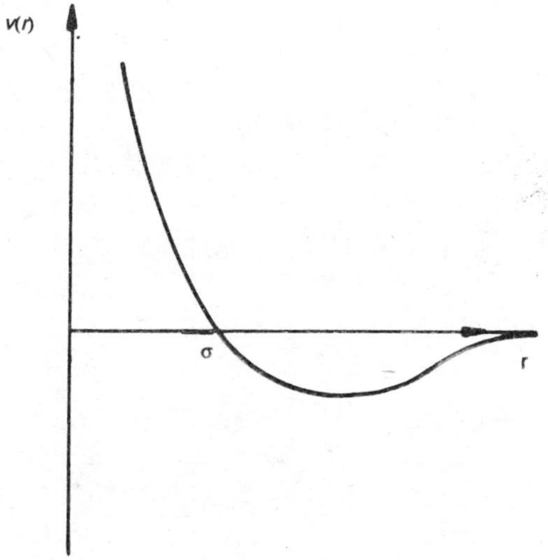

Fig. 5.1: $v(r)$ vs. r.

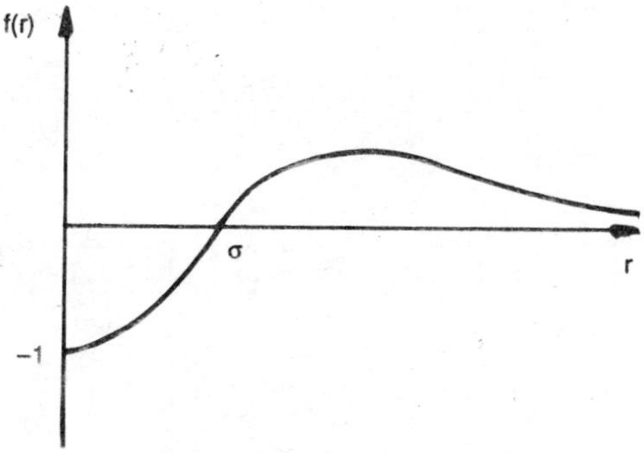

Fig. 5.2: $f(r)$ vs. r.

$$Z = \frac{1}{N!}\int \frac{d^{3N} p_i\, d^{3N} r_i}{h^{3N}}\, e^{-\sum \frac{p_i^2}{2mkT}}\, e^{-\sum_{\text{pairs}} v(r_{ij})/kT}$$

$$= \frac{V^N}{N!}\left(\frac{2\pi mkT}{h^2}\right)^{3N/2} \frac{1}{V^N}\int d^{3N} r_i\, e^{-\sum_{\text{pairs}} \frac{v(r_{ij})}{kT}}$$

$$= \frac{V^N}{N!}\left(\frac{2\pi\, mkT}{h^2}\right)^{3N/2} \frac{1}{V^N}\int d^{3N} r_i \prod_{\text{pairs}} (1 + f(r_{ij}))$$

$$= \frac{V^N}{N!}\left(\frac{2\pi\, mkT}{h^2}\right)^{3N/2} \frac{1}{V^N}\int d^{3N} r_i \prod_{\text{pairs}} (1 + f_{ij}) \qquad \text{... (5.1.3)}$$

where in the last but one line of Eq. (5.1.3), we have used the definition of $f(r)$ from Eq. (5.1.1) and introduced the notation $f(r_{ij}) = f_{ij}$ in the last line. The free energy is consequnetly

$$F = -kT \ln Z = -NkT \ln V - \frac{3NkT}{2} \ln \frac{2\pi\, mkT}{h^2}$$

$$+ N \ln N - N - kT \ln \frac{1}{V^N}\int d^3 r_i \prod_{\text{pairs}} (1 + f_{ij}) \qquad \text{... (5.1.4)}$$

and the pressure P follows as

$$P = -\left(\frac{\partial F}{\partial V}\right)_T = \frac{NKT}{V} + kT\frac{\partial}{\partial V}\ln\frac{1}{V^N}\int d^3 r_i \prod_{\text{pairs}}(1 + f_{ij})$$

$$= \frac{NkT}{V}\left[1 + \frac{V}{N}\frac{\partial}{\partial V}\ln\frac{1}{V^N}\int d^3 r_i \prod_{\text{pairs}}(1 + r_{ij})\right]$$

$$= \frac{NkT}{V}\left[1 + V\frac{\partial \overline{F}}{\partial V}\right] \qquad \text{...(5.1.5)}$$

where

$$\overline{F} = \frac{1}{N}\ln\frac{1}{V^N}\int d^3 r_i \prod_{\text{pairs}}(1 + f_{ij}) \qquad \text{... (5.1.6)}$$

At this stage, we have an exact result. If $v_{ij} = 0$, $f_{ij} = 0$ and hence $\overline{F} = 0$, giving back the ideal gas law. Since $\overline{F}$ cannot be exactly obtained for any non-zero v_{ij}, we try to get a handle on the problem by assuming that v_{ij} is small and hence f_{ij} is small. Accordingly, we retain only the O (f) terms in Eq. (5.1.6) and find

$$\overline{F} = \frac{1}{N}\ln\frac{1}{V^N}\int d^{3N} r_i \left(1 + \prod_{\text{pairs}} f_{ij}\right)$$

$$= \frac{1}{N}\ln\frac{1}{V^N}\left[V^N + \frac{N^2 V^{N-1}}{2}\int f(r)\, d^3 r\right]$$

$$= \frac{1}{N}\ln\left[1 + \frac{N^2}{2V}\int f(r)\, d^3 r\right] \qquad \text{... (5.1.7)}$$

where we have approximated the number of pairs $N(N-1)/2$ by $N^2/2$ for large N. Evaluating the derivative $\partial\overline{F}/\partial V$ to O (f), we find

$$\frac{\partial \overline{F}}{\partial V} = -\frac{N}{2V^2}\int f(r)\, d^3 (r) \qquad \text{... (5.1.8)}$$

The equation of state [Eq. (5.1.5)] now becomes

$$\frac{PV}{NkT} = 1 - \frac{N}{2V} \int f(r)\, d^3 r \qquad \text{... (5.1.9)}$$

A phenomenological equation of state for the real gas is Van der Waals equation which reads

$$\left(P + \frac{a}{V^2}\right)(V - b) = NkT \qquad \text{... (5.1.10)}$$

where a and b are constants for a given gas. We can rewrite this equation of state as

$$P = \frac{NkT}{V-b} - \frac{a}{V^2} = \frac{NkT}{V}\left(1 + \frac{b}{V}\right) - \frac{1}{V^2} + O\left(\frac{1}{V^3}\right)$$

$$= \frac{NkT}{V} + \frac{bNkT - a}{V^2} + \ldots$$

or

$$\frac{PV}{NkT} = 1 + \left(b - \frac{a}{NkT}\right)\frac{1}{V} + \ldots \qquad \text{... (5.1.11)}$$

To see how this compares with Eq. (5.1.9), we need to have an expression for $f(r)$. We make the approximation

$$f(r) = \begin{cases} -1 \text{ for } r < \sigma \\ -(1 - e^{-\beta v(r)}) \text{ for } r > \sigma \end{cases}, \qquad \text{... (5.1.12)}$$

leading to

$$\int f(r)\, d^3 r = -\frac{4}{3}\pi\sigma^3 - 4\pi \int_\sigma^\infty r^2 \left(1 - e^{-\beta v(r)}\right) dr$$

and

$$\frac{PV}{NkT} = 1 + \frac{2\pi}{3}\frac{N}{V}\sigma^3 - \frac{2\pi N}{V}\int_\sigma^\infty r^2 \left(e^{-\beta v(r)} - 1\right) dr$$

$$\cong 1 + \frac{2\pi}{3}\frac{N}{V}\sigma^3 - \frac{2\pi N}{VkT}\int_{\sigma}^{\infty}\{-v(r)\}\, r^2\, dr \qquad \text{... (5.1.13)}$$

for weak potentials, i.e. $v << kT$. Note that $-\int_{\sigma}^{\infty} v(r)\, r^2\, dr$ is positive definite since beyond σ, $v(r)$ is attractive. Thus as an expansion in inverse powers of V, Eq. (5.1.13) has the same structure as Van der Waals equation to the lowest order, when we identify

$$b = \frac{2}{3}\pi N\sigma^3 \qquad \text{... (5.1.14a)}$$

and

$$a = 2\pi N^2 \int_{\sigma}^{\infty}\{-v(r)\} r^2 dr \qquad \text{... (5.1.14b)}$$

If we could retain all powers of f, i.e. considered arbitrarily strong potentials, we would generate an equation of state of the form

$$\frac{P}{nkT} = 1 + A_1 n + A_2 n^2 + \ldots \qquad \text{... (5.1.15)}$$

where $n = N/V$ is the number density. The various $A_1, A_2 \ldots$ etc. are known as virial coefficients. What we have generated is the first virial coefficient A_1. The interesting feature is that A_1 has a zero at a finite temperature. This fact has been experimentally verified.

5.2 Linked Cluster Expansion

This is a systematic way of obtaining the equation of state of a gas. The technique is useful as it shares many features of the diagrammatic techniques of many-body physics, but the end result is in a sense disappointing. It is incapable of providing an equation of state which correctly describes the high density behaviour and hence cannot describe a phase transition. In that sense, the phenomenological equation of state of Van der Waals is superior.

We return to Eq. (5.1.3) and write it as

$$Z = \frac{1}{N!}\left(\frac{2\pi mkT}{h^2}\right)^{3N/2} Z_N = \frac{V}{N!}\frac{1}{\lambda^{3N}} Z_N, \qquad \text{... (5.2.1)}$$

where $\lambda = h/\sqrt{2\pi mkT}$ is the De Broglie wavelength, and

$$Z_N = \int d^3 r_1\, d^3 r_2 \ldots d^3 r_n \prod_{\text{pairs}} \left(1 + f_{ij}\right)$$

$$= \int d^3 r_1\, d^3 r_2 \ldots d^3 r_N \,(1 + f_{12} + f_{13} + f_{23} + \ldots$$

$$+ f_{12}f_{13} + f_{12}f_{23} + \ldots + f_{12}f_{23}f_{31} + \ldots\Big) \qquad \text{... (5.2.2)}$$

The various terms in the brackets in Eq. (5.2.2) can be given a diagrammatic representation by setting up the following rules:

i) Each particle is represented by a circle. So an N-particle situation will involve N circles.
ii) We represent f_{pq} by a line linking the p th circle (i.e. circle representing the p th particle) to the qth circle.
iii) In any given diagram some circles may be linked and some may be free. Every linkage contributes to f while each unlinked circle contributes unity.

As an example consider $N = 3$, for which Eq. (5.2.2) can be explicitly written as,

$$Z_3 = \int d^3 r_1\, d^3 r_2\, d^3 r_3 \Big[1 + f_{12} + f_{23} + f_{13} + f_{12}f_{13} + f_{12}f_{23}$$

$$+ f_{13}f_{23} + f_{12}f_{23}f_{13}\Big]$$

$$= Z_{30} + Z_{31} + Z_{32} + Z_{33} \qquad \text{... (5.2.3)}$$

where

$$Z_{30} = \int d^3 r_1\, d^3 r_2\, d^3 r_3$$

$$Z_{31} = \int d^3 r_1\, d^3 r_2\, d^3 r_3 \Big[f_{12} + f_{23} + f_{13}\Big]$$

$$Z_{32} = \int d^3 r_1\, d^3 r_2\, d^3 r_3 \Big[f_{12}f_{13} + f_{13}f_{23} + f_{12}f_{23}\Big]$$

$$Z_{33} = \int d^3 r_1 \, d^3 r_2 \, d^3 r_3 f_{12} f_{23} f_{13}$$

and the diagrammatic representations are shown in Fig. 5.3.

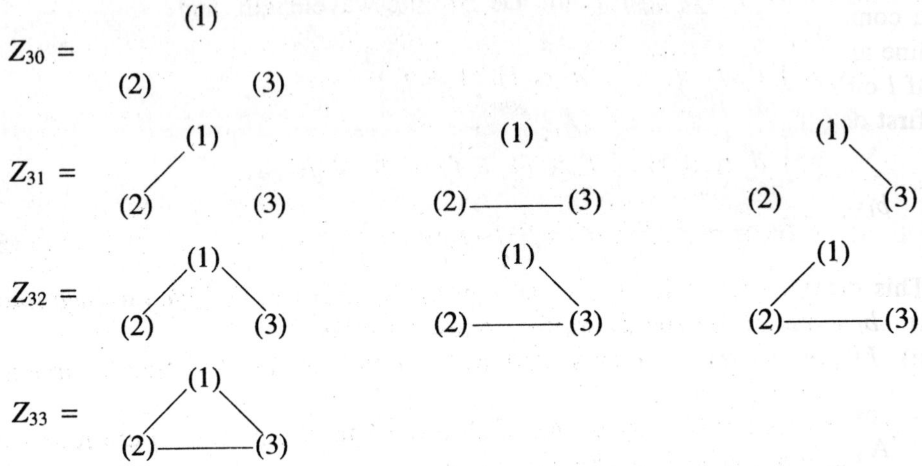

Fig. 5.3: The three-particle graphs

A graph with N circles is called a N-particle graph. From the above, it is clear that for $N = 3, Z_3$ [of Eq. (5.2.3)] = sum of all *distinct* three-particle graphs. An example of graphs which are not distinct is shown in Fig. 5.4.

Fig. 5.4: Non-distinct 3-particle graphs.

For arbitrary N,

Z_N = Sum of all distinct N-particle graphs ... **(5.2.4)**

For a given N, any graph, in general, can be viewed as a product of several subgraphs. An example for $N = 8$ is shown in Fig. 5.5.

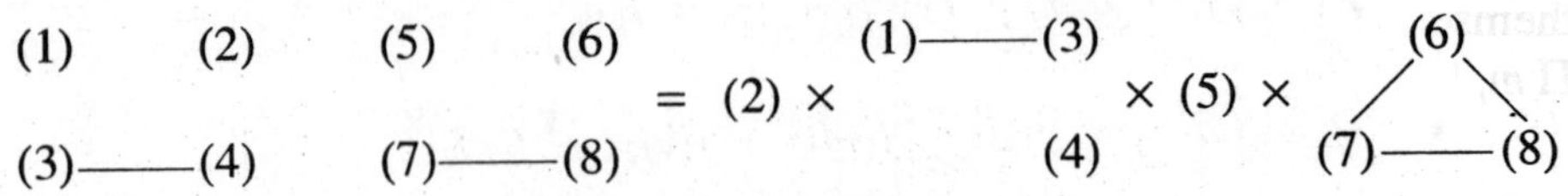

Fig. 5.5: Factoring of graphs.

The factoring shown in Fig. 5.5 leads to the contribution

$$\left[\int d^3 r_2\right]\left[d^3 r_5\right]\left[\int d^3 r_1\, d^3 r_3\, d^3 r_4 f_{34}\right]\times\left[\int d^3 r_6\, d^3 r_7\, d^3 r_8 f_{67} f_{78} f_{68}\right]$$

from that particular graph. It should be noted that each factor corresponds to a connected graph—connected in the sense that each circle is connected to a line and thus directly or indirectly connected to every other circle in the graph. If l circles are involved in a connected graph, then it is called an l-cluster. We first define a dimensionless number b_l associated with each cluster integral by

$$b_l = \lambda^{\frac{1}{3l-3}} \cdot \frac{1}{V} \cdot \frac{1}{l\,!} \times [\text{Sum of all possible } l\text{-clusters}] \qquad \text{... (5.2.5)}$$

This definition ensures that

i) b_l is dimensionless

ii) $\underset{V \to \infty}{Lt}\; b_l$ is finite. This is true because all the f_{ij}'s have a finite range and only the integration over the centre of gravity produces a factor of V.

A general N-particle graph will be composed of l-clusters where l may range from 1 to N. In a given situation, an l-cluster occurs m_l times and we must have $\Sigma\, m_l \cdot l = N$. Now, the set of numbers $\{m_l\}$ is not unique, e.g. if $N = 3$, we can have $l = 1, m_l = 3$; or $l = 1, m_l = 1$ and $l = 2, m_l = 1$; or $l = 3$, $m_l = 1$. Thus, there are three ways of satisfying $\sum m_l \cdot l = 3$. These three ways are the three types of nontrivial graphs (Z_{31}, Z_{32}, Z_{33}) in Fig. 5.3.

Consider an arbitrary N-particle graph with a given distribution of m_l. The magnitude of the integral $(\{m_l\})$ associated with this graph is

$$\prod_{l=1}^{N} \{b_l \lambda^{3l-3}\, V\, l\,!\}^{m_l} \qquad \text{... (5.2.6)}$$

We now need to know in how many different ways these l-clusters with the distribution $\{m_l\}$ can be formed. Clearly all N particles can be permuted among themselves and thus there are $N!$ possibilities. But not all these permutations can generate distinct graphs. Clearly the permutation of the m_l l-clusters among themselves cannot generate new graphs. That cuts the number down by $\prod_l m_l\,!$ · Further, within the same cluster, the l-particles can be permuted among themselves in $l!$ ways without producing a distinct graph and since there are m_l l-clusters, the number of such redundant permutations is $(l\,!)^{m_l}$. Thus the total number of distinct ways P in which a given distribution $\{m_l\}$ is obtained is

$$P(\{m_l\}) = \frac{N\,!}{\prod_l (l\,!)^{m_l} \prod_l m_l\,!} \qquad \text{... (5.2.7)}$$

The contribution of this distribution $\{m_l\}$ to Z_N is PI and can be written as

$$Z_N(\{m_l\}) = \frac{N\,!}{\prod_l l\,!^{m} l \prod_l \cdot m\,!} \cdot \prod_l \left(b_l\,\lambda^{31-3}\,Vl\,!\right)^{m_l}$$

$$= \prod_l \frac{N\,!}{m_l\,!}\left(\frac{V}{\lambda^3}\right)^{m_l} \lambda^{3N}\, b_l^{m_l} \qquad \text{... (5.2.8)}$$

The partition function is

$$Z_N = \sum_{\{m_l\}} Z_N\,(\{m_l\}) \qquad \text{... (5.2.9)}$$

with the constraint $\Sigma\, l\, m_l = N$.

It is this constraint which makes the evaluation of the sum impossible. Hence we resort to the grand canonical ensemble where the partition function can be written as

$$Q = \sum_{\varepsilon,N} e^{-\beta\varepsilon}\, e^{\mu N\beta}$$

$$= \sum_N e^{\mu N\beta}\, Z(N)$$

$$= \sum_N e^{\mu\beta N} \frac{1}{N\,!} \frac{1}{\lambda^{3N}} \cdot Z_N$$

$$= \sum_N e^{\beta\mu N} \sum_{\{m_l\}} \prod_l \frac{3N}{m_l\,!}\left(\frac{V}{l^3}\right)^{m_l} \times b_l^{m_l}$$

$$= \sum_{N=0}^{\infty} \sum_{\{m_l\}} \prod_{l=1}^{N} \frac{1}{m_l\,!} \left(\frac{V\, b_l\, e^{\beta\mu l}}{\lambda^3} \right)^{m_l}$$

where we have made use of Eqs. (5.2.1) for $Z(N)$ which is just the Z of that equation and Eqs. (5.2.8) and (5.2.9). Now the various sums over the set $\{m_l\}$ are freed from the constraint of fixed N and consequently we can perform the unrestricted sum over each m_l for a given l and finally perform the product over l. Thus

$$Q = \prod_l \sum_{m_1} \frac{1}{m_1!} \left(\frac{V\, b_1\, e^{\mu\beta l}}{\lambda^3} \right)^{m_1} \sum_{m_2} \frac{1}{m_2\,!} \left(\frac{V\, b_2\, e^{\mu\beta l}}{\lambda^3} \right)^{m_2} \cdots$$

$$= \prod_l e^{\left(V b_l e^{\mu \beta l}\right)/\lambda^3} \qquad \ldots \textbf{(5.2.11)}$$

The equation of state follows from the relation,

$$PV = kT \ln Q = kT \sum_l \frac{V}{\lambda^3}\, b_l\, e^{\mu l/kT} \qquad \ldots \textbf{(5.2.12)}$$

once μ is obtained from

$$N = \frac{\partial}{\partial \mu} (PV) \Big|_T = \frac{V}{\lambda^3} \sum l\, b_l\, e^{\mu l/kT} \qquad \ldots \textbf{(5.2.13)}$$

elimination of μ between Eqs. (5.2.12) and (5.2.13) yields the equation of state—a relation between P, V, N and T. In principle, one needs to obtain the relation between P, V, N and T for arbitrtary N and V and then take the thermodynamic limit. Needless to say, this is virtually impossible. If we take the limit of $N \to \infty$ and $V \to$ with $n = N/V$ finite, then Eqs. (5.2.12) and (5.2.13) become

$$\frac{P}{kT} = \frac{1}{\lambda^3} \Sigma\, b_l\, e^{\,\mu l/kT} \qquad \ldots \textbf{(5.2.14a)}$$

and

$$n = \frac{1}{\lambda^3} \Sigma l\, b_l\, e^{\frac{\mu l}{kT}} \qquad \text{... (5.2.14b)}$$

where the b_l's are finite numbers in this limit. We can obtain a series solution for Eqs. (5.2.14a) and (5.2.14b) as

$$\frac{P}{nkT} = 1 + A_1 n + A_2 n^2 + \ldots \qquad \text{... (5.2.15)}$$

which is the virial expansion introduced in Sec. 5.1. This expansion essentially treats n as small and hence from Eq. (5.2.14b), we get

$$n = \frac{1}{\lambda^3} (b_1 e^{\mu} + 2b_2\, e^{2\mu} + \ldots)$$

$$= \frac{e^{\mu} b_1}{\lambda^3} \left(1 + \frac{2b_2}{b_1} e^{\mu} + \frac{3b_3}{b_1} e^{2\mu} \ldots\right) \qquad \text{... (5.2.16)}$$

and a perturbative solution for the chemical potential yields as the first approximation

$$\frac{e^{\mu} b_1}{\lambda^3} = n \qquad \text{... (5.2.17a)}$$

and

$$\frac{e^{\mu} b_1}{\lambda^3} = n \left(1 - 2 \frac{b_2}{b_1} \lambda^3 n\right) + O(n^3) \qquad \text{... (5.2.17b)}$$

as the second approximation. Now, the pressure can be expressed from Eq. (5.2.14) as

$$\frac{P}{kT} = \frac{b_1}{\lambda^3} e^{\mu} + \frac{b_2}{\lambda^3} e^{2\mu} +$$

$$= n - \frac{2b_2}{b_1^2} \lambda^3 n^2 + \frac{b_2}{b_1^2} \lambda^3 n^2 + O(n^3)$$

$$= n \left[1 - \frac{b_2}{b_1^2} \lambda^3 n + O(n^2)\right]$$

or

$$\frac{P}{nkT} = \left(1 - \frac{b_2}{b_1^2}\lambda^3 n + O(n^2)\right) \qquad \text{... (5.2.18)}$$

Thus the first virial coefficient is given by $(b_2/b_1^2)\ \lambda^3$. Referring to the definition of b_l (Eq. 5.2.5), it is evident that $b_1 = 1$ and

$$b_2 = \frac{1}{\lambda^3}\cdot\frac{1}{V}\cdot\frac{1}{2}\cdot \times \text{ (All possible 2-clusters)} \qquad \text{... (5.2.19)}$$

But there can be only one kind of 2-cluster and that has the contribution $\int d^3 r_1\, d^3 r_2 f(r_{12}) = V \int f(r)\, d^3 r$, so that Eq. (5.2.18), becomes

$$\frac{P}{nkT} = 1 - \frac{1}{2}\, n \int f(r)\, d^3 r + O(n^2) \qquad \text{... (5.2.20)}$$

exactly identical to Eq. (5.1.9). Thus, what we have generated now is a systematic way of evaluating the different terms of the virial expansion of Eq. (5.1.15). However, this constraint of going to the thermodynamic limit before obtaining the equation of state makes the virial expansion useless at high densities and hence for studying phase transitions. Phenomenological equations of state that are due to Van der Waals are consequently very useful.

5.3 Van der Waals Equation of State

For one mole of a real gas, the equation of state proposed by Van der Waals (Eq. 5.1.10) can be written as

$$P = \frac{RT}{V-b} - \frac{a}{V^2} \qquad \text{... (5.3.1)}$$

Clearly $P \to \infty$ as $V \to b$ and $P \to O$ as $V \to \infty$ for a fixed value of T. The intermediate behaviour is not monotonic as extrema exist at

$$\frac{RT}{(V-b)^2} = \frac{2a}{V^3}$$

or

$$V^3 - \frac{2a}{RT}V^2 + \frac{4ab}{RT}V - \frac{2ab^2}{RT} = 0 \qquad \text{... (5.3.2)}$$

This gives rise to three roots for V, one of which is smaller than b and hence irrelevant for our purpose. Of the two relevant ones, one is a maximum, the other a minimum. As the temperature is raised, the separation between the maximum and the minimum decreases and if we consider the isotherm at $T = T_c$, the maximum and the minimum merge and there is only one point at which $\partial P/\partial V = 0$. It is obvious that this limiting situation must be a point of inflection and $\partial^2 P/\partial V^2$ must be zero. If the molar volume is V_c at $T = T_c$, then the second derivative of Eq. (5.3.1) yields

$$\left.\frac{\partial^2 P}{\partial V^2}\right|_{T=T_c} = 0 \text{ or } \frac{RT_c}{(V_c - b)^3} = \frac{3a}{V_c^4} \qquad \text{... (5.3.3)}$$

and together with Eq. (5.3.2) at $T = T_c$, $V = V_c$ (critical point) leads to

$$V_c = 3b, \; RT_c = \frac{8a}{27b} \qquad \text{... (5.3.4)}$$

The above result could also be obtained in a different way. We could have asked for the condition where the two roots of the cubic come together, i.e. Eq. (5.3.2) has the structure $(V-\alpha)^2(V-\beta) = 0$. This yields Eq. (5.3.4) once more. Yet another way of arriving at Eq. (5.3.4) is to consider how many values of V yield the same pressure on a given isotherm, i.e. we cast Eq. (5.3.1) itself as a cubic in V for a fixed T and P and determine the three roots. All three roots are real for $T < T_c$, as it must according to our previous discussion of the maximum, minima and the behaviour at $V = b$ and $V = \infty$. For $T > T_c$, the relation between P and V has got to be monotonic and at $T = T_c$, when there is an extremum as a point of inflection, the three roots of V must come together and we should have $(V - V_c)^3 = 0$. The condition that the cubic in V has this structure again yields Eq. (5.3.4).

In the light of the above discussion it is fairly obvious what the different isotherms should look like and they are shown in Fig. 5.6.

For $T > T_c$, the isotherms are like a standard vapour with $\partial P/\partial V < 0$ everywhere. It is for $T < T_c$, in the region between the minimum and maxi-

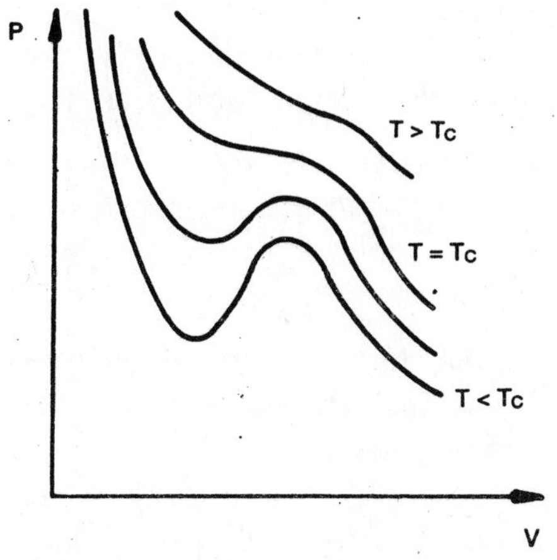

Fig. 5.6: Isotherms for Van der Waals equation.

mum that Van der Waals equation produces an unphysical situation by yielding $\partial P/\partial V > 0$. For the equation of state to make sense for $T < T_c$, it can no longer represent a homogeneous substance as we have assumed so far—it must represent two phases in equilibrium. To see whether this is the case we must construct the free energies of phases which are in thermal and mechanical equilibrium, i.e. same values of T and P. As discussed before there are three such candidates shown explicitly in Fig. (5.7) as the points A, B and C. Which of them would be the favoured one is determined by a consideration of the free energy; the lowest free energy state would be favoured.

The free energy has to be calculated by integrating along an isotherm

$$F = -\int P dV = -\int dV \left[\frac{RT}{V-b} - \frac{a}{V^2}\right] = -RT \ln (V-b) - \frac{a}{V} + F_o \qquad \text{... (5.3.5)}$$

and the result is shown in Fig. (5.7) with the three points A, B and C indicated on the curve. The solid curve is the result of integration. The dashed line joins the tangents at A at C. It is on this dashed line that the free energy of mixed phase composed of A and C will be lying and as can be seen from Fig. (5.7), the mixed phase has a lower free energy than the homogeneous phase shown by the solid line. Thus, between A and C, there is a coexistence of two phases and Van der Waals equation of state shows a phase transition as T is lowered below T_c. For $T < T_1$, between the points A and C, the pressure will remain constant as the molar volume of phase A changes to the molar volume of phase C. This is a first-order transition and the latent heat involved can be calculated from the Clausius-Clapeyron relation. The limiting transition occurs at $T = T_c$. Here there is no latent heat involved, but as one passes T_c from above, one passes from a one-phase (vapour) region to a two-phase coexistence state. The compressibility is infinite at $T = T_c$. The transition at $T = T_c$ is a second-order transition characterized by absence of latent heat in the transition and

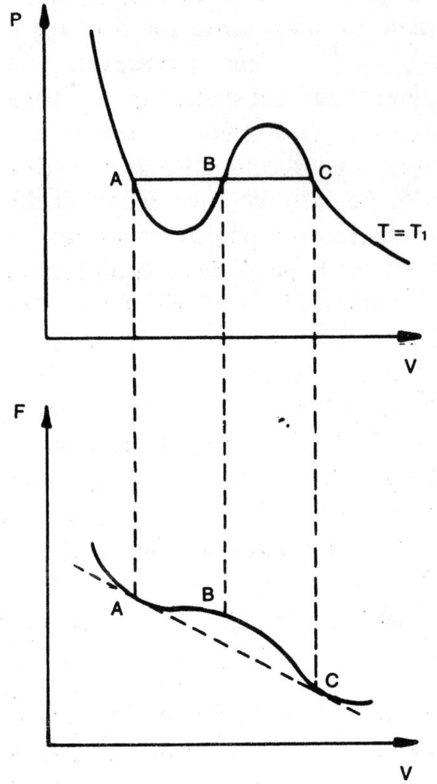

Fig. 5.7: Van der Waals isotherm and the two phase region.

singularities in thermodynamic quantities. At a temperature $T > T_c$, we determine the compressibility as

$$\frac{1}{\chi_T} = -\frac{\partial P}{\partial V}\bigg)_T = \frac{RT}{(V-b)^2} - \frac{2a}{V^3} \quad \text{... (5.3.6)}$$

Along the line of critical volume $V = V_c = 3b$, if we change the temperature, then

$$\frac{1}{\chi_T} = \frac{RT}{4b^2} - \frac{2a}{27b^3} = \frac{R}{4b^2}\left[T - \frac{8a}{27b}\right]$$

$$= \frac{R}{4b^2}\left(T - T_c\right)$$

or

$$\chi_T \propto \left(T - T_c\right)^{-1} \quad \text{... (5.3.7)]}$$

Thus the susceptibility diverges in a characteristic fashion as T approcahes T_c and while the absolute magnitude of the susceptibility is dependent on the nature of the substance (reflected in b and T_c), the nature of the singularity at T_c, i.e. $(T-T_c)^{-1}$, is universal. As the temperature is lowered towards the critical temperature along the critical isochore, the susceptibilities of all gases approach infinity, the behaviour being described by $(T - T_c)^{-1}$. This universality shown by the Van der Waals gas is quite remarkable and, as we will see later, is a standard feature of all second-order phase transitions.

We can explore yet another singular behaviour pertaining to the Van der Waals gas close to T_c but below it. As we remarked before, at T_c the maximum and the minimum come together causing a merging of two of the roots of Eq.

(5.3.2). If $T < T_c$, then the two roots will be displaced slightly from V_c and if one calls $T = T_c(1-\varepsilon)$ and the two roots which were $V = V_c = 3b$, are now at $V_{1,2} = V_c(1 \pm \delta)$, then clearly the structure of Eq. (5.3.2) to O (ε) will be

$$V^3 - \frac{2a}{RT_c}V^2 + \frac{4\,abV}{RT_c} - \frac{2ab^2}{RT_c} + \left(\frac{2aV_c^{\,2}}{RT_{cc}} - \frac{4abV_c}{RT_c} + \frac{2ab^2}{RT_c}\right) = 0$$

$$V^3 - \frac{2a}{RT}V^2 + \frac{4abV}{RT} - \frac{2ab^2}{RT} = -\,\varepsilon\, 27\, b^3 \qquad \text{... (5.3.8)}$$

Now setting $V = V_c(1+\delta)$ and working to O (δ^2)

$$V_c^3\left(1+3\delta+\delta^2\right) - \frac{2a}{RT_c}V_c^2\left(1+2\delta+\delta^2\right) + \frac{4abV_c}{RT_c}(1+\delta) - \frac{2ab^2}{RT_c} = -27b^3\varepsilon$$

$$\text{or } \delta\left[3V_c^3 - \frac{4aV_c^2}{RT_c} + \frac{4\,abV_c}{RT_c}\right] + \delta^2\left(V_c^3 - \frac{2a}{RT_c}V_c^2\right)$$

$$= -\,27\,b^3\varepsilon \qquad \text{... (5.3.9)}$$

But

$$3V_c^3 - \frac{4aV_c^3}{RT_c} + \frac{4\,abV_c}{RT_c} = b^3\left(81 - \frac{27\times 9}{2} + \frac{27\times 3}{2}\right) = 0$$

and

$$V_c^3 - \frac{2a}{RT_c}\cdot V_c^2 = b^3\left[27 - 2\times\frac{27}{4}\times 9\right] = -\frac{135}{4}\,b^3$$

or

$$\delta^2 = \frac{4}{5}\varepsilon \qquad \text{... (5.3.10)}$$

Consequently, for $T \cong T_c$ but below it, the difference between the two molar volumes is given by

$$\Delta V = V_1 - V_2 = 2V_c\,\delta = V_c\,\frac{4}{\sqrt{5}}\,\varepsilon^{1/2} \qquad \text{... (5.3.11)}$$

The difference in densities between the two phases is

$$\Delta\rho \propto \Delta V \left(= V_c\, \varepsilon^{1/2}\right) \propto \left(T_c - T\right)^{1/2} \qquad \text{... (5.3.12)}$$

The singular behaviour [i.e. $(T_c - T)^{1/2}$] is once again universal, while the actual $\Delta\rho$ will depend on the material properties. It should be noted that $\Delta\rho = O$ in the one-phase region $(T > T_c)$ and is non-zero only in the two-phase region $(T < T_c)$ and hence is an important characteristic of this second-order phase transition taking place at $T = T_c$. It is an example of what in general is called an order parameter for the second-order transition—a quantity which is zero above the transition and non-zero below. The analogue is the spontaneous magnetization in the paramagnetic-ferromagnetic transition.

5.4 Existence of Phase Transitions

The analysis of the Van der Waals gas in the last section showed that phase transitions lead to singularities in the thermodynamic functions. It is natural to ask how the partition function which appears to be an analytic function of its arguments develops a singularity. The answer lies in the limit $V \to \infty$. As we take the limit $V \to \infty$, singularities may develop as the limit function of a sequence of analytic functions need not be analytic.

Considering the grand partition function Q, we have

$$\frac{P}{kT} = \frac{1}{V} \ln Q \qquad \text{... (5.4.1a)}$$

and

$$n = \frac{1}{V}\, z \frac{\partial}{\partial z} \ln Q \qquad \text{... (5.4.1b)}$$

where $z = e^{\mu/kT}$. To have the possibility of singularities, we must take the thermodynamic limit $V \to \infty$ and write

$$\frac{P}{kT} = \operatorname*{Lt}_{V \to \infty} \frac{1}{V} \ln Q \qquad \text{... (5.4.2a)}$$

$$n = \underset{V \to \infty}{\text{Lt}} \frac{1}{V} z \frac{\partial}{\partial z} \ln Q \qquad \textbf{(5.4.2b)}$$

with n fixed. In Eq. (5.4.2b) the limit and the differentiation can be interchanged only if the convergence is uniform.

It was shown by Yang and Lee that the phase transitions are controlled by the zeros of the grand partition function in the complex z-plane. A transition occurs if a zero approaches the real axis in the limit $V \to \infty$. The two theorems governing the transition are:

Theorem 1: The limit $F_\infty(z) = \underset{V \to \infty}{\text{Lt}} \frac{1}{V} \ln Q$ exists for all $z > 0$ and is a continuous non-decreasing function of z and independent of the shape of the confining volume so long as the area of the surface that encloses the volume increases no faster than $V^{2/3}$ as the volume is increased.

Theorem 2: If R is a region of the complex z-plane containing a segment of the real axis where there are no zeros of Q, then in this region the quantity $\frac{1}{V} \ln Q$ converges uniformly to its limit as $V \to \infty$. The limit is analytic for all z in R.

A thermodynamic phase is defined by the values of z described in Theorem 2. In this region uniform convergence implies that limits can be interchanged in Eq. (5.4.2b) and we have

$$\frac{P(z)}{kT} = F_\infty(z) \qquad \textbf{... (5.4.3a)}$$

and

$$n = z \frac{\partial}{\partial z} F_\infty(z) \qquad \textbf{... (5.4.3b)}$$

The equation of state can be obtained by graphical elimination of z (see Fig. 5.8).

If the entire real axis is contained in R, then there is only one thermodynamic phase possible. Note that the fact that $n(z)$ is a non-decreasing function of z follows from

$$z \frac{\partial}{\partial z} \{n(z)\} = z \frac{\partial}{\partial z}\left(z \frac{\partial}{\partial z} \frac{1}{V} \ln Q\right) = \langle n^2 \rangle - \langle n \rangle^2 \geq 0 \qquad \textbf{... (5.4.4)}$$

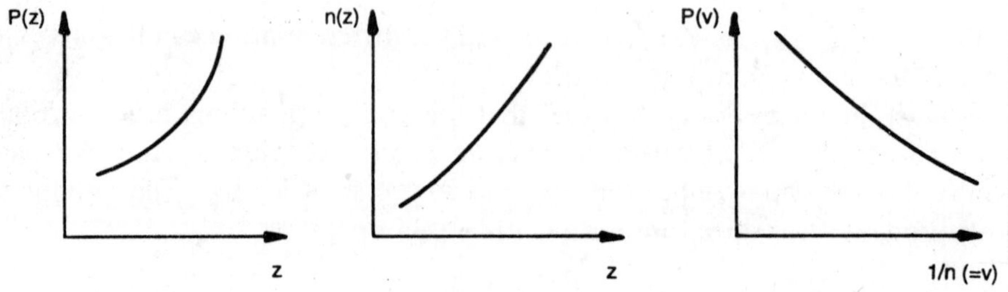

Fig. 5.8: No zero of the partition function on the real axis and the corresponding single phase equation of state.

If we now consider a situation where a zero of the grand partition function approaches a point z_0 on the real axis, then there are two thermodynamic phases, corresponding to distinct regions R_1 and R_2 (separated by z_0) where the conditions of Theorem 2 hold separately. At $z = z_0$, $F_\infty(z)$ and hence $P(z)$ are continuous by Theorem 1, but the derivative may be discontinuous and that leads to the two different phases for $z < z_0$ and $z > z_0$. Now since $n(z)$ is a non-decreasing function of z, it will have to make a finite jump at the point where $\partial P/\partial z$ is discontinuous, corresponding to a first-order transition. The situation is shown in Fig. 5.9. The similarity of the P vs $1/n$ plot with the isotherm of the Van der Waals gas is obvious.

5.5 Correlation Functions

For dense gases and liquids a particularly important quantity is the structure function or the correlation function or what is also known as the distribution function. The distinguishing feature of these systems is the interaction between particles and if $V(\mathbf{r}_1, \mathbf{r}_2, \ldots \mathbf{r}_N)$ be the general potential energy of the N-particle system, the probability of finding a particle at $\mathbf{r}_1$, another at $\mathbf{r}_2$, a third at $\mathbf{r}_3$, and so on for all the N particles is given by the canonical distribution function $1/Z_N \, e^{-(1/kT)V(\mathbf{r}_1, \ldots \mathbf{r}_2 \ldots \mathbf{r}_N)}$.

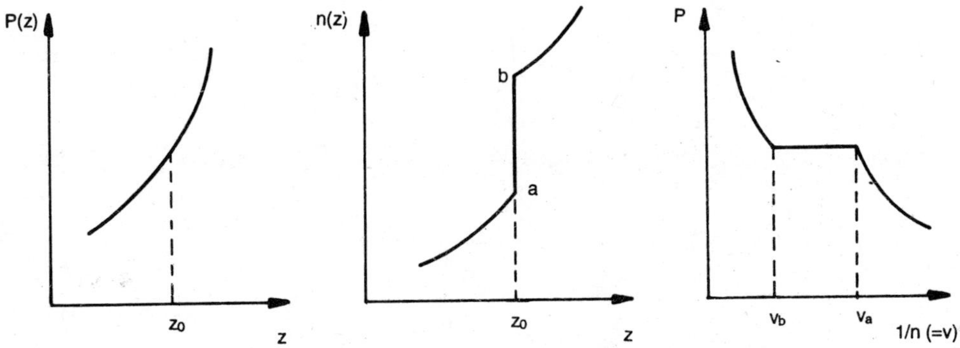

Fig. 5.9: A first order transition.

To obtain the probability of finding a particle at the point $\mathbf{r}_1$, we need to integrate over all the other coordinates $\mathbf{r}_2, \mathbf{r}_3, \ldots \mathbf{r}_N$ and thus the one-particle distribution function $n_1(\mathbf{r}_1)$ is defined as

$$n_1(\mathbf{r}_1) = \frac{N}{Z_N} \int d^3 r_2, \ldots d^3 r_N \, e^{-\frac{V(\mathbf{r}_1, \mathbf{r}_2, \ldots \mathbf{r}_n)}{kT}} \qquad \ldots (5.5.1)$$

The two-particle distribution function or the two-point correlation function $n_2(\mathbf{r}_1, \mathbf{r}_2)$ is similarly defined by integrating out over the N–2 remaining coordinates.

$$n_2(\mathbf{r}_1, \mathbf{r}_2) = \frac{N(N-1)}{Z_N} \int d^3 r_3 \ldots d^3 r_N \, e^{-V(\mathbf{r}_1, \mathbf{r}_2, \ldots \mathbf{r}_N)/kT} \qquad \ldots (5.5.2)$$

If we expect translational invariance and rotational invariance, then $n_2(\mathbf{r}_1, \mathbf{r}_2) = n_2(r_{12})$, where $r_{12} = |\mathbf{r}_1 - \mathbf{r}_2|$. As $r_{12} \to \infty$, $n_2 \to n_1^2$ since at large distances the particles are essentially independent of each other.

The immediate question is whether one can find a differential equation for $n_2(\mathbf{r}_1, \mathbf{r}_2)$ which would lead to a procedure for finding $n_2(\mathbf{r}_1, \mathbf{r}_2)$, if the potential is known. Accordingly, one writes (assuming a pairwise interaction for the remainder of the section).

$$n_2(r_1, r_2) = \frac{N(N-1)}{Z_N} \int e^{-\beta \sum\limits_{i \neq 1} V(r_{i1}) - \beta \sum\limits_{\substack{j \neq 1 \\ i \neq j, 1}} V(r_{ij})} d^3 r_3 \, d^3 r_4 \ldots d^3 r_N \qquad \ldots (5.5.3)$$

and differentiating with respect to $\mathbf{r}_1$ obtains,

$$\nabla_1 n_2(\mathbf{r}_1, \mathbf{r}_2) = -\frac{N(N-1)}{Z_N}\beta \int \left[\nabla_1 V(r_{12}) + \sum_{i \neq 1,2} \nabla_1 V(r_{1i})\right]$$

$$\times e^{-\beta \sum_{pairs} V} d^3 r_3 d^3 r_4 \ldots d^3 r_n$$

$$= -\beta \left[\nabla_1 V(r_{12}) n_2(r_{12}) + \frac{N(N-1)}{Z_N} \int \sum_{i \neq 1,2} \nabla_1 V(r_{1i}) \right.$$

$$\left. \times e^{-\beta \sum_{pairs} V} d^3 r_3 d^3 r_4 \ldots d^3 r_N \right] \qquad \ldots (5.5.4)$$

We now define the three-point correlation function or the three-particle distribution function as

$$n_3(\mathbf{r}_1, \mathbf{r}_2, \mathbf{r}_3) = \frac{N!}{(N-3)!} \frac{1}{Z_N} \int d^3 r_4 \ldots d^3 r_N \, e^{-\beta \sum_{pairs} V(r_{ij})} \qquad \ldots (5.5.5)$$

and recognizing that the second term on the right hand side of Eq. (5.5.4) can be written as

$$\frac{N(N-1)(N-2)}{Z_N} \int \nabla_1 \mathbf{V}(r_{13}) d^3 r_3 \int d^3 r_4 \ldots d^3 r_N \times e^{-\beta \sum_{pairs} V},$$

we arrive at

$$\nabla_1 n_2(r_{12}) = -\beta \left[\left(\nabla_1 V(r_{12})\right) n_{12} + \int d^3 r_3 \nabla_1 V\left(r_{13}\right) n_3(r_1, r_2, r_3)\right] \qquad \ldots (5.5.6)$$

This is the Born-Green equation and it illustrates a problem typical of all non-linear systems. The two-point correlation function n_2 is related to the three-point correlation function n_3. If we now want to find an equation for n_3, we would find that it would be involving n_4 and so on. This is the closure problem which is found in turbulence, or in any system described by a non-linear Langevin equation and so on. An approximate solution is to truncate the hierarchy at some point by invoking a relation between a higher correlation and

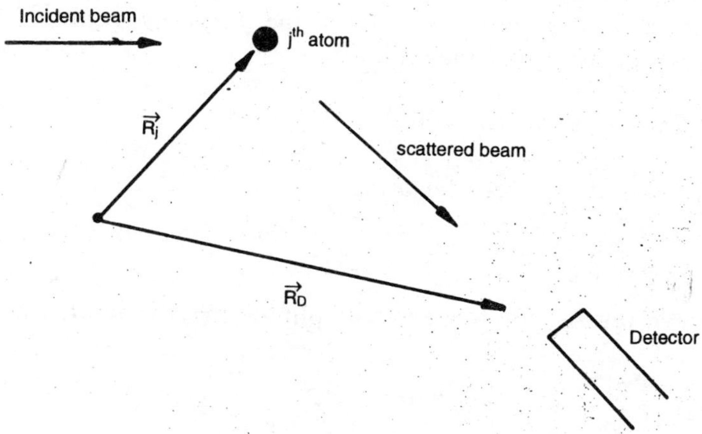

Fig. 5.10: The scattering geometry.

lower correlations. One such approximation is that of Kirkwood, which assumes that

$$n_3(r_1, r_2, r_3) = n_2(r_1, r_2) \times n_2(r_1, r_3) \times n_2(r_2, r_3)/n_1^3 \quad \text{... (5.5.7)}$$

Substituting for $n_3(r_1, r_2, r_3)$ from Eq. (5.5.7) in Eq. (5.5.6), we may solve for n_2—still a formidable task but nevertheless solvable.

The two-point correlation function can be determined fairly accurately experimentally by neutron or X-ray scattering from the gas or liquid. The typical experimental arrangement is shown in Fig. 5.10.

The scattering amplitude from the jth atom (position vector $\mathbf{R_j}$)is

$$S_j = a(\theta)\, e^{i\,\mathbf{k_{in}}\cdot\mathbf{R_j}}\, \frac{e^{ikR_{jD}}}{R_{jD}} \quad \text{... (5.5.8)}$$

In the above $\mathbf{k_{in}}$ is the momentum of the incoming beam and R_{jD} is the distance between the jth atom and the detector with

$$\mathbf{R_{jD}} = \mathbf{R_D} - \mathbf{R_j} \quad \text{... (5.5.9a)}$$

The collision is elastic and hence the magnitude k of the momentum of the incoming beam is the same as that of the outgoing beam. Thus, (since $\mathbf{k}_{out}$ and $\mathbf{R}_{jD}$ are in the same direction)

$$kR_{jD} = \mathbf{k} \cdot \mathbf{R}_{jD} = \mathbf{k}_{out} \cdot \left(\mathbf{R}_D - \mathbf{R}_j\right) \qquad \text{... (5.5.9b)}$$

and since the detector is very far away,

$$R_{jD} \cong R_D \qquad \text{... (5.5.9c)}$$

With the above results, the scattering amplitude may be written as

$$S_j \cong a(\theta) \frac{e^{i\mathbf{k}_{out} \cdot \mathbf{R}_D}}{R_D} e^{i\mathbf{R}_j \cdot (\mathbf{k}_{in} - \mathbf{k}_{out})}$$

$$= a(\theta) \frac{e^{i\mathbf{k}_{out} \mathbf{R}_D}}{R_D} e^{i\mathbf{q} \cdot \mathbf{R}_j} \qquad \text{... (5.5.10)}$$

where $\mathbf{q} = \mathbf{k}_{in} - \mathbf{k}_{out}$ is the scattering wavevector. For the total amplitude, we have to sum over all j and the intensity I is then obtained from the absolute magnitude of ΣS_j, i.e.

$$I(\theta) = \left|\sum S_j(\theta)\right|^2 = \left| a(\theta) \frac{e^{ik \cdot \mathbf{R}_D}}{R_D} \sum e^{i\mathbf{q} \cdot R_j}\right|^2$$

$$= |a(\theta)|^2 \frac{1}{R_D^2} |G|^2 \qquad \text{... (5.5.11)}$$

where

$$|G|^2 = \sum_i e^{i\mathbf{q} \cdot \mathbf{R}_i} \sum_j e^{-i\mathbf{q} \cdot \mathbf{R}_j}$$

$$= \sum_{ij} e^{i\mathbf{q}.(\mathbf{R}_{ij})}$$

For a liquid or gas, we need to average this $|a|^2$ over the probability of finding scatterers at $\mathbf{R}_i$ and $\mathbf{R}_j$ and hence the measured quantity is

$$\langle |a|^2 \rangle = \sum_{\text{pairs}} \int d^3 R_1 \ldots d^3 R_N \frac{1}{Z_N} e^{-\beta \sum_{\text{pairs}} V(R_{ij})} e^{+iq \cdot \mathbf{R}_{ij}}$$

$$= \frac{N(N-1)}{2} \int d R_1^3 \ldots d^3 R_N \frac{e^{iq \cdot \mathbf{R}_{12}}}{Z_N} e^{-\beta \sum_{\text{pairs}} V(R_{ij})}$$

$$= \int d^3 R_1 \, d^3 R_2 \, e^{iq \cdot \mathbf{R}_{12}} \, n_2 \left(\mathbf{R_1}, \mathbf{R_2}\right)$$

$$= \int d^3 R_{cm} \, d^3 R_{12} \, e^{iq \cdot \mathbf{R}_{12}} \, n_2 \left(\mathbf{R}_{12}\right)$$

$$= \frac{4\pi V}{q} \int r \, n_2(r) \sin qr \, dr \qquad \ldots \textbf{(5.5.13)}$$

where r is the magnitude or $\mathbf{R}_{12}$ and $\mathbf{R}_{cm} = (\mathbf{R}_1 + \mathbf{R}_2)/2$.

In arriving at the final form we have assumed translational and rotational invariance which makes $n_2(\mathbf{R}_1, \mathbf{R}_2)$ a function of R_{12} alone. Thus, measuring the scattering intensity at different values of θ (i.e. different q) would allow one to invert the relation expressed in Eq. (5.5.13) and obtain the correlation function $n_2(r)$.

5.6 Ising Model

This is a model of interacting magnetic moments at various points in space. For convenience, we imagine a definite lattice in space (linear lattice in one dimension, square lattice in two dimensions, cubic in three dimensions and so on) and put classical spins (essentially magnetic moment vectors) on each site. Each spin has the same magnitude but can have different orientations. In the Ising model, the spins can only point up or down. The interaction is short ranged—specifically we shall take them only to extend to nearest neighbours and it is such that the parallel alignment of spins is favoured. The energy function of this interacting system is taken to be

$$E = -\sum_{\substack{\text{nearest} \\ \text{neighbour}}} J_{ij} S_i S_j \qquad \ldots \textbf{(5.6.1)}$$

where S_i is the spin at site i and can have two values + 1 and – 1. J_{ij} is positive and exists only if i and j are two nearest neighbour positions. A simplifying assumption is to take $J_{ij} = J$ independent of the sites and then we have the standard Ising model energy expression

$$E = -J \sum_{\text{nn}} S_i S_j \qquad \text{... (5.6.2)}$$

If the whole system is put in an uniform external magnetic field h, then an additional term in the energy expression is required depicting the interaction of the spins with the magnetic field and we have

$$E = -J \sum_{\text{nn}} S_i S_j - h \sum S_i \qquad \text{... (5.6.3)}$$

Exact soluiion of the model with $h = 0$ is straightforward in one dimension and difficult in two dimensions. In three dimensions, the model cannot be exactly solved. With $h \neq 0$, the solution in one dimension is again straightforward and in two dimensions somewhat more involved. We will present the exact solution in one dimension and present an approximate method for solving the problem in any dimensions.

The important question that one has to address is whether the model would show spontaneous magnetization (i.e. magnetization in the limit of $h \to 0$) at a finite temperatue. Certainly at high temperatures the average magnetic moment would be zero, because the free energy would be dominated by the entropy term. But as the temperature is lowered, the energy term begins to dominate and it becomes increasingly difficult to turn the spin, so that it is possible that at low temperatures the spins will be aligned causing a phase transition from a state of zero average magnetization (paramagnet) to a state of non-zero average magnetization (ferromagnet).

5.6.1 Exact Solution in One Dimension

We need to evaluate the partition function

$$Z = \sum_{\{S_i\}} e^{\beta J \sum_i S_i S_{i+1}} \qquad \text{... (5.6.4)}$$

in the case of zero external field. This can be done by a trick. The product $S_i S_{i+1}$ can only acquire two values +1 or –1. So we define a new set of variables η_i which are defined by

$$\eta_i = \left.\begin{array}{l} +1 \text{ if } S_i S_{i+1} = 1 \\ -1 \text{ if } S_i S_{i-1} = 1 \end{array}\right\} \quad \text{... (5.6.5)}$$

The partition function can be written in terms of these variables as

$$Z = \sum_{\substack{\eta_1 = \pm 1 \\ \eta_2 = \pm 1}} e^{\beta J \sum \eta_i} = \prod_i \sum_{\eta_i = \pm 1} e^{\beta J \eta_i} = (2 \cosh \beta J)^N \quad \text{... (5.6.6)}$$

The free energy is $F = -kT \ln Z = -NkT \ln \cosh \beta J$.

In the presence of an external magnetic field, the above trick fails and we resort to a technique which has very general applicability. We consider the problem with N sites and evaluate the partition function without summing over the spin S_N. Then,

$$Z\left(S_N\right) = \sum_{S_1, S_2 \ldots S_{N-1}} e^{\Sigma \beta J S_i S_{i+1} + h \beta \Sigma S_i} \quad \text{... (5.6.7)}$$

We now put another site $(N + 1)$ and perform the sum over the spin at site N to write

$$Z\left(S_{N+1}\right) \sum_{S_N} = e^{\beta J S_N S_{N+1}} Z_N(S_N)\, e^{\beta h (S_N + S_{N+1})} \quad \text{... (5.6.8)}$$

Now, S_N can acquire two values +1 and –1 and hence we can write Z_N as a two-conponent vector $Z_N(+)$ and $Z_N(-)$. Clearly the relation between Z_{N+1} and Z_N is given by

$$Z_{N+1}(+)' = e^{\beta J + 2\beta h} Z_N(+) + e^{-\beta J} Z_N(-)$$

$$Z_{N+1}(-) = e^{-\beta J} Z_N(+) + e^{\beta J - 2\beta h} Z_N(-) \quad \text{... (5.6.9)}$$

or in matrix form

$$Z_{N+1}^{\alpha} = \sum_{\beta} T_{\alpha\beta} Z_N^{\beta}, \qquad \text{... (5.6.10)}$$

where $T_{\alpha\beta}$, called the transfer matrix for obvious reasons, is given by

$$T = \begin{bmatrix} e^{\beta J + 2\beta h} & e^{-\beta J} \\ e^{-\beta J} & e^{\beta J - 2\beta h} \end{bmatrix} \qquad \text{... (5.6.11)}$$

Now T is a real symmetric matrix and hence can be diagonalized and its eigenvectors will span the two-dimensional vector space we are dealing with. If we call the two eigenvalues λ_1 and λ_2 and the associated eigenvectors $\begin{pmatrix} \alpha_1 \\ \alpha_2 \end{pmatrix}$ and $\begin{pmatrix} \beta_1 \\ \beta_2 \end{pmatrix}$, then clearly we can write

$$\begin{pmatrix} Z_1^{+} \\ Z_1^{-} \end{pmatrix} = c_1 \begin{pmatrix} \alpha_1 \\ \alpha_2 \end{pmatrix} + c_2 \begin{pmatrix} \beta_1 \\ \beta_2 \end{pmatrix} \qquad \text{... (5.6.12)}$$

From Eq. (5.6.10) it is obvious that

$$\begin{pmatrix} Z_{N+1}^{+} \\ Z_{N+1}^{-} \end{pmatrix} = (T)^N \mathbf{Z}_1 = \lambda_1^N c_1 \begin{pmatrix} \alpha_1 \\ \alpha_2 \end{pmatrix} + \lambda_2^N c_2 \begin{pmatrix} \beta_1 \\ \beta_2 \end{pmatrix} \qquad \text{... (5.6.13)}$$

and the partition function

$$Z_{N+1} = Z_{N+1}^{+} + Z_{N+1}^{-} = \lambda_1^N \left(c_1 \alpha_1 + c_1 \alpha_2\right) + \lambda_2^N \left(c_2 \beta_1 + c_2 \beta_2\right) \qquad \text{... (5.6.14)}$$

In the thermodynamic limit only the larger eigenvalue survives and if $\lambda_1 > \lambda_2$, we have for $N >> 1$

$$Z_{N+1} = \lambda_1^N \left(c_1 \alpha_1 + c_1 \alpha_2\right) \qquad \text{... (5.6.15)}$$

and the free energy

$$F = -NkT \ln \lambda_1 \qquad \text{... (5.6.16)}$$

In the present case the two eigenvalues of T of Eq. (5.6.11) are

$$\lambda_{1,2} = e^{\beta J} \cosh 2\beta h \pm \left[e^{-2\beta J} + e^{2\beta J} \sinh^2 2\beta h\right]^{1/2} \qquad \text{... (5.6.17)}$$

and hence

$$F = -N \ln\left[e^{\beta J} \cosh 2\beta h + \sqrt{e^{-2\beta J} + e^{2\beta J} \sinh^2 2\beta h}\right] \quad \text{... (5.6.18)}$$

The magnetization M is given by

$$M = -\frac{\partial F}{\partial h} = N \frac{2\beta\, e^{\beta J} \sinh 2\beta h}{\sqrt{e^{-2\beta J} + e^{2\beta J} \sinh^2 2\beta h}} \quad \text{... (5.6.19)}$$

As $h \to 0$,

$$M \cong 4\beta^2 N e^{2\beta J} h \quad \text{... (5.6.20)}$$

showing that there can be no spontaneous magnetization as $h \to 0$ unless $T \to 0$. The susceptibility is given by

$$\chi = \left.\frac{\partial M}{\partial h}\right|_{h=0} = \frac{4N}{(kT)^2} e^{2J/kT} \quad \text{... (5.6.21)}$$

Thus as $T \to 0$, the susceptibility shows a singularity. Hence a phase transition occurs in the one-dimensional Ising model at $T = 0$. Later (Part III) we will show that in two dimensions a phase transition occurs at a finite temperature.

5.6.2 Approximate Method for Any Dimensions

We will present the technique for $h = 0$, the generalization to $h \neq 0$ is straightforward. In this approximation, we first write the energy expression as $E = -J \Sigma S_i S_j = -\sum_i h_i S_i$, where $h_i = \sum_{j \neq i} S_j$ is the field at the site i due to all the other spins. At this point there is no approximation—it comes when we say that h_i is the same at every site i and equal to some average $h = \langle h_i \rangle = J \langle m \rangle$, m being the average magnetization. The energy expression is now

$$E = -h \sum S_i \quad \text{... (5.6.22)}$$

and the partition function can be trivially evaluated. But h is an unknown quantity. So it has to be consistently fixed by determining the magnetization m from the partition function. We now have an explicit equation for m and

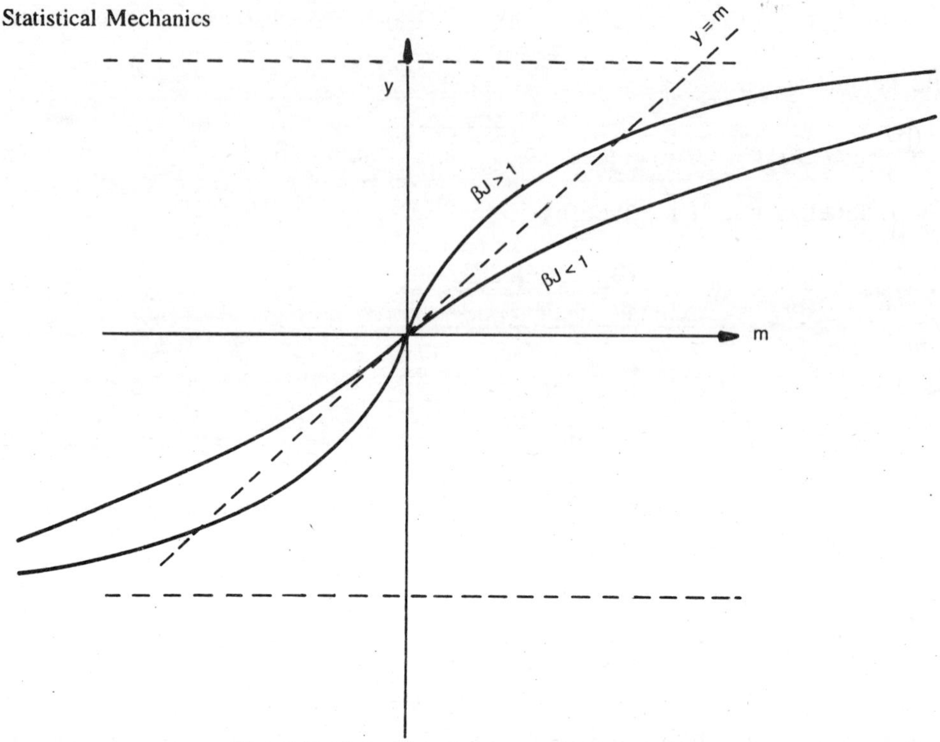

Fig. 5.11: Solution of the mean field equation.

the approximation is operationally complete. This is called the 'mean field approximation' and has a wide range of applicability as a first crack at a difficult problem concerning interactions. For the energy of Eq. (5.6.22), the mean magnetization m is

$$m = \left(e^{\beta h} - e^{-\beta h}\right)\left(e^{\beta h} + e^{-\beta h}\right)$$

$$= \tanh \beta h = \tanh \beta J m \qquad \text{... (5.6.23)}$$

The desired equation for m has been obtained. The solution has to be obtained graphically as shown in Fig. 5.11. The dashed line is the plot $y = m$ and the solid curves correspond to $y = \tanh \beta J m$ for various values of βJ. Clearly for $\beta J < 1$, there is only one solution, namely $m = 0$. However for $\beta J > 1$, there is a second solution with $m \neq 0$.

It m is very small, in other words βJ greater than but very close to unity, the hyperbolic tangent in Eq. (5.6.23) can be expanded to give

$$m = \beta J m - (\beta J)^3\, m^3/3 + \ldots$$

or

$$\frac{(\beta J)^3}{3} m^3 = m (\beta J - 1) \quad \text{... (5.6.24)}$$

For $\beta J < 1$, clearly the only solution is $m = 0$, but for $\beta J > 1$, one has the non-zero value of m given by

$$\begin{aligned} m^2 &= \frac{3}{(\beta J)^3} (\beta J - 1) \\ &= 3 \left(\frac{\beta}{\beta_c} - 1 \right) \\ &= 3 \left(\frac{T_c - T}{T} \right) \cong 3 \left(\frac{T_C - T}{T_c} \right), \end{aligned} \quad \text{... (5.6.25)}$$

where $\beta_c J = 1$ defines the critical temperature below which a solution with $m \neq 0$ exists.

That this is the favoured phase for $T < T_c$, can be established only after the free energy has been constructed. This will be dealt with in one of the problems at the end of the chapter. For the moment, we accept that when it exists, the solution with $m \neq 0$, is the stable solution and hence we have a second-order phase transition (no latent heat) at $T = T_c$, with $m = 0$ (paramagnet) above T_c and $m \neq 0$ (ferromagnet) below T_c. Thus the magnetization is the order parameter for the transition and its temperature dependence just near the transition is given by

$$m \propto \left(T_c - T \right)^{1/2}, \quad \text{... (5.6.26)}$$

Once again a singular behaviour and a universal behaviour independent of the material that is getting magnetized are seen. In fact, the even wider universality may be noticed by comparing with Eq. (5.3.12) describing the order parameter for the Van der Waals gas near the critical point. There too the behaviour is $(T_c - T)^{1/2}$.

The approximation described above is independent of spatial dimensionality. Hence this approximation would lead to a phase transition at finite temperature even in one dimension which we have seen above that it does not happen. Thus the approximation can be quite faulty. However, if a transition is known to occur, then this approximation is a good way of describing it. As a rule of thumb, it is a bad approximation in low dimensions but good in higher dimensions.

Problems

5.1 The pair correlation function $D(\mathbf{r}_1, \mathbf{r}_2)$ of the system of particles is defined as follows:
$D(\mathbf{r}_1, \mathbf{r}_2)\, d^3 r_1\, d^3 r_2$ = probability of simultaneously finding a particle in the volume element $d^3 r_1$ about $\mathbf{r}_1$ and a particle in the volume element $d^3 r_2$ about $\mathbf{r}_2$.
For a gas of classical particles with pairwise interaction (repulsive at very short range and attractive but decaying fast at long range), find the correlation function to the lowest non-zero order in the interaction. The density of the gas is very low.

5.2 An ensemble of 10^{10} weakly interacting spinless particles, each with the mass of an electron, are identical in appearance but obey classical statistics. They are confined in a critical box which is 10^{-6} cm on an edge. Each particle undergoes a potential interaction with the box which is of two sorts. One is attractive and leads to a bound state well localized near the centre of the box and having an energy of –1 eV. The other interaction is a strong repulsive one which prevents the particles from escaping through the walls of the box. Find at what temperature the pressure in the box is 1 atm?

5.3 Long molecules, each of length l are constrained to lie at equal separations along a one-dimensional chain. The molecules are capable of two orientations—parallel or perpendicular to the chain. The nearest neighbour interaction energy is $-\varepsilon$ if one is parallel and the other perpendcular and $+\varepsilon$ if both are perpendicular to the chain length. If the total number of molecules is N, find the average length of the chain as a function of temperature.

5.4 Consider the one-dimensional Ising model at temperatures close to $T = 0$. At $T = 0$, all the spins are aligned. For very low temperatures, there will be a few breaks in the aligned state. The following figure shows a situation with two breaks.

$$\uparrow \uparrow \uparrow \uparrow \downarrow \downarrow \downarrow \downarrow \downarrow \downarrow \uparrow \uparrow \uparrow \uparrow \uparrow \uparrow \uparrow$$

Each break costs an energy J. The statistical mechanics of the Ising model for $T=0$ can be described as the statistical mechanics of a dilute gas of breaks (non-interacting). Use this model to determine the correlation function of the Ising model as $T \rightarrow 0$.

5.5 Consider the approximate treatment for the magnetic transition carried out in Sec. 5.6. For $T < T_c$ show by explicit calculation of the free energy that the state with $m \neq 0$ will be favoured.

Chapter Six

Quantum Statistics

6.1 Introduction

Our discussion so far has treated all particles as classical particles obeying Newton's laws. This is valid if the temperature is relatively high so that the De Broglie wavelength is much smaller than the relevant length scales or the density is low enough so that inter-particle separation is much greater than the atomic size.

We now remove the restriction of low density and high temperatures which means that the De Broglie wavelength ($\approx h/\sqrt{mkT}$) becomes of the order of $n^{-1/3}$ ($n = N/V$). The quantum nature of the particles is now going to become effective and the necessary restrictions have to be taken into account. For a system of identical particles this restriction amounts to the symmetry under interchange of particles in the quantum mechanical many-body wavefunction. Since two successive interchanges of any two particles (say in single particle states i and j) reproduces the same many-body state (so far as probabilites go), a particular exchange either keeps the state unaltered (symmetric wavefunction) or introduces a change of sign (antisymmetric wavefunction). If the wavefunction changes sign, then clearly there cannot be two particles in the same quantum state i, since under exchange the same wavefunction would be reproduced but with changed sign. This can only happen if the wavefunction vanishes identically. Thus Pauli's exclusion principle is obeyed if the wavefunction is antisymmetric and particles with such wavefunctions are called fermions. If the wavefunction is symmetric, then there is no restriction on the number of particles in a particular state and such particles are called bosons. The spin statistics theorem relates the intrinsic spin of a particle to its symmetry property and thus half integer spin particles are fermions while integer spin ones are bosons. Electrons and nucleons are fermions having spin 1/2, while pions (spins 0), photons (spin 1) etc. are bosons.

6.2 Quantum Distribution Functions

We consider a quantum particle capable of existing in any one of a series of single particle quantum states labelled by i (i = 1, 2, 3,...). The energy of the i th state is ε_i.

A collection of such non-interacting particles in equilibrium at a temperature T is considered and we want to find the average number $\langle n_i \rangle$ of particles in the i th state. If the number of particles is n_i in any given configuration, then the total number of particles is

$$N = \sum n_i \qquad \text{... (6.2.1)}$$

and the total energy is

$$E = \sum \varepsilon_i \, n_i \qquad \text{... (6.2.2)}$$

(In this energy expression, we are making use of the vital assumption of the absence of interaction among the particles). To find the average value of n_i, it is best to consider one of the standard ensembles, evaluate the partition function and thence obtain $\langle n_i \rangle$. The grand canonical ensemble is best suited for our purpose since in that case neither E nor N need to be held constant. Unconstrained sums are easiest to do and hence we evaluate the grand canonical partition function

$$Q = \sum_{E,N} e^{-\beta (E - \mu N)} \qquad \text{... (6.2.3)}$$

where μ is the chemical potential. Using Eqs. (6.2.1) and (6.2.2),

$$Q = \sum_{\varepsilon_i,\, n_i} e^{-\beta \sum_i (\varepsilon_i n_i - \mu n_i)} \qquad \text{... (6.2.4)}$$

where it is convenient to do the sum over n_i first and write

$$Q = \prod_{\varepsilon_i} Q^{(i)} = \prod_{\varepsilon_i} \left[\sum_{n_i} e^{-\beta n_i(\varepsilon_i - \mu)} \right] \quad \text{... (6.2.5)}$$

If the particles are fermions, then each n_i is restricted to be zero or one and

$$Q_F^{(i)} = \sum_{n_i} e^{-\beta(\varepsilon_i - \mu) n_i} = 1 + e^{\beta(\varepsilon_i - \mu)} \quad \text{... (6.2.6)}$$

For bosons on the other hand, n_i can range from zero to infinity and we have

$$Q_B^{(i)} = \sum_{n_i = 0}^{\infty} e^{-\beta(\varepsilon_i - \mu) n_i} = \left[1 - e^{-\beta(\varepsilon_i - \mu)}\right]^{-1} \quad \text{... (6.2.7)}$$

Hence, the average number of particles with energy ε_i per quantum state is obtained using an expression similar to. Eq. (3.3.14).

6.2.1 Fermions

For fermions

$$\langle n_i \rangle = kT \frac{\partial}{\partial \mu} \ln Q_F^{(i)}$$

$$= \frac{1}{e^{\beta(\varepsilon_i - \mu)} + 1} \quad \text{... (6.2.8)}$$

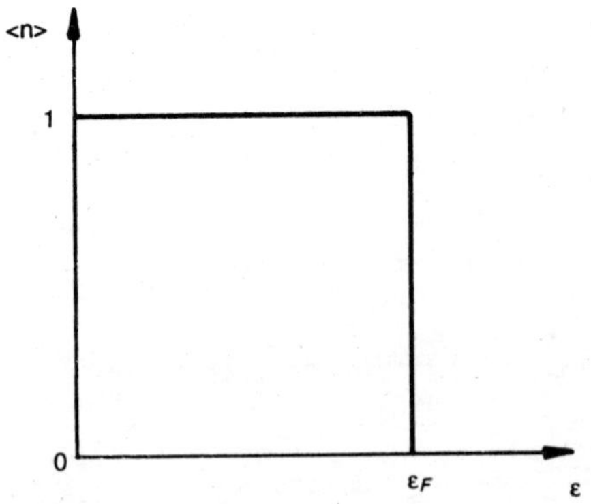

Fig. 6.1: Fermi distribution at T = 0.

It is instructive to picture the average occupation per state for different temperatures. The most important case is $T = 0$, i.e. $\beta = \infty$. In this case for $\varepsilon_i < \mu$, $e^{\beta(\varepsilon_i - \mu)} \rightarrow 0$ and hence there is on the average one particle state in this energy range. On the other hand, for $\varepsilon_i > \mu$, $e^{\beta(\varepsilon_i - \mu)} \rightarrow \infty$ and hence $\langle n_i \rangle \rightarrow 0$. Thus all the particles are accommodated in the states below the chemical potential μ and the distribution is pictured in Fig. 6.1.

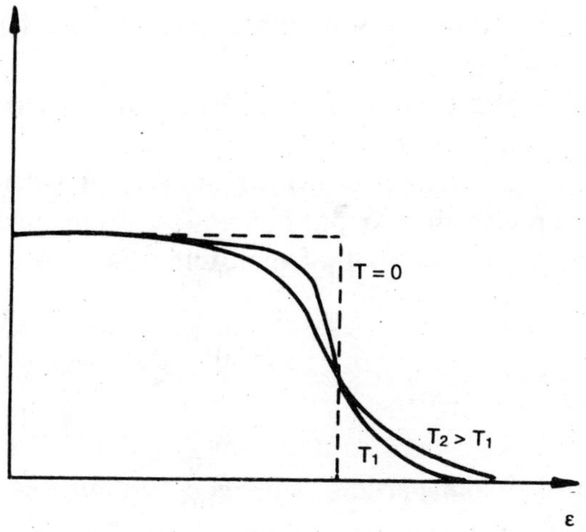

Fig. 6.2: Fermi distribution for $T \neq 0$.

The chemical potential μ at $T = 0$ is called the fermi energy ε_F. For temperatures $T > 0$, the distribution develops a tail for $\varepsilon_i > \mu$ with $\langle n_i \rangle = 1/2$ at $\varepsilon_i = \mu$. As the temperature increases, the tail becomes more and more pronounced as shown in Fig. 6.2.

6.2.2 Bosons

In the case of bosons

$$\langle n_i \rangle = kT \frac{\partial}{\partial \mu} \ln Q_B^{(i)} = \frac{1}{e^{\beta(\varepsilon_i - \mu)} - 1} \quad \text{... (6.2.9)}$$

Here the important point to note is that since $\langle n_i \rangle$ must exist even for $\varepsilon_i = 0$, one must have $e^{-\beta\mu} \geq 1$ or $\mu \leq 0$. Thus the chemical potential for bosons must be less than or equal to zero. Turning back to Eq. (6.2.7), it should now be clear that this is the same condition that the infinite sum should converge for all $\varepsilon_i \geq 0$. For those bosons which do not obey particle conservation (photons, phonons, etc.), the chemical potential is identically zero. This is due to the fact that the total number of particles N is now a variable which is fixed in equilibrium by the condition of minimum free energy and hence

$\mu = \left(\frac{\partial F}{\partial N}\right)_{V,T} = 0$. For those bosons where the chemical potential vanishes at a finite temperature, the phenomenon of Bose condensation occurs as will be shown in a subsequent chapter.

We end this section by writing down the number of particles between energy ε and $\varepsilon + d\varepsilon$. The particles are taken to be free (i.e. no interaction and no confining potential) and hence the number of quantum states per unit energy interval denoted by $dg/d\varepsilon$ is

$$\frac{dg}{d\varepsilon} = (2s + 1)\,\frac{2\pi V}{h^3}\,(2m)^{3/2}\,\varepsilon^{1/2} \qquad \text{... (6.2.10)}$$

where s is the spin of the particle and hence number of particles $n(\varepsilon)\,d\varepsilon$ in the energy range $d\varepsilon$ is

$$n(\varepsilon)\,d\varepsilon = (2s + 1)\,\frac{2\pi V}{h^3}\,(2m)^{3/2}\,\frac{\varepsilon^{1/2}}{\varepsilon^{\beta(\varepsilon-\mu)} \pm 1}\,d\varepsilon, \qquad \text{... (6.2.11)}$$

where the upper sign hold for fermions and the lower sign for bosons.

6.3 Equation of State

Before establishing the equation of state, we will prove a relationship which is independent of statistics—this relates the pressure volume product to the total energy. Using the standard result of grand canonical ensemble, we have

$$PV = kT \sum_i \ln Q^{(i)}_{F,B} = \pm\, kT\,(2s + 1)\,\frac{2\pi\, V\,(2m)^{3/2}}{h^3} \times$$

$$\int_0^\infty \ln\left(1 \pm e^{-(\varepsilon - \mu)/kT}\right)\varepsilon^{1/2}\,d\varepsilon \qquad \text{... (6.3.1)}$$

where the upper sign holds for fermions while the lower sign holds for bosons. Integrating by parts the integral on the right hand side of Eq. (6.3.1), we arrive at

$$PV = \pm\, kT\ (2s+1)\ \frac{2\pi V(2m)^{3/2}}{h^3} \left\{ \left[\frac{2}{3}\varepsilon^{3/2} \ln\left(1 \pm e^{-(\varepsilon-\mu)/kT}\right) \Big|_0^\infty \right] \right.$$

$$\left. + \frac{2}{3} \int_0^\infty \left[\frac{\varepsilon^{3/2}\ (\pm)\ e^{-\frac{\varepsilon-\mu}{kT}}}{1 \pm e^{-(\varepsilon-\mu)/kT}} \cdot \frac{d\varepsilon}{kT} \right] \right\}$$

$$= \frac{2}{3}\ (2s+1)\ \frac{2\pi V\,(2m)^{3/2}}{h^3} \int_0^\infty \frac{\varepsilon^{3/2} d\varepsilon}{e^{\frac{\varepsilon-\mu}{kT}} \pm 1}$$

$$= \frac{2}{3} \int_0^\infty \varepsilon\ n(\varepsilon)\ d\varepsilon = \frac{2}{3}\ E \qquad \dots \textbf{(6.3.2)}$$

where E is the average energy of the system. The above result, as should be clear from the derivation, holds irrespective of whether the particles are fermions or bosons—in fact by comparing with Eqs. (4.1.7) and (4.1.8) it can be seen that it holds for the classical ideal gas as well. This is expected from Eq. (6.3.2) since it is a result true regardless of values of T and μ, and hence must be valid at large negative values of μ, which corresponds to the classical ideal gas. To obtain the equation of state, we need in addition to Eq. (6.3.1), an equation for the total number of particles N,

$$N = \frac{\partial}{\partial\mu}(PV) = \pm\,(2s+1)\ \frac{2\pi V}{h^3}\ (2m)^{3/2} \int_0^\infty \frac{(\pm)\,\varepsilon^{1/2}\ d\varepsilon}{e^{(\varepsilon-\mu)/kT} \pm 1} \qquad \dots \textbf{(6.3.3)}$$

Our purpose will be to eliminate the chemical potential between Eqs. (6.3.1) and (6.3.3) and obtain a relation of the form

$$\frac{PV}{NkT} = 1 + \sum_{l=1}^{\infty} A_l\, n^l \qquad \dots \textbf{(6.3.4)}$$

where $n = N/V$.

We proceed by assuming that μ is negative and hence a binomial expansion of $(1 \pm e^{-\beta(\varepsilon-\mu)})^{-1}$ is possible. This allows us to write Eq. (6.3.3) as

$$N = (2s+1)\ \frac{2\pi V}{h^3} \cdot (2m)^{3/2} \int_0^\infty \frac{\varepsilon^{1/2}\ d\varepsilon}{e^{(\varepsilon-\mu)\,kT} \pm 1}$$

$$= (2s+1)\cdot\frac{2\pi V}{h^3}(2m)^{3/2}\int_0^\infty e^{(\mu-\varepsilon)/kT}\Big[1 \pm e^{(\mu-\varepsilon)/kT}$$

$$+ e^{2(\mu-\varepsilon)/kT} \pm e^{3(\mu-\varepsilon)/kT} + \dots\Big]\,e^{1/2}\,d\varepsilon$$

$$= (2s+1)\frac{2\pi V}{h^3}(2m)^{3/2}\int e^{(\mu-\varepsilon)/kT}\sum(\pm)^n e^{n(\mu-\varepsilon)/kT}\,\varepsilon^{1/2}\,d\varepsilon$$

$$= (2s+1)\frac{2\pi V}{h^3}(2mkT)^{3/2} e^{\mu/kT}\sum_{n=0}^{\infty}(\pm)^n\, e\frac{\mu n/kT}{(n+1)^{3/2}}\Gamma\left(\frac{3}{2}\right)$$

$$= (2s+1)\,V\cdot\left(\frac{2\pi m\,kT}{h^2}\right)^{3/2} e^{\mu/kT}\sum_{n=0}^{\infty}\frac{(\pm)^n e^{\mu n/kT}}{(n+1)^{3/2}} \qquad \dots \textbf{(6.3.5)}$$

Identifying the de Brodlie wavelength as $\lambda_d = = \dfrac{h}{(2\pi mkT)^{1/2}}$, we can write Eq. (6.3.5) as

$$e^{\mu/kT}\sum_{n=0}^{\infty}\frac{(\pm)^n e^{\mu n/kT}}{(n+1)^{3/2}} = \frac{n}{2s+1}\lambda_d^3 = n\lambda^3 \qquad \dots \textbf{(6.3.6)}$$

(we have written $\lambda_d = (2s+1)^{1/3}\lambda$).

Now, if $n\lambda^3 << 1$, which is attained if the density is very low (small n) or λ is very small (high T), then the left hand side of Eq. (6.3.6) must be much smaller than unity and hence we must have $e^{\mu/kT} << 1$, which requires μ to be large and negative. We proceed by writing Eq. (6.3.6) as

$$e^{\mu/kT}\left\{1 \pm \frac{e^{\mu/kT}}{2^{3/2}} + \frac{e^{2\mu/kT}}{3^{3/2}} + \cdots\right\} = n\lambda^3 \qquad \dots \textbf{(6.3.7)}$$

Expanding

$$e^{\mu/kT} = n\,\lambda^3 + \sum_{l=2}^{\infty} B_l\,(n\,\lambda^3)^l \qquad \dots \textbf{(6.3.8)}$$

and inserting in Eq. (6.3.7),

$$n\lambda^3\left[1 + B_1(n\lambda^3) + \ldots\right]\left\{1 \pm \frac{n\lambda^3}{2^{3/2}}\left(1 + B_1 n\lambda^3 + \ldots\right) + \right.$$

$$\left.\frac{1}{3^{3/2}}(n\lambda^3)^2 + \ldots\right\} = n\lambda^3 \qquad \ldots \text{(6.3.9)}$$

Matching terms of equal power of $n\lambda^3$ on either side of Eq. (6.3.9), we find

$$B_1 = \pm\frac{1}{2^{3/2}} \qquad \ldots \text{(6.3.10)}$$

by working to $O((n\lambda^3)^2)$. Higher-order matchings would lead to a systematic determination of the higher-order coefficients. To this order, we have

$$e^{\frac{\mu}{kT}} = n\lambda^3\left[1 \pm \frac{n\lambda^3}{2^{3/2}} + \ldots\right] \qquad \ldots \text{(6.3.11)}$$

To find the equation of state, we make use of Eq. (6.3.2) and obtain

$$\frac{PV}{NkT} = \frac{2}{3}\frac{E}{NkT} = \frac{2}{3}(2s+1)\frac{2\pi V(2m)^{3/2}}{h^3 NkT}\int\frac{\varepsilon^{3/2}\,d\varepsilon}{e^{(\varepsilon-\mu)/kT} \pm 1}$$

$$= \frac{2}{3}(2s+1)\frac{2\pi V(2m)^{3/2}}{h^3 NkT}\int_0^\infty e^{(\mu-\varepsilon)/kT}$$

$$\left[1 \pm e^{(\mu-\varepsilon)/kT} + e^{2(\mu-\varepsilon)/kT} + \ldots\right]\varepsilon^{3/2}\,d\varepsilon$$

$$= \frac{2}{3}(2s+1)\frac{2\pi(2mkT)^{3/2}V}{Nh^3}e^{\mu/kT} \times$$

$$\sum_{n=0}^{\infty}\frac{e^{\frac{\mu n}{kT}}(\pm)^n x^{3/2}e^{-x}}{(n+1)^{5/2}}\,dx \quad (\text{writing } x = (n+1)\varepsilon)$$

$$= (2s+1)\frac{V}{N}\left(\frac{2\mu\, mkT}{h^2}\right)^{3/2} e^{\frac{\mu}{kT}}\sum_{n=0}^{\infty}\frac{(\pm)^n e^{\mu n/kT}}{(n+1)^{5/2}}$$

$$= \frac{e^{\mu/kT}}{n\lambda^3} \sum_{n=0}^{\infty} (\pm)^n \frac{e^{\mu n/kT}}{(n+1)^{5/2}}$$

$$= \frac{e^{\mu/kT}}{n\lambda^3} \left(1 \pm \frac{e^{\mu/kT}}{2^{5/2}} + \cdots\right) \quad \text{... (6.3.12)}$$

Using the fugacity expansion from Eq. (6.3.11), we arrive at the equation of state

$$\frac{PV}{NkT} = \left(1 \pm \frac{n\lambda^3}{2^{3/2}} + \ldots\right)\left(1 \mp \frac{n\lambda^3}{2^{5/2}} + \ldots\right)$$

$$= 1 \pm \frac{n\lambda^3}{2^{3/2}} \mp \frac{n\lambda^3}{2^{5/2}} - O\left((n\lambda^3)^2\right)$$

$$= 1 \pm \frac{n\lambda^3}{2^{5/2}} + O\left[(n\lambda^3)^2\right] \quad \text{... (6.3.13)}$$

Clearly, we have obtained an equation of state in the desired form of Eq. (6.3.4), with

$$A_1 = \pm \frac{1}{2^{5/2}} \quad \text{... (6.3.14)}$$

The correction to the classical ideal gas now is, as expected, quantum mechanical in nature and determined by the De Broglie wavelength. For high temperatures and low densities, when quantum effects are expected to be unimportant, the classical equation of state is obtained.

6.4 Entropy of the Quantum Ideal Gas

We will illustrate the calculation of the entropy in the microcanonical ensemble. This will illustrate the use of Boltzmann's equation $S = k \ln \Gamma\ (E,N)$ in the quantum mechanical case. The energy E is now almost fixed (allowing for only a very small spread) and the total number of particles N is fixed as well. As usual, the non-interacting particles are enclosed in an extremely large box which causes the momentum to be quantized and thus the energy levels ε_i (labelled by momentum) are very closely spaced. If there are n_i particles in the level ε_i, then

$$\sum n_i = N \quad \text{(fixed)} \qquad \text{... (6.4.1a)}$$

$$\sum \varepsilon_i n_i = E \text{ (fixed)} \qquad \text{... (6.4.1b)}$$

First, we need to find out the number of ways in which the total number of particles can be arranged to give n_i particles in the i th level. We call this $W(\{n_i\})$. Clearly,

$$\Gamma(E,N) = \sum_{\{n_i\}} W(\{n_i\}) \qquad \text{... (6.4.2)}$$

To find $W(\{n_i\})$, we need to follow a standard trick. The energy ε_i is taken to be the mean energy of a bunch of levels centred around ε_i and g_i in number. These g_i levels are clustered together forming a cell whose occupation number is the number n_i introduced. We now consider $w(n_i, g_i)$, the number of ways in which n_i particles are distributed among the g_i states in the cell. Each arrangement within a cell is independent of the arrangements within the other cells and hence

$$W(\{n_i\}) = \prod_i w\left(n_i, g_i\right) \qquad \text{... (6.4.3)}$$

The determination of $w(n_i, g_i)$ is statistics dependent.

i) *Fermions:* To each level g_i there can be at most one particle. So n_i of the g_i levels will be occupied in a given arrangement. The total number of arrangements are obviously equal to the number of ways in which n_i things can be chosen from a total g_i and hence

$$w^{\mathrm{F}}\left(n_i, g_i\right) = \frac{g_i\,!}{n_i\,!\left(g_i - n_i\right)!} \qquad \text{... (6.4.4)}$$

ii) *Bosons:* To each level there can be upto n_i particles in this case. We consider the $g_i - 1$ partitions of the g_i levels and together with the n_i particles, we have a collection of $n_i + g_i - 1$ objects. The $g_i - 1$ partitions can be placed anywhere interspersed between the n_i particles and the total number of configurations obtained therefore will give w^B (g_i, n_i). The total number of arrangements is obviously the total number of permutations of the $(n_i + g_i - 1)$ objects, which is equal to $(n_i + g_i - 1)$! But of these the n_i identical particles and the $g_i - 1$ identical partitions can arrange themselves in n_i ! and $(g_i - 1)$! ways respectively without giving a new arrangement. So the number of independent configurations is

$$w^B \left(g_i, n_i\right) = \frac{(n_i + g_i - 1)\,!}{n_i\,!\,(g_i - 1)\,!} \quad \text{... (6.4.5)}$$

The calculation of Γ is through the relation of Eq. (6.4.2) which yields

$$\Gamma^F = \sum_{\{n_i\}} \prod_i w^F$$

$$= \sum_{\{n_i\}} \prod_i \frac{g_i\,!}{n_i\,!\,(g_i - n_i)\,!} \quad \text{... (6.4.6a)}$$

and

$$\Gamma^B = \sum_{\{n_i\}} \prod_i \frac{(n_i + g_i - 1)\,!}{n_i\,!\,(g_i - 1)\,!} \quad \text{... (6.4.6b)}$$

The summation over $\{n_i\}$ keeping the constraints of Eqs. (6.4.1a) and (6.4.1b) is impossible and hence we replace the sum by the single term $\{n_i\} = \{\bar{n}_i\}$, where $\{\bar{n}_i\}$ is that particular set which maximizes W under the given constraints. It is more convenient to maximize $\ln W$ and hence we have using Stirling's approximation, $\ln n\,! = n \ln n - n$,

$$\ln w^F = \sum_i \left[\ln g_i\,! - \ln n_i\,! - \ln (g_i - n_i)\,! \right]$$

$$= \sum_i \Big[g_i \ln g_i - g_i - n_i \ln n_i + n_i - (g_i - n_i)$$

$$\times \ln (g_i - n_i) + g_i - n_i \Big]$$

$$= \sum_i \{ g_i [\ln g_i - \ln (g_i - n_i)] - n_i [\ln n_i - \ln (g_i - n_i)] \}$$

Differentiation with respect to n_i leads to

$$\delta \ln w^{F} = \sum_i \left[\frac{(g_i - n_i)}{g_i - n_i} \delta n_i + \ln (g_i - n_i) \delta n_i - \delta n_i - \ln n_i \cdot \delta n_i \right]$$

$$= \sum_i \ln \left(\frac{g_i}{n_i} - 1 \right) \delta n_i \qquad \text{... (6.4.7)}$$

Similarly,

$$\ln w^{B} = \sum_i \Big[(n_i + g_i - 1) \ln (n_i + g_i - 1) - (n_i + g_i - 1)$$

$$- n_i \ln n_i + n_i - (g_i - 1) \ln (g_i - 10 + g_i - 1 \Big]$$

$$= \sum_i \Big[(n_i + g_i - 1) \ln (n_i + g_i - 1) - n_i \ln n_i$$

$$- (g_i - 1) \ln (g_i - 1) \Big]$$

leading to

$$\delta (\ln w^{B}) = \sum_i \left(\ln (n_i + g_i - 1) \delta n_i - \ln n_i \cdot \delta n_i \right)$$

$$= \sum_i \ln \left(1 + \frac{g_i - 1}{n_i} \right) \delta n_i \qquad \text{... (6.4.8)}$$

To take into account the constraints of Eqs. (6.4.1a) and (6.4.1b), we need to write them in the differential form

$$\sum_i \delta n_i = 0 \qquad \text{... (6.4.9a)}$$

$$\sum_i \varepsilon_i \delta n_i = 0 \qquad \text{... (6.4.9b)}$$

and introducing Lagrange multipliers λ_1 and λ_2, we have for fermions

$$\sum_i \left(\ln \left(\frac{g_i}{n_i} - 1 \right) - \lambda_1 - \lambda_2 \varepsilon_i \right) \delta n_i = 0 \qquad \text{... (6.4.10)}$$

and for bosons

$$\sum_i \left(\ln \left(1 + \frac{g_i - 1}{n_i} \right) - \lambda_1 - \lambda_2 \varepsilon_i \right) \delta n_i = 0 \qquad \text{... (6.4.11)}$$

The maximizing set $\{\bar{n}_i\}$ for fermions satisfy

$$\frac{g_i}{\bar{n}_i} = 1 + e^{\lambda_1 + \lambda_2 \varepsilon_i}$$

or

$$\bar{n}_i = \frac{g_i}{e^{\lambda_1 + \lambda_2 \varepsilon_i} + 1}, \qquad \text{... (6.4.12)}$$

while for bosons

$$\bar{n}_i = \frac{g_i}{e^{\lambda_1 + \lambda_2 \varepsilon_i} - 1}, \qquad \text{... (6.4.13)}$$

where we have assumed $g_i >> 1$. Equations (6.4.12) and (6.4.13) are exactly identical to the grand canonical results expressed in Eq. (6.2.11), once we have identified $\lambda_2 = \beta$, $\lambda_1 = \beta\mu$ and $g_i = g(\varepsilon)d\varepsilon = (2s+1)\frac{2\pi V}{h^3}(2m^{3/2}\varepsilon^{1/2})\,d\varepsilon$. Our strategy is to write

$$\Gamma^{F,B} = W^{F,B}(\{\bar{n}_i\}), \qquad \text{... (6.4.14)}$$

which can be safely done if the fluctuation in $\bar{n}_i$ is small. Hence for self consistency, we will need to demonstrate that $\langle \bar{n}_i^2 \rangle - \langle \bar{n}_i \rangle^2$ vanishes in the thermodynamic limit.

Before doing so we complete the calculation of the entropy by writing

$$\Gamma^F = \prod_i \frac{g_i\,!}{\bar{n}_i\,!\,(g_i - \bar{n}_i)\,!} \qquad \text{... (6.4.15)}$$

and

$$\Gamma^B = \prod_i \frac{(\bar{n}_i + g_i - 1)\,!}{\bar{n}_i\,!\,(g_i - 1)\,!} \qquad \text{... (6.4.16)}$$

which lead to

$$S^F = k\left[\sum_i g_i \ln g_i - \bar{n}_i \ln \bar{n}_i - (g_i - \bar{n}_i)\ln (g_i - \bar{n}_i)\right]$$

$$= k\sum_i \{g_i\,[\ln g_i - \ln (g_i - \bar{n}_i)] - \bar{n}_i\,[\ln \bar{n}_i - \ln (g_i - \bar{n}_i)]\}$$

$$= k\sum_i \left\{\bar{n}_i \ln\left(\frac{g_i}{\bar{n}_i} - 1\right) - g_i \ln\left(1 - \frac{\bar{n}_i}{g_i}\right)\right\} \qquad \text{... (6.4.17)}$$

and

$$S^B = k\sum_i \left\{\bar{n}_i \ln\left(1 + \frac{g_i}{\bar{n}_i}\right) + g_i \ln\left(1 + \frac{\bar{n}_i}{g_i}\right)\right\} \qquad \text{... (6.4.18)}$$

To conclude our argument, we need to show that the fluctuations in n_i, are negligible. We proceed by noting that (see Eqs. (6.2.8) and (6.2.9))

$$\langle n_i \rangle = -\frac{\partial}{\partial \varepsilon_i} kT \ln Q^{(i)} \qquad \text{... (6.4.19)}$$

Differentiating,

$$\frac{\partial}{\partial \varepsilon_i}\langle n_i \rangle = -kT\frac{\partial}{\partial \varepsilon_i}\left(\frac{1}{Q^{(i)}}\frac{\partial Q^{(i)}}{\partial \varepsilon_i}\right)$$

$$= -kT\left[\frac{1}{Q^{(i)}}\frac{\partial^2 Q^{(i)}}{\partial \varepsilon_i^2} - \left(\frac{1}{Q^i}\frac{\partial Q^{(i)}}{\partial \varepsilon_i}\right)^2\right] \qquad \text{... (6.4.20)}$$

We note that the average occupation can be written as

$$<n_i> = \frac{1}{Q} \sum e^{\mu \beta \Sigma n_i - \beta \Sigma \varepsilon_i n_i} n_i$$

$$= -kT \frac{\partial}{\partial \varepsilon_i} \ln Q = -kT \frac{\partial}{\partial \varepsilon_i} \ln Q' .$$

while

$$<n_i^2> = \frac{1}{Q} \Sigma e^{\mu \beta \Sigma n_i - \beta \Sigma \varepsilon_i n_i} n_i^2 .$$

$$= \frac{(kT)^2}{Q} \frac{\partial^2}{\partial \varepsilon_i^2} Q = \frac{(kT)^2}{Q^{(i)}} \frac{\partial^2}{\partial \varepsilon_i^2} Q^{(i)} \qquad \text{... (6.4.21)}$$

Using the above result in Eq. (6.4.20), we obtain,

$$<(\delta n_i)^2> = <n_i^2> - <n_i>^2 = -kT \frac{\partial}{\partial \varepsilon_i} <n_i>$$

$$= -kT \frac{\partial}{\partial \varepsilon_i} \left[e^{\beta(\varepsilon_i - \mu)} \pm 1 \right]^{-1}$$

$$= \frac{e^{\beta(\varepsilon_i - \mu)}}{(e^{\beta(\varepsilon_i - \mu)} \pm 1)^2} = <n_i> \pm <n_i>^2 \qquad \text{... (6.4.22)}$$

(once again, the upper sign is for fermions and the lower sign for bosons).

In the above form, the fluctuations in n_i are not necessarily small. However, as we consider the infinite volume limit, the level distribution becomes continuous and instead of a single level i we consider a group of levels clustered around the level i so that

$$n_i = \sum_k n_k$$

leading to,

$$<n_i^2> - <n_i>^2 = <(\Sigma n_k)^2> - <\Sigma n_k>^2$$

$$= <\sum_{m,l} n_l n_m> - <\sum_l n_l> <\sum_m n_m> \qquad \text{... (6.4.23)}$$

If the suffixes m and l are unequal, the right hand side of Eq. (6.4.23) vanishes which we prove by noting the following:

$$\frac{\partial}{\partial \varepsilon_l} <n_m> = -kT \frac{\partial}{\partial \varepsilon_l}\left(\frac{1}{Q^{(m)}} \frac{\partial Q^{(m)}}{\partial \varepsilon_m}\right)$$

$$= -kT \frac{\partial^2}{\partial \varepsilon_l \, \partial \varepsilon_m} \ln Q = -kT \frac{\partial}{\partial \varepsilon_l}\left(\frac{1}{Q} \frac{\partial Q}{\partial \varepsilon_m}\right)$$

$$= -kT\left[\frac{1}{Q} \frac{\partial^2 Q}{\partial \varepsilon_l \, \partial \varepsilon_m} - \frac{1}{Q^2} \frac{\partial Q}{\partial \varepsilon_l} \frac{\partial Q}{\partial \varepsilon_m}\right]$$

$$= -\beta\left[<n_l n_m> - <n_l><n_m>\right]$$

For $m \neq l$, $\frac{\partial}{\partial \varepsilon_l} <n_m> = 0$ and hence the result. Using this result, we can write Eq. (6.4.23) as

$$<n_i^2> - <n_i>^2 = \sum_k \left(<n_k^2> - <n_k>^2\right) \qquad \text{... (6.4.24)}$$

From Eq. (6.4.22), we can write

$$\sum_k \left(<n_k^2> - <n_k>^2\right) = \sum_k \left(<n_k> \pm <n_k>^2\right)$$

$$= <n_i> \pm \sum_k <n_k>^2$$

so that

$$<n_i^2> - <n_i>^2 = <n_i> \pm \sum_k <n_k>^2 \qquad \text{... (6.4.25)}$$

which is the desired result. The right hand side is proportional to the volume V. While the first term on the right hand side is automatically so, the second term too is proportional to the volume as well which appears in the measure of integration as $Vd^3k/(2\pi)^3$. Thus the relative fluctuation $\frac{(<n_i^2> - <n_i>^2)}{<n_i>^2}$ is

$O\left(\frac{1}{V}\right)$, and in the infinite volume, the fluctuations vanish and the entropy expressions of Eqs. (6.4.17) and (6.4.18) are perfectly valid.

Problems

6.1 Prove that in 2-dimensions the specific heat $C_v(N,T)$ of an ideal non-relativistic gas of spinless fermion is identical to the specific heat of a corresponding gas of bosons for all values of N and T.

Chapter Seven

Fermi Distribution: Examples

7.1 Degenerate Fermi Gas

A collection of independent fermions is called degenerate if it is at very low temperatures or if its density is very high. The characteristic of the degenerate Fermi gas is that the distribution function $f(\varepsilon) = 1$ if $\varepsilon < \mu$ and zero if $\varepsilon > \mu$. This is because the temperature of the gas can be considered to be absolute zero and the limiting form of the Fermi distribution discussed in Sec. 6.2 obtains. We will consider spin half fermions so that the factor $2s+1 = 2$ and we have

$$n(\varepsilon)\,d\varepsilon = 2 \cdot 2\pi V \left(\frac{2m}{h^2}\right)^{3/2} \varepsilon^{1/2}\,d\varepsilon, \text{ for } \varepsilon < \mu$$

$$= 0 \text{ for } \varepsilon > \mu \qquad \text{... (7.1.1)}$$

If the total number of fermions is N, then

$$N = \int_0^{\mu} n(\varepsilon)\,d\varepsilon = 4\pi V \left(\frac{2m}{h^2}\right)^{3/2} \frac{2}{3} \cdot \mu^{3/2}$$

$$= \frac{8\pi}{3} V \left(\frac{2m}{h^3}\right)^{3/2} \varepsilon_F^{3/2} \qquad \text{... (7.1.2)}$$

where we identify the chemical potential μ at absolute zero as the Fermi energy ε_F. In terms of the number density $n = N/V$, the Fermi energy can be written as

$$\varepsilon_F = \left(\frac{3n}{8\pi}\right)^{2/3} \frac{h^2}{2m} \qquad \text{... (7.1.3)}$$

The Fermi temperature T_F is defined as $\varepsilon_F = kT_F$, while the Fermi velocity v_F is defined as $v_F = (2\varepsilon_F/m)^{1/2}$ and the Fermi momentum as $P_F = mv_F$. Clearly the Fermi temperature is proportional to the two thirds power of density and hence a high density Fermi gas has a high Fermi temperature. The Fermi gas is degenerate if its temperature T is much lower than T_F—this is generally the criterion for determining whether the limiting form of the distribution for the degenerate gas is applicable in a given situation or not. The total energy of the degenerate Fermi gas (non-relativistic) is

$$E = \int_0^{\varepsilon_F} \varepsilon\, n\, (\varepsilon)\, d\varepsilon = 4\pi V \left(\frac{2m}{h^2}\right)^{3/2} \int_0^{\varepsilon_F} \varepsilon^{3/2}\, d\varepsilon$$

$$= \frac{8}{3}\pi V \left(\frac{2m}{h^2}\right)^{3/2} \varepsilon_F^{3/2}\, \frac{3}{5}\, \varepsilon_F$$

$$= \frac{3}{5} N \varepsilon_F \qquad \text{... (7.1.4)}$$

The pressure exerted by the Fermi gas is found from the general expression $P = \frac{2}{3}\, E/V$ and hence

$$P = \frac{2}{3} n \cdot \varepsilon_F = \frac{2}{5}\, nkT_F \qquad \text{... (7.1.5)}$$

Thus, one pictures the degenerate Fermi gas as being at absolute zero with all the levels below $\varepsilon = \varepsilon_F$ occupied, the occupancy factor being $(2s + 1)$, where s is the spin of the fermions. The number of fermions and the zero point kinetic energy coming from the uncertainty principle allows the gas to exert pressure, which is given by Eq. (7.1.5). The intense pressure which the degenerate Fermi gas can exert manifests itself in the form of the white dwarf stars, which we deal with in the next section.

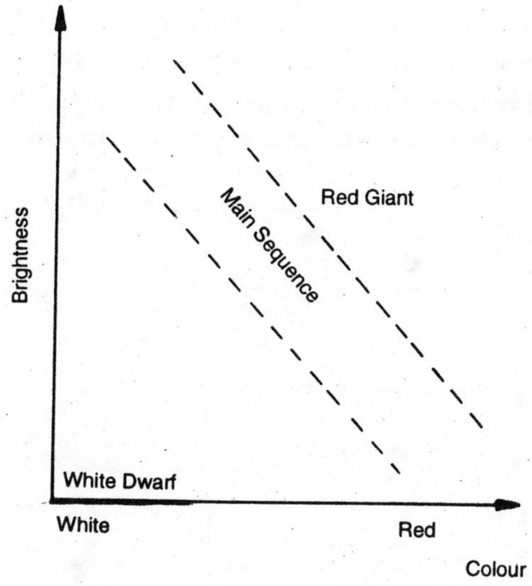

Fig. 7.1: Russel-Hertzprung plot.

7.2 White Dwarfs

The brightness of a star diminishes as the colour becomes redder and in a brightness against colour plot (known as Russell-Hertzprung diagram (Fig. 7.1) most of the stars lie in a strip. Certain stars are brighter than their red colour warrants—these are the red giants, while certain others are dimmer than their white colour would suggest—these are the white dwarfs. The reason for their dimness is the absence of hydrogen. The mass of the star is provided by the helium nuclei while the stability is provided by the ionized electrons. The mass of the star would try to drive it towards gravitational collapse—this is prevented by the zero point pressure exerted by the electrons which can be considered a degenerate Fermi gas. We consider the Fermi gas of electrons to consist of N particles. The number of nucleons is then $2N$ and the mass of the star is approximately $2N\, m_n$ where m_n is the nucleonic mass. Knowing the approximate mass of the star, we can estimate N and from the approximate size estimate, the volume V. Thus the density $n = N/V$ is known and the Fermi temperature of the star can be calculated. It is roughly 10^{11} K >> 10^7 K, the temperature of the star. Hence the electron gas can be taken to be at zero temperature and thus completely degenerate.

As explained above, the role of the nucleons is to provide the mass while the role of the electrons is to exert the pressure. The typical energy of the electrons is much larger than the rest mass of the electrons and hence we need to use the relativistic expression for energy in calculating the pressure of the degenerate electron gas:

$$E = V \cdot \frac{8\pi}{h^3} \int_0^{P_F} \sqrt{\left(p^2 c^2 + m^2 c^4\right)}\, p^2\, dp$$

$$= V \cdot \frac{8\pi}{h^3} \cdot c \int_0^{P_F} \left(1 + \frac{m^2 c^2}{2 p^2}\right) \cdot p^3\, dp$$

$$= V \cdot \frac{8\pi}{h^3} \cdot c \left(\frac{P_F^{\,4}}{4} + \frac{m^2c^2}{4} P_F^{\,2} + \ldots\right)$$

$$= V \cdot \frac{8\pi}{h^3} \cdot \frac{m^4 c^5}{4} \left(X_F^4 + X_F^2 + \ldots\right), \qquad \text{... (7.2.1)}$$

where $X_F = P_F/mc$. The pressure can be obtained by differentiating E with respect to V, as

$$P = -\frac{\partial E}{\partial V} = -\frac{2\pi}{h^3} m^4 c^5 \left(X_F^4 + X_F^2 + V \frac{\partial X_F^2}{\partial V}\left(2X_F^2 + 1\right) + \ldots\right) \qquad \text{... (7.2.2)}$$

Now,

$$V \cdot \frac{\partial}{\partial V} X_F^2 = 2V \frac{\partial}{\partial V} \frac{\varepsilon_F}{mc^2} = \frac{2}{mc^2} V \cdot \frac{\partial \varepsilon}{\partial V}$$

$$= \frac{2}{mc^2} V \cdot \frac{\partial}{\partial V} \left(\frac{h^2}{2m} \left(\frac{3}{8\pi} \frac{N}{V}\right)^{2/3}\right)$$

$$= -\frac{4}{3} \frac{m\varepsilon_F}{m^2 c^2} = -\frac{2}{3} X_F^2 \qquad \text{... (7.2.3)}$$

and hence Eq. (7.2.2) becomes

$$P = -\frac{2\pi m^4 c^5}{h^3}\left(X_F^4 + X_F^2 - \frac{4}{3} X_F^4 - \frac{2}{3} X_F^2 + \dots\right)$$

$$= \frac{2\,\pi m^4 c^5}{3h^3}\left(X_F^4 - X_F^2 + \dots\right) \qquad \dots (7.2.4)$$

In terms of the mass and radius of the star, we can write

$$M = 2N\, m_n\,, \; V = \frac{4}{3}\pi R^3$$

leading to

$$2m_n \cdot n = \frac{3M}{4\pi\, R^3} \text{ or } n = \frac{3M}{8\pi m_n} \cdot R^3$$

and

$$X_F = \overline{M}^{1/3}/\overline{R} \qquad \dots (7.2.5)$$

where.

$$\overline{M} = \frac{9\pi}{8}\frac{M}{m_n} \text{ and } \overline{R} = R/\left(\frac{h}{mc}\right) \qquad \dots (7.2.6)$$

In terms of $\overline{M}$ and $\overline{R}$, we can write the pressure as

$$P = K\left(\frac{\overline{M}^{4/3}}{\overline{R}^4} - \frac{\overline{M}^{2/3}}{\overline{R}^2}\right) \qquad \dots (7.2.7)$$

where

$$K = \frac{2\pi}{3}\, mc^2 \left(\frac{mc}{h}\right)^3 \qquad \dots (7.2.8)$$

The equilibrium radius is found by equating the work done against the pressure to the gravitational potential energy. The gravitational potential energy of a system of total mass M and radius R is

$$V(R) = -\frac{\alpha GM^2}{R} = -\alpha G\left(\frac{8\,m_n}{9\pi}\right)^2 \frac{mc}{h}\frac{\overline{M}^2}{\overline{R}} \qquad \text{... (7.2.9)}$$

where α is a number of order unity and can be calculated only when the density is known as a function of the radial distance. The work done against the pressure is

$$W = \int_{\infty}^{R} P4\pi r^2 dr = 4\pi\left(\frac{h}{mc}\right)^3 \int_{\infty}^{\overline{R}} P(\overline{r})\ 4\pi\,\overline{r}^2\ d\overline{r} \qquad \text{... (7.2.10)}$$

which must be equated to the potential energy $V(R)$. Differentiating the right hand sides of Eqs. (7.2.9) and (7.2.10)

$$4\pi\,\overline{R}^2\ P(\overline{R}) = \alpha\,G\left(\frac{8m_n}{9\pi}\right)^2\left(\frac{mc}{.h}\right)^4 \frac{\overline{M}^2}{\overline{R}^2} \qquad \text{... (7.2.11)}$$

or

$$K\left(\frac{\overline{M}^{4/3}}{\overline{R}^4} - \frac{\overline{M}^{2/3}}{\overline{R}^2}\right) = K'\frac{\overline{M}^2}{\overline{R}^4}$$

where we have substituted for $P(\overline{R})$ from Eq. (7.2.7) and defined $K' = \alpha g/4\pi\left(\frac{8m_n}{9\pi}\right)^2\left(\frac{mc}{h}\right)^4.$

Rearranging,

$$\frac{1}{\overline{R}^4}\left(\frac{K}{K'}\ \overline{M}^{4/3} - \overline{M}^2\right) = \frac{K}{K'}\cdot\frac{\overline{M}^{2/3}}{\overline{R}^2}$$

or

$$\overline{R}^2 = \overline{M}^{2/3}\left(1 - \frac{\overline{M}^{2/3}}{\overline{M}_0^{\ 2/3}}\right) \qquad \text{... (7.2.12)}$$

where

$$\overline{M}_0^{\ 2/3} = K/K' = \frac{2\pi}{3}\ mc^2\left(\frac{mc}{h}\right)^3 \Big/ \frac{\alpha G}{4\pi}\left(\frac{8m_n}{9\pi}\right)^2\left(\frac{mc}{h}\right)^4$$

$$= \frac{2\pi}{3} \cdot 4\pi \left(\frac{1}{2\pi}\right)^3 \cdot \left(\frac{9\pi}{8}\right)^2 \frac{1}{\alpha} \frac{hc}{Gm_n^2}$$

$$= \frac{27\pi}{64\alpha} \quad \frac{hc}{Gm_n^2} \qquad \text{... (7.2.13)}$$

Clearly for $\overline{R}$ as given by Eq. (7.2.12) to exist, we must have $\overline{M} < \overline{M}_0$. Thus $\overline{M}_0$ is the upper limit on the mass of a white dwarf. The dimensionless number $hc/Gm_n^2 \approx 10^{39}$. The dimensional mass M_0 corresponing to $\overline{M}_0$ is (for $\alpha \approx 1$)

$$M_0 = \frac{8}{9\pi} m_n \cdot \overline{M}_0 \approx 10^{33} \text{ g}, \approx M_{\text{sun}} \qquad \text{... (7.2.14)}$$

For an accurate determination of M_0, one must have a proper estimate of α and when this is taken into account $M_0 \approx 1.4\, M_{\text{sun}}$—an upper limit on the white dwarf mass known as the Chandrasekhar limit.

7.3 Specific Heat of an Electron Gas

In the last two sections we dealt with a gas of fermions at $T = 0$. We now consider the case of finite temperature when it is possible for the fermions to populate states with energy above the chemical potential μ and consequently the chemical potential is now going to be a function of the total number of particles N and the temperature T. It is this temperature-dependent chemical potential which we want to determine. The most convenient way to proceed is to use Eq. (6.3.2) and write

$$PV = \frac{2}{3} E = \frac{2}{3} \cdot \frac{2V}{4\pi^2} \left(\frac{2m}{h^2}\right)^{3/2} \int_0^\infty \frac{\varepsilon^{3/2}\, d\varepsilon}{e^{\beta(\varepsilon - \mu)} + 1} \qquad \text{... (7.3.1)}$$

(Note that for a spin s fermion, the factor 2 associated with V in the above expression will be changed to $(2s + 1)$. To evaluate the integral note that

$$\int_0^\infty \frac{\varepsilon^{3/2}\, d\varepsilon}{e^{\beta(\varepsilon - \mu)} + 1} = (kT)^{5/2} \int_{-\beta\mu}^\infty \frac{(x + \mu\beta)^{3/2}}{e^x + 1} = (kT)^{5/2}\, I\, (\mu\beta)$$

and $\mu\beta = \alpha$ ... (7.3.2)

where

$$I(\alpha) = \int_{-\alpha}^{\infty} \frac{(x + \alpha)^{3/2}}{e^x + 1} \qquad \text{... (7.3.3)}$$

We first note that an exact evaluation of $I(\alpha)$ is impossible and one has to settle for a perturbative calculation. Having done the $T = 0$ calculation in the previous sections, we now wish to extend those results to $T \neq 0$ but small nonetheless. This means $\alpha = \mu/kT >> 1$ and our approximate evaluation of $I(\alpha)$ will deal with the range where $\alpha \longrightarrow \infty$. Thus,

$$I(\alpha) = \int_{-\alpha}^{\infty} \frac{(x+\alpha)^{3/2}}{e^x+1} dx = \int_{-\alpha}^{0} \frac{(x+\alpha)^{3/2}}{e^x+1} dx + \int_{0}^{\infty} \frac{(x+\alpha)^{3/2}}{e^x+1} dx$$

$$= \int_{0}^{\alpha} \frac{(\alpha - x)^{3/2}}{1+e^{-x}} dx + \int_{0}^{\infty} \frac{(x + \alpha)^{3/2}}{e^x + 1} dx$$

$$= \int_{0}^{\alpha} \frac{e^x}{1+e^x} (\alpha - x)^{3/2} dx + \int_{0}^{\infty} \frac{(\alpha + x)^{3/2}}{e^x + 1} dx$$

$$= \int_{0}^{\alpha} (\alpha - x)^{3/2} dx - \int_{0}^{\alpha} \frac{(\alpha - x)^{3/2}}{1 + e^x} dx + \int_{0}^{\infty} \frac{(\alpha + x)^{3/2}}{e^x + 1} dx$$

$$= \frac{2}{5}\alpha^{5/2} + \int_{0}^{\infty} \frac{(\alpha + x)^{3/2} - |(\alpha - x)|^{3/2}}{e^x + 1} + \int_{\alpha}^{\infty} \frac{|(\alpha + x)|^{3/2}}{e^x + 1} dx \qquad \text{... (7.3.4)}$$

The last term on the right hand side in $O(e^{-\alpha})$ and hence negligible. The second term can be expanded as

$$(\alpha + x)^{3/2} - (\alpha - x)^{3/2} = \alpha^{3/2} + \frac{3}{2}\alpha^{1/2} x + \ldots - \left(\alpha^{3/2} - \frac{3}{2}\alpha^{1/2} x + \ldots\right)$$

$$= 3\,\alpha^{1/2}\, x + O\left(\frac{1}{\alpha^{1/2}}\right) \qquad \text{... (7.3.5)}$$

Evaluation of $I(\alpha)$ to two-term accuracy ($\alpha^{5/2}$ and $\alpha^{1/2}$), then requires the evaluation of the definite integral $I_1 = \int\limits_0^\infty \frac{x\,dx}{e^x + 1}$, which we can write as,

$$I_1 = \int\limits_0^\infty x\, e^{-x}\left[1 + e^{-x}\right]^{-1} dx$$

$$= \int\limits_0^\infty x\, e^{-x} \sum_{n=0}^{\infty} (-)^n\, e^{-nx}\, dx$$

$$= \sum_{n=0}^{\infty} \int\limits_0^\infty x\, e^{-(n+1)x}\, (-)^n\, dx$$

$$= \sum_{n=0}^{\infty} (-)^n \frac{1}{(n+1)^2} \int\limits_0^\infty y\, e^{-y}\, dy$$

$$= \sum_{n=0}^{\infty} \frac{(-1)^n}{(n+1)^2} = \frac{1}{1^2} + \frac{1}{3^2} + \frac{1}{5^2} - \left(\frac{1}{2^2} + \frac{1}{4^2} + \frac{1}{6^2}\right)$$

$$= \frac{1}{1^2} + \frac{1}{2^2} + \frac{1}{3^2} + \ldots - 2\left(\frac{1}{2^2} + \frac{1}{4^2} + \frac{1}{6^2} + \ldots\right)$$

$$= \zeta(2) - 2 \cdot \frac{1}{2^2}\left(\frac{1}{1^2} + \frac{1}{2^2} + \frac{1}{3^2} + \ldots\right)$$

$$= \zeta(2) - \frac{1}{2}\zeta(2) = \frac{1}{2}\zeta(2) \qquad \text{... (7.3.6)}$$

where

$$\zeta(n) = \frac{1}{1^n} + \frac{1}{2^n} + \frac{1}{3^n} + \ldots \qquad \text{... (7.3.7)}$$

From any text on mathematical physics it can be verified that $\zeta(2) = \pi^2/6$ and hence the required integral $I_1 = \pi^2/12$. Now using Eqs. (7.3.4) and (7.3.5), we have

$$I(\alpha) = \frac{2}{5}\alpha^{5/2} + \frac{\pi^2}{4}\alpha^{1/2} + O\left(\frac{1}{\alpha^{1/2}}\right) \qquad \text{... (7.3.8)}$$

and from Eqs. (7.3.1) and (7.3.2), we obtain,

$$PV = \frac{V}{3\pi^2}\left(\frac{2m}{h^2}\right)^{3/2}(kT)^{5/2}\left[\frac{2}{5}\left(\frac{\mu}{kT}\right)^{5/2} + \frac{\pi^2}{4}\left(\frac{\mu}{kT}\right)^{1/2} + \ldots\right]$$

$$PV = \frac{V}{3\pi^2}\left(\frac{2m}{h^2}\right)^{3/2}\left[\frac{2}{5}\left(\mu^{5/2} + \frac{\pi^2}{4}(kT)^2\mu^{1/2}\right)\right] \qquad \text{... (7.3.9)}$$

Differentiating with respect to μ yields N

$$N = \frac{V}{4\pi^2}\left(\frac{2m}{h^2}\right)^{3/2}\left[\mu^{3/2} + \frac{\pi^2}{8}(kT)^2\frac{1}{\mu^{1/2}} + \ldots\right] \qquad \text{.. (7.3.10)}$$

As $T \to 0$, the first term correctly reproduces the zero temperature result of Eq. (7.1.2). The Fermi energy ε_F is given by Eq. (7.1.3) and eliminating N/V in favour of ε_F, we find

$$\mu = \varepsilon_F\left[1 + \frac{\pi^2}{8}\left(\frac{kT}{\mu}\right)^2 + \ldots\right]^{-2/3}$$

$$\cong \varepsilon_F\left[1 - \frac{\pi^2}{12}\left(\frac{kT}{\varepsilon_F}\right)^2 + \ldots\right] \qquad \text{... (7.3.11)}$$

to the leading order in the temperature T. Note that this result does go in the expected direction in that, for very high temperatures, the chemical potential is negative (see Chapter 6) and thus the first correction to the positive zero temperature chemical potential is to decrease it.

To find the specific heat, we first need to find the entropy. This is best done by noting PV is $2/3\,E$ and thus from Eq. (7.3.10)

$$E = \frac{V}{2\pi^2}\left(\frac{2m}{h^2}\right)^{3/2}\left[\frac{2}{5}\,\mu^{5/2} + \frac{\pi^2}{4}\,(kT)^2\,\mu^{1/2} + \ldots\right]$$

$$= \frac{V}{2\pi^2}\left(\frac{2m}{h^2}\right)^{3/2}\left[\frac{2}{5}\,\varepsilon_F^{5/2}\left(1 - \frac{5}{2}\cdot\frac{\pi^2}{12}\left(\frac{kT}{\varepsilon_F}\right)^2 + \ldots\right) + \frac{\pi^2}{4}\,(kT)^2\,\varepsilon_F^{1/2} + \ldots\right]$$

$$= \frac{V}{2}\left(\frac{2m}{h^2}\right)^{3/2}\left[\frac{2}{5}\,\varepsilon_F^{5/2} + \frac{\pi^2}{6}\,(kT)^2\,\varepsilon_F^{1/2} + \cdots\right] \qquad \ldots\ \textbf{(7.3.12)}$$

Differentiating with respect to T at constant volume

$$C = \frac{\partial E}{\partial T} = \frac{V}{2\pi^2}\cdot\left(\frac{2m}{h^2}\right)^{3/2}\cdot\frac{\pi^2}{3}\cdot k^2\,\varepsilon_F^{1/2}\,T + \ldots$$

$$= N\cdot 3\pi^2\left(\frac{h^2}{2m}\right)^{3/2}\frac{1}{\varepsilon_F^{3/2}}\,\frac{1}{2\pi^2}\left(\frac{2m}{h^2}\right)^{3/2}\frac{\pi^2}{3}\,k\,(kT)\,\varepsilon_F^{1/2} + \ldots$$

$$= \frac{\pi^2}{2}\,Nk\cdot\frac{kT}{\varepsilon_F} + \ldots \qquad \ldots\ \textbf{(7.3.13)}$$

Thus, as $T \to 0$, the specific heat of the electron gas is given by

$$C = \frac{\pi^2}{2}\,Nk\,\frac{kT}{\varepsilon_F}$$

which vanishes linearly with T as T tends to zero. This is to be contrasted with the T^3-dependence of the contribution of the vibrations of the nuclei forming the crystal lattice. The total specific heat for a metal will be a combination of that obtained above in Eq. (7.3.14) and the lattice part obtained in Eq. (4.5.12). At very low temperatures the linear part dominates while at higher temperatures

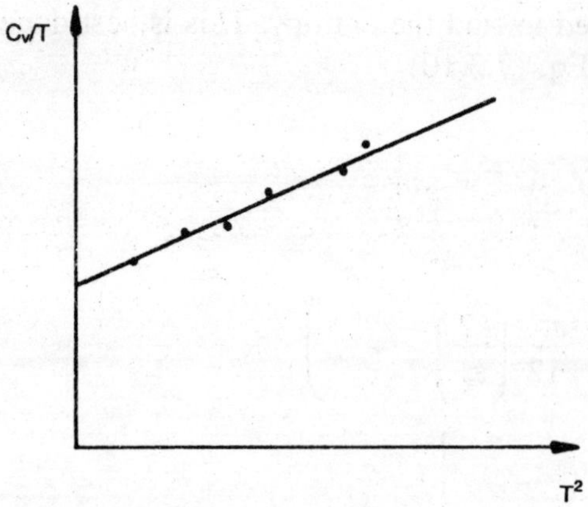

Fig. 7.2: Measured specific heat of copper at low temperatures showing the linear behavior of the electron gas and the additional T^3 contribution of the lattice vibrations.

the T^3 effect is more prominent. The results are in surprisingly good agreement [Fig. (7.2)] with experiments for a wide variety of metals in spite of the fact that effects like presence of impurities, periodic structure of the lattice, etc. have been ignored. In the next section, the effect of the periodic structure of the lattice will be discussed.

7.4 One-dimensional Metal—Effect of Periodic Lattice Structure

To begin with we would like to find out the allowed energy levels ε of an electron moving in a weak periodic potential $U(x)$. We assume the periodicity parameter to be a, so that

$$U(x + a) = U(x) \qquad \text{... (7.4.1)}$$

Further, we assume that the length L of the sample is such that

$$\int_0^L U(x)\, dx = 0 \qquad \text{... (7.4.2)}$$

The energy levels ε are to be obtained by a perturbative treatment of the Schrödinger equation

$$H\psi = \left[-\frac{h^2}{2m}\frac{\partial^2}{\partial x^2} + U(x) \right] \psi(x) = \varepsilon\ \psi(x) \qquad \text{... (7.4.3)}$$

where the potential $U(x)$ is the perturbation. The unperturbed energies are those of the free particle,

$$\varepsilon = \frac{p^2}{2m} \qquad \text{... (7.4.4)}$$

where p is the momentum and the wavefunctions are the suitably normalized plane wave solutions $\psi_p = \frac{1}{\sqrt{L}} e^{ipx/h}$. The first-order energy shift E_1 is given by the standard formula

$$E_1 = \int_0^L \psi_P^* \, U(x) \, \psi_p \, dx = \frac{1}{L} \int_0^L U(x) \, dx = 0 \qquad \text{... (7.4.5)}$$

However, the above formula for the first-order energy shift is true for non-degenerate states alone and clearly there are two wavefunctions $\psi_{\pm p} = 1/\sqrt{L} \, e^{\pm ipx/h}$ belonging to the same energy level $\varepsilon = p^2/2m$ and accordingly we must deal with the degenerate state perturbation theory which will yield non-trivial results whenever the potential can cause transitions between the two degenerate states, i.e. whenever the matrix element

$$M_{12} = \int \psi_p^* \, U(x) \, \psi_p \, dx \qquad \text{... (7.4.6)}$$

will be non-zero. Expanding the periodic potential in a Fourier series as

$$U(x) = \sum_{n=0}^{\infty} U_n \, e^{inqx} + c.c. \qquad \text{... (7.4.7)}$$

where

$$q = \frac{2\pi}{a} \qquad \text{... (7.4.8)}$$

(a being the lattice periodicity introduced in Eq. (7.4.1)) it is easy to see that $M_{12} = 0$ whenever

$$2p = \pm \, nq\hbar \quad \text{or} \quad p = \pm \, n \frac{\pi}{a} \hbar \qquad \text{... (7.4.9)}$$

We will concentrate on the $n = 1$ case for the present. The energy levels when p is in the vicinity of $(\pi/a)\hbar$, i.e. $p = (\frac{\pi}{a} + k)\hbar$ can be found by diagonalizing the 2×2 matrix whose diagonal elements are $\int \psi_p^* \, H \, \psi_p \, dx$ and

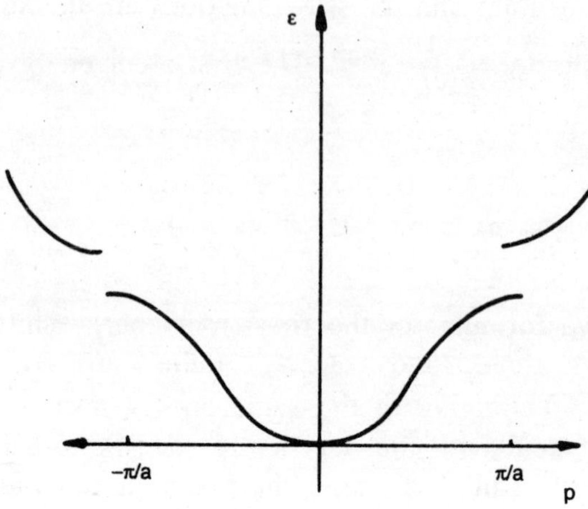

Fig. 7.3: The discontinuity in the ε vs p plot.

$\int \psi_{-p}^{*} H \psi_{-p}\, dx$ (which are clearly equal to $\left[\left(\frac{\pi}{a} \pm k\right)\right]^{2} \hbar^{2}$ and the off diagonal elements are $\int \psi_p^{*} H \psi_{-p}\, dx$ and $\int \psi_{-p}^{*} H \psi_p\, dx$ (which from Eq. (7.4.6) are clearly seen to be U_1 and U_1^{*}). Thus the energy levels ε are found from

$$\text{Det}\begin{bmatrix} \frac{1}{2m}\left(\frac{\pi}{a} + k\right)^{2} \hbar^{2} - \varepsilon & U_1 \\ U_1^{*} & \frac{1}{2m}\left(\frac{\pi}{a} - k\right)^{2} \hbar^{2} - \varepsilon \end{bmatrix} = 0$$

or

$$\varepsilon^{2} - \varepsilon \frac{\hbar^{2}}{m}\left(\frac{\pi^{2}}{a^{2}} + k^{2}\right) + \frac{\hbar^{2}}{4m^{2}}\left(\frac{\pi^{2}}{a^{2}} - k^{2}\right)^{2} - |U_1|^{2} = 0$$

$$\varepsilon_{1,2} = \frac{\hbar^{2}}{2m}\left(\frac{\pi^{2}}{a^{2}} + k^{2}\right) \pm \left(\frac{\pi^{2} k^{2} \hbar^{4}}{a^{2} m^{2}} + |U_1|^{2}\right)^{1/2} \qquad \text{... (7.4.10)}$$

At $k = 0$, i.e. $p = \pm (\pi/a)\,\hbar$, there is an energy gap given by $\Delta\varepsilon = \varepsilon_1 - \varepsilon_2 = 2\,U_1$. The situation is shown in Fig. 7.3. Far from

$p = \pm (\pi/a)\hbar$ the free particle result in valid to first order in U, while at $p = \pm (\pi/a)\hbar$ there is a discontinuity in ε with the approach to the two limiting values being quadratic in nature as is obvious from Eq. (7.4.10).

While we have restricted the above discussion to $n = 1$, the same phenomenon occurs for all n and whenever $\varepsilon = n^2\pi^2\hbar^2/2ma^2$, we have a discontinuity in the ε vs. p plot.

We now address the question regarding the position of the Fermi level in this one dimensional system. If the total number of electrons is N, then we have for a system of length L,

$$N = \frac{2L}{h}\int_{-p_F}^{p_F} dp = \frac{4Lp_F}{h} = \frac{2L}{\pi\hbar}\,p_F \qquad \dots (7.4.11)$$

Now, if there is one electron per site (sites come at separations of a), then $N = L/a$ and

$$p_F = \frac{\pi}{2a}\cdot\hbar \qquad \dots (7.4.12)$$

The Fermi level is far away from the energy gap and we have the usual metallic properties associated with the electron gas, which is to say that it is possible to excite electrons across the Fermi level with a small amount of thermal energy and thus electrical and thermal conduction is feasible.

On the other hand, if there are two electrons per site (the maximum allowable), then $N = 2L/a$ and

$$p_F = \frac{\pi}{a}\cdot\hbar \qquad \dots (7.4.13)$$

The Fermi level is now right in the middle of the energy discontinuity and this can have very significant consequences.

One of the quantities that is strongly affected is the density of states. There are no states at $p = p_F$ for ε lying between $\varepsilon_F + |U_1|$ and $\varepsilon_F - |U_1|$. We want the density of states as ε approaches $\varepsilon_F - |U_1|$. The general formula for the density of states $n(\varepsilon)$ in one dimension can be written as

$$n(\varepsilon) = \frac{2L}{h}\int dp\,\delta(\varepsilon - \varepsilon_p) \qquad \dots (7.4.14)$$

which is the trivial statement that if we are to convert the basic density of states $2L\, dp/h$ to the energy variable, then we must integrate over all momenta which give the same energy. In using this relation in the present case, we note that it is most convenient to set the zero of the energy at ε_F and work in terms of small deviations Δ from ε_F, i.e.

$$\varepsilon = \varepsilon_F + \Delta \qquad \text{... (7.4.15)}$$

For the momentum, we work in terms of small deviation from $\pi\hbar / a$. Further, remembering $\Delta << \varepsilon_F$, Eq. (7.4.14) leads to

$$n(\varepsilon) \cong \frac{2L}{h} \frac{|\Delta|}{\sqrt{\Delta^2 - |U_1|^2}} \qquad \text{... (7.4.16)}$$

for $\Delta > U_1$. We can now determine the energy difference $E(T) - E(0)$ for $T << T_F$ by the following argument. At $T = 0$, all the electrons occupy the states with $\varepsilon < \varepsilon_F$, while for $T > 0$, there are some electrons in states $\varepsilon > \varepsilon_F$. For every electron raised above the Fermi level, there is a hole created in the erstwhile filled Fermi sea. If we measure the energy from the Fermi level ε_F, i.e. $\varepsilon' = \varepsilon_F + \Delta$ (Eq. (7.4.15)) then the absence of a particle with energy ε' from the Fermi sea corresponds to a hole of energy $\Delta' = \varepsilon_F - \varepsilon'$. Thus for $T > 0$, we have a gas of holes with energy lying between 0 and ε_F and a gas of particles (electrons) with energy greater than ε_F. At $T = 0$, this gas disappears. For $T << T_F$, this gas will have vanishingly small energy and will be determined by the distribution in the vicinity of the Fermi level. The probability of finding a hole with energy $\Delta' = \varepsilon_F - \varepsilon'$ is clearly $1 - f(\varepsilon') = (e^{\beta(\mu - \varepsilon')} + 1)^{-1} \cong [e^{\beta(\varepsilon_F - \varepsilon')} + 1]^{-1} = (\varepsilon^{\beta\Delta'} + 1)^{-1}$, which is the same as the distribution of particles. For a given energy Δ $(= \varepsilon - \varepsilon_F)$, the distribution of holes is $(e^{\beta\Delta} + 1)^{-1}$ and the distribution of particles is also $(e^{\beta\Delta} + 1)^{-1}$ for $T << T_F$ and the energy of this gas of particles and holes is determined to a first approximation by the density of states at the Fermi level. The density of states is the same for electrons and holes and hence

$$E(T) - E(0) = \int_0^\infty \Delta \left(e^{\beta\Delta} + 1\right)^{-1} n(\varepsilon_F)\, d\Delta \qquad \text{... (7.4.17)}$$

In this case, the density of states near $\varepsilon = \varepsilon_F$ is given by Eq. (7.4.16) and hence

$$E(T) - E(0) = \frac{2L}{h}\left(\frac{2m}{\varepsilon_F}\right)^{1/2} \int_0^\infty \frac{\Delta^2}{\sqrt{\Delta^2 - |U_1|^2}} \frac{d\Delta}{e^{\beta\Delta} + 1}$$

$$\cong \frac{2L}{h}\left(\frac{2m}{\varepsilon_F}\right)^{1/2}\left(\frac{\pi}{\Delta^2}\right)^{1/2} U_1^{3/2}\,(kT)^{1/2}\, e^{-U_1/kT} \qquad \text{... (7.4.18)}$$

The leading term in the specific heat will accordingly be

$$C(T) \cong \frac{2L}{h}\sqrt{2m/\varepsilon_F}\,\sqrt{\pi/\Delta^2}\, e^{-U_1/kT}\, U_1^{5/2}\,(kT)^{-3/2}\, k \qquad \text{... (7.4.19)}$$

This is a very different specific heat from the normal situation where $C \propto T$. In this case, where there is a gap at the Fermi level, the specific heat is much smaller (it approaches zero exponentially as $T \to 0$) and this is as it should be because there is a barrier of height $2|U_1|$ formed at the point where the particle-hole gas is created and thus temperatures of the order $kT \cong |U_1|$ are required for the formation of particle-hole pairs. The conductivity is determined by this gas of particles and holes. For $\Delta >> kT$, the formation of this gas is strongly suppressed and one has an insulator, while for smaller Δ the behaviour is that of a semiconductor.

We now compare the ground state energy ($T = 0$) of the system with and without the energy gap. Clearly,

$$E(U_1) - E(U_1 = 0) = \int_{-\varepsilon_F}^{U_1} [n(U_1) - n(U_1 = 0)]\,\Delta\, d\Delta \qquad \text{... (7.4.20)}$$

where $n(U_1)$ is the function exhibited in Eq. (7.4.16).

Note that $n(U_1)$ differs from $n(U_1 = 0)$ only near $\Delta = 0$ and thus while $n(U_1)$ is expressed by Eq. (7.4.16), $n(U_1 = 0)$ is to be evaluated at the Fermi surface.

Thus to the lowest order of accuracy, we can write,

$$E(U_1) - E(U_1 = 0) \cong \frac{2L}{h}\left(\frac{2m}{\varepsilon_F}\right)\int_{-\varepsilon_F}^{U_1}\left(\frac{\Delta^2}{\sqrt{\Delta^2 - |U_1|^2}} - \Delta\right) d\Delta$$

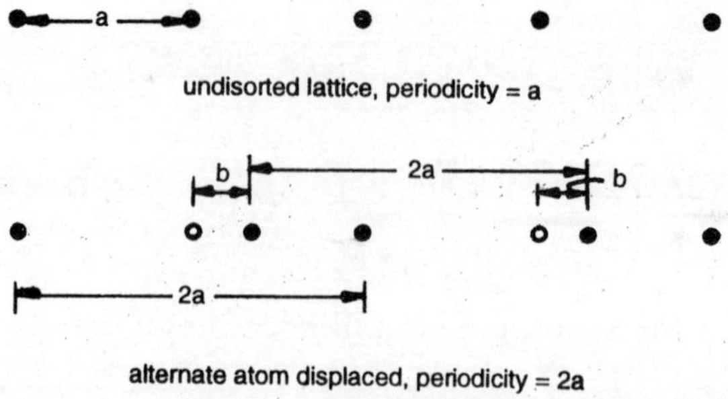

Fig. 7.4: Lattice distortion.

$$= \frac{2L}{h}\left(\frac{2m}{\varepsilon_F}\right)^{1/2}\left[\int_{-\varepsilon_F}^{U_1}\left[\sqrt{\Delta^2-|U_1|^2}-\Delta\right]d\Delta + |U_1|^2\int_{-\varepsilon_F}^{U_1}\frac{d\Delta}{\sqrt{\Delta^2-|U_1|^2}}\right]$$

$$\cong \frac{2L}{h}\left(\frac{2m}{\varepsilon_F}\right)^{1/2}|U_1|^2\left[\int_{-\varepsilon_F}^{U_1}\frac{d\Delta}{\sqrt{\Delta^2-|U_1|^2}}-\frac{1}{2}\int_{-\varepsilon_F}^{U_1}\frac{d\Delta}{\Delta}\right]$$

$$\cong -\frac{L}{h}\left(\frac{2m}{\varepsilon_F}\right)^{1/2}|U_1|^2 \ln\left(\varepsilon_F/|U_1|\right) \qquad \text{... (7.4.21)}$$

Thus for $N = 2L/a$, when there is a gap at the Fermi surface, the system with the gap has a substantially lower energy.

We now return to a lattice where $N = L/a$. The Fermi level in this case is far removed from the energy gap and the usual metallic properties will be found. But the interaction between the electron and lattice can now deform the lattice. Accordingly, we assume that every other atom in the lattice suffers a displacement b, so that the periodicity of the lattice is now $2a$ instead of a as shown in Fig. 7.4.

Now, there is a weak energy discontinuity at $p = \pi/2a$ which is the position of the Fermi level with $N = L/a$. This discountinuity is weak since the lattice is still almost periodic with period a but the small deformation b induces a period $2a$, i.e. we can write the potential $U(x, b)$ as

$$U(x,b) = U(x,0) + b\,U'(x,0) + \ldots$$

where $U(x,0)$ is our previous potential with period a while $U'(x,0)$ is the potential with period $2a$. The strength of the potential term causing the discountinuity at the Fermi surface is bU_1'. If we consider the energetics, then to achieve a deformation b will require an energy $1/2\ kb^2$, where K is an elastic constant. Opening up a gap at the Fermi level, however, lowers the energy by the amount shown in Eq. (7.4.21) where we now have to replace U_1 by bU_1'. So the change in energy achieved by distorting the lattice is

$$\Delta E\ (b) = \frac{1}{2}\ Kb^2 - \frac{L}{h}\ \sqrt{2m/\varepsilon_F}\ b^2\ |U_1'|^2\ \ln \frac{\varepsilon_F}{b\,|U_1'|} \qquad \text{... (7.4.22)}$$

Clearly for small b, the second term will dominate over the first and it is energetically favourable for the lattice to deform in such a manner that a gap opens up at the Fermi surface. Thus electron-lattice interaction in one dimension causes the lattice to deform structurally in the above manner and the one-dimensional metal loses its metallic property because of the gap at the Fermi surface. This instability is known as the Peierl's instability. It does not occur in two or three dimensions where the effect of the periodicity of the lattice on the electrons shows up in the form of charge density waves.

7.5 Emission of Electrons from a Metal Surface

We consider a metal in equilibrium with a dilute electron gas at very low temperatures. The periodic distribution of the ions which produces the periodic potential in which the electrons move is interrupted at the surface giving rise to a potential barrier which needs to be overcome if an electron is to leave the surface. We model this barrier by giving the electron at the bottom of the conduction band an energy –V. The maximum kinetic energy of an electron is, to a good approximation at the low temperatures concerned, equal to the Fermi energy ε_F. The minimum energy required to eject an electron from the metal is clearly

$$W = V - \varepsilon_F \qquad \text{... (7.5.1)}$$

which is called the work function of the metal. Now to calculate the chemical potential of this electron gas inside the metal, we note that the Fermi distribution for particles having momentum between $\mathbf{p}$ and $\mathbf{p} + d^3\mathbf{p}$ is

$[e^{\beta(p^2/2m - V - \mu)} + 1]^{-1}$ and hence at low temperatures ($\beta \rightarrow \infty$), the standard calculation of the Fermi energy yields

$$\varepsilon_F = V + \mu \qquad \text{... (7.5.2)}$$

Comparing Eqs. (7.5.1) and (7.5.2), we have

$$\mu = -W \qquad \text{... (7.5.3)}$$

inside the metal.

Outside the metal, the density of the electron gas is very low and hence this corresponds to a situation where the Fermi distribution will reduce to the classical distribution and thus the number of electrons between **p** and **p** + d^3**p** will be given by $e^{-\beta(p^2/2m - \mu_0)}$, where μ_0 is the chemical potential outside the metal. If the total number of eletrons is N and the volume V, then outside the metal.

$$N = 2 \cdot V \frac{4\pi}{h^3} \int_0^\infty p^2\, e^{(-p^2/2mkT)}\, e^{\mu_0/kT}\, dp \qquad \text{... (7.5.4)}$$

(spin 1/2 of the electron has been taken into account), giving rise to the electron density

$$n = 2 \cdot \left(\frac{2\pi mkT}{h^2}\right)^{3/2} e^{\mu_0/kT} \qquad \text{... (7.5.5)}$$

For equilibrium between the metal and the electron gas, the chemical potentials must be equal and hence

$$\mu_0 = \mu = -W \qquad \text{... (7.5.6)}$$

Thus the number density of the electrons outside the metal is given by

$$n = 2 \cdot \left(\frac{2\pi mkT}{h^2}\right)^{3/2} e^{-W/kT} = \frac{e^{-W/kT}}{\lambda^3} \qquad \text{... (7.5.7)}$$

The equilibrium that is achieved here is a dynamical one in that in every unit time interval, a certain number of electrons strike the metal surface and a fraction of them are absorbed. To keep the number of electrons fixed, an equal number of electrons must emerge from the metal surface per unit area per unit time. This gives the emission current from the metal surface. The number of electrons with momentum between p_z and $p_z + dp_z$ (the surface is $z = 0$) and density given by Eq. (7.5.7) striking a unit area of the metal in unit time, is clearly $\frac{2n\lambda^3}{h^3} \int_{-\infty}^{\infty} dp_x \int_{0}^{\infty} dp_y \int_{0}^{\infty} dp_z\, v_z e^{-p^2/2mkT}$ and hence the total number of electrons, N, striking the metal surface per unit area per unit time is

$$\overline{N} = 2\,\frac{e^{-W/kT}}{h^3} \int_0^{\infty} \frac{p_z}{m}\, e^{\frac{-p_z^2}{2mkT}}\, dp_z \int_0^{\infty} e^{\frac{-p_x^2}{2mkT}}\, dp_x \int_{-\infty}^{\infty} e^{-\frac{p_y^2}{2mkT}}\, dp_y$$

$$= 2 \cdot \frac{2mkT\,\pi}{h^2} \cdot \frac{kT}{h}\, e^{-W/kT} \qquad \text{... (7.5.8)}$$

If the reflection coefficient at the metal surface is r, then the number of electrons absorbed on the metal surface per unit area per unit time is $(1 - r)\; 4\pi\, mkT/h^2 \cdot (kT/h)\; e^{-W/kT}$ and this must equal the emission rate. The emission current (I) is consequently given by

$$I = \frac{4\,\pi\, m\,(kT)^2}{h^3}\,(1 - r)\; e^{-W/kT} \qquad \text{... (7.5.9)}$$

This thermionic current is associated with the work of Richardson and Dushman. The factor $e^{-W/kT}$ reflects the depletion of the current due to the effect of the potential barrier.

7.6 Correlations in a Fermi Gas

The two-point distribution function $n\,(\mathbf{r}_1, \mathbf{r}_2)$ was defined in Chapter 5 and evaluated in the case when the classical particles have a pairwise interaction. For independently moving particles it should be easy to see that

$$n\,(\mathbf{r}_1, \mathbf{r}_2) = n^2 + n\;\delta\;(\mathbf{r}_1 - \mathbf{r}_2) \qquad \text{... (7.6.1)}$$

For fermions which are otherwise free, the correleation function $n\,(\mathbf{r}_1, \mathbf{r}_2) = n\,(\mathbf{r}_{12})$ does not have the trivial form of Eq. (7.6.1) since the two-point function has to keep track of the fact that fermions obey the exclusion principle and hence there cannot be two fermions in the same state. This restriction is best implemented in momentum space where we define the Fourier transform of $n(\mathbf{r}_{12})$ as $C(\mathbf{k})$ given by

$$C(\mathbf{k}) = \int d^3 r_{12}\, e^{-i\mathbf{k}\cdot\mathbf{r}_{12}}\, n(\mathbf{r}_{12}) \qquad \text{... (7.6.2)}$$

Now $C(\mathbf{k})$ represents the particle scattering amplitude with momentum transfer $\mathbf{k}$ as we had seen in Sec. 5.5. For fermions this amplitude is

$$C(\mathbf{k}) \propto \frac{2}{V} \sum_{\mathbf{p}} f(\mathbf{p})\,[1 - f(\mathbf{p}+\mathbf{k})] \qquad \text{... (7.6.3)}$$

where $f(\mathbf{p}) = [e^{(\varepsilon(p)-\mu)\beta} + 1]^{-1}$ (the factor 2 is the spin degeneracy). It describes a process where a particle comes with momentum $\mathbf{p}$, exchanges momentum $\mathbf{k}$ with the scatterer and goes off with momentum $\mathbf{p} + \mathbf{k}$. The factor $[1 - f(\mathbf{p}+\mathbf{k})]$ ensures that there are no particles with momentum $\mathbf{p} + \mathbf{k}$ in the final state. The correlation function is coordinate space is found by inverting Eq. (7.6.2) as

$$n(\mathbf{r}_{12}) = \int \frac{d^3 k}{(2pi)^3} \cdot C(\mathbf{k}) = n\,\delta(\mathbf{r}_{12}) - 2\,|f(\mathbf{r}_{12})|^2 \qquad \text{... (7.6.4)}$$

where

$$f(\mathbf{r}) = \frac{1}{V} \sum_{p} f(\mathbf{p})\, e^{i\mathbf{p}\cdot\mathbf{r}} = \frac{I(\mathbf{r})}{2\pi^2 r} \qquad \text{... (7.6.5)}$$

with

$$I(\mathbf{r}) = \int_0^\infty dp \frac{p \sin pr}{e^{(\varepsilon-\mu)\beta} + 1}.$$

The integral $I(r)$ can be found in Vol. 5 of *Theoretical Physics* by Landau and Lifshitz and is given by

$$I(\mathbf{r}) = \frac{\partial}{\partial r}\left[\frac{\sin(p_F r)}{2\zeta \sinh(r/2\zeta)}\right] \qquad \text{... (7.6.6)}$$

where

$$2\zeta = P_F / mkT \text{ and } \mu = p_F^2 / 2m$$

As $T \to 0$, the length scale ζ becomes extremely large and we can always take $r << \zeta$. In this limit,

$$I(r) \cong \frac{\partial}{\partial r}\left(\frac{1}{r} \sin p_F r\right) \qquad \text{.... (7.6.7)}$$

and

$$C(r) \cong n\,\delta(\mathbf{r}) - \frac{p_F^4}{2\pi^4 r^2}\left[\frac{\cos p_F r}{p_F r} - \frac{\sin p_F r}{(p_F r)^2}\right]^2 \qquad \text{... (7.6.8)}$$

We note that for large r, there is no correlation length, the correlation falls of as r^{-4}. Also to be noted is the appearance of the periodic factors $\cos^2 p_F r$ etc. which implies a spatial periodicity with period $2\pi/2p_F$. This is an effect of the existence of a sharp Fermi surface. The oscillations in the correlation function are related to the Friedel oscillations in the electron gas. For $T > 0$, there is a finite correlation length ζ. For $\zeta >> p_F^{-1}$ and $r/\zeta >> 1$, we find

$$C(r) \sim \frac{p_F^2 e^{-r/\zeta}}{2\pi^5 \zeta^2 r^2} \sin^2 p_F r \qquad \text{... (7.6.9)}$$

which represents an exponential decay modulated by a spatial periodicity of π/p_F.

Problems

7.1 For a two-dimensional gas of non-interacting fermions find the chemical potential μ at any temperature T in terms of the number density n. Check to see that as $T \to 0$, the chemical potential correctly reduces to the two-dimensional free energy and that as $T \to \infty$, the appropriate classical limit is obtained.

7.2 Show that for a three-dimensional electron gas, the ratio of constant pressure specific heat C_p to the constant volume specific heat C_v is given by

$$\frac{C_p}{C_v} - 1 = \frac{\pi^2}{3}\left(\frac{kT}{\varepsilon_F}\right)^2 + \dots,$$

for $kT << \varepsilon_F$.

7.3 Show that for $kT << \varepsilon_F$, if the number of electrons changes by one at fixed temperature, the chemical potential changes by $\Delta\mu = [V g(\varepsilon_F)]^{-1}$, where $g(\varepsilon)$ is the density of states. Consequently, show that the maximum change in the occupation probability of a level is $\Delta f = \varepsilon/6kT \cdot 1/N$. This shows that the distribution function is insensitive to a small change in the number of electrons.

7.4 Find the equation of state of a relativistic completely degenerate electron gas (energy momentum relation is $E^2 = p^2c^2 + m^2c^4$).

7.5 A sufficiently dense neutrino (rest mass zero) star may be treated as a degenerate gas of relativistic fermions. Find the relation between mass and radius for the equilibrium of such a star.

7.6 Find the specific heat of a degenerate extreme relativistic electron gas.

7.7 For photoelectric effect the escape condition for electrons is $\left(\frac{p_z^2}{2m} + h\nu\right) > w$. Show that for $h(\nu - \nu_0) >> kT$, where $h\nu_0 = w - \mu \approx w - \varepsilon_F = \varphi$, the current density J is

$$J \approx \frac{2\pi me}{h}(\nu - \nu_0)^2$$

Chapter Eight

Electrons in a Magnetic Field

8.1 Introduction

In this chapter we will deal with various situations that can arise in an electron gas in the presence of a magnetic field. To do so we must first deal with the energy levels of a system of otherwise free electrons put in a uniform magnetic field **H** taken to be in the z-direction. The Hamiltonian can be obtained by replacing the momentum **p** by $\mathbf{p} + e\mathbf{A}/c$, where $-e$ $(e > 0)$ is the charge of the electron and hence the Schrödinger equation that needs to be solved is

$$\frac{1}{2m}\left(\mathbf{p} + \frac{e\mathbf{A}}{c}\right)^2 \psi = E\,\psi \qquad \text{... (8.1.1)}$$

We now have to fix a gauge. It should be noted that the Schrödinger equation above is gauge invariant in that the transformation $\mathbf{A}(\mathbf{r}) \to \mathbf{A}(\mathbf{r}) - \nabla\alpha(\mathbf{r})$, $\psi(\mathbf{r}) \to e^{-ie/\hbar c\ \alpha(\mathbf{r})}\psi(\mathbf{r})$, leaves Eq. (8.1.1) unchanged. We fix the gauge by choosing $A_x = -Hy$ and $A_y = A_z = 0$ (Landau gauge).

In the Landau gauge, Eq. (8.1.1) becomes

$$\left[\frac{1}{2m}\left(p_x - \frac{eH_y}{c}\right)^2 + \frac{p_y^2}{2m} + \frac{p_z^2}{2m}\right]\psi = E\psi \qquad \text{... (8.1.2)}$$

In the z-direction the motion is that of a free particle with energies given by $E' = p_z^2/2m$, appropriately quantized in a one dimensional box in that direction. In the x-y plane, the energy eigenvalue equation is

$$\left[\frac{1}{2m}\left(p_x - \frac{eH_y}{c}\right)^2 + \frac{p_y^2}{2m}\right] F(x,y) = \varepsilon F(x, y) \qquad \text{... (8.1.3)}$$

We note that p_x is clearly a constant of motion and hence for a given value of p_x ($p_x = 2\pi\hbar n_x / L$ is the appropriate quantization), we can write Eq. (8.1.3) as

$$\left[\frac{p_y^2}{2m} + \frac{1}{2} m \left(\frac{eH}{mc}\right)^2 \left(y - y_0\right)^2\right] f(y) = \varepsilon f(y) \qquad \text{... (8.1.4)}$$

where $y_0 = cp_x/eH$. The energy eigenvalues are clearly

$$\varepsilon = \left(n + \frac{1}{2}\right)\frac{eH\hbar}{mc} \qquad \text{... (8.1.5)}$$

and the two-dimensional wavefunction is

$$F(x, y) = \frac{1}{\sqrt{L_x}} e^{ik_x \cdot x} N_n e^{-\frac{eH(y-y_0)^2}{\hbar c}} H_n\left(\frac{eH(y-y_0)}{\hbar c}\right)$$

where H_n (ζ) (ζ in this case being $eH(y - y_0)/hc$) is the Hermite polynomial of order n and N_n the appropriate normalization factor and $k_x = p_x / \hbar$.

The above energy levels are known as Landau levels and since the operator p_x is a constant of motion they must be degenerate. To find the degeneracy, we note that the energy eigenvalue is independent of k_x and so k_x can take all the possible values that ensure $y_0 = (\hbar c/eH) \cdot k_x$ does not exceed the size L_y of the system in the y-direction. With $k_x = 2\pi n_x/L_x$, $y_0 = (hc/eHL_x) \cdot n_x$ and $\Delta y_0 = (hc/eHL_x)\ \Delta n_x$, the degeneracy g, which is clearly the maximum possible value of Δn_x is given by

$$g = (\Delta n_x)_{\max} = \frac{eHL_x}{hc} (\Delta y)_{\max}$$

$$= e HL_x L_y/hc = eAH/hc \qquad \text{... (8.1.7)}$$

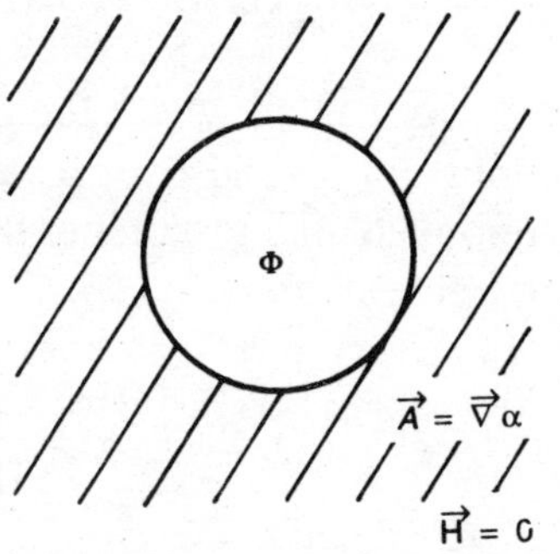

Fig. 8.1: $\vec{H} = 0$ everywhere in the plane except in a circular region where there is a magnetic flux Φ.

Thus the degeneracy of a Landau level is proportional to the area of the x–y plane (this corresponds to the fact that the trajectory can be centred around any point in the x–y plane) and the applied magnetic field H.

The Landau levels with their degeneracy (independent of the level) are the only things required for a study of the magnetic properties but to conclude this section it is worth pointing out certain features of our result. Imagine a plane with a circular hole in the centre (Fig. 8.1). There is a certain amount of magnetic flux $\mathbf{\Phi}$ passing through the hole but everywhere in the plane we have $\mathbf{H} = 0$. If $\mathbf{H} = 0$, then the vector potential $\mathbf{A}$ must be pure gauge, i.e. we can write

$$\mathbf{A} = \nabla \alpha \qquad \text{... (8.1.8)}$$

It must be noted that it is not possible to carry out a gauge transformation that reduces $\mathbf{A}$ to zero everywhere since we must have for any closed path C about the centre in the plane

$$\int_C \mathbf{A} \cdot d\mathbf{l} = \int_S (\nabla \times \mathbf{A}) \cdot d\mathbf{S} = \int_S \mathbf{H} \cdot d\mathbf{S} = \mathbf{\Phi} \qquad \text{... (8.1.9)}$$

where S is an open surface whose boundary curve is C and hence the surface has to intersect the flux $\mathbf{\Phi}$. We can write down the solution for α as

$$\alpha = \frac{\Phi\, \theta}{2\pi} \qquad \text{... (8.1.10)}$$

We now consider the solution of the Schrödinger equation for an electron in the plane with the boundary condition that the wave-function has to vanish in the hole. Because of the gauge invariance of the Schrödinger equation, we can try to remove $\mathbf{A}$ from the equation by the transformation $\mathbf{A} \to \mathbf{A} - \nabla\alpha$ but then the wavefunction undergoes the transformation $\psi \to \psi e^{-ie\alpha/\hbar c} = \psi\, e^{-i\,\theta\, e\Phi/hc}$, which makes it discontinuous in space in general, i.e. when

$\theta \to \theta + 2\pi$, we do not regain the same wavefunction. There are two ways out of this difficulty.

i) If the wavefunction ψ is localized and hence can vanish where the discontinuity is to occur (i.e. the wavefunction does not spread over a range of $\theta = 2\pi$), then the difficulty is trivially removed. This can be true for electrons trapped by impurities but not for free electrons.

ii) If the flux Φ is such that

$$\frac{e\Phi}{hc} = n \qquad \text{... (8.1.11)}$$

where n is an integer. In this case changing θ by 2π does not affect anything since $e^{2n\pi} = 1$. The vector potential can now indeed be gauged away and the electron does not know about the flux. This is the flux quantization in superconductivity.

8.2 Magnetic Properties: $T = 0$

Before calculating the magnetization and susceptibility, let us recall the thermodynamic relations involved. The fundamental relation is

$$TdS = dE - PdV - \mu dN + MdH \qquad \text{... (8.2.1)}$$

where the term MdH indicates that if the external field is infinitesimally increased the total magnetic energy of the system is decreased and hence positive magnetic work is done by the system (if $M > 0$). In terms of the potential $\tilde{F} = E - TS - \mu N$, it is clear that

$$\left(\frac{\partial \tilde{F}}{\partial H}\right)_{V,\mu,T} = -M \qquad \text{... (8.2.2)}$$

As shown in Chapter 3, $\tilde{F} = -PV$ and hence we also have

$$M = \frac{\partial (PV)}{\partial H}\bigg|_{V,\mu,T} = \frac{2}{3}\left.\frac{\partial E}{\partial H}\right)_{V,\mu,T} \qquad \text{... (8.2.3)}$$

We now need to point out there are two sources of magnetism—the intrinsic magnetic moment of the electrons coming from their spin and the magnetic moment arising out of the orbital motion. We will treat them separately. To deal with the intrinsic magnetic moment, we note that the spin-up electrons have the magnetic energy $-\sigma H$, where $\sigma = |e|\hbar/2mc$ is the Bohr magneton, while spin-down ones have the magnetic energy σH. Accordingly, the chemical potential μ of the spin-up electrons becomes $\varepsilon_F - \sigma H$, while the potential for the spin down ones is $\varepsilon_F + \sigma H$, where ε_F is th Fermi energy defined in Eq. (7.1.3). Clearly the total potential $\tilde{F}$ can be written as

$$\tilde{F} = \frac{1}{2}\tilde{F}\left(\varepsilon_F + \sigma H\right) + \frac{1}{2}\tilde{F}\left(\varepsilon_F - \sigma H\right)$$

$$= \tilde{F}\left(\varepsilon_F\right) + \frac{1}{2}(\sigma H)^2 \frac{\partial^2 \tilde{F}}{\partial \varepsilon_F^2} + \ldots$$

$$= \tilde{F}\left(\varepsilon_F\right) - \frac{1}{3}(\sigma H)^2 \frac{\partial^2 E}{\partial \varepsilon_F^2} + \ldots \qquad \ldots \textbf{(8.2.4)}$$

Recalling that $E = 3/5\, N\, \varepsilon_F = 8\pi V/3\, (2m/h^2)^{3/2}\, \varepsilon_F^{5/2}$, Eq. (8.2.4) reduces to (note that E is to be a function of V and ε_F for Eq. (8.2.3) to be applicable)

$$E = E(\varepsilon_F) + 3\pi V \left(\frac{2m}{h^2}\right)^{3/2} (\sigma H)^2\, \varepsilon_F^{1/2} + \ldots \qquad \ldots \textbf{(8.2.5)}$$

From Eq. (8.2.3), the zero-field spin susceptibility is given by

$$\chi_{sp} = \left(\frac{\partial M}{\partial H}\right)_{H=0} = \frac{2\sigma^2}{3}\frac{\partial^2 E}{\partial H^2} = 4\pi V \left(\frac{2m}{h^2}\right)^{3/2} \sigma^2\, \varepsilon_F^{1/2},$$

resulting in a susceptibility per unit volume as

$$\frac{\chi_{sp}}{V} = \frac{\sigma^2 m\, p_F}{\pi^2 \hbar^3} \qquad \ldots \textbf{(8.2.6)}$$

The positive value of the susceptibility indicates that it is paramagnetic. This phenomenon of the paramagnetic response of the electron gas to an external field because of its intrinsic spin is known as Pauli paramagnetism.

To treat the orbital contribution, we note that the total energy of a particular state is $E = (n + 1/2)\, eH\hbar/mc + p_z^2/2m$ and that the grand partition function Q for a system of fermions is given by ($\omega_c = eH/mc$)

$$\ln Q = \sum_{\text{states}} \ln \left(1 + e^{-(E-\mu)/kT}\right)$$

$$= 2\,\frac{|e|HA}{hc}\,\frac{L}{h}\sum_{n=0}^{\infty}\int_{-\infty}^{\infty} dp_z \,\ln\left\{1 + e^{-[(2n+1)\frac{\hbar\omega_c}{2} + \frac{p_z^2}{2m} - \mu]\beta}\right\}$$

$$= 2\,\frac{|e|\,HV}{h^2 c}\sum_{n=0}^{\infty}\int_{-\infty}^{\infty} dp_z \,\ln\left\{1 + e^{-[(2n+1)\frac{\hbar\omega_c}{2} + \frac{p_z^2}{2m} - \mu]\beta}\right\} \quad \text{... (8.2.7)}$$

Note that the degeneracy for the motion in the plane is $|e|HA/hc$ for every n and the number of states between p_z and $(p_z + dp_z)$ is Ldp_z/h while the spin degeneracy fator is 2. The combination leads to Eq. (8.2.7). The potential $\tilde{F} = -kT \ln Q$ and hence

$$\tilde{F} = 2\sigma H \sum_{n=0}^{\infty} f\,[\mu - (2n+1)\,\sigma H] \quad \text{... (8.2.8)}$$

with

$$f(\mu) = -\frac{kTmv}{2\pi^2\hbar^3}\int_{-\infty}^{\infty}\ln\left[1 + e^{\left(\mu - \frac{p_z^2}{2m}\right)\beta}\right] dp_z \quad \text{... (8.2.9)}$$

To evaluate the sum we have to make use of the Euler-McLaurin sum formula which gives

$$\sum_{n=1}^{\infty} f(n+a) \cong \int_a^{\infty} F(x)\,dx - \frac{1}{2}F(a) - \frac{1}{12}F'(a) \quad \text{... (8.2.10)}$$

Setting $a = 1/2$, we find

$$\sum_{n=0}^{\infty} F\left(n+\frac{1}{2}\right) \cong \int_0^{\infty} F(x)\,dx - \int_0^{1/2} F(x)\,dx + \frac{1}{2}F\left(\frac{1}{2}\right) - \frac{1}{12}F'\left(\frac{1}{2}\right) \quad \text{... (8.2.11)}$$

If we approximate $F(x)$ by $F(0) + xF'(0)$ for $0 < x < 1/2$, then we obtain

$$\sum_{n=0}^{\infty} F\left(n+\frac{1}{2}\right) \cong \int_0^{\infty} F(x)\,dx - \frac{1}{2}F(0) - \frac{F'(0)}{8} + \frac{1}{2}F\left(\frac{1}{2}\right) - \frac{1}{12}F'\left(\frac{1}{2}\right)$$

$$= \int_0^{\infty} F(x)\,dx + \frac{1}{2}\frac{1}{2}F'(0) - \frac{1}{8}F'(0) - \frac{1}{12}F'(0)$$

$$= \int_0^{\infty} F(x)\,dx + \frac{1}{24}F'(0) \qquad \ldots \textbf{(8.2.12)}$$

Thus, in view of Eq. (8.2.12), the sum in Eq. (8.2.8) becomes

$$\sum_{n=0}^{\infty} f\left(\mu - \left(n+\frac{1}{2}\right)2\sigma H\right) = \int_0^{\infty} f(\mu - 2\sigma Hx)\,dx$$

$$+ \frac{1}{24}\frac{\partial}{\partial n} f(\mu - 2n\sigma H)\Big|_{n=0}$$

$$= \frac{1}{2\sigma H}\int_{-\infty}^{\mu} f(y)\,dy - \frac{2\sigma H}{24}\frac{\partial f}{\partial \mu} + O(H^2) \qquad \ldots \textbf{(8.2.13)}$$

where $y = \mu - 2n\sigma Hx$

The potential $\tilde{F}$ is now given by (of Eq. (8.2.9))

$$\tilde{F} = \int_{-\infty}^{\mu} f(y)\,dy - \frac{(2\sigma H)^2}{24}\frac{\partial f}{\partial \mu} \qquad \ldots \textbf{(8.2.14)}$$

In all the above manipulations we have always been working with weak magnetic fields so that terms higher than quadratic in H are being dropped. In fact, in spite of the fact that we are dealing with very low temperatures, the magnetic field is such that $\sigma H \ll kT$. The passage from Eq. (8.2.11) to Eq. (8.2.12) depends on the fact that the change in F in one step $n \to n+1$ is small and this is guaranteed by the condition $\sigma H \ll kT$. If this condition is violated, then new features appear which are taken up in Sec. 8.4.

Turning to Eq. (8.2.14), we note that the first term is independent of H. Consequently it is the potential $\tilde{F}(\mu)$ in zero magnetic field. Thus

$$\tilde{F}(\mu) = \int_{-\infty}^{\mu} f(y)\, dy, \qquad \text{... (8.2.15)}$$

giving on differentiation

$$\frac{\partial \tilde{F}}{\partial \mu} = f(\mu)$$

and transforming Eq. (8.2.14) into the form

$$\tilde{F} = \tilde{F}(\mu) - \frac{\sigma^2 H^2}{6} \frac{\partial^2 \tilde{F}}{\partial \mu^2} \chi \qquad \text{... (8.2.16)}$$

Noting that the orbital susceptibility $\chi_{orb} = -\partial^2 \tilde{F}/dH^2$, we have

$$\chi_{orb} = -\frac{\partial^2 \tilde{F}}{\partial H^2} = \frac{\sigma^2}{3} \frac{\partial^2 \tilde{F}}{\partial \mu^2} = -\frac{2\sigma^2}{9} \frac{\partial^2 E}{\partial \mu^2}$$

$$= -\frac{1}{3} \chi_{sp} \qquad \text{... (8.2.17)}$$

Thus the orbital motion gives rise to a diamagnetic response (negative susceptibility) which has one third the strength of the paramagnetic spin susceptibility. The combined response of the electron gas at $T = 0$ is paramagnetic with the susceptibily $2/3\ \chi_{sp}$.

It should be noted that it was not necessary to split the calculation into two parts. The total energy due to spin and orbital motion can be written as $\varepsilon_{tot} = p_z^2/2m + (n + 1/2)\hbar\,\omega_c \pm \sigma H$. This set of energy eigenvalues can be written as

$$\varepsilon_{tot} = \frac{p_z^2}{2m} + 2n\sigma H, \; n = 0, 1, 2, \qquad \text{... (8.2.18)}$$

where $n = 0$ occurs once and all the non-zero values of n occur twice. Hence the density of states is the usual $(2eHV/\hbar^2 c)\ dp_z$ for $n \neq 0$ but only half as much for $n = 0$. The potential is then given by the sum

$$\tilde{F} = 2\sigma H \left\{ \frac{1}{2} \tilde{F}(\mu) + \sum_{n=1}^{\infty} f(\mu - 2\sigma\, nH) \right\} \qquad \text{... (8.2.19)}$$

It is evaluated by using

$$\sum_{n=1}^{\infty} F(n) = \int_0^{\infty} F(x) - \frac{1}{2} F(0) - \frac{1}{12} F'(0) \qquad \text{... (8.2.20)}$$

obtained from Eq. (8.2.10) by setting $a = 0$ for $\sigma H << kT$.

8.3 Magnetism of the Electron Gas: $T >> T_F$

At high temperatures, the distribution function behaves like the classical Maxwell-Boltzmann distribution and hence the appropriate limiting form has to be used. To treat the paramagnetism coming from the intrinsic spin first, we turn to Eq. (8.2.4) and note that it has to be written in terms of the chemical potential μ for arbitrary temperatures. Accordingly,

$$\tilde{F} = \frac{1}{2} [\tilde{F}(\mu + \sigma H) + \tilde{F}(\mu - \sigma H)]$$

$$= \tilde{F}(\mu) + \frac{1}{2} (\sigma H)^2 \frac{\partial^2 \tilde{F}}{\partial \mu^2} + \dots$$

$$= \tilde{F}(\mu) - \frac{1}{3} (\sigma H)^2 \frac{\partial^2 E}{\partial \mu^2} + \dots \qquad \text{... (8.3.1)}$$

If we have the distribution appropriate to the Maxwell-Boltzmann form, then

$$E \cong \frac{3}{2} NkT = 3V \cdot \left(\frac{2\pi\, mkT}{\hbar^2} \right)^{3/2} e^{\mu/kT}\, kT, \qquad \text{... (8.3.2)}$$

where $e^{\mu/kT} << 1$ (this is what reduces the Fermi distribution $[e^{(\varepsilon - \mu)/kT} + 1]^{-1}$ to approximately $e^{-(\varepsilon - \mu)/kT}$) and thus,

$$\left(\frac{\partial^2 E}{\partial \mu^2}\right) = 3\,\frac{V}{kT}\left(\frac{2\pi\, mkT}{\hbar^{\;2}}\right)^{3/2} e^{\mu/kT}$$

$$= \frac{3}{2}\cdot\frac{N}{kT} \qquad \text{... (8.3.3)}$$

In the above, the expression for number density in the classical limit (see Chapter 6) has been repeatedly used. From Eq. (8.3.1), we find

$$\chi_{sp} = \frac{2}{3}\,\sigma^2\,\frac{\partial^2 E}{\partial \mu^2} \cong \frac{N\sigma^2}{kT}\quad (T >> T_F) \qquad \text{... (8.3.4)}$$

Thus the Curie law for paramagnetic susceptibility ($\chi \propto 1/T$) is recovered.

We now turn to the orbital contribution and note that for small fields, i.e. $\sigma H << kT$, the result of Eq. (8.2.16) is valid for all temperatures and hence $\chi_{orb} = -\ 1/3\ \chi_{sp}$, when the appropriate high temperature approximations are made. In fact, if we compare Eq. (8.2.16) with the second step in Eq. (8.2.4), then we note that for all temperatures

$$\chi_{orb} = -\,\frac{1}{3}\,\chi_{sp} \qquad \text{... (8.3.5)}$$

so long as the field is vanishingly small. If the field becomes large, i.e. the criterion $\sigma H << KT$ is no longer satisfied, then the susceptibility depends on the applied field and Eq. (8.3.5) is no longer applicable. We study this situation in the next section.

8.4 De Hass-Van Alphen Effect: $T = 0$

We now consider the effect of a finite field at $T = 0$. Clearly in this limit the sum formula of Eq. (8.2.12) is no longer valid as the $f(x)$ of Eq. (8.2.8) need not change by a small amount when n goes to $n + 1$. This will happen when the argument approaches zero and it is under these conditions that it is essential that $\sigma H << kT$ (the criterion that ensures a small H) be valid. Since this can not be ensured for $T = 0$ we do not use Eq. (8.2.20) in this section. We calculate the total energy E of N electrons confined to a plane with single particle energies being given by Eq. (8.1.5)

$$\varepsilon_j = \left(j + \frac{1}{2}\right) 2\sigma H \qquad \text{... (8.4.1)}$$

The degeneracy factor (including the spin degeneracy)

$$g = 2\ eAH/hc = N.H/H_0 \qquad \text{... (8.4.2)}$$

where N is the number of particles and $H_0 = nhc/2e$, with $n = N/A$ being the number density of the electrons.

At $T = 0$, all the particles tend to settle in the lowest energy states and clearly if $H > H_0$, $g > N$ and hence all the electrons can be accomodated in the lowest Landau level. Hence if $H > H_0$,

$$\therefore E_0 = N\sigma H \qquad \text{... (8.4.3)}$$

while for $H < H_0$, not all particles can be accomodated in the ground Landau level. Suppose that H is such that j th Landau level is completely filled while some spill over to the $(j + 1)$th level.

Clearly then

$$(j + 1)\, g < N < (j + 2)\ g$$

or

$$\frac{1}{j + 2} < \frac{H}{H_0} < \frac{1}{j + 1} \qquad \text{... (8.4.4)}$$

For H in this interval, the energy is

$$E = g \sum_{i=0}^{j} (2j + 1)\ \sigma H + [N - g\,(j + 1)]\ (2j + 3)\,\sigma H$$

$$= \sigma H\ [g\,j\,(j+1) + (j+1)g - g\,(j+1)\ (2j+3) + N\,(2j+3)]$$

$$= \sigma H\ [N\ (2j+3) - g\,(j+1)\ (j+2)]$$

$$= \sigma H \left[N\ (2j+3) - NH/H_0\ (j+1)\ (j+2)\right]$$

$$= N\ \sigma H_0 \left[(2j+3)\ \frac{H}{H_0} - (j+1)\ (j+2) \left(\frac{H}{H_0}\right)^2\right] \qquad \text{... (8.4.5)}$$

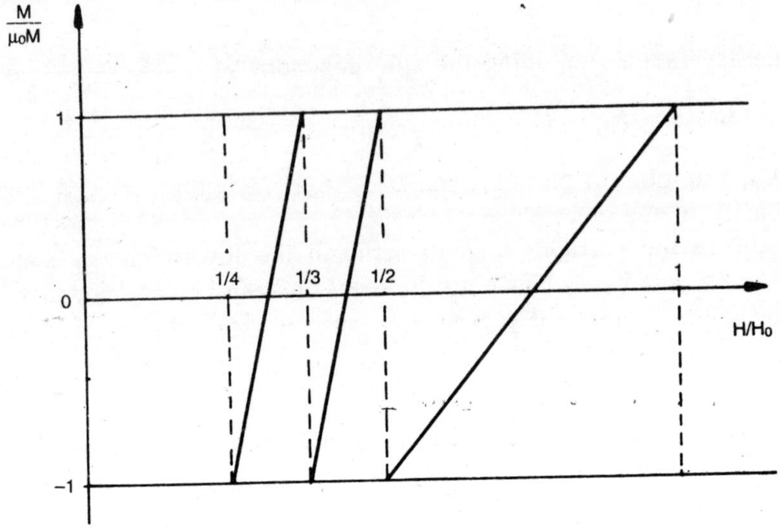

Fig. 8.2: Oscillations in magnetization.

Writing $x = H/H_0$, we summarize the above results as

$$E = \begin{cases} N\mu_0 H_0 x & \text{if } x > 1 \\ N\mu_0 H_0 \left[(2j+3)\,x - (j+1)\,(j+2)x^2\right] & \text{if } \dfrac{1}{j+2} < x < \dfrac{1}{j+1} \end{cases}$$

(8.4.6)

The magnetization can be calculated from the relation $M = -\,\partial/\partial x\ (E/H_0)$ and susceptibility from $\chi = \partial\,(M/H_0)/\partial x$. Hence

$$M = \begin{cases} -N\mu_0 & \text{if } x > 1 \\ -N\mu_0\left[2j+3 - 2(j+1)\,(j+2)x\right] \end{cases}$$

$$\text{if } \frac{1}{j+2} < x < \frac{1}{j+1} \quad \text{... (8.4.7)}$$

and

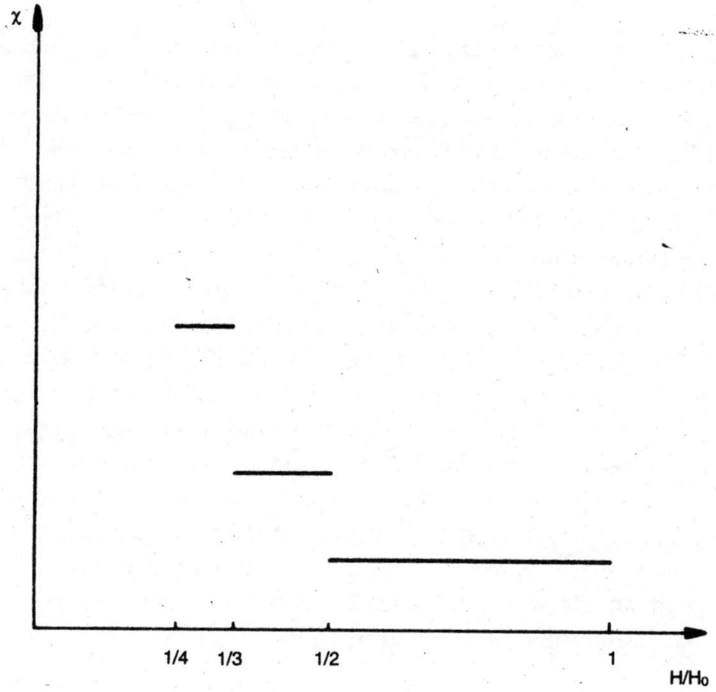

Fig. 8.3: The susceptibility at $T = 0$ as a function of H/H_0.

$$\chi = \begin{cases} 0 & \text{if } x > 1 \\ 2\,(j+1)\,(j+2)\,N\mu_0/H_0 & \text{if } \dfrac{1}{j+2} < x < \dfrac{1}{j+1} \end{cases} \qquad \text{... (8.4.8)}$$

As shown in Fig. 8.2, the magnetization M in units of $\mu_0 N$ oscillates between $+1$ and -1 as H is varied in the range $0 < H < H_0$ and the grows linearly. The susceptibility exhibits steps as a function of H decreasing to a zero value for $H \geq H_0$ (Fig. 8.3). The oscillations in magnetization were discovered experimentally in a metal by De Haas and Van Alphen in 1931 and predicted for an electron gas by Landau in 1930.

8.5 Quantum Hall Effect

Hall effect deals with the situation of crossed electric and magnetic fields. We imagine a situation with the electric field **E** in the x-direction and the magnetic field **H** in the z-direction. The electrons which acquire a velocity in the x-direction due to the electric fields are deflected in the y-direction due to the Lorentz force arising from the x-velocity and the magnetic field along the z-axis. This sets up a current in the y-direction. We will immediately specialize to a two-dimensional situation embedded in a three-dimensional space with the magnetic field chosen in the z-direction and the velocity of the charge carriers (electrons in our case) confined to a two-dimensional plane. In the absence of the magnetic field, we would have the conductivity given by σ_0, where $\mathbf{j} = \sigma_0 \mathbf{E}$. A crude estimate of σ_0 can be obtained by assuming in accordance with the Pauli principle that only the electrons near the Fermi surface take part in the conduction process and that between successive collisions each electron travels a mean distance l_0 in time τ_0 ($l_0 \equiv \tau_0 v_F$). After a collision the velocity can be in any direction and on the average can be taken to be zero. Before the next collision, the electron will pick up a velocity $\Delta \mathbf{v} = -e\,\mathbf{E}\,\tau_0/m$ due to its motion in an electric field and the current is obtained by adding the effect of all such electrons and hence

$$\mathbf{j} = -en\,\mathbf{v} = \frac{e^2 n\,\tau_0}{m}\,\mathbf{E} = \sigma_0\,\mathbf{E} \qquad \text{... (8.5.1)}$$

where n is the free electron density. Thus the conductivity σ_0 is given by $e^2\,n\,\tau_0/m$ or $e^2\,nl_0/p_F$. The above picture is purely classical—quantum effects usually change m to m^* and affect the value of τ_0, leaving the general form of Eq. (8.5.1) unchanged. In the presence of a magnetic field **H**, the Lorentz force is a given by $-e\,\mathbf{v} \times \mathbf{H}/c$ and hence (c is the velocity of light)

$$\mathbf{j} = \sigma\,[\mathbf{E} - \mathbf{j} \times \mathbf{H}/nec] \qquad \text{... (8.5.2)}$$

The resistivity tensor ρ_{ij} defined as $E_l = \Sigma_m\,\rho_{lm}\,j_m$ is read off from Eq. (8.5.2) as

$$\rho = \begin{pmatrix} \sigma_0^{-1} & H/nec \\ -H/nec & \sigma_0^{-1} \end{pmatrix} \qquad \text{... (8.5.3)}$$

The conductivity tensor is obtained by inversion and we find

$$\sigma_{xx} = \sigma_0/(1 + \omega_c^2 \tau_0^2) \qquad \text{... (8.5.4a)}$$

and

$$\sigma_{xy} = \frac{nec}{H} + \frac{1}{\omega_c \tau_0} \sigma_{xx}, \qquad \text{... (8.5.4b)}$$

where $\omega_c = eH/mc$.

The quantum Hall effect concerns the resistivity and conductivity tensors of a two dimensional electron gas at very low temperatures and in high magnetic fields. The results are surprising in that the conductivity tensor has the form

$$\sigma = \begin{pmatrix} 0 & -ne^2/h \\ ne^2/h & 0 \end{pmatrix} \qquad \text{... (8.5.5)}$$

where n is an integer or a fraction of the form p/q.

The direct resistivity is zero and the Hall resistivity acquires the form $\rho_{xy} = h/ne^2$. As the magnetic field is varied and the resistivity measured, the Hall resistivity shows a plateau with the value h/ne^2 wherever the filling fraction ν of the lowest Landau level is in the vicinity of the number n. If n_0 is the number density of the charge carriers, then from the degeneracy of the Landau levels, we have

$$\nu = \frac{hcn_0}{eH} \qquad \text{... (8.5.6)}$$

and whenever ν acquires a value near n with changing H, one observes a plateau in ρ_{xy} with ρ_{xx} simultaneously dropping down to zero. This situation is shown in Fig. 8.4.

The vanishing of σ_{xx} makes the system dissipationless and hence related to superfluidity and superconductivity. The vanishing of σ_{xx} is essential in understanding the quantization of σ_{xy}.

To set up a naive picture for the integer quantum Hall effect, so called when n is an integer ($n = p/q$ with q odd describes the fractional quantum Hall effect), we imagine varying H and arriving at a value for which $\nu = 1$, i.e. the lowest Landau level is completely filled. Then the Fermi level must lie between two Landau levels and hence one would require a finite amount of energy to make new states available to the electrons. Scatterings would become difficult, τ_0 would tend to infinity and consequently, we would have $\rho_{xx} = \rho_{yy} = 0$ and ρ_{xy} quantized as seen. This would occur for all integral fill-

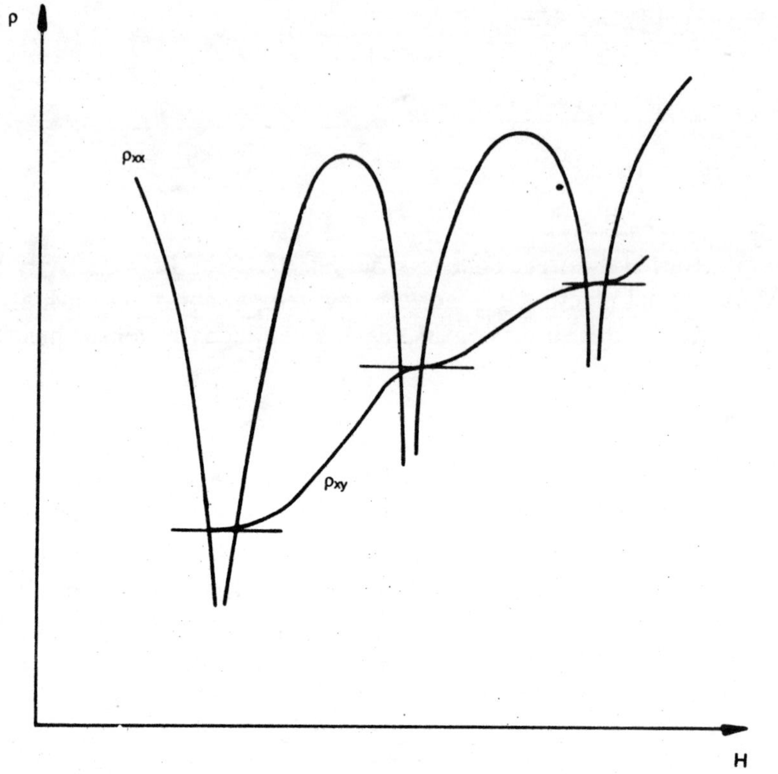

Fig. 8.4: Resistivities as functions of magnetic field.

ing fractions and we could have an explanation for the integral effect. However, this can only explain the quantized values of σ_{xy} and dissipationless situation at certain discrete values of H. Experimentally what one observes are plateaus in the conductivity (or resistivity) vs H plot (see Fig. 8.4). The observed plateaus, as we shall see in the following, can only be explained on the basis of the inevitable impurities present in a real sample.

We now present an argument for the quantization of the Hall conductivity which is based on gauge invariance and hence has the potential of being quite general. This discussion follows Halperins's formulation of Laughlin's argument. The geometry will be taken to be that of a disc with a hole punched in the centre (any convenient geometry can be chosen as all evidence points to the fact that the quantum Hall effect is independent of size, shape, connectivity or edge conditions in the sample). The usual magnetic field **H** is present in

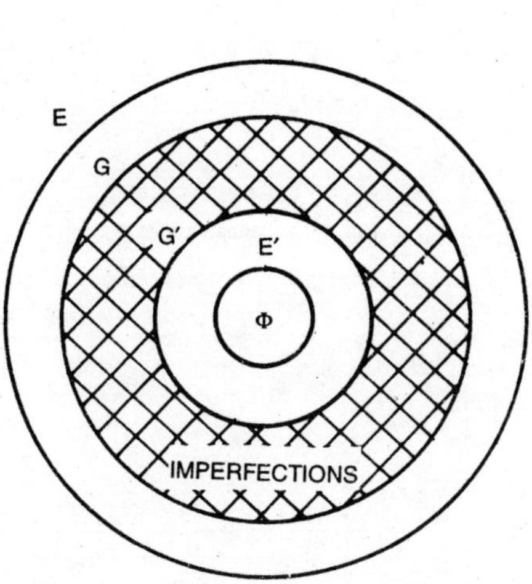

Fig. 8.5: The geometry for the discussion of resistivity quantization.

the z-direction, but in addition through the hole in the disc we imagine the existence of a variable flux Φ (confined to the hole). As shown in Fig. 8.5, the sample is a disc of realistic region with impurities, the boundaries being designated by GG′. the regions EG and $G'E'$ are completely devoid of impurities and act as guard rings for the impurities. We imagine that the total number of electrons is such that exactly n Landau levels are filled in the guard rings. Since there are no impurities in the guard rings, the Fermi level must lie between two Landau levels in the guard rings. The guard rings serve to relate the conducting properties of the realistic region to that of the ideal region.

We now consider changing the flux Φ adiabatically. There will exist a circumferential electric field **E** everywhere in the disc now, such that if we take any closed contour surrounding the hole, we must have

$$\frac{1}{c}\frac{d\Phi}{dt} = \frac{1}{c}\oint \frac{\partial \mathbf{H}}{\partial t} \cdot d\mathbf{S} = \oint_C \mathbf{E} \cdot d\mathbf{l} = \oint_C \rho_{mn}\, j_m\, dl_n \qquad \text{... (8.5.7)}$$

where ρ_{mn} is the resistivity tensor.

Let the loop C be in one of the guard rings (say EG). The resistivity is now given by $\rho_{xx} = 0$ and $\rho_{xy} = h/ne^2$ and if the flux change is $\Delta\Phi = \Phi_0 = hc/e$, then we have (a circumferential **E** field in this case means a Hall current in the radial direction)

$$\frac{h}{e} = \frac{1}{c} \cdot \Delta\Phi = \int dt \oint_C j_r \times \frac{h}{ne^2} dl$$

or

$$ne = \int dt \oint_C j_r\, dl$$

$$= \text{Charge transferred across the loop } C \qquad \text{... (8.5.8)}$$

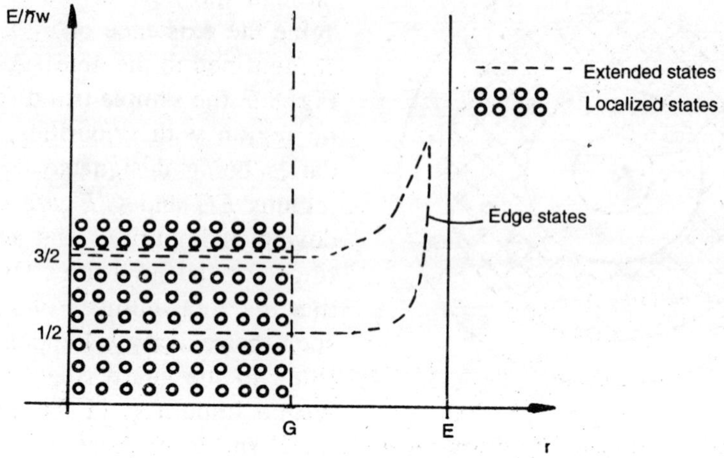

Fig. 8.6: Energy levels of the ring system vs. radius.

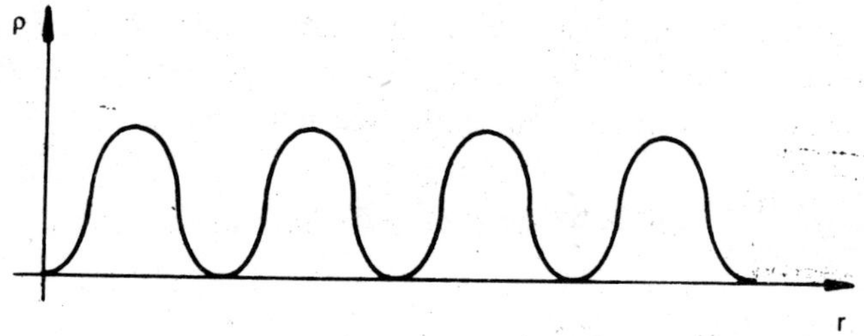

Fig. 8.7: Electron density as a function of radius in the impurity-free region.

The electrons transported across *EG* must have come from the edge *E* (say) and will cross into the sample at *G*. At a physical edge such as *E*, there is a quasi continuum of edge states for each Landau level as illustrated in Fig. 8.6. These edge states will be filled to the Fermi level. We now explain how the current flows across the guard ring. The electron density across the region *EG* with a completely filled Landau level will appear as shown in Fig. 8.7.

As we start changing the flux, the electrons have to respond to the change, until the change in flux becomes $\Delta\Phi = \Phi_0$. Gauge invariance implies that if $\Delta\Phi = \Phi_0$, the effect of the vector potential can be removed by a gauge transformation and thus the electron distribution has to look exactly as it did before the flux change was turned on. So the distribution of Fig. 8.7 has to be regained for $\Delta\Phi = \Phi_0$, which can occur if an integral number of electrons have moved across the region *EG*. This movement, as expected, results in an infinitesimal change of energy. Now assume that there is no dissipation in the sample *GG′*. This will happen if the Fermi level lies in the region of the localized states in Fig. 8.6. This assumption is equivalent to saying that the current is normal to the electric field or else the diagonal component of the conductivity tensor vanishes. Now if there is no build-up of charge in *GG′* or its edges, the electrons delivered to the sample must be transferred through it and the conducting behaviour of the sample must be identical to the behaviour of the perfect system with integrally filled Landau levels. Consequently, the quantized Hall conductivity will be observed in the sample.

The above conclusion will be violated if there are no electrons or additional electrons transferred across GG′. This would imply the existence of states in the interior of the sample or at the boundary with the guard ring, which could accept or provide the electrons. Since an adiabatic process is capable of transferring electrons into or from these states, they must be near the Fermi level.

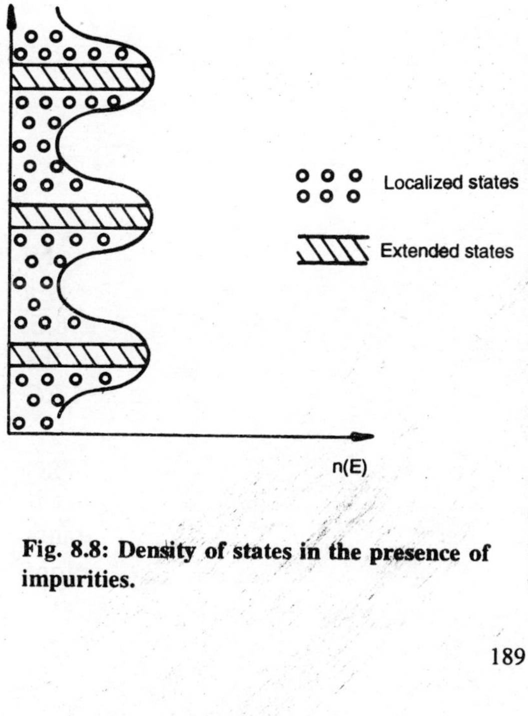

Fig. 8.8: Density of states in the presence of impurities.

The only states which are affected by the flux Φ are the extended states which have an appreciable amplitude on the entire path circling the hole. The localized states, as discussed in Sec. 8.1, are not affected by the flux changes and hence their occupation or nature cannot change. The existence of these localized states is essential to the understanding of the plateaus. In the presence of the impurities, the density of states near each Landau level is broadened and the situation is as shown in Fig. 8.8.

Now as the magnetic field is decreased (i.e. the density is increased) the localized states graudally fill up without changing the occupation of the extended states and hence with no change in the Hall conductance. Thus, over a range of magnetic field values, while the Fermi level E_F lies between the core of the extended states in two successive Landau levels, the Hall conductivity is a constant and it is only when E_F crosses the core of the extended states on the next Landau level that the conductivity jumps to the next plateau.

Thus if the Fermi level lies in a mobility gap, so that the diagonal conductivity vanishes, the Hall conductivity gets quantized. The trivial case, $n = 0$, corresponds to an insulator, If the impurity is weak enough and the impurity region connects smoothly to an impurity-free region, the non-trivial cases n = 1, 2, 3,... etc. are realized and we have the integral quantum Hall effect.

The fractional quantum Hall effect requires the inclusion of the electron-electron interaction for its explanation and cannot be couched in the simple terms given above. The nature of the electronic states in the guard ring would depend on the Coulomb repulsion of electrons and to explain the fractional Hall effect in the above terms one would have to show that for the interacting case, a transfer of fractional number of electrons occurs if the flux changes by an integer number of flux quanta. This particular formulation of the fractional effect is not yet available, although Laughlin's picture of the incompressible quantum fluid when the filling fraction ν takes on certain values like 1/3, 2/3, 2/5, etc. gives a satisfactory picture of the phenomenon.

Chapter Nine

Bose-Einstein Distribution: Examples

9.1 Bose-Einstein Condensation

In Chapter 6, we dealt with the dilute Bose gas where $\lambda^3 n << 1$. In this section, we turn to the other limit where the quantum effects are all important. We shall deal with spin 0 particles for simplicity and $2s + 1 = 1$ in the density of states of Eq. (6.2.10). The fundamental equation which we want to explore in detail is

$$N = V \,.\, 2\pi \left(\frac{2m}{h^2}\right) \int_0^\infty \frac{\varepsilon^{1/2}\, d\varepsilon}{e^{(\varepsilon-\mu)\beta} - 1} \qquad \text{... (9.1.1)}$$

For the dilute Bose gas, the appropriate analysis begins with the assumption $e^{-\mu\beta} >> 1$. In the case we want to discuss now, $e^{-\mu\beta}$ will be near its minimum possible value, namely unity. At high temperatures $e^{-\mu\beta} >> 1$ and its value decreases as we lower the temperature. If $e^{-\mu\beta}$ should become equal to unity at some temperature T_c, Eq. (9.1.1) does not have a solution for temperatures $T < T_c$, since $e^{-\mu\beta}$ cannot be smaller than unity. Separating the $\varepsilon = 0$ term, we write Eq. (9.1.1) as

$$N = N_0(\mu) + 2\pi\, V \left(\frac{2m}{h^2}\right)^{3/2} \int_0^\infty \frac{\varepsilon^{1/2}\, d\varepsilon}{e^{(\varepsilon-\mu)\beta} - 1} \qquad \text{... (9.1.2)}$$

where $N_0(\mu)$ is the ground state occupancy given by $N_0(\mu) = e^{\beta\mu}/(1 - e^{\beta\mu})$. Writing $\beta\varepsilon = x$, Eq. (19.1.2) becomes

$$N = N_0(\mu) + 2\pi V \left(\frac{2mkT}{h^2}\right)^{3/2} \int_0^\infty \frac{x^{1/2}\, dx}{e^{-\mu\beta} \cdot e^x - 1} \qquad \text{... (9.1.3)}$$

The difficulty mentioned above becomes transparent from Eq. (9.1.3). If at $T = T_c$, $e^{\mu\beta} = 1$, then Eq. (9.1.1) becomes

$$N = 2\pi V \left(\frac{2mkT_c}{h^2}\right)^{3/2} \int_0^\infty \frac{x^{1/2}\, dx}{e^x - 1} \qquad \text{... (9.1.4)}$$

For $T < T_c$ the prefactor of the integral has to decrease and hence the integral itself has to increase to keep the total number of particles fixed at N. The integral can increase only if $e^{-\mu\beta} < 1$ for $T < T_c$ but such values of $e^{-\mu\beta}$ are not allowed in Bose-Einstein statistics. In the presence of $N_0(\mu)$, we arrive at a solution to this problem by maintaining $\mu = 0$ for $T < T_c$ and letting $N_0(\mu)$ vary to keep Eq. (9.1.3) satisfied with $e^{-\mu\beta} = 1$. Writing $e^{\mu\beta} = z$, we can write Eq. (9.1.3) as

$$N = N_0 + 2\pi V \left(\frac{2mkT}{h^2}\right) \int_0^\infty \frac{z\, x^{1/2} e^{-x}}{1 - z\, e^{-x}}\, dx$$

$$= N_0 + 2\pi V \left(\frac{2mkT}{h^2}\right) \int_0^\infty \sum_{n=0}^{\infty} z^{n+1}\, x^{1/2}\, e^{-(n+1)x}\, dx$$

$$= N_0 + V \left(\frac{2\pi mkT}{h^2}\right)^{3/2} \int_{n=0}^{\infty} \frac{z^{n+1}}{(n+1)^{3/2}}$$

$$= N_0 + V \left(\frac{2\pi mkT}{h^2}\right)^{3/2} \zeta_{3/2}(z) \qquad \text{... (9.1.5)}$$

where the function

$$\zeta_{3/2}(z) = \sum_{n=0}^{\infty} \frac{z^{n+1}}{(n+1)^{3/2}} \qquad \text{... (9.1.6)}$$

is clearly a monotonically increasing function of z for $z > 0$ and rises from 0 for $z = 0$ to $\zeta_{3/2}(1) = 2.612$ at $z = 1$ (the derivative is discontinuous at $z = 1$). We now write Eq. (9.1.5) as

$$N_0 = N - V\left(\frac{2\pi mkT}{h^2}\right)^{3/2} \zeta_{3/2}(z) \qquad \text{... (9.1.7)}$$

and for $0 \leq z \leq 1$, it is clear that the highest T at which $N_0 \neq 0$ is given by T_c which satisfies

$$\left(\frac{2\pi mkT_c}{h^2}\right)^{3/2} \zeta_{3/2}(1) = \frac{N}{V} = n \qquad \text{... (9.1.8)}$$

For $T < T_c$, z having already reached unity cannot change any more and hence the particle density in the ground state has to change according to

$$n_0(T) = \frac{N_0}{V} = \frac{N}{V} - \left(\frac{2\pi mkT}{h^2}\right)^{3/2} \zeta_{3/2}^{(1)}$$

$$= \frac{N}{V} - \frac{N}{V}\left(\frac{T}{T_c}\right)^{3/2} = n\left(1 - \frac{T^{3/2}}{T_c^{3/2}}\right) \qquad \text{... (9.1.9)}$$

The ground state, i.e. $\mathbf{p} = 0$ ($\varepsilon_p = 0$) now has a finite fraction of particles (in the thermodynamic limit). We can imagine that this Bose gas has two possible phases—one in which the particles are spread thinly over all the different energy levels and another phase in which a finite density of particles is obtained in the ground state which is now macroscopically occupied, while over the rest of the states the particles are thinly spread. The transformation from one phase to another occurs at $T = T_c$ and results in a ground state condensate for $T < T_c$, with the condensate accounting for all the particles in the system at $T = 0$.

To get a better understanding of the transition, we calculate the thermodynamic quantities for the two phases.

i) $T < T_c$. In this range, the total energy is obtained by noting that $\mu = 0$ and only the particles outside the condensate contribute, i.e. in Eq. (9.1.3), we do not have any contribution from N_0 to the energy. Consequently,

$$E = 2\pi V \left(\frac{2mkT}{h^2}\right)^{3/2} kT \int_0^{\infty} \frac{x^{3/2}}{e^x - 1}\, dx$$

$$= 2\pi V \left(\frac{2mkT}{h^2}\right)^{3/2} kT \frac{3}{2} \pi^{1/2} \zeta_{5/2}$$

(this is obtained by following the same sequence of steps that was used for the second term on the right hand side of Eq. (9.1.5)

$$= \frac{3}{2} V \left(\frac{2\pi mkT}{h^2}\right)^{3/2} kT\, \zeta_{5/2}$$

$$= \frac{3}{2} NkT \left(\frac{T}{T_c}\right)^{3/2} \zeta_{5/2} / \zeta_{3/2}$$

$$= \frac{3}{2} \cdot \frac{1.341}{2.312} NkT \left(\frac{T}{T_c}\right)^{3/2}$$

$$= 0.770\, NkT \left(\frac{T}{T_c}\right)^{3/2} \qquad \text{... (9.1.10)}$$

The equation of state follows as:

$$PV = \frac{2}{3} E = 0.513\, NkT \left(\frac{T}{T_c}\right)^{3/2}$$

or

$$P = \left(\frac{2\pi mk}{h^2}\right)^{3/2} kT^{5/2} \zeta_{5/2} \qquad \text{... (9.1.11)}$$

Differentiation the energy E yields the specific heat at constant volume as

$$C_v = 1.925\ Nk \left(\frac{T}{T_c}\right)^{3/2} \qquad \text{... (9.1.12)}$$

The entropy is obtained by integrating C/T and hence

$$S = 1.282\ Nk \left(\frac{T}{T_c}\right)^{3/2} \qquad \text{... (9.1.13)}$$

ii) $T > T_c$: Now $N_0 = 0$ and z has to be fixed from Eq. (9.1.3), which yields

$$N = V \left(\frac{2\pi mkT}{h^2}\right)^{3/2} \zeta_{3/2}(z) \qquad \text{... (9.1.14)}$$

The energy follows from Eq. (9.1.3) as

$$E = 2\pi V \left(\frac{2mkT}{h^2}\right)^{3/2} \int_0^\infty \frac{x^{3/2}\,dx}{\frac{1}{z}\,e^x - 1}$$

$$= \frac{3}{2} V \left(\frac{2\mu kT}{h^2}\right) \zeta_{5/2}(z)\ kT \qquad \text{... (9.1.15)}$$

To find the entropy we note that in the grand canonical ensemble

$$E - TS - \mu N = -PV = -\frac{2}{3} E$$

or $TS = \frac{5}{3} E - \mu N$

$$\text{or} \quad S = \frac{5}{2} kV \left(\frac{2\pi mkT}{h^2}\right)^{3/2} \zeta_{5/2}(z) - Nk \ln z \qquad \text{... (9.1.16)}$$

The specific heat can be obtained by differentiating the energy at constant volume .

$$C = \frac{15}{4} kV \left(\frac{2\pi mkT}{h^2}\right)^{3/2} \zeta_{5/2}(z) + \frac{3}{2} VkT \left(\frac{2\pi mkT}{h^2}\right)^{3/2} \times \frac{d}{dz} \zeta_{5/2}(z) \cdot \frac{dz}{dT}$$

... (9.1.17)

To find dz/dT at constant V and N, we differentiate Eq. (9.1.14) to write

$$0 = \frac{d}{dT}\left(\frac{N}{V}\right) = \frac{d}{dT}\left(\frac{2\pi mkT}{h^2}\right)^{3/2} \zeta_{3/2}(z)$$

or

$$\frac{3}{2} T^{1/2} \zeta_{3/2}(z) + T^{3/2} \frac{d}{dz} \zeta_{3/2}(z) \cdot \frac{dz}{dT} = 0$$

or

$$\frac{dz}{dT} = -\frac{3}{2T}\left[\frac{d}{dz} \zeta_{3/2}(z)\right]^{-1} \cdot \zeta^{3/2}(z) \qquad \text{... (9.1.18)}$$

We now find from the definition of Eq. (9.1.6)

$$z \cdot \frac{d}{dz} \zeta_{5/2}(z) = \zeta_{3/2}(z) \qquad \text{... (9.1.19a)}$$

and

$$z \cdot \frac{d}{dz} \zeta_{3/2}(z) = \zeta_{1/2}(z) \qquad \text{... (9.1.19b)}$$

Consequently,

$$\frac{1}{z}\frac{dz}{dT} = -\frac{3}{2T}\frac{\zeta_{3/2}(z)}{\zeta_{1/2}(z)} \qquad \text{... (9.1.20)}$$

(using Eq. (9.1.19b) in Eq. (9.1.18))
and Eq. (9.1.17) becomes

$$C = \frac{15}{4} kV \left(\frac{2\pi mkT}{h^2}\right)^{3/2} \zeta_{5/2}(z) - \frac{9}{4} Nk \frac{\zeta_{3/2}(z)}{\zeta_{1/2}(z)} \qquad \text{... (9.1.21)}$$

If we compare the thermodynamic quantities E, S and C above and below the transition, then they are continuous at $T = T_c$. However, we might note the fact that in the low temperature phase, i.e. for $T < T_c$, the pressure is independent of volume and hence studying isotherms could be interesting. If

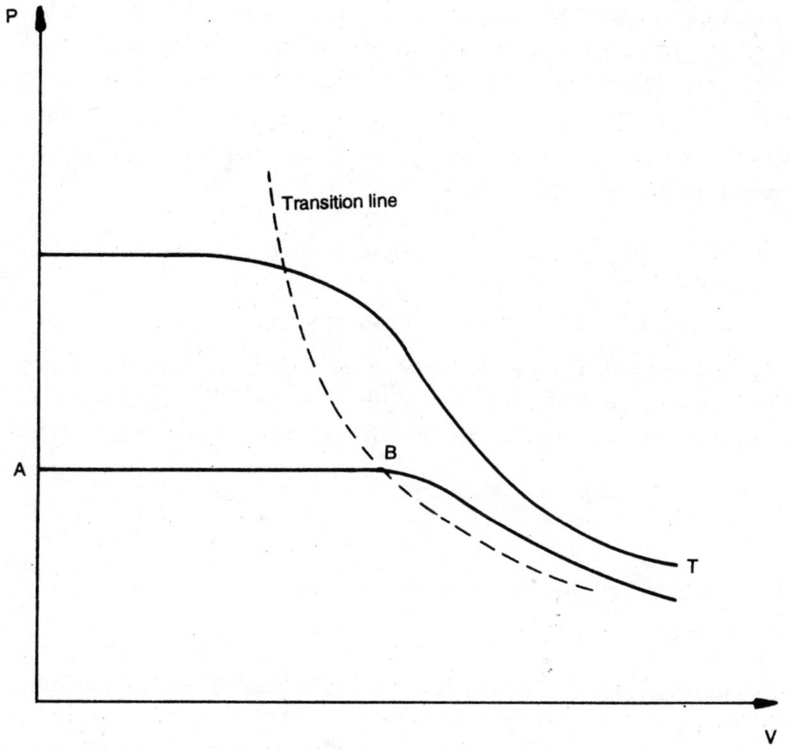

Fig. 9.1: Isotherms for the Bose gas. The dashed line corresponds to the condensation.

we hold T fixed and vary V starting from large values to smaller ones (going from dilute gas to the degenerate gas) then at a critical V ($= V_c$) which satisfies

$$\left(\frac{2\pi mkT_c}{h^2}\right)^{3/2} \zeta_{3/2}(1) = \frac{N}{V_c} \qquad \text{... (9.1.22)}$$

the condensation phenomenon will set in and for $V < V_c$, there will be a finite condensate N, determined by

$$N_0 = N \left(1 - \frac{V}{V_c}\right) \qquad \text{... (9.1.23)}$$

In the P-V diagram at a fixed T (isotherms shown in Fig. 9.1), the pressure is unchanging over the range $0 < V < V_c$ and then decreases with increasing V.

For higher temperatures, the critical V_c occurs at smaller and smaller values. The flat part of each curve can be thought of as the coexistence region for the two phases—a standard feature of the first order transition. Further the pressure is determined solely by the temperature in this region, which is characteristic of the vapour pressure and from Eq. (9.1.11), we find the slope of the vapour pressure curve to be

$$\frac{dP}{dT} = \frac{5}{2}\,\zeta_{5/2}(1)\left(\frac{2\pi mkT}{h^2}\right)^{3/2}\cdot k \qquad \text{... (9.1.24)}$$

Now for an isotherm at the temperature T, the specific volumes are zero and V_c/N for the condensate and the gas phase respectively. Using the Clausius-Clapeyron relation, we can now get the latent heat of the transition per particle as

$$L = T\left(\frac{V_c}{N} - 0\right)\cdot\frac{dP}{dT} = T\,\frac{V_c}{N}\cdot\frac{dP}{dT}$$

$$= \frac{5}{2}\cdot\frac{\zeta_{5/2}(1)}{\zeta_{3/2}(z)}\cdot kT \qquad \text{... (9.1.25)}$$

Thus the condensation has certain features of a first-order transition.

Without trying to classify this transition as first order or second order, we will now go to T slightly above T_c and imagine approaching T_c from above. Clearly the chemical potential will be very close to zero for $T \equiv T_c$ and we want to determine the exact manner in which μ approaches zero as $\delta T = (T - T_c)$ tends to zero. To do this we need to analyze Eq. (9.1.20) near $z = 1$ and $T = T_c$. However, $\zeta_{1/2}(z)$ is divergent as $z \to 1$ and hence we need to go back to Eq. (9.1.14) which says that for fixed N and V,

$$T^{3/2}\,\zeta_{3/2}(z) = \text{constant} \qquad \text{... (9.1.26)}$$

Taking differentials near $z = 1$ and $T = T_c$

$$\frac{3}{2}\,\zeta_{3/2}(1)\,\frac{\delta T}{T_c} = -\frac{d\,\zeta_{3/2}(z)}{dz}\Big|_{z=1}\,\delta z \qquad \text{... (9.1.27)}$$

Since $\left.\dfrac{d\zeta_{3/2}(z)}{dz}\right|_{z=1}$ diverges at $z = 1$, it is proper to start with the definition of derivatives and write

$$\frac{d}{dz}\,\zeta_{3/2}(z)\Bigg|_{z=1}\delta z = \zeta_{3/2}(1+\delta z) - \zeta_{3/2}^{(1)} \qquad \text{... (9.1.28)}$$

An integral representation of Eq. (9.1.28) is clearly

$$\frac{d}{dz}\,\zeta_{3/2}(z)\Bigg|_{z=1}\delta z = \frac{1}{\Gamma(3/2)}\int_0^\infty dx\,x^{1/2}\left[\frac{1}{\dfrac{e^x}{1+\delta z}-1} - \frac{1}{e^x-1}\right]$$

$$= \frac{1}{\Gamma\left(\frac{3}{2}\right)}\int_0^\infty \frac{dx\,x^{1/2}\sigma z\,e^x}{(e^x-1)\left(\dfrac{e^x}{1+\delta z}-1\right)} \qquad \text{... (9.1.29)}$$

We note two features of Eq. (9.1.29).

i) As $\delta \to 0$, the integral is completely dominated by the small-x regime and hence to get the leading term in δz, it suffices to linearize e^x.

ii) The integral is well defined only for negative δz, which is as it should be since the maximum value of z (i.e. $z = 1$) is achieved at $T = T_c$ and z has to decrease as T increases above T_c.

In accordance with (ii) we set $\delta z = -\varepsilon$ $(-\varepsilon > 0)$ and find by linearizing e^x,

$$\varepsilon\,\frac{d}{dz}\,\zeta_{3/2}(z)\,|_{z=1} \cong \frac{\varepsilon}{\Gamma(3/2)}\int_0^\infty \frac{dx}{x^{1/2}}\cdot\frac{1}{x+\varepsilon} = \frac{\pi\varepsilon^{1/2}}{\Gamma(3/2)} \qquad \text{... (9.1.30)}$$

Returning to Eq. (9.1.29), we find

$$\varepsilon^{1/2} = \frac{3}{2}\,\frac{\zeta_{3/2}(1)\,\Gamma(3/2)}{\pi}\,\frac{\delta T}{T_c} \qquad \text{... (9.1.31)}$$

Now $z = e^{\mu/kT}$ leads to $\delta z = z\,[\frac{\delta\mu}{kT} - \frac{\mu\,\delta T}{kT\ \ T}]$ and near $T = T_c$, $\mu = 0$, $z = 1$, we have $\delta z \sim \delta\mu/kT_c$. Combined with Eq. (9.1.30), this leads to

$$\delta\mu = \frac{9}{4}\,kT_c\cdot\left(\frac{\zeta_{3/2}(1)\,\Gamma(3/2)}{\pi}\right)^2\left(\frac{\delta T}{T_c}\right)^2 \qquad \text{... (9.1.32)}$$

The slope of the chemical potential is consequently discontinuous at $T = T_c$. This leads to interesting consequences for the slope of the specific heat. We turn to the energy expression, viewing it as a function of the independent variables z and T and can expand as a Taylor series to $O(\delta T)^2$ near $T = T_c$ and $z = 1$ as (N and V are held fixed)

$$E(T, z) = E(T_c, 1) + \left(\frac{\delta E}{\delta T}\right)_{T_c, 2} \delta T + \left(\frac{\delta E}{\delta z}\right)_{T_c, 1} \delta z + \frac{1}{2}\left(\frac{\delta^2 E}{\delta T^2}\right)_{T_c, 1} (\delta T)^2 \quad \text{... (9.1.33)}$$

We do not need any further terms since δz is $O(\delta T)^2$ from Eq. (9.1.31). The slope of the specific heat at $T = T_c$, is clearly $\delta^2 E/\delta(\delta T)^2$ at $T = T_c$ and $z = 1$ and using Eqs. (9.1.31) and (9.1.33), the slope can be easily shown to be

$$\left(\frac{dC_v}{dT}\right)_{T \to T_c} + = \frac{Nk}{T_c}\left[\frac{45}{8}\frac{\zeta_{5/2}(1)}{\zeta_{3/2}(1)} - \frac{27}{4}\left(\frac{\zeta_{3/2}(1)\Gamma(3/2)\,2}{\pi}\right)\right]$$

$$\cong -0.776\,\frac{Nk}{T_c} \quad \text{... (9.1.34)}$$

For T approaching T_c from below, Eq. (9.1.12), shows the slope to be

$$\left(\frac{dC_v}{dT}\right)_{T \to T_c^-} = \frac{2.89Nk}{T_c} \quad \text{... (9.1.35)}$$

The specific heat, thus, has a maximum at $T = T_c$ and a slope discontinuity of 3.66 Nk/T. The behaviour of C_v as a function of T is shown in Fig. 9.2.

It should be emphasized that the above picture is valid for a non-interacting Bose gas and any interaction, however weak, may alter the picture drastically.

We now comment on the procedure followed in Eq. (9.1.3), when the number of particles residing in the ground state, i.e. $p = 0$, were singled out. We could have split off any finite number of terms without changing the value of the integral, i.e. we could equally well have written

$$N = N_0 + N_1 + V\left(\frac{2m}{h^2}\right)^{3/2} \int \frac{\varepsilon^{1/2} d\varepsilon}{e^{\beta(\varepsilon - \mu)} - 1}$$

where

$$N_1 = \frac{1}{\frac{1}{z} e^{\beta\varepsilon_1} - 1} \quad \text{and} \quad N_0 = \frac{z}{1 - z}$$

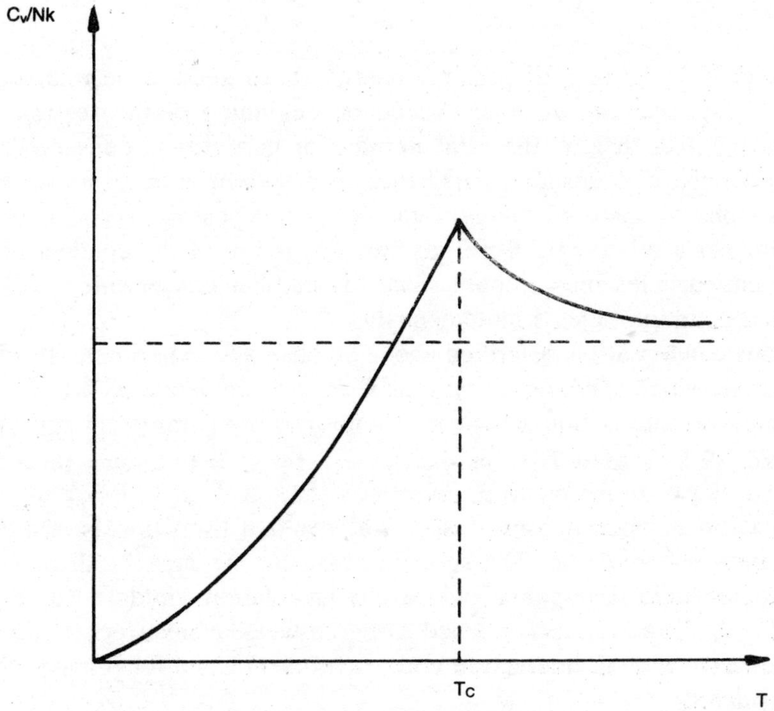

Fig. 9.2: Specific heat around the condensation point T_c.

If this is as reasonable as removing N_0 alone from the sum, then we do not have a consistent procedure for establishing that particles have definitely condensed in the **p** = 0 state. We consider the gas to be in a box of volume V, which we ultimately take to be infinity. Now as $z \to 1, N_1 \to (e^{\beta\varepsilon_1} - 1)^{-1}$ and since the energies of a particle in a box are proportional to $V^{-2/3}$, it is clear that for large $V, N_1 \sim V^{2/3}$. Note that N_1 $(z = 1)$ is the maximum value of N_1 (z) under the restriction $z \leq 1$ and hence what is true for N_1 $(z = 1)$ is true for all z. Thus the corresponding particle density $N_1/V \sim V^{-1/3}$ and hence must vanish in the thermodyamic limit. On the other hand, in a finite volume V, the chemical potential adjusts itself to provide a finite limit for

$$\underset{\substack{z \to 1 \\ V - \infty}}{\text{Lt}} \quad \frac{z}{V} \cdot \frac{1}{1-z}$$

and thus only N_0 among all possible energy states need to be retained.

To end this section, we note that at the beginning of the section, we laid stress on the fact that N, the total number of particles is conserved. This is essential for the condensation and hence in a system with no conservation of numbers—photon gas or a phonon gas—it is not possible to have Bose condensation. For a relativistic Bose gas, the discussion of the condensation must take the antiparticles into account since the number conservation will involve particles and antiparticles simultaneously.

Can this condensation described above be observed experimentally. The only Bose system which remains a liquid at very low temperatures where quantum effects predominate is liquid helium. If one puts the parameters appropriate to ^{4}He in Eq. (9.1.8), then T_c turns out to be 3.14 K. It is known that ^{4}He does undergo a phase transition to a superfluid state at $T \pm 2.17$ K and it is to a certain extent an open question as to whether that transition is related to the condensation phenomenon. The specific heat near the superfluid transition has a near logarithmic divergence, while our calculation yields a finite specific heat at $T = T_{c.}$ However, as remarked earlier, a weakly imperfect (i.e. individual molecules have a weak interaction with each other) gas might change the situation considered.

9.2 Black-body Radiation

While it is debatable whether the Bose-Einstein distribution has anything to do with the real life transition in ^{4}He, there no doubt that it is a completely accurate description of the equilibrium state of radiation with matter. The radiation in thermal equilibrium is called black body radiation and the radiation may be regarded as a gas consisting of photons. The equations of the radiation field are governed by Maxwell's equations which are linear and hence the electromagnetic oscillations which constitute radiation can be written as a set of harmonic oscillators and the energy quanta of these oscillators are the photons. Since the governing equations are linear and the description in terms of simple harmonic oscillators exact, there is no interaction among the photons. The photons have integral intrinsic angular momentum (spin) and hence the gas of photons in equilibrium will be correctly described by the Bose-Einstein distribution at a given temperature.

If the radiation is not in vacuum but in a material medium, then properties of the medium come into play and it is possible that Maxwell's equations may not be linear any more and in this case, for the condition of an ideal Bose gas to be approximately fulfilled, one must have the interaction between the matter and radiation to be weak. This is true all across the radiation spectrum in gases except for frequencies in the vicinity of the absorption lines of the material. For high density matter, the interaction may not be weak.

Since the energy ε of a photon with frequency ω and mometum $\mathbf{k}$ is given by

$$\varepsilon = \hbar\,\omega = \hbar\,c\,k \qquad \text{... (9.2.1)}$$

where c is the velocity of light, we have to find the density of states in terms of k or ω. We also need to find the chemical potential μ. If the particle number is not conserved the chemical potential has got to be zero. This is because the average number of particles in the system will be determined by the condition of equilibrium, namely the condition minimizing the free energy F with respect to all the variables in the system. This leads to $\delta F/\delta N = 0$ but since $\mu = \delta F/\delta N$, we have the condition of vanishing chemical potential. Thus the number of photons with energy $\omega(k)$ is

$$n(\omega) = \left(e^{\hbar\omega/kT} - 1\right)^{-1} \qquad \text{... (9.2.2)}$$

We now need to specify the number of states with momentum between $\mathbf{k}$ and $\mathbf{k} + d\mathbf{k}$. To do so we note that the spatial and temporal dependence of the electric field corresponding to the photon will have the plane waveform determined by $e^{i(\mathbf{k}\cdot\mathbf{r}-\omega t)}$. Now periodic boundary conditions on the electric field in a cube of size L will force the usual quantization

$$\mathbf{k} = \frac{2\pi}{L}\cdot\mathbf{n} \qquad \text{... (9.2.3)}$$

where $\mathbf{n} = n_x\hat{i} + n_y\hat{j} + n_z\hat{k}$, n_x, n_y and n_z being integers and thus the number of states between $\mathbf{k}$ and $\mathbf{k} + d^3\mathbf{k}$ follow the free-particle argument of Chapter 4 exactly, and can be written as $V/(2\pi)^3 \cdot 4\pi k^2 dk$. Now the spin degeneracy has to be included and although $s = 1$ would lead to $2s + 1 = 3$, we note that the electromagnetic oscillations are transverse and hence there can be only two polarization directions (the two transverse directions). Thus the spin degeneracy is 2 and the number of photons with energy between $\hbar\,\omega$ and $\hbar\,(\omega + d\omega)$ can be written as

$$N(\omega)d\omega = \frac{8\pi V}{(2\pi)^3 c} \frac{\omega^2\, d\omega}{e^{\hbar\omega\beta} - 1} \quad \text{... (9.2.4)}$$

The total energy of the radiation at a temperature T is

$$E = \int_0^\infty \hbar\,\omega\, N(\omega)\, d\omega = \frac{8\pi V\hbar}{(2\pi)^3 c^3} \int_0^\infty \frac{\omega^3 d\omega}{e^{\beta\hbar\omega} - 1}$$

$$= \frac{8\pi V\hbar}{(2\pi)^3 (\beta\hbar\,)^4 c^3} \int_0^\infty \frac{x^3 dx}{e^x - 1} \quad \text{(where } x = \beta\,\hbar\,\omega)$$

$$= 8\pi \left(\frac{1}{hc}\right)^3 (kT)^4 \sum_{n=0}^{\infty} \frac{1}{(n+1)^4}\, \Gamma(4)$$

$$= 48\pi \left(\frac{1}{hc}\right)^3 V\,(kT)^4\, \zeta_4\,(1)$$

$$= \frac{48\pi^5}{90}\, V\,.\,kT \left(\frac{kT}{hc}\right)^3 = \frac{8}{15}\,\pi^2\, VkT \left(\frac{\pi kT}{hc}\right)^3 \quad \text{... (9.2.5)}$$

The specific heat at constant volume is consequently,

$$C_v = \left.\frac{\partial E}{\partial T}\right)_V = \frac{32}{15}\,\pi^2 \left(\frac{\pi kT}{hc}\right)^3 \quad \text{... (9.2.6)}$$

What we would like to verify experimentally is the T^4-law of Eq. (9.2.5) or more importantly the actual distribution $E(\omega, T)$ which we define by

$$E(T)/V = \int_0^\infty E(\omega,\, T)\; dT \quad \text{... (9.2.7)}$$

leading to

$$E\,(\omega,T) = \frac{\hbar}{\pi^2 c^3} \cdot \frac{\omega^3}{e^{\beta\hbar\omega} - 1} \quad \text{... (9.2.8)}$$

To test the above relation we imagine a small hole through which the black body radiation communicates with the outside world. Now, the amount of energy of frequency ω coming out per unit area per unit time in a direction θ to the normal is the amount of energy of frequency ω contained in a volume element of height $c\cos\theta$ and base of area unity, while the probability of coming out in the solid angle $d\Omega = 2\pi \sin\theta \, d\theta$ is $d\Omega/4\pi$ (isotropy). Hence the amount of energy coming out per unit area per unit time is obtained by integrating over the hemisphere and hence

$$I(\omega, T) = \int_{\text{hemisphere}} \frac{d\Omega}{4\pi} \, c \cos\theta E(\omega,T)$$

$$= \frac{c}{4} E(\omega,T) \qquad \text{... (9.2.9)}$$

while the total energy emitted, u, is given by

$$u = \int_0^\infty d\omega \; I(\omega,T) = \sigma T^4 \qquad \text{... (9.2.10)}$$

where

$$\sigma = \frac{\pi^2 k^4}{60 \, (\hbar c)^3} \qquad \text{... (9.2.11)}$$

is called the Stefan-Boltzmann constant.

The measurement of both $I(\omega,T)$ and $I(T)$ can be carried out and provides an excellent verification of the Bose distribution for photons. The shape of spectral curves of $I(\omega,T)$ vs ω at different temperatures is shown in Fig. 9.3.

Note that there is a characteristic maximum in the spectrum $I(\omega,T)$ and as the temperature increases, the maximum shifts to higher frequencies. This is known as Wien's displacement law.

Finally, consider a body surrounded by black body radiation. The amount of radiation energy incident per unit time on a surface area dS at an angle θ to a chosen direction is clearly $cE(\omega, T) \cos\theta \, dS \, d\omega$. If the absorptive power of the body is $A(\theta,\omega)$ at this angle and frequency then the amount of energy absorbed per unit time is $c\,A(\theta, \omega)\, E(\omega, T) \cos\theta \; dS \; d\omega$. We consider fluorescence and scattering to be absent. Also radiation is merely absorbed by the

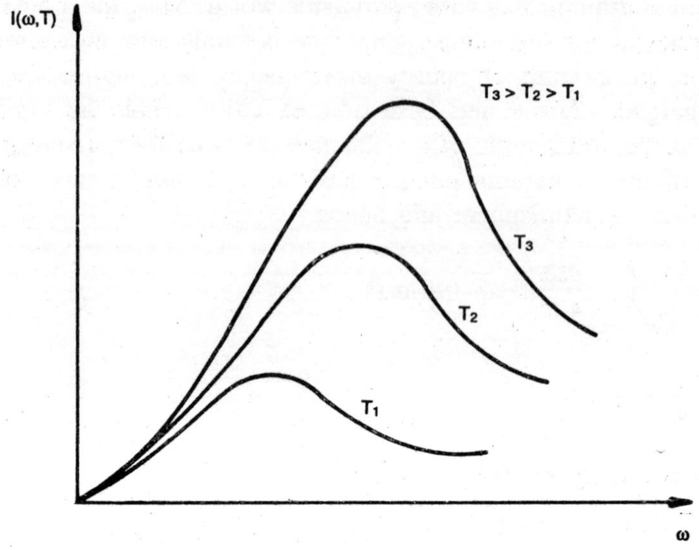

Fig. 9.3: The intensity as a function of ω for different temperatures.

body—it does not pass through. Now if the emissivity of the surface is $J(\theta, \omega)$, then the amount of energy emitted per unit area per unit time is $J(\theta, \omega)\ dS\ d\omega$ and detailed balance immediately yields

$$J(\theta, \omega)/A(\omega, \theta) = c\,E(\omega, T)\cos\theta \qquad \text{... (9.2.12)}$$

independent of the properties of the body—this is Kirchoff's law. For a black body $A(\omega, \theta) = 1$ and the actual emissive power is independent of the material property.

9.3 Superfluidity of ^{4}He

Helium has several remarkable properties. It remains a liquid even at the lowest temperatures under normal pressures. This is because of the inert gas nature which causes the interaction between helium atoms to be very weak. In spite of this weak interaction the other inert gases do solidify. Helium, however, because of its very low mass has a high degree of zero-point kinetic energy and the combination of low mass and weak interatomic forces keeps it a liquid at $T \cong 0$. Thus the two isotopes of helium— ^{4}He and ^{3}He are the only liquids which can show quantum features—all other liquids freeze into solids before the temperature can become low enough for the De Broglie wavelength to become of the order of the atomic spacing.

The liquefaction of ^{4}He occurs at $T \cong 5$ K. If this liquid is further cooled, then a further change occurs at $T = T_\lambda = 2.17$ K. Below T_λ, the liquid boils without bubbling—its thermal conductivity becomes infinite! In addition to the apparently infinite thermal conductivity, its viscosity, under certain conditions, also appears to be zero. In particular, if it flows through a narrow capillary at speeds less than a critical velocity v_c, then it shows no viscosity (this narrow capillary is called a superleak). The narrower the capillary, higher the critical velocity. However, if the viscosity is measured by the rotating cylinder technique (a rotating can placed in a liquid bath experiences a drag due to the viscosity of the liquid), the measured value goes down with decreasing temperature. This behaviour of viscosity is explained by a model which assumes that below T_λ, liquid helium is an intimate mixture of two liquids—a normal fluid which exhibits viscosity and a superfluid which does not exhibit any viscosity. The superleak allows only the superfluid to flow.

A further assumption that the superfluid carries no entropy helps explain the mechano-caloric effect. If two containers of ^{4}He, A and B, are maintained at the same temperature but different pressures and connected by a superleak (see Fig. 9.4), then the temperature of the container at the lower pressure falls and a temperature difference ΔT is maintained which is related to the pressure difference ΔP as $\Delta T = \Delta P/\rho s$, where s is the specific entropy. Conversely, if A and B are at the same pressure but different temperatures, then on being connected by a superleak a pressure difference ΔP develops, given by $\Delta P = \rho s \, \Delta T$ as stated before. If one of the containers (the one at lower temperature) is a thin tube, then the pressure head causes a fountain of liquid helium to emerge.

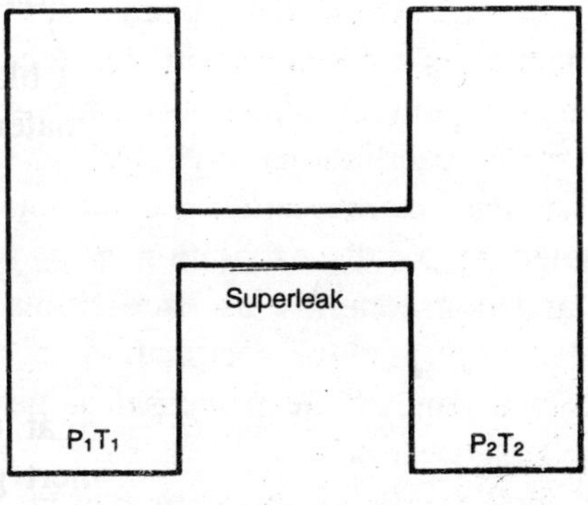

Fig. 9.4: Flow caused by temperature gradient.

Yet another interesting property of superfluid helium is its ability to creep up the side of the beaker in which it is kept. For all liquids Van der Waals forces of attraction between beaker and liquid molecules cause the formation of a layer of liquid whose thickness decreases with height. In all liquids, except He in its superfluid phase, the layer formation is inhibited by the small but finite temperature differences between the wall and the liquid. If the wall is at a higher temperature the layer forms droplets and falls back into the liquid. For $T < T_\lambda$, the infinite thermal conductivity prevents any temperature difference and hence the liquid layer creeps up the wall and escapes from the beaker. The above properties clearly show that for $T < T_\lambda$, the liquid is in a different thermodynamic state and this state is called He II. The transition at $T = T_\lambda$ is second order (no latent heat) and is called the lambda transition due to the behaviour of the specific heat, which becomes infinitely large at $T < T_\lambda$, and the specific heat vs temperature curve has the appearance of the letter lambda (λ) (see Fig. 9.5).

It should be noted that ^{3}He does not exhibit the above features unless the temperature is lowered to milli kelvins. The prime difference between ^{3}He and ^{4}He being one of statistics— ^{3}He is half integral spin, while ^{4}He has spin zero—we expect that the superfluid property of ^{4}He is due to the fact that it obeys Bose statistics. The main purpose of this section is to bring out the connection between superfluidity and the Bose distribution.

To conclude this qualitative introduction to the properties of He II, we note that the picture of He II as being composed of two kinds of fluids leads to a prediction about wave propagation. If we call ρ_n and ρ_s the densities of the normal and superfluid components of He II, then the actual density $\rho = \rho_s + \rho_n$. Generally wave propagation implies oscillations in the density ρ (sound waves), but in this case we can have out of phase oscillations in ρ_s and ρ_n so that ρ need not vary. However, the oscillations in ρ_s and ρ_n will propagate in the form of waves and these oscillations are called second sound. Since the total entropy of He II is determined by ρ_n, the oscillation in ρ_n will manifest itself as oscillations in entropy and consequently as oscillations in temperature. The prediction and subsequent experimental verification of the existence of these temperature oscillations was one of the principal achievements of the "two-fluid" theory.

Information regarding the energy spectrum (i.e. relation between energy and momentum) of a collection of helium atoms can be had from the study of low temperature specific heat. It is experimentally observed that for very low temperatures, the specific heat is proportional to T^3 (the specific heat of a

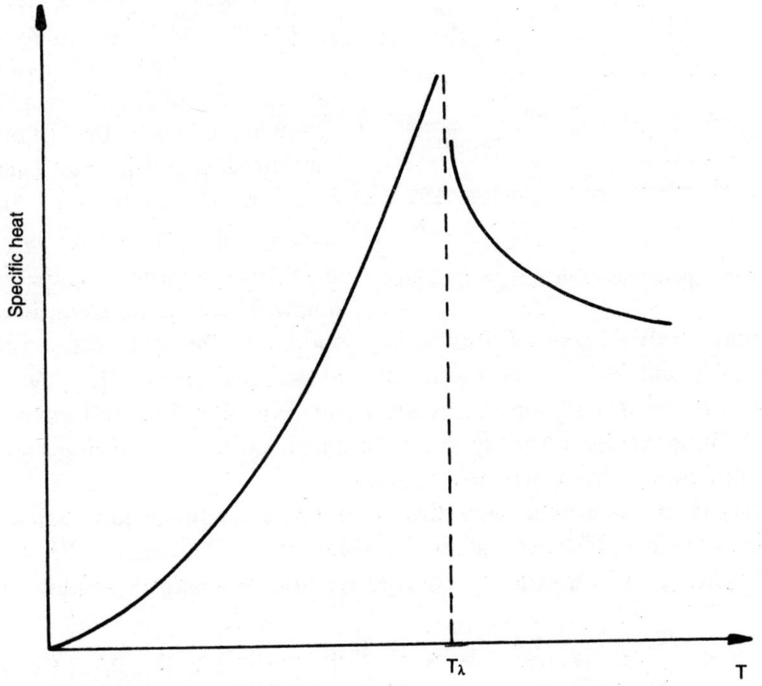

Fig. 9.5: Specific heat near the superfluid transition.

phonon gas), while for relatively higher temperatures there is an additional contribution having the form $e^{-\Delta/kT}$. The very low temperature properties are determined by the lowest momentum values and the T^3- law in this regime suggests that for very small p, the dispersion is phonon-like, i.e.

$$E = \hbar\,\omega = \hbar\,c\,k = c\,p\;(p\text{ small}) \qquad \text{... (9.3.1)}$$

The exponential behaviour at higher temperature suggests that for higher momenta (which are sampled at higher temperatures), the dispersion relation has the form

$$E = \hbar\,\omega = \Delta + \frac{(p-p_0)^2}{2m^*} \quad (p \cong p_0) \qquad \text{... (9.3.2)}$$

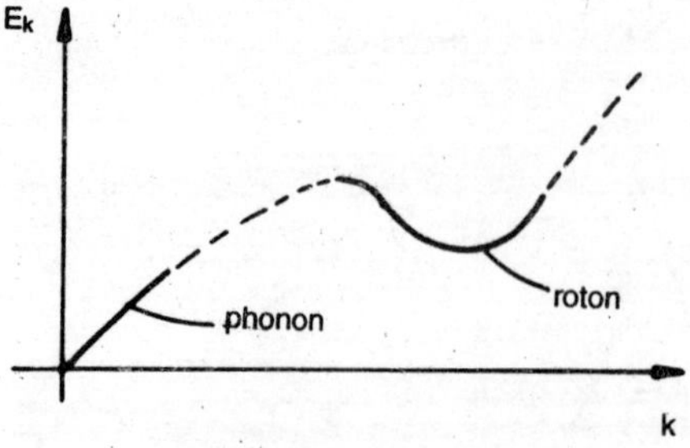

Fig. 9.6: The spectrum according to Landau.

(that this leads to the desired specific heat is the subject of one of the problems). The total specific heat is the sum of those obtained separately on the basis of Eq. (9.3.1) and (9.3.2). Comparing with the experimentally measured specific heat, Landau could find the values of the fitting parameters c, Δ, p_0 and m^*. The value of c thus obtained is the velocity of sound at low temperatures and the experimental confirmation of this value vindicates the essential correctness of Eqs. (9.3.1) and (9.3.2). However, the above dispersion [Eq. (9.3.2)] yields only parts of the E vs p curve, as shown in Fig. 9.6. The full curve cannot be obtained theoretically, although it can be inferred from scattering measurements by a relationship which we now discuss.

We begin by assuming that the fluid has an equilibrium density $\bar{\rho}$ when its velocity is zero. The excitations cause a density fluctuation $\delta\rho$ and provide a small velocity $\mathbf{v}$. The energy can be written to quadratic order as

$$E = \bar{E} + \int \psi\,(\mathbf{r})\,\delta\rho\,d^3r + \frac{1}{2}\int \Phi\,(\mathbf{r}-\mathbf{r}')\,\delta\rho\,(\mathbf{r})\,\delta\rho\,(\mathbf{r}')\,d^3\mathbf{r}\,d^3r'$$

$$+ \frac{1}{2}\int \bar{\rho}v^2\,(\mathbf{r})\,d^3\,\mathbf{r} \qquad \text{... (9.3.3)}$$

where $\Psi\,(\mathbf{r}')$ and $\varphi\,(\mathbf{r},\,\mathbf{r},\,)$ are to be thought of as coefficients for expansion in powers of $\rho\,(r)$.

The fact that the unperturbed fluid is isotropic and homogeneous makes $\varphi\,(\mathbf{r},\,\mathbf{r}')$ a function of $(\mathbf{r}-\mathbf{r}')$ alone and also make $\Psi(\mathbf{r})$ a constant. The constraint on the total mass remaining unchanged implies $\int \delta\rho\,(\mathbf{r})d^3\,r = 0$ and thus we have

$$E = \bar{E} + \frac{\bar{\rho}}{2}\int v^2\,(\mathbf{r})\,d^3r + \frac{1}{2}\int \varphi\,(\mathbf{r}-\mathbf{r}')\,\delta\rho\,(\mathbf{r})\,\delta\rho\,(\mathbf{r}')\,d^3\,rd^3\,r' \qquad \text{... (9.3.4)}$$

where $\bar{E}$ is determined by $\bar{\rho}$ alone. The continuity equation provides a constraint in the form

$$\frac{\partial \rho}{\partial t} + \nabla \cdot (\rho \mathbf{v}) = 0 \qquad \text{... (9.3.5)}$$

which to the correct order of consistency can be written as

$$\frac{\partial}{\partial t}\, \delta\rho + \bar{\rho}\, \nabla \cdot \mathbf{v} = 0 \qquad \text{... (9.3.6)}$$

To incorporate the costraint of Eq. (9.3.6) in Eq. (9.3.4), we write down the energy expression in momentum space by using the Fourier transforms

$$v(\mathbf{r}) = \int v(\mathbf{k})\, e^{i\mathbf{k}\cdot\mathbf{r}} \frac{d^3k}{(2\pi)^3} \qquad \text{... (9.3.7a)}$$

$$\delta\rho(\mathbf{r}) = \int \delta\rho(\mathbf{k})\, e^{i\mathbf{k}\cdot\mathbf{r}} \frac{d^3k}{(2\pi)^3} \qquad \text{... (9.3.7b)}$$

and

$$\varphi(\mathbf{r}-\mathbf{r}') = \int \varphi(\mathbf{k})\, e^{i\mathbf{k}\cdot(\mathbf{r}-\mathbf{r}')} \frac{d^3k}{(2\pi)^3} \qquad \text{... (9.3.7c)}$$

The energy becomes

$$E = \bar{E} + \frac{\bar{\rho}}{2} \sum_k \mathbf{v}(\mathbf{k}) \cdot v(-\mathbf{k}) + \frac{1}{2} \sum_k \varphi(\mathbf{k})\, \delta\rho(\mathbf{k})\, \delta\rho(-\mathbf{k}) \qquad \text{... (9.3.8)}$$

In momentum space, Eq. (9.3.6) becomes

$$\delta\dot{\rho}(\mathbf{k}) = -i\mathbf{k}\, \bar{\rho} \cdot \mathbf{v}(\mathbf{k}) \qquad \text{... (9.3.9)}$$

Substituting for $v(\mathbf{k})$ in Eq. (9.3.8), bearing in mind that small oscillations in a liquid are longitudinal [$\mathbf{v}(\mathbf{k})$ is parallel to $\mathbf{k}$], we have

$$E - \bar{E} = \frac{1}{2} \cdot \sum_k \frac{|\delta\dot{\rho}(k)|^2}{\bar{\rho}\, k^2} + \frac{1}{2} \sum_k \varphi(k)\, |\delta\rho(k)|^2 \qquad \text{... (9.3.10)}$$

where use has been made of the fact that $\delta\rho$ is real implies $\delta\rho(k)^* = \delta\rho(-k)$ and similarly for $\mathbf{v}(\mathbf{k})$. The right hand side of Eq. (9.3.10) corresponds to a set of non-interacting oscillators (one for each k value) with frequency $\Omega(k)$ given by

$$\Omega^2 (k) = \bar{\rho}\ \varphi (k)\ k^2 \qquad \text{... (9.3.11)}$$

and accordingly the energy spectrum for each k value is given by

$$E (k) = \left(n + \frac{1}{2}\right) \hbar\ \Omega(k) \qquad \text{... (9.3.12).}$$

where n is an integer or zero. The ground state energy of the system is given by $n = 0$ and hence

$$E_{\text{ground}} = \bar{E} + \sum_k \frac{1}{2} \hbar\ \Omega(k) \qquad \text{... (9.3.13)}$$

The second term on the right hand side is the zero-point energy and we have

$$\frac{1}{2} \hbar\ \Omega (k) = \frac{1}{2\bar{\rho}\, k^2} < |\delta\dot{\rho}\ (k)\ |^2 > + \frac{1}{2}\ \varphi (k)\ < |\delta\rho\ (k)\ |^2 >$$

$$= \varphi (k)\ <|\ \delta\rho\ (k)|^2> \qquad \text{... (9.3.14)}$$

where we have made use of the virial theorem (angular brackets denote expectation values) for the simple harmonic oscillator which asserts that the two expectation values in the first line of Eq. (9.3.14) are equal. The excitation energy $\varepsilon\ (k)$ is clearly $\hbar\ \Omega\ (k)$ and hence

$$\varepsilon\ (p) = \varepsilon(k) = \hbar\ \Omega\ (k) = \hbar\ \frac{(k)^2}{\Omega\ (k)} = \frac{\hbar^2\ \bar{\rho}\,\varphi\ (k)\ k^2}{2\varphi(k)\ <|\ \delta\rho(k)\ |^2>}$$

$$= \frac{\hbar^2\ \bar{\rho}\, k^2}{2 < |\ \delta\rho(k)\ |^2>} = \frac{p^2}{2S\ (p)} \qquad \text{... (9.3.15)}$$

where $p = \hbar k$ is the momentum and $S(p) = <|\ \delta\rho\ (k)\ |^2 >/\ \bar{\rho}$ is the Fourier transform of the density correlation function or the structure factor

$$S(\mathbf{r} - \mathbf{r}\ ') = <\delta\rho\ (r)\ \delta\rho\ (r')\ >/\ \bar{\rho} \qquad \text{... (9.3.16)}$$

As we have seen before (see Chapter 5), the structure factor can be experimentally determined from neutron scattering) and thus the excitation curve can be experimentally obtained using Eq. (9.3.15). The curve shown in Fig. 9.7 does

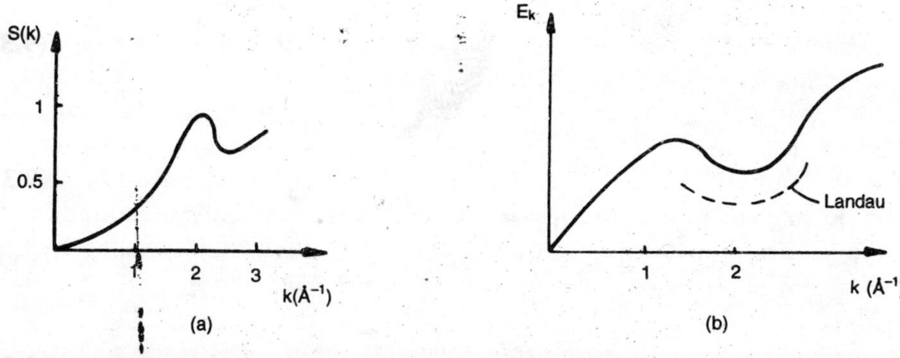

Fig. 9.7: (a) Experimental structure factor $S(k)$. (b) Using $S(k)$ to obtain E_k. Solid curve. agrees qualitatively with Landau's curve but is off in magnitude.

have a linear behaviour near $p = 0$ and a minimum at p_0 — features which were present in Landau's theory shown in Fig. (9.6).

An excitation curve of the kind shown in Fig. 9.7 does lead to superfluidity. This can be understood by considering He II flowing down a capillary with velocity $\mathbf{V}$. The kinetic energy $E = 1/2\ MV^2$. For the fluid to be normal there will be momentum exchange with the capillary walls and hence excitations will be created in the fluid. In a superfluid such excitations cannot be formed and consequently there is no momentum or energy transfer and hence no viscosity. If an excitation of momentum $\mathbf{p}$ is created then the fluid velocity changes to $\mathbf{V}$, given by

$$\mathbf{V}' = \mathbf{V} - \mathbf{p}/M \qquad \text{... (9.3.17)}$$

The energy of the fluid is

$$E = \frac{1}{2} MV'^2 = \frac{1}{2} M(\mathbf{V} - \mathbf{p}/M)^2 \cong \frac{1}{2} MV^2 - \mathbf{V} \cdot \mathbf{p} = E - \mathbf{V} \cdot \mathbf{p} \qquad \text{... (9.3.18)}$$

If the excitation of momentum $\mathbf{p}$ has an energy $\varepsilon(p)$, then the energy of the fluid must have decreased by $\varepsilon(p)$. If the maximum value of $\mathbf{V} \cdot \mathbf{p}$ which is pV, is less than $\varepsilon(p)$ then the excitation cannot be formed and hence we have a critical velocity V_c given by

$$V_c = \varepsilon(p)/p\,|_{\min} \qquad \text{... (9.3.19)}$$

If the excitation curve gives a non-zero minimum value of $\varepsilon(p)/p$ then we expect that no excitations can be formed in the system for flow velocities lower than V_c and superfluidity will be observed. Clearly if $\varepsilon(p)/p$ vanishes at the origin i.e. as $p \to 0$, we will never get a finite V_c. Hence if superfluidity is to be observed, the low lying excitations must be phonons—the quanta for sound waves. The above semi-phenomenological discussion of superfluidity does not invoke the question of statistics at any point. The lesson from the above discussion is that the low-p excited states can only be sound waves for superfluidity and this restriction on the excited states ought to be a consequence of the Bose statistics—this is the final issue that we want to address.

The importance of Bose statistics for superfluidity was made plausible by Feynman who considered the problem of constructing wavefunctions for a collection of N helium atoms interacting via a pairwise potential, attractive when the atoms are separated and repulsive if the atoms touch. The argument proceeds by noting the following facts:

i) Since the Hamiltonian is real, the wavefunctions can be chosen to be real.
ii) The ground state wavefunction has no nodes.
iii) The excited state wavefunction must be orthogonal to the ground state wavefunction and must have one node.

As a starting hypothesis, we can picture the ground state wavefunctions is to be constant if the atoms are separated and zero if they touch. We denote the function by $\varphi(\mathbf{r}_1, \mathbf{r}_2 \ldots \mathbf{r}_N)$. If the first excited state wavefunction $\psi(\mathbf{r}_1, \mathbf{r}_2 \ldots \mathbf{r}_N)$, then $\int d^3r_1 \ldots d^3r_N \, \varphi(\mathbf{r}_1 \ldots \mathbf{r}_N) \, \psi(\mathbf{r}_1 \ldots \mathbf{r}_N)|$ has to be equal to zero. If we call a given distribution of $\mathbf{r}_1, \mathbf{r}_2 \ldots \mathbf{r}_N$ a configuration, then we note that over approximately half the configurations ψ must be positive (we call these configurations A) and over the other half it must be negative (we call these B). If ψ is not a phonon state then the change from A to B cannot alter the density and if it has to be a low energy state then the change must involve large distances that is large scale movements of atoms must be necessary to change A to B.

Let us divide the system into two volumes left and right—if there are more atoms to the left we have the configuration A (let us say) and if there are more to the right we have B (Fig. 9.8). Now, to convert A to B, we would need to move atoms from left to right (large scale displacements), but in the process the density will change to the reverse (high on left and low on right) (Fig. 9.8.) Thus these wavefunctions are phonon-like. If the density is not to vary, then for an A atom moved to the right, a B atom has to move to the left and the small changes in neighbouring positions, etc. are ironed out to leave the

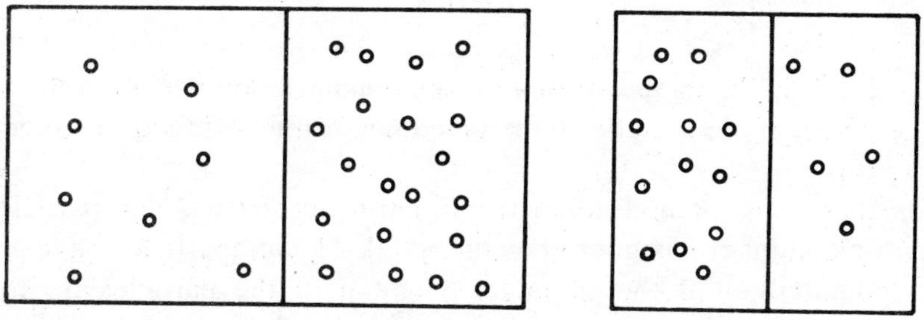

Fig. 9.8: Excited state configuration.

density unchanged. This too appears to be a large displacement and thus a low energy non-phonon state has been produced. However, this conclusion is erroneous due to Bose statistics.

The atoms are identical and obey Bose statistics, so an exchange of two atoms does not change the wavefunction. Consequently, instead of the two large distance changes, we first imagine interchanging atoms of A type that takes us close to the final configuration in the B-form and then moving them a small distance to get the final configuration. So the change from A to B requires very small movements and costs a lot of energy. The exchange would change the sign of the wavefunction if the identical particles were fermions and the above argument fails. Thus the low-lying states are phonon-like as a consequence of He atoms obeying Bose statistics.

Problems

9.1 Consider a B.E. gas of particles whose energy (ε)–momentum (p) relation is $\varepsilon = \alpha p^s$, where α is a constant and s is a variable index (usual case $s = 2$) . In a D-dimensional space, find the boundary in the s-D plane which demarcates the no-condensation region from that in which B.E. condensation occurs at a finite temperature.

9.2 For a B.E. gas in a gravitational field g, show that the condensation sets in at

$$T_c \approx T_c^{(o)} \left[1 + \frac{8}{9} \frac{1}{\Gamma(3/2)} \left(\frac{m\pi g L}{k T_c^{(o)}} \right)^{1/2} \right]$$

where $T_c^{(o)}$ is the temperature at which condensation occurs in the absence of g and L is the length of the container in the z-direction (direction of gravity).

9.3 Consider a gas of non-interacting relativistic bosons. The particle + antiparticle number has got to be conserved. The antiparticles have a chemical potential $-\mu$ if the chemical potential of the particles is μ. Write down the expression for the "number of particles" with energy $\varepsilon = (p^2c^2 + m^2c^4)^{1/2}$ (note that if "particle" count of the particles is +1, that of antiparticles is – 1). Discuss the B.E. condensation in this system, explaining clearly what is the criterion for the condensation to set in. If $mc^2 << kT_c$, find an expression for the critical temperature T_c in terms of the fundamental constants.

9.4 The Landau model treats superfluid helium at low temperatures ($\leq$ 1 K) as a weakly excited quantum sytem in which deviations from the ground state (T = 0 K) are described in terms of a "gas of elementary excitations" of the system against a quiescent background. The gas of excitations corresponds to the normal fluid and the background to the superfluid. Landau suggested the excitation spectrum shown in Fig. 9.6 on phenomenological grounds. For $p << p_0$, the spectrum is linear and hence gives a specific heat $\propto T^3$ at very low temperatures. For $p \approx p_0$, Landau called the excitations rotons and gave them the frequency spectrum

$$\varepsilon_{(p)} = \Delta + (p - p_0)^2 / 2\mu$$

The rotons are bosons and their number indefinite. Show that the specific heat of a roton gas is

$$C_r = R \left\{ \frac{3}{4} + \frac{\Delta}{kT} + \left(\frac{\Delta}{kT} \right)^2 \right\} e^{-\Delta/kT}$$

9.5 The most visible "super" quality of the fluidity of liquid ^{4}He is its ability to crawl up walls and out of a containing vessel. The reason is that it coats the wall with a thick film. At height z above the free liquid surface (in equilibrium with its own vapour), the film adhering to a vertical wall has a thickness proportional to $z^{-1/2}$. Derive this result by the following steps:

a) Suppose the temperature is such that the superfluid fraction is f. Consider an element of the adhering film at height z, of thickness d, vertical dimension Δz and unit width in the other direction. Treat the superfluid component in this volume element as a free Bose gas. Write down its energy, with contribution from potential energy due to gravity and kinetic energy due to its being confined to a thickness d.

b) Minimize the energy with respect to d.

Chapter Ten

Superconductivity

10.1 Introduction

In this chapter, we will give a fairly phenomenological picture of conventional superconductivity. Generally metals which are not very good conductors at room temperatures become superconductors when the temperature is reduced to 5–10K. We give a qualitative picture of these superconductors in view of the interest generated of late in superconductivity at considerably higher temperatures (~100K) in certain ceramic compounds. The mechanism of that superconductivity is supposed to be different from the mechanism that we will be discussing in the next section. The characteristic properties of a superconductor like tin are:

i) It behaves as if there is no D.C. electrical resistivity. Current, if set up in superconductors does not seem to decay.
ii) It behaves as a perfect diamagnet. A sample in thermal equilibrium in an applied magnetic field carries surface currents. These currents give rise to an additional field that exactly cancels the field in the interior of the specimen.
iii) The superconductor behaves as if there was an energy gap of width 2Δ centred around the Fermi level in the set of allowed one-electron levels. The gap increases as the temperature decreases and levels off to a maximum value of $\Delta(0)$ at very low temperatures.

The transition to a superconducting state is sharp in bulk specimen. Below a critical temperature T_c the superconducting properties appear, the most dramatic of which is the apparent vanishing of the D.C. resistance. Above T_c, the resistivity has the form $\rho(T) = \rho_0 + BT^5$, the constant term appearing from impurity and defect scattering and the term in T^5 arising from phonon scattering. Below T_c, these mechanisms are ineffective in dissipating the current and the resistivity abruptly drops to zero. The superconductivity can be destroyed by large magnetic fields, if the current exceeds a critical value (the size of the critical current is determined by the geometry of the sample) or by applying an A.C. electric field at sufficiently high frequencies.

Consider a superconductor at a temperature below the critical temperature T_c. As a magnetic field is applied a certain amount of energy is expended to establish the magnetic field of the screening currents so that the magnetic field inside the specimen is cancelled out. If the applied field is large enough it will be energetically favourable for the system to revert back to the normal state, allowing the field to penetrate. Although the normal state has a higher free energy than the superconducting state below T_c in zero field the increase in free energy will be compensated for by the decrease in magnetic energy as the screening currents disappear. The manner in which field penetration occurs can depend upon the material for the same geometry and it has been seen that materials can behave in two distinct ways:

i) Below a critical field $H_c(T)$ that increases as T decreases further below T_c, there is no penetration at all. For $H > H_c$, the entire sample reverts to the normal state and the field penetrates perfectly. These are called the type-I materials.

ii) Below a lower critical field $H_{c1}(T)$, there is no penetration of the field at all, while above an upper critical field $H_{c2}(T)$, the sample reverts back to the normal state with full penetration of the field. For $H_{c1} < H < H_{c2}$, there is partial penetration of the flux, and the sample develops a rather complicated microscopic structure of both normal and superconducting regions, known as the mixed state. In the mixed state (Abrikosov state), the flux penetrates the sample in the form of thin filaments of flux. Within each filament the field is high and the material not superconducting. Outside the core of the filaments, the material remains superconducting. Circulating around each filament is a vortex of screening current. These are called type-II materials.

At low temperatures, the specific heat of a metal is of the form $AT + BT^3$, where the linear term is due to electronic excitations and the cubic term is the lattice contribution. For superconductors below T_c, this form is significantly altered. The linear term is absent and instead replaced by a term whose temperature dependence is of the form $e^{-\Delta/kT}$. This is the characteristic behaviour of a system whose excited states are separated from the ground state by an energy of order Δ. Experiments show that the energy gap is of the order kT_c.

Further evidence of the existence of a gap in the energy spectrum can be had from tunneling experiments. The conduction electrons in a superconductor and a normal metal can be brought into thermal equilibrium by placing them in close contact such that there is just a thin insulating layer separating them. In thermal equilibrium, the chemical potential of the two metals will be equal and if a potential difference is now applied, the chemical potential of one will be raised with respect to the other and if both metals are in the normal state, a current will immediately flow. This is the tunnelling current. However, if one of the metals is normal and the other superconducting, the potential difference has to be raised to at least Δ before a tunneling current begins to flow.

Acoustic attenuation and A.C. electrical conductivity also provide evidence for the existence of the gap. In the next section, we will give, following Weisskopf, a qualitative description of the phenomenon of superconductivity.

10.2 Pairing Theory: Qualitative

Consider the metal as a cubic lattice of positive ions filled with a degenerate gas of free electrons at zero temperature. The lattice distance d is of the order of a few Bohr radii. If ε_A is an atomic energy of a few electron volts, then clearly

$$\varepsilon_A \sim \frac{e^2}{d} = \frac{\hbar^2}{md}\cdot\frac{me^2}{\hbar^2} = \frac{\hbar^2}{m}\cdot\frac{d}{d^2 a_B}$$

$$= \frac{\hbar^2}{md^2} f \sim \frac{\hbar^2}{md^2}. \qquad \text{... (10.2.1)}$$

where $f = d/a_B$ is a number of order unity though always somewhat greater than unity. The maximum momentum and hence velocity of the electrons is possessed by those electrons at the top of the Fermi surface

$$p_F \sim \hbar / d$$

$$v_F \sim \hbar / md \qquad \text{... (10.2.2)}$$

The mechanics of the lattice can be described as that of independent oscillation of mass M and frequency given by the Debye frequency ω_D. The potential energy of the displacement (typically d) is

$$\frac{1}{2} M \omega_D^2 \left(\frac{d}{2}\right)^2 \sim \varepsilon_A \sim \frac{\hbar^2}{m(d/2)^2}$$

$$\omega_D = b \frac{\hbar}{d^2} \frac{1}{\sqrt{mM}} \sim \sqrt{m/M}\ \varepsilon_A/\hbar \qquad \text{... (10.2.3)}$$

The number b is 8 in our estimate and indeed lies between 5 and 10 for most metals.

We now describe the effect of the motion of the electron on the lattice. Consider an electron at the Fermi surface. It spends a time $\tau \sim d/v_F$ near the ion. During this time the momentum transferred to the ion is approximately

$$\Delta_p \sim \tau . \frac{e^2}{d^2} \sim e^2 / dv_F \qquad \text{... (10.2.4)}$$

This makes the ion travel towards the electron and the displacement suffered can be estimated as

$$\delta \sim \Delta p / M \omega_D$$

$$\sim \frac{e^2}{dv_F} \frac{1}{M} \sqrt{mM/\hbar}\ d^2$$

$$= \sqrt{m/M} \frac{d}{a_B} \cdot d$$

$$\sim \sqrt{m/M}\ d \qquad \text{... (10.2.5)}$$

The ions therefore are displaced a distance δ towards the path of the electron. They do not remain in this displaced position for ever but return after a time of the order of ω_D^{-1}. It should be noted that we are at $T = 0$ and the displacement δ does not spread over the lattice in the form of sound waves. The displacement is a part of the quantum state of the electron in a lattice, different from that of the free electron. In moving through the lattice, the electron behaves like a quasi particle. The fact that the ion will return after a time ω_D^{-1} implies that the length of the region over which there are displaced ions associated with an electron is

$$l \sim v\tau \sim v/\omega_D \sim \sqrt{M/m} \cdot d \qquad \text{... (10.2.6)}$$

The displacement of the ion towards the electron implies that in this tube of length l and diameter d the potential is different from that we would find in an undeformed lattice. The potential in the absence of the deformation is e^2/d and since the separation has decreased by a distance δ, we must have a decrease in the potential by the amount

$$\Delta U = -\frac{e^2}{d^2}\,\delta \sim -\frac{e^2}{d}\sqrt{m/M} \qquad \text{... (10.2.7)}$$

A second electron entering this tube sees a screening of the field of the first electron and hence feels an effective attraction. Only the electrons near the Fermi surface can make use of such a weak interaction—the electrons below the Fermi surface are 'frozen' in that they cannot easily change their quantum states. In order to feel the effect of the potential tube fully, the electrons must meet "head-on", i.e. to say the closest distance of approach should be equal to or less than d, since d is of the order of wavelength of the electrons at the Fermi surface. This means the electrons must be in a relative S-wave state. In a higher angular momentum state, the impact parameter will be higher and the electrons would miss their mutual potential tubes.

The mutual potential then has the form (relative to the usual background)

$$V(r) \sim = -\sqrt{m/M}\,\frac{e^2}{d}\,\delta_{LO} \qquad r < 1 \qquad \text{... (10.2.8)}$$

$$\cong 0\ ,\ \text{for} \quad r < 1$$

where δ_{LO} is the δ-function that decrees that only the S-wave state is important. The two electrons must have opposite spins to satisfy Pauli principle in the S-wave state. Thus, we have the result that two electrons of opposite momenta and spins feel a weak attraction for each other at the Fermi surface.

It is this attraction that gives rise to bound states on top of the Fermi distribution. If we visualize the potential of Eq. (10.2.8) to be a trough (in D = 3) of depth U and radial extension l, then indeed there would be bound states with energy

$$\varepsilon \approx -U + \frac{\pi^2\hbar^2}{2ml^2} \qquad \text{... (10.2.9)}$$

if $U >> \hbar^2/ml^2$. Although the potential of Eq. (10.2.8) would satisfy this criterion, it is not obvious that a bound state would exist. This is because the two electrons are at the Fermi surface and have a relative kinetic energy of $2\varepsilon_F$ since $\varepsilon_F >> U$, it can be argued that they would be in the continuum and hence unbound. However, it should be noted that because of the Pauli principle, this state would be the lowest possible state and hence the above argument may not be valid. The effect, as it turns out, is that a bound state is formed although the binding energy is much weaker than that expected on the basis of Eq. (10.2.9). The potential of Eq. (10.2.8) decreases the energy of a pair of electrons at the Fermi surface from $2\varepsilon_F$ to $(2\varepsilon_F - \Delta)$ if the total momentum is zero and we are in the S-wave state (see Sec. 10.4 for the relevant demonstration using the Schrödinger equation), where

$$\Delta = 2\hbar\,\omega_D \cdot e^{-\zeta \cdot 2\pi\,\varepsilon_F/e^2 d} \qquad \text{... (10.2.10)}$$

where ζ is a numerical constant of order unity.

The pair of electrons is called a Cooper pair and Δ is their binding energy. If we define,

$$\gamma = e^{-\zeta\, 2\pi\varepsilon_F\, d/e^2} \qquad \text{... (10.2.11)}$$

then $\gamma \sim 10^{-2}$ to 10^{-3} $(\varepsilon_F\, d/e^2 \sim 1)$ and we can write

$$\frac{\Delta}{\varepsilon_A} = 2\sqrt{m/M}\;\gamma \qquad \text{... (10.2.12)}$$

making Δ roughly $10^{-4} \sim 10^{-5}$ times smaller than atomic energies. Thus, $\Delta \sim 10^{-4}$ eV which corresponds to a temperature of a few degrees Kelvin.

Since the wavefunction of a Cooper pair is an S-wave, the motion it describes is the periodic back and forth movement of the pair covering a relative distance ρ between them. For distances $r > \rho$, the wavefunction goes to zero much faster than r^{-1}. This is quite analogous to the motion of two nucleons in a deuteron or the electron and positron in the ground state of positronium. If the momentum spread of the wavefunction around the Fermi momentum (i.e. approximately the range of momentum of the free particle states which have to be superposed to give the wavefunction of the Cooper pair) is Δ_p, then the energy spread would be

$$\Delta\varepsilon \cong \Delta p \cdot p_F/m \qquad \text{... (10.2.13)}$$

and since $\Delta\varepsilon$ is of the order of the binding energy Δ,

$$\frac{\Delta p}{p_F} \sim m\Delta/p_F^2 \sim \Delta/\varepsilon_F << 1 \qquad \text{... (10.2.14)}$$

Hence the momentum spread is very small and thus for $r << \rho$, the pair wavefunction is almost sin $(\rho_F\, r/h)$. For $r > \rho$, the wavefunction vanishes due to interference of the waves with different p. The "size" ρ of the pair can then be estimated as

$$\rho \sim \frac{\hbar}{\Delta p} \sim \frac{\hbar\, p_F}{m\Delta} \sim \frac{p_F^2\, d}{m\Delta} \sim \frac{\varepsilon_F}{\Delta} \cdot d >> d \qquad \text{... (10.2.15)}$$

The wavefunction is thus tremendously spread out which is a characteristic of the low binding energy. The large extension of the wavefunction is responsible for the absence of the influence of electrostatic repulsion on the binding. For distances somewhat larger than d, the electron interactions are screened out by the electron gas. Since $\rho >> d$, the effect of the repulsion which is confined to distances smaller than a few d, changes the wavefunction only near the centre and the extension of the binding potential being $l >> d$, the binding energy is largely unaffected. The length ρ is called the coherence length and can be of the order of 10^{-4} cm.

We now turn to the electron gas as a whole. Cooper pairs are formed only among those electrons that lie in the uppermost part of the Fermi distribution, not lower than Δ below the surface ε_F in energy, or Δp below p_F in momentum. Electrons lying lower than that cannot make use of the empty states to form a Cooper pair wavefunction. All Cooper pairs have momentum $p = 0$ and spin $S = 0$. That puts them in the same quantum state. This does not violate any statistics since the two-electron system obeys Bose statistics.

The number n' of Cooper pairs can be estimated from the number of electrons n (number per unit volume) which would be of $O\left(\frac{1}{d^3}\right)$

$$n' = n\frac{\Delta}{\varepsilon_F} \sim \frac{\Delta}{d^3\varepsilon_F} \sim \frac{1}{d^2\rho} \qquad \text{... (10.2.16)}$$

The distance D between Cooper pairs is

$$D \sim \left(\frac{1}{n'}\right)^{1/3} \sim d\left(\frac{\rho}{d}\right)^{1/3} \sim (30\text{–}40)\, d \qquad \text{... (10.2.17)}$$

The size of a Cooper pair is much larger than that and hence the Cooper pairs must necessarily overlap. However, it is important to note that two Cooper pair wavefunctions are still orthogonal. The Cooper pair wavefunction is also orthogonal to the wavefunctions of the free electrons which are more than Δ below the Fermi level. The interaction between the electrons discussed above thus changes the Fermi distribution to an ensemble of Cooper pairs among the n' electrons in the Δ-layer and of free electrons below the Δ-layer.

We next demonstrate that a finite amount of energy of the order of Δ is required to reach the first excited state of this ensemble. For a free-electron Fermi gas, the excited states are infinitely near to the ground state—they are the states in which one or more electrons are moved into a level slightly higher than the Fermi level. These excited states carry electric current. The existence of a finite energy required to reach the first excited state in the present case can be seen from the fact that if a free electron is to be taken outside the Fermi surface, it needs to be given an energy of O (Δ) and if an electron from the Δ-layer is to be taken across the Fermi level, the pair has to be broken requiring an energy of order Δ. To complete this argument, one has to understand what prevents a simple Cooper pair from changing its motion from the momentum state $p = 0$ to a very small momentum p' without breaking up. If $p'^2/m \ll \Delta$, then this would represent an excited state of the ensemble with energy much lower than Δ. However, this is impossible since it would require that the free electron states making up the Cooper pair change to contain pairs of free electron levels for which the total momentum is p'. Since all electrons in the Δ-layer are paired, the change of motion of one pair would affect all the other pairs. The only way in which one Cooper pair can change its momentum from $p = 0$ to p' is by having all the Cooper pairs change simultaneously and that would cost an energy much larger than Δ.

One must now consider how an ensemble of electrons with zero total momentum $p = 0$ ever produces a current and even a supercurrent without any resistance at all. It might seem that in a state $p = 0$, there is no net motion of charge. The reason that a current can be carried is that in the presence of a magnetic field, the velocity is no longer proportional to the momentum. The velocity is obtained from the kinetic energy by differentiation with respect to the momentum p and since the kinetic energy in the presence of a magnetic field characterized by a vector potential is $\frac{1}{2m}\left(\mathrm{p} - \frac{e\mathrm{A}}{c}\right)^2$ we find for the velocity,

$$\mathbf{v} = \frac{1}{m}(\mathbf{p} - e\,\mathbf{A}/c) \qquad \text{... (10.2.18)}$$

The current density produced by the ensemble of electrons is

$$\mathbf{j} = e\,\Sigma\,\mathbf{v}_i = \frac{e}{m}\Sigma\,p_i - \frac{e^2}{mc}\,n\mathbf{A}$$

$$= -\frac{e^2 n}{mc}\,\mathbf{A} \qquad \text{... (10.2.19)}$$

since $\Sigma\, p_i$ = total momentum = 0 and n is the total number of electrons per unit volume. We can write

$$\mathbf{j} = -\frac{e^2 n}{mc}\,\mathbf{A} = -\frac{\mathbf{A}}{\lambda^2} \qquad \text{... (10.2.20)}$$

where

$$\lambda = \left(\frac{m\,c^2}{n\,e^2}\right)^{1/2} = \left(\frac{m\,c^2 d}{e^2}\right)^{1/2} d = \left(\frac{d}{a_B}\right)^{1/2} \frac{\hbar\,c}{e^2}\,d \qquad \text{... (10.2.21)}$$

Now, the fine structure constant $e^2/\hbar c \sim 1/137$ and hence the length λ is of the order of hundred times the lattice distance.

The current is supported because of the energy gap. The current above is just the induction current set up when the magnetic field is switched on. The kinetic energy (T) increases when the field is switched on

$$T = \frac{1}{2m} \sum_i \left(\mathbf{p}_i - \frac{e\mathbf{A}}{c} \right)^2$$

$$= \frac{1}{2m} \Sigma p_i^2 + \frac{e^2 n}{2mc^2} A^2 \qquad \text{... (10.2.22)}$$

($\Sigma\ \mathbf{p}_i \cdot \mathbf{A} = 0$ since $\Sigma\ \mathbf{p}_i = 0$). If the metal were non-superconducting, this increase of kinetic energy would be compensated by changing the momentum of each electron by $e\mathbf{A}/c$ so that the kinetic energy reverts to its original value within a short relaxation time. In the superconductor, the Cooper pairs at the crust prevent such a redistribution—momentum redistribution costs energy and hence the situation with altered kinetic energy has to persist and thus the current does not decay.

However, it must be pointed out that a gap in the spectrum is not the sole criterion for persistent current. Had that been so, insulators would have shown superconducting behaviour as well. The reason an insulator does not behave in a similar fashion is because the free particle kinetic energy expression is no longer correct at the top of a band. If we consider a one-dimensional situation for simplicity, then $T(p)$ is like $p^2/2m$ for the lower part of the band. Near the upper end of the band the function $T(p)$ has a point of inflection and the tangent is horizontal at the upper end, i.e. $dT/dp = 0$ at $p = \pm\ p_{max}$. Now, in the absence of the magnetic field

$$j = e\Sigma \frac{dT}{dp} = e \int_{-p_{max}}^{p_{max}} \frac{dT}{dp} \cdot g(p)\ dp = 0 \qquad \text{... (10.2.23)}$$

since $g(p)$ the density of states is an even function and dT/dp is an odd function of p.

In the presence of a magnetic field, we have the kinetic energy as some function of $(p - eA/c)$ and expanding as a power series of the potential A, which is usually small

$$T(p - eA/c) = T(p) - eA/c \frac{dT}{dp} + \ldots$$

The current is

$$j = e \int_{-p_{max}}^{p_{max}} \frac{dT}{dp} g(p)\, dp = e \int_{-p_{max}}^{p_{max}} \frac{dT(p)}{dp} g(p)\, dp$$

$$- \frac{e^2 A}{c} \int_{-p_{max}}^{p_{max}} \frac{d^2T}{dp^2} g(p)\, dp$$

$$= - \frac{e^2 Ag}{c} \int_{-p_{max}}^{p_{max}} \frac{d^2T}{dp^2} dp$$

$$= - \frac{e^2 Ag}{c} \left. \frac{dT}{dp} \right|_{-p_{max}}^{p_{max}} = 0 \qquad \text{... (10.2.24)}$$

We have used the fact that a one-dimensional density of states is p-independent and that at the top of the band $dT/dp = 0$. Thus, there is no current in the presence of a magnetic field.

We now show that the magnetic field does not penetrate into the superconductor. The current being given by Eq. (10.2.24), we have the magnetic field as being given by

$$\nabla \times \mathbf{B} = \mathbf{j} = - \frac{\mathbf{A}}{\lambda^2}$$

Another curl operation leads to

$$\nabla^2 \mathbf{B} = \mathbf{B}/\lambda^2$$

showing that the field dies down exponentially from the surface.

Consider an infinite superconductor with a plane surface. In a rectangular coordinate system, the space $X > 0$ is filled with superconducting metal, the plane defined by $X = 0$ is the surface. Let us imagine that there is a magnetic field B_0 for $X > 0$ in the y-direction. Inside the material

$$B_y = B_0 \, e^{-X/\lambda} \quad (X > 0)$$

and $B_x = B_z = 0$. The corresponding vector potential is

$$A_z = - \lambda \, B_0 \, e^{-X/\lambda} \quad (X > 0)$$

with $A_x = A_y = 0$

$$j_z = -A_z/\lambda^2 = \frac{B_0}{\lambda}\, e^{-X/\lambda} \quad (X > 0)$$

with $j_x = j_y = 0$. Thus there is a current in the z-direction but confined within a depth λ from the top. The total current flowing across a rectangle of depth λ and width S in the y-direction is clearly $B_0/\lambda \cdot \lambda S = S B_0$ (we have replaced $e^{-X/\lambda}$ by 1 for $X < \lambda$ and by zero for $X > \lambda$).

Finally consider a superconducting wire with its axis in the z-direction, with a circular magnetic field B_0 around the wire, parallel to its surface and perpendicular to the z-axis. We conclude from the preceding discussion that there will be a current j in the wire flowing in the z-direction of strength

$$j = 2\pi R B_0$$

The current is restricted to a depth λ from the surface of the wire. This current produces a circular magnetic field B_0 at the radius R and the situation is self-consistent and stable. The current produces the field B_0 which, in turn, causes the same current to appear in the "frozen" electron gas, kept in the ground state by the Cooper pair crust at the top of the Fermi surface, so long as $kT < \Delta$. The electrons at a depth greater than λ from the surface do not move as there is no magnetic field there, while the electrons within a depth λ move keeping the total momentum zero. This state of affairs does not continue for all values of B_0. As the magnetic field is increased, the kinetic energy of the electrons increases, and if this increment can break the Cooper pairs, then the electrons will redistribute and get rid of the excess energy by losing the kinetic energy, i.e. stopping the current. If we consider a unit area on the surface of the wire, then the additional kinetic energy in a volume $V = \lambda \cdot 1$ is

$$\Delta T = V \frac{ne^2}{2mc^2} A^2 = V \frac{A^2}{2\lambda^2}$$

The current will be stable if ΔT is less than the total energy G of the binding of Cooper pairs

$$G = V \Delta n' = V \cdot \Delta \cdot n \cdot \Delta/\varepsilon_F$$

where n' is the number of Cooper pairs per unit volume. The critical magnetic field B_{cr} is given by $B_{cr} \sim A_{cr}/\lambda$ and hence

$$\frac{V}{2} B_{cr}^{2} \sim V \cdot n\, \Delta^2/\varepsilon_F$$

$$B_{cr}^{2} \sim 2n\, \Delta^2/\varepsilon_F$$

With $\Delta/\varepsilon_F \sim 10^{-4}$ and $\varepsilon_F \sim 5$ eV, B_{cr} turns out to be around 300 G. A superconductor with a critical field of this order is termed a type I superconductor.

This essentially concludes the brief survey of superconductivity. We have not attempted to touch upon the topics of flux quantization or Josephson tunneling. Our focus has been on the pairing mechanism of conventional superconductivity. Accordingly, in the two remaining sections of this chapter we will attempt to quantify the attractive interaction and demonstrate that in the presence of the attraction, two electrons near the Fermi surface will always form a bound state.

10.3 Origin of the Attractive Interaction

In this section, we attempt to quantify the heuristic argument about the attractive interaction between the two electrons. To do so, we work in terms of the dielectric constant and obtain the interaction between two electrons in momentum space in the form

$$V(k, \omega) = \frac{e^2}{\varepsilon(k, \omega)\, k^2} \qquad \text{... (10.3.1)}$$

where $\varepsilon(k, \omega)$ is the momentum and frequency dependent dielectric constant. The effect that we want to incorporate is the strong reaction of the lattice to the electron and the screening of the electron charge by the ion cloud. It is this screening which is capable of overcompensating the electron charge. A second electron coming along may now see a positive charge and feel attracted.

To calculate $\varepsilon(k, \omega)$, we consider an external charge $\delta\rho$, having the Fourier transform

$$\delta\rho(r) = \delta\rho \cos(k \cdot r)\, e^{i\omega t} + c.c. \qquad \text{... (10.3.2)}$$

If the screening charge formed in response to $\delta\rho$ is $\rho(r)$, with the Fourier transform

$$\rho(r) = \rho \cos(k \cdot r)\ e^{i\omega t} + c.c. \qquad \text{... (10.3.3)}$$

then the dielectric constant $\varepsilon(k, \omega)$ is

$$\varepsilon(k, \omega) = \frac{\delta\rho}{\rho + \delta\rho} \qquad \text{... (10.3.4)}$$

We need to calculate the response charge ρ, which will have contributions from ions as well as electrons denoted by ρ_i and ρ_e respectively. Clearly

$$\rho = \rho_i + \rho_e$$

The total charge sets up the potential V, which satisfies

$$\nabla^2 V = -(\delta\rho + \rho_i + \rho_e) \qquad \text{... (10.3.5)}$$

In the resulting electric field $\mathbf{E} = -\nabla V$, the ions move according to the equation of motion

$$M\ \mathbf{J}_i = nZe^2\ \mathbf{E} \qquad \text{... (10.3.6)}$$

where J_i is the ionic current, n the number density of ions and z the ionic charge. The ionic current satisfies the continuity equation

$$\frac{\partial \rho_i}{\partial t} + \nabla \cdot \mathbf{J}_i = 0 \qquad \text{... (10.3.7)}$$

Combining Eqs. (10.3.5)–(10.3.7), we get

$$\frac{\partial^2 \rho_i}{\partial t^2} = \frac{nZe^2}{M} \nabla^2 V = -\frac{nZe^2}{M}(\delta\rho + \rho_i + \rho_e)$$

In Fourier space,

$$\omega^2\ [\rho_i(\mathbf{k},\omega)] = \omega_i^2\ [\delta\rho(\mathbf{k},\omega) + \rho_i(\mathbf{k},\omega) + \rho_e(\mathbf{k},\omega)] \qquad \text{... (10.3.8)}$$

where $\omega_i^2 = \dfrac{nZe^2}{M}$. (Typically, $\omega_i \approx 10^{13}/\text{s}$)

One now needs to know the response of the electrons alone. The characteristic frequencies involved with electrons are i) the plasma frequency $\omega_e = \omega_i \sqrt{M/m}$ and ii) the frequency associated with the Fermi energy

$\omega_F = \varepsilon_F / h$. Both ω_e and ω_F are much greater than ω_i and hence for frequencies ω close to ω_i, the electron gas is virtually static. We calculate the response in what is known as the Thomas-Fermi approximation. We recall from Chapter 7 that the Fermi energy of an electron gas is proportional to $n^{2/3}$, where n is the number density. We now assume that in an external potential $V(r)$, the Fermi energy is raised to $\varepsilon_F + eV(r)$ and the resulting density $n_e(r)$ is proportional to $[\varepsilon_F + eV(r)]^{3/2}$. We write $n_e(r) = ne + \rho_e(r)$ and expanding in $\rho_e(r)$ and $eV(r)$

$$\frac{\rho_e(r)}{-ne} = \frac{3}{2}\frac{eV}{\varepsilon_F} \qquad \text{... (10.3.9)}$$

Using Eq. (10.3.5) in momentum space

$$\rho_e(k) = -\frac{k_s^2}{k^2}(\rho + \delta\rho) \qquad \text{... (10.3.10)}$$

where

$$k_s^2 = \frac{3}{2}\frac{ne^2}{\varepsilon_F}$$

Combing Eqs. (10.3.8) and (10.3.10),

$$\rho = \rho_i + \rho_e = \left(\frac{\omega_i^2}{\omega^2} - \frac{k_s^2}{k^2}\right)(\rho + \delta\rho) \qquad \text{... (10.3.11)}$$

This leads to

$$\varepsilon(k,\omega) = \frac{\omega^2(k_s^2 + k^2) - \omega_i^2 k^2}{\omega^2 k^2}$$

Writing,

$$\omega_k^2 = \omega_i^2 \frac{k^2}{k^2 + k_s^2}$$

$$\frac{1}{\varepsilon(k,\omega)} = \frac{\omega^2 k^2}{(k_s^2 + k^2)[\omega^2 - \omega_k^2]}$$

$$= \frac{k^2}{k^2 + k_s^2} \left[1 + \frac{\omega_k^2}{\omega^2 - \omega_k^2} \right] \quad \text{... (10.3.12)}$$

For $\omega \leq \omega_k$, $\varepsilon (k,\omega)$ can become negative signalling an attraction.

10.4 Cooper Instability

In this section, we demonstrate using the Schrödinger equation, the formation of the bound state of two electrons at the Fermi surface through a weak attractive interaction. We begin with a degenerate gas of N free electrons at $T = 0$, with the Fermi energy given by ε_F and the Fermi momentum by k_F. Two electrons at the Fermi surface are taken to interact with a weak attractive intereaction. Their centre of mass is at rest. The two-particle Schrödinger equation takes the form

$$-\frac{h^2}{2m} (\nabla_1^2 + \nabla_2^2)\, \psi + V(\mathbf{r}_1 - \mathbf{r}_2)\, \psi = E\, \psi \quad \text{... (10.4.1)}$$

We will measure the energy relative to the Fermi energy and hence write $E = 2\varepsilon_F + \varepsilon_b$. Further, it will be convenient to work in the momentum space and hence we introduce the momentum space wavefunction $\varphi (k)$ as

$$\psi (\mathbf{r}_1, \mathbf{r}_2) = \psi (\mathbf{r}_1 - \mathbf{r}_2) = \int \varphi (k)\; e^{i\mathbf{k}.(\bar{r}_1 - \bar{r}_2)} \frac{d^3k}{(2\pi)^3} \quad \text{... (10.4.2)}$$

Translation invariance implies that $\psi (\bar{r}_1, \bar{r}_2)$ is a function of $\mathbf{r}_1 - \mathbf{r}_2$ above. The filled Fermi sphere immediately puts the following vital restriction on $\varphi (k)$

$$\varphi (\mathbf{k}) = 0 \text{ for } |\, k \,| < k_f \quad \text{... (10.4.3)}$$

The states inside the Fermi sphere are not available to the two electrons because of the exclusion principle. In momentum space, Eq. (10.3.1) becomes

$$\frac{h^2}{2m} (k^2 + k^2)\; \varphi (\mathbf{k}) + \sum_{\mathbf{k}'} V_{\mathbf{k}\mathbf{k}'}\, \varphi (\mathbf{k}') = (2\varepsilon_F + \varepsilon_b)\, \varphi (\mathbf{k})$$

or,

$$\left(\frac{h^2}{m}k^2 - 2\varepsilon_F - \varepsilon_b\right)\varphi(\mathbf{k}) = -\sum_{\mathbf{k}'} V_{\mathbf{kk}'}\,\varphi(\bar{k}') \qquad \text{... (10.4.4)}$$

We now have to take a form for $V_{\mathbf{kk}'}$. As seen qualitatively in Sec. 10.2 the attractive interaction has a long range in coordinate space. Accordingly, it is short ranged in momentum space and exists over a range $\hbar\omega_D$ (ω_D being the Debye frequency—ν_m of Sec. 4.5) above the Fermi surface.

In that range, we take it to be a constant V_0, so that

$$V_{\mathbf{kk}'} = -V_0 \text{ if } \mathbf{k} < \mathbf{k}', \text{ such that } \varepsilon_F < \frac{h^2k^2}{2m}, \frac{h^2k'^2}{2m} < \varepsilon_F + \hbar\,\omega_D \qquad \text{... (10.4.5)}$$

The negative sign comes from the fact that it is an attraction. We can now write Eq. (10.4.4) as

$$\varphi(\mathbf{k}) = \frac{V_0 C}{\frac{h^2k^2}{m} - 2\varepsilon_F - \varepsilon_b} \qquad \text{... (10.4.6)}$$

where C is a constant given by

$$C = \sum_{k'} \varphi(\bar{\mathbf{k}}'), \qquad \text{... (10.4.7)}$$

the summation being over the range where $V_{\mathbf{k},\mathbf{k}'} \neq 0$. From Eqs. (10.4.6) and (10.4.7)

$$C = \sum_{\mathbf{k}} \varphi(\mathbf{k}) = V_0\, C\sum_{\mathbf{k}} \frac{1}{\frac{h^2k^2}{m} - 2\varepsilon_F - \varepsilon_b} \qquad \text{... (10.4.8)}$$

The sum over **k**, brings in a phase space factor $\frac{4\pi k^2\,dk}{(2\pi)^3} = N(\varepsilon)\,d\varepsilon$, where $N(\varepsilon)$ is the density of one electron level for a given spin. In terms of the energy variable measured from the Fermi level $\left(\varepsilon = \frac{h^2k^2}{2m} - \varepsilon_F\right)$ Eq. (10.4.8) becomes

$$1 = V_0 \int_0^{h\omega_D} \frac{N(\varepsilon)\, d\varepsilon}{2\varepsilon - \varepsilon_b} \qquad \text{... (10.4.9)}$$

Over the narrow range of integration shown in the above equation, $N(\varepsilon)$ is approximately constant and equal to $N(0)$—the density of states at the Fermi level. Thus,

$$1 = \frac{N(0)\, V_0}{2} \ln \left(1 - \frac{2h\omega_D}{\varepsilon_b}\right) \qquad \text{... (10.4.10)}$$

leading to

$$\varepsilon_b = \frac{-2\hbar\, \omega_D\, e^{-\frac{2}{N(0)V_0}}}{1 - e^{-\frac{2}{N(0)V_0}}} \qquad \text{... (10.4.11)}$$

which in the weak coupling limit $N(\omega)\; V_0 << 1$, becomes

$$\varepsilon_b \approx -\, 2h\omega_D\, e^{-\frac{2}{N(0)\, V_0}} \qquad \text{... (10.4.12)}$$

The negative sign indicates that the electrons form a bound state. The bound state is formed regardless of how small the potential V_0 is. This is the unusual feature which comes from the blocking off of a certain region of momentum space by the filled Fermi sphere. It should be noted that Eq. (10.4.12) cannot be obtained in perturbation theory.

Chapter 11

Boltzmann Transport Equation

11.1 Introduction

The macroscopic state of a system in equilibrium is characterized by a few parameters. The state of a gas contained in a box is determined by its pressure, volume and temperature while the temperature, magnetization and volume characterize a piece of magnetic material. In an equilibrium state, these characterizing quantities do not change with time. What is more, if an unforced system happens to be in a non-equilibrium state, the general tendency of the system is to approach an equilibrium state. In the case of a magnetic material at very high temperatures, we know from observation that the equilibrium state is one in which the system shows no net magnetization. The material is composed of individual molecules which carry magnetic moments, all having the same magnitude but capable of pointing in any direction. The zero magnetization at high temperatures is produced by an equal probability of the molecules to orient themselves in all directions. Now suppose that one forcibly produces a state of non-zero magnetization, for example by applying a strong external magnetic field and then removing it. The state with finite magnetization thus produced is a non-equilibrium state. If one watches it over a period of time, the state evolves and eventually the equilibrium state with zero magnetization is achieved. This equilibration is brought about by the magnetic interaction between the individual molecules which causes the reorientation of the moments.

As another example we may consider one mole of a gas confined to occupy half of a given volume V at a maintained temperature T. If the constraint, e.g. a partition which divides V into two parts, is removed then we have a non-equilibrium situation where the whole volume V is now available to the gas at temperature T, but its presssure it nearly double than that would be obtained in an equilibrium situation. Once again, if one waits for the system to evolve, the collisions between the gas molecules help bring about equilibrium and the final state is characterized by the temperature T, volume V and the pressure appropriate to that obtained from the thermodynamic equation of state. The equilibrium is brought about by interaction. Collisions are caused only because the molecules interact with each other (however weakly) and can be understood only on the microscopic scale.

The above examples make it clear that the equilibrium state is one which does not change with time. This is where the question of time scales comes in. It is virtually impossible to talk about a state of absolute equilibrium—one that would never change with time—it is proper to talk about a time τ_s for a particular state to change and the time τ_o over which the state is being observed. The observed state can be considered in equilibrium only if $\tau_s >> \tau_o$. A trivial example is an amount of water in a beaker which for most practical purposes can be taken to be in equilibrium and concepts of equilibrium thermodynamics applied to it (e.g. at any given instant its temperature can be measured). If, however, one observes over a periods of days, the water level in the beaker will fall due to evaporation and hence over this prolonged period of observation the state is not an equilibrium one. The other point to note it that the approach to equilibrium is a consequence of interactions at a microscopic level and any attempt to describe the approach to equilibrium must proceed at the microscopic level. For a gas of N molecules in a volume V, this elaborate description was achieved by Boltzmann and this is what we will describe in the next few chapters.

11.2 Distribution Function

As discussed in the previous section, the approach to equilibrium can be understood on a microscopic level. If we consider N molecules of a gas in a volume V, then N is of the order of Avogadro number which is 10^{23} and hence specifying the dynamics of all the N particles is virtually impossible task. One has to opt for a coarser description and this involves specifying the number

$dN(\bar{r}, \bar{p}, t)$ of molecules lying within a momentum range **p** to **p** + d**p** and a spatial range r to **r** + d**r**. A distribution fuction f(**r**, **p**, t) is the density in phase space defined by

$$dN = f\, d^3r\, d^3p \qquad \text{... (11.2.1)}$$

with the normalization

$$N = \int dN = \int f(\mathbf{r}, \mathbf{p}, t)\, d^3r\, d^3p \qquad \text{... (11.2.2)}$$

The volume element $d^3r\, d^3p$ in the coordinate and momentum space is an infinitesimal when compared to macroscopic dimensions, but certainly large compared to atomic dimensions and thus we expect the function f (related to the physical density) to be smooth if we are not concerned about atomic dimensions. We can generally expect the function f to the spatially uniform (especially if external forces are absent) and then the integration over d^3r produces the total volume and we have

$$N = V \int f(\mathbf{p}, t)\, d^3p \qquad \text{... (11.2.3)}$$

Thus $\int f(\mathbf{p}, t)\, d^3p$ is the time-dependent particle density.

Equilibrium corresponds to the $\underset{t \to \infty}{\text{Lt}}\ f(\mathbf{r}, \mathbf{p}, t)$ if it exists, whence $f(\mathbf{r}, \mathbf{p}, t)$ presumably becomes independent of time. The task of the theory is to construct an equation of motion for the function f. This, in principle, is a much simpler task than solving the $6N$ Hamilton's equation of motion associated with the N particles. This simplifying point of view where the microscopic dynamics is given a coarser description is a sort of bridge between the microscopic and macroscopic description. As was seen in Chapter 2, it is the equilibrium distribution function that, together with a linking hypothesis, allows for the calculation of all thermodynamic properties of all systems. In this chapter we intend to obtain the equation for the time development of f(**r**, **p**, t).

The expected structure of the time development equation is

$$\frac{\partial f}{\partial t} = Lf \qquad \text{... (11.2.4)}$$

where L is an operator which can be of any sort—in general non-linear. The equilibrium distribution f_0 corresponds to

$$\frac{\partial f_0}{\partial t} = Lf_0 = 0 \qquad \text{... (11.2.5)}$$

which is a fixed point of the time development of f. The equilibrium is stable only if a small perturbation from the equilibrium state leads to a time development where the perturbation dies away or keeps oscillating about the zero value. This dynamics of the perturbation can be obtained from a linear stability analysis where one sets $f = f_0 + \delta f$ and linearizes the right hand side of Eq. (11.2.4) about f_0 to write

$$\frac{\partial}{\partial t}\,\delta f = K\,\delta f \qquad \text{... (11.2.6)}$$

where K is now a linear operator. The solution can lead to oscillatory behaviour which generally corresponds to wave propagation or relaxational behaviour which leads to damping coefficients like viscosity, thermal conductivity, etc. We now proceed to find the equation of motion for f.

11.3 In the Absence of Collisions

In the simplest situation, all interaction between the molecules is ignored. An external force $\mathbf{F}$ is supposed to act which may be uniform or may depend on position. If the position and momentum of the ith particle at time t is $\mathbf{r}_i$ and $\mathbf{p}_i$ respectively, and if the external force at the position of the ith particle is $\mathbf{F}_i$, then in an interval δt, the position and momentum are altered to

$$\mathbf{r}'_i = \mathbf{r}_i + \frac{\mathbf{p}_i}{m}\,\delta t \qquad \text{... (11.3.1a)}$$

$$\mathbf{p}'_i = \mathbf{p}_i + \mathbf{F}_i\,\delta t \qquad \text{... (11.3.1b)}$$

where $\mathbf{r}'_i$ and $\mathbf{p}'_i$ are respectively the position and momentum at the time $t + \delta t$. If we consider a volume element $d^3\mathbf{r}\;d^3p$ about $\mathbf{r}$ and $\mathbf{p}$, then the number of particles in this volume element is

$$dN = f(\mathbf{r}, \mathbf{p}, t)\,d^3r\,d^3p \qquad \text{... (11.3.2)}$$

In a time δt, the volume element $d^3r\,d^3p$ changes to the element $d^3r'\,d^3p'$ where $\mathbf{r}'$ and $\mathbf{p}'$ are given by Eqs. (11.3.1a) and (1.3.1b). In the absence of collisions, all the dN particles in $d^3r\, d^3p$ at time t are transferred to the volume $d^3r'\, d^3p'$ at time $t + \delta t$. The distribution function is now given by

$$dN = f(\mathbf{r}', \mathbf{p}', t + \delta t)\, d^3r'\, d^3p' \qquad \text{... (11.3.3)}$$

From Eqs. (11.3.2) and (11.3.3), we see

$$f(\mathbf{r}, \mathbf{p}, t)\, d^3 r\, d^3 p = f(\mathbf{r}', \mathbf{p}', t + \delta t)\, d^3r'\, d^{3\prime} p' \qquad \text{... (11.3.4)}$$

To see how $d^3r\, d^3p$ is related to $d^3r'\, d^3p'$, we need only consider the element $dxdp_x$. Here x, y and z are the components of $\mathbf{r}$ and p_x, p_y, p_z are the components of $\mathbf{p}$. Since

$$x' = x + \frac{p_x}{m}\, \delta t$$

and

$$p'_x = p_x + F_x\, \delta t$$

$$dx'\, dp'_x = \begin{vmatrix} \dfrac{\partial x'}{\partial x} & \dfrac{\partial x'}{\partial p_x} \\ \dfrac{\partial p'_x}{\partial x} & \dfrac{\partial p'_x}{\partial p_x} \end{vmatrix} dx\, dp_x = \begin{vmatrix} 1 & \delta t/m \\ \dfrac{\partial F_x}{\partial x}\delta t & 1 \end{vmatrix} dx\, dp_x$$

$$= \left[1 + \frac{1}{m}\frac{\partial F_x}{\partial x}(\delta t)^2\right] dx\, dp_x \qquad \text{... (11.3.5)}$$

Dropping all terms of $O\,((\delta t)^2)$, we arrive at

$$dx'\, dp'_x = dx\, dp_x \qquad \text{... (11.3.6)}$$

Identical arguments lead to $dy'\, dp'_y = dy\, dp_y$ and $dz\, dp'_z = dz\, dp_z$, so that $d^3r\, d^3p' = d^3r\, d^3p$, and consequently Eq. (11.3.4) becomes

$$f(\mathbf{r}, \mathbf{p}, t) = f(\mathbf{r}', \mathbf{p}', t + \delta t)$$

$$= f(\mathbf{r}, \mathbf{p}, t) + \frac{\partial f}{\partial t}\partial t + \frac{\partial f}{\partial \mathbf{r}} \cdot \frac{\mathbf{p}}{\mathbf{m}}\delta t + \frac{\partial f}{\partial \mathbf{p}} \cdot \mathbf{F}\, \delta t$$

or

$$\frac{\partial f}{\partial t} + \frac{\mathbf{p}}{\mathbf{m}} \cdot \frac{\partial f}{\partial \mathbf{r}} + \mathbf{F} \cdot \frac{\partial f}{\partial \mathbf{p}} + 0 \qquad \text{... (11.3.7)}$$

Thus in the absence of collisions, the distribution function satisfies a linear evolution equation. In the next section we consider the effect of collisions.

11.4 In the Presence of Collisions

Interactions between molecules which lead to collisions can cause molecules to leave and enter the volume element $d^3r\,d^3p$ that we have been considering. The gas is considered to be sufficiently dilute so that only binary interactions need to be considered. At large distances attractive interaction is sufficiently weak so that we can ignore it. At close range, namely at a separation of the order of molecular diameter, the interaction is very strongly repulsive so that the entire effect of the interaction between two molecules manifests itself in the form of elastic collisions. In what follows, we will ignore all quantum effects, i.e. the de Broglies wavelength $\lambda \sim \hbar / \sqrt{mk_BT}$ is much smaller than the typical intermolecular separation $(V/N)^{1/3}$. For a dilute gas this condition is easily satisfied. In the presence of collisions, if molecules enter or leave the volume element at the rate $|\partial f/\partial t|_{coll}$, then Eq. (11.3.7) becomes

$$\frac{\partial f}{\partial t} = \frac{\partial f}{\partial t} + \frac{\mathbf{p}}{m} \cdot \frac{\partial f}{\partial \mathbf{r}} + \mathbf{F} \cdot \frac{\partial f}{\partial \mathbf{p}} = \left.\frac{\partial f}{\partial t}\right|_{coll} = \left.\frac{\partial f}{\partial t}\right|_{in} - \left.\frac{\partial f}{\partial t}\right|_{out} \qquad \text{... (11.4.1)}$$

where $|\partial f/\partial t|_{in}$ is the contribution from collisions which cause molecules to enter the volume $d^3r\,d^3p$ and $|\partial f/\partial t|_{out}$ is the contribution from collisions which cause the molecules to leave the volume $d^3r\,d^3p$.

We consider molecules with velocity $\mathbf{v}$ to collide elastically with molecules of velocity $\mathbf{v}_1$ and acquire velocities $\mathbf{v}_2$ and $\mathbf{v}_3$ respectively after the collision. This process causes loss of molecules from the volume element $d^3r\,d^{3}p$. The reverse process where a molecule of velocity $\mathbf{v}_2$ strike a molecule of velocity $\mathbf{v}_3$ producing velocities $\mathbf{v}$ and $\mathbf{v}_1$ after the collision. causes an increase in the number of molecules in the volume $d^3r\,d^3p$. The first process will be characterized by the cross section $\sigma(\mathbf{v}, \mathbf{v}_1 \rightarrow \mathbf{v}_2, \mathbf{v}_3)$ and the second by the cross-section $\sigma(\mathbf{v}_2, \mathbf{v}_3 \rightarrow \mathbf{v}, \mathbf{v}_1)$. So far as the spatial part is concerned the whole process occurs in the volume d^3r. The number of molecules with velocity between $\mathbf{v}$ and $\mathbf{v} + d\mathbf{v}$ in the volume d^3r is $f(\mathbf{r}, \mathbf{v}, t)\, d^3r\, d^3v$ (Instead of the momentum it is simpler to consider the velocity as the variable in what follows). The number of possible collisions per unit time per unit area of these molecules with those with velocity $\mathbf{v}_1$ is determined by their own number and the number of molecules with velocity between $\mathbf{v}_1$ and $\mathbf{v}_1 + d\mathbf{v}_1$ lying in the physical volume with unit cross-sectional area and height $|\mathbf{v} - \mathbf{v}_1|$. The number of such molecules is $f(\mathbf{r}, \mathbf{v}_1, t)\, |\mathbf{v} - \mathbf{v}_1|\, d^3 v$, and hence the number of possible collisions per unit time (the incident flux) is $f(\mathbf{r}, \mathbf{v}, t) f(\mathbf{r}, \mathbf{v}_1, t)\, |\mathbf{v} - \mathbf{v}_1|\, d^3r\, d^3v\, d^3v_1$.

At this point, an assumption creeps in almost unnoticed. The two varieties of molecules—those with velocity $\mathbf{v}$ and those with velocity $\mathbf{v}_1$—are uncorrelated in the sense that the joint probability distribution of these molecules has been taken as the product of the individual probability distributions. This lack or correlation is an assumption—generally called the assumption of molecular chaos. This assumption is at the heart of the success of statistical mechanics and as seen is Chapter 2, the molecular dynamics is such that the lack of correlation is a plausible dynamical state.

The elastic collisions change the direction of relative velocity of the two molecules from the incident $\mathbf{v} - \mathbf{v}_1$ to the final $\mathbf{v}_2 - \mathbf{v}_3$. In a solid angle $d\Omega$ about the final direction for a given $\mathbf{v}$, $\mathbf{v}_1$, $\mathbf{v}_2$ and $\mathbf{v}_3$, the number of particles scattered per unit time is the number of particles leaving the element $d^3r\, d^3v$ in that direction and is given by

$$dN_{out} = |\mathbf{v} - \mathbf{v}_1| f(\mathbf{r}, \mathbf{v}, t) f(\mathbf{r}, \mathbf{v}_1, t)\, \sigma(\mathbf{v}, \mathbf{v}_1 \rightarrow \mathbf{v}_2, \mathbf{v}_3)\, d\Omega\, d^3r\, d^3v\, d^3v_1 \quad \text{... (11.4.2)}$$

The total number leaving the element $d^3r\, d^3v$ per unit time as obtained by integrating over all possible $\mathbf{v}_1$ and solid angle is

$$N_{out} = \int_{\mathbf{v}_1, \Omega} |\mathbf{v} - \mathbf{v}_1| f(\mathbf{r}, \mathbf{v}, t) f(\mathbf{r}, \mathbf{v}_1, t)\, \sigma(\mathbf{v}, \mathbf{v}_1 \rightarrow \mathbf{v}_2, \mathbf{v}_3)\, d\Omega\, d^3v\, d^3v_1 d^3r \quad \text{... (11.4.3)}$$

The reverse collision, i.e. molecules with velocities $\mathbf{v}_2$ and $\mathbf{v}_3$ colliding to give rise to final velocities $\mathbf{v}$ and $\mathbf{v}_1$ is obtained by arguments identical to those given above and lead to

$$N_{in} = \int_{\mathbf{v}_2, \mathbf{v}_3} |\mathbf{v}_2 - \mathbf{v}_3| f(\mathbf{r}, \mathbf{v}_2, t) f(\mathbf{r}, \mathbf{v}_3, t)\ \sigma\ (\mathbf{v}_2, \mathbf{v}_3 \rightarrow \mathbf{v}, \mathbf{v}_1)\ d\Omega\, d^3v_2 d^3v_3 d^3r \quad \text{... (11.4.4)}$$

We now note the following equalities:

i) the collision is elastic and hence the magnitude of the relative velocity of approach is equal to the magnitude of the relative velocity of receding, i.e.

$$|\mathbf{v} - \mathbf{v}_1| = |\mathbf{v}_2 - \mathbf{v}_3| \quad \text{... (11.4.5)}$$

ii) in terms of the centre of mass velocity $\mathbf{v}_{cm} = \frac{1}{2}(\mathbf{v}_1 + \mathbf{v})$ and the relative velocity $\mathbf{v}_{rel} = \mathbf{v} - \mathbf{v}_1$, the volume element $d^3v_1 d^3v = d^3v_{cm}\, d^3\, v_{rel}$ as the Jacobian of the transformation is unity

$$d^3\, v_{cm}\, d^3\, v_{rel} = \begin{vmatrix} \dfrac{\partial v_{cm}}{\partial v} & \dfrac{\partial v_{cm}}{\partial v_l} \\ \dfrac{\partial v_{rel}}{\partial v} & \dfrac{\partial v_{rel}}{\partial v} \end{vmatrix} d^3vd^3\, v_1 \begin{vmatrix} \frac{1}{2} & + & \frac{1}{2} \\ 1 & - & 1 \end{vmatrix} d^3vd^3v_1$$

Similarly $d^3v_2\, d^3v_3 = d^3v_{cm}\, d^3v'_{rel}$. The centre of mass velocity is unchanged and hence d^3v_{cm} is unaltered. The volume element $d^3v'_{rel}$ is equal to the element d^3v_{rel} since the elastic collisions simply rotates the direction of $\mathbf{v}_{rel}$ and hence the magnitude of a volume element about $\mathbf{v}_{rel}$ is unchanged. Thus,

$$d^3v_1\, d^3v = d^3v_{cm}\, d^2v_{rel} = d^3v_{cm}\, d^3v'_{rel} = d^3v_2\, d^3v_3 \quad \text{... (11.4.6)}$$

iii) The dynamics is inviarant under time reversal, rotation and reflection. The time reversal invariance implies

$$\sigma\ (\mathbf{v}_2, \mathbf{v}_3 \rightarrow \mathbf{v}, \mathbf{v}_1) = \sigma\ (-\mathbf{v}, -\mathbf{v}_1 \rightarrow -\mathbf{v}_2, -\mathbf{v}_3) \quad \text{... (11.4.7)}$$

To exploit the invariances under rotation and reflection, we consider a rotation by 180° about the z-axis which sends v_x to $-v_x$ and v_y to $-v_y$. Now a reflection about the x—y plane takes v_z to $-v_z$ and thus under the combined operation $\mathbf{v} \rightarrow -\mathbf{v}$. Invariance under rotation and reflection consequently implies

$$\sigma(-\mathbf{v}, -\mathbf{v}_1 \rightarrow -\mathbf{v}_2, -\mathbf{v}_3) = \sigma(\mathbf{v}, \mathbf{v}_1 \rightarrow \mathbf{v}_2, \mathbf{v}_3)$$

and hence

$$\sigma(\mathbf{v}_2, \mathbf{v}_3 \to \mathbf{v}, \mathbf{v}_1) = \sigma(\mathbf{v}, \mathbf{v}_1 \to \mathbf{v}_2, \mathbf{v}_3) \qquad \text{... (11.4.8)}$$

Using Eqs. (11.4.5), (11.4.6) and (11.4.8), we can rewrite Eq. (11.4.4) as

$$N_{\text{in}} = \int_{\mathbf{v}_1, \Omega} |\mathbf{v} - \mathbf{v}_1| f(\mathbf{r}, \mathbf{v}_2, t)\, f(\mathbf{r}, \mathbf{v}_3, t)\, \sigma(\mathbf{v}, \mathbf{v}_1 \to \mathbf{v}_2, \mathbf{v}_3)\, d\Omega\, d^3v\, d^3v_1 d^3r \qquad \text{... (11.4.9)}$$

Thus the number of extra molecules ending up per unit time in the volume element $d^3r\, d^3v$ due to collision is

$$N_{\text{in}} - N_{\text{out}} = \int_{\mathbf{v}_1, \Omega} |\mathbf{v} - \mathbf{v}_1|\, \sigma(\mathbf{v}, \mathbf{v}_1 \to \mathbf{v}_2, \mathbf{v}_3) \Big[f(\mathbf{r}, \mathbf{v}_2, t) f(\mathbf{r}, \mathbf{v}_3, t)$$

$$- f(\mathbf{r}, \mathbf{v}_1, t)\, f(\mathbf{r}, \mathbf{v}, t)\Big]\, d\Omega\, d^3v_1\, d^3r\, d^3v$$

$$= \left|\frac{\partial f}{\partial t}\right|_{\text{coll}} d^3r\, d^3v \qquad \text{... (11.4.10)}$$

and Eq. (11.4.1) becomes

$$\frac{\partial f}{\partial t} + \mathbf{v} \cdot \frac{\partial f}{\partial \mathbf{r}} + \frac{\mathbf{F}}{m} \cdot \frac{\partial f}{\partial \mathbf{v}} = \int |\mathbf{v} - \mathbf{v}_1| \times \sigma(\mathbf{v}, \mathbf{v}_1 \to \mathbf{v}_2, \mathbf{v}_3)$$

$$\times \Big[f(\mathbf{r}, \mathbf{v}_2, t)\, f(\mathbf{r}, \mathbf{v}_3, t) - f(\mathbf{r}, \mathbf{v}_1, t) f(\mathbf{r}, \mathbf{v}, t)\Big] \times d\Omega\, d^3v_1 \qquad \text{... (11.4.11)}$$

This is the Boltzmann transport equation, which gives the time evolution of the distribution function f and is of the form $\partial f / \partial t = Lf$ that we had anticipated in Eq. (11.2.4).

We now find the equilibrium distribution f_0 in the absence of any external force **F**. Since the right hand side of Eq. (11.4.11) cannot produce any spatial dependence in f, in the absence of any external force, the distribution function f cannot depend on **r** and thus the equilibrium distribution for which $\partial f_0/dt = 0$, has to satisfy

$$\int d^3v_1\, d\Omega\ \sigma(\mathbf{v}_1, \mathbf{v} \rightarrow \mathbf{v}_2, \mathbf{v}_3)\ \{f_0(\mathbf{v}_2) f_0(\mathbf{v}_3) - f_0(\mathbf{v}_1) f_0(\mathbf{v})\} = 0$$

or

$$f_0(\mathbf{v}_2) f_0(\mathbf{v}_3) = f_0(\mathbf{v})\ f_0(\mathbf{v}_1) \qquad \text{... (11.4.12)}$$

The product of the distribution function is collision invariant and hence

$$\ln f_0(\mathbf{v}) + \ln f_0(\mathbf{v}_1) = \text{Collision invariant} \qquad \text{... (11.4.13)}$$

Now, in an elastic collision the additive invariants are momentum and kinetic energy:

$$\mathbf{v} + \mathbf{v}_1 = \mathbf{v}_2 + \mathbf{v}_3 = \text{Collision invariant}$$

$$v_1^2 + v^2 = v_2^2 + v_3^2 = \text{Collision invariant}$$

Consequently

$$\ln f_0(\mathbf{v}) = \alpha + \boldsymbol{\beta} \cdot \mathbf{v} + \gamma v^2$$

$$= A_0 - A(\mathbf{v} - \mathbf{v}_0)^2 \qquad \text{... (11.4.14)}$$

where α, $\boldsymbol{\beta}$ and γ are constants and so are A_0, A and $\mathbf{v}_0$ expressible in terms of α, $\boldsymbol{\beta}$ and γ. We can write the above as

$$f_0(v) = C e^{-A(\mathbf{v} - \mathbf{v}_0)^2} \qquad \text{... (11.4.15)}$$

and choosing the origin of $\mathbf{v}$ as $\mathbf{v}_0$ without any loss of generality

$$f_0(v) = C\ e^{-Av^2} \qquad \text{... (11.4.16)}$$

where C is a constant. The constants A and C are related because of the normalization

$$N = \int f_0(v)\, d^3r\, d^3v = VC \int_0^{\infty} e^{-Av^2}\, d^3v$$

$$= 4\pi \cdot V C \int_0^{\infty} v^2 \, e^{-Av^2} \, dv$$

$$= \left(2\pi \, VC/A^{3/2}\right) \times \Gamma\left(\frac{3}{2}\right)$$

$$= \left(\frac{\pi}{A}\right)^{3/2} V C$$

or

$$\left(\frac{\pi}{A}\right)^{3/2} C = \frac{N}{V} = n, \qquad \text{... (11.4.17)}$$

where n is the number density.

To make further progress, we can calculate the pressure exerted by the gas enclosed in a cubical volume V. To obtain the pressure we can choose a particular face (say xy) and calculate the number of molecules with the z-component of velocity between v_z and $v_z + dv_z$ incident per unit time per unit area on this face. Clearly, the number of such molecules is

$$dN = C \int_{-\infty}^{\infty} dv_x \int_{-\infty}^{\infty} dv_y \, e^{-A(v_x^2 + v_y^2)} \, e^{-Av_z^2} \, dv_z \times 1 \times v_z \qquad \text{... (11.4.18)}$$

Every time such a molecule hits the x–y plane and is reflected the change in momentum is $2mv_z$ and hence the force per unit area.

$$dF = 2mv_z \, dN \qquad \text{... (11.4.19)}$$

To get the total force per unit area and hence the pressure p, we integrate over v_z from 0 to ∞ and get using Eq. (11.4.18),

$$P = 2m \, C \int_{-\infty}^{\infty} dv_x \int_{-\infty}^{\infty} dv_y \, e^{-A(v_x^2 + v_y^2)} \int_0^{\infty} v_z^2 \, e^{-Av_z^2} \, dv_z$$

$$= \frac{2mC}{A} \left\{\Gamma\left(\frac{1}{2}\right)\right\}^2 \frac{1}{A^{1/2}} \cdot \frac{1}{2A} \cdot \Gamma\left(\frac{3}{2}\right)$$

$$= \frac{m}{2A}\left(\frac{\pi}{A}\right)^{3/2} C = \frac{m}{2A}\, n \qquad \text{... (11.4.20)}$$

In equilibrium the pressure P of the gas is related to its volume V and temperature T. The only interaction that we have assumed is elastic collision between the gas molecules and that is a fair approximation to an ideal gas. For an ideal gas, the relation between P, V and T is

$$PV = N k T$$

where N is the number of molecules and k is a constant known as Boltzmann's constant with the numerical value 1.38×10^{-16} erg / K.
Thus

$$P = n k T \qquad \text{... (11.4.21)}$$

for the ideal gas and comparing with Eq. (11.4.20)

$$A = \frac{m}{2kT} \qquad \text{... (11.4.22)}$$

The equilibrium distribution can then be written as

$$f_0(v) = n\left(\frac{m}{2\pi k T}\right)^{3/2} e^{-\frac{mv^2}{2kT}} \qquad \text{... (11.4.23)}$$

which gives the density of molecules under equilibrium condition with velocity between $\mathbf{v}$ and $\mathbf{v} + d\mathbf{v}$. This is the familiar Maxwell-Boltzmann distribution for velocities in a non interacting gas of classical particles.

Chapter 12

Consequences of the Transport Equation

12.1 Introduction

In the previous chapter, we have obtained the evolution equation for the distribution function f and used it to obtain the equilibrium distribution function. In this chapter our focus will be on setting up a non-equilibrium distribution and obtaining its time evolution. We will begin with an equilibrium distribution and give it a small perturbation. The resulting perturbation will evolve in time according to the linearized version of Eq. (11.4.11). The full transport equation [Eq. (11.4.11)] is virtually impossible to solve because of the non-linear collision integral, but for small perturbations about the equilibrium state the non-linear term can be linearized and under such circumstances the approach to equilibrium can be analytically studied. For attaining equilibrium the collision term is essential and hence the decay of the perturbation to the equilibrium state is possible only if the collision term is included in the transport equation. If we want to consider a simple situation without the collision term, then a small perturbation to the equilibrium state can oscillate about the equilibrium distribution if a net unbalanced force is generated on distortion of the equilibrium. If we ignore collisions and also the presence of external forces, there is no possibility of a non-equilibrium distribution ever approaching the equilibrium distribution. As an example of this, we may consider a non-equilibrium distribuition where the particles have their velocities distributed in a Maxwell-

Boltzmann fashion but are all concentrated at $r = 0$ at time $t = 0$. Thus, the distribution at $t = 0$ is

$$f(\mathbf{r}, \mathbf{v}, 0) = N\left(\frac{m}{2\pi kT}\right)^{3/2} e^{-\frac{mv^2}{2kT}} \delta^3(r) \qquad \text{... (12.1.1)}$$

The evolution occurs without collision and no interactions between the particles which can produce an external force. The transport equation takes the form

$$\frac{\partial f}{\partial t} + \mathbf{v} \cdot \frac{\partial f}{\partial \mathbf{r}} = 0 \qquad \text{... (12.1.2)}$$

with the solution

$$f(\mathbf{r}, \mathbf{v}, t) = g(\mathbf{r} - \mathbf{v}t, \mathbf{v}) \qquad \text{... (12.1.3)}$$

where g is an arbitrary function, which can be fixed by the initial condition:

$$f(\mathbf{r}, \mathbf{v}, 0) = g(\mathbf{r}, \mathbf{v}) = N\left(\frac{m}{2\pi\, kT}\right) e^{-\frac{mv^2}{2kT}} \delta^3(r) \qquad \text{.... (12.1.4)}$$

Having thus fixed $g(\mathbf{r}, \mathbf{v})$, we find from Eq. (12.1.3)

$$f(\mathbf{r}, \mathbf{v}, t) = N\left(\frac{m}{2\pi\, kT}\right)^{3/2} e^{-\frac{mv^2}{2\,kT}} \delta^3(\mathbf{r} - \mathbf{v}t) \qquad \text{... (12.1.5)}$$

The number of particles $n(\mathbf{r}, t)d^3r$ lying between $\mathbf{r}$ and $\mathbf{r} + d\mathbf{r}$ is obtained by integrating over all velocities and thus

$$n(\mathbf{r}, t)d^3 r = \int_v d^3v\, f(\mathbf{r}, \mathbf{v}, t)\, d^3 r$$

$$= N\left(\frac{m}{2\pi kT}\right)^{3/2} d^3 r \int e^{-\frac{mv^2}{2kT}} \delta^3(\mathbf{r} - \mathbf{v}t)\, d^3 v$$

$$= N\left(\frac{m}{2\pi kT}\right)^{3/2} \frac{1}{t^3} e^{-\frac{mr^2}{2kTt^2}} d^3r$$

using the standard rules of δ-function integration. This leads to

$$n(\mathbf{r}, t) = N\left(\frac{m}{2\pi kT}\right)^{3/2} \frac{1}{t^3}\, e^{-\frac{mr^2}{2kTt^2}} \qquad \text{... (12.1.6)}$$

As time evolves the particles concentrated at $r = 0$ start spreading out because of the initial velocities and ultimately diffuse out over all space. As expected, there is no equilibrium distribution due to the absence of ineractions.

12.2 Interaction without Collisions

We consider a collection of electrons spread out on a two-dimensional surface. A sample of such a two-dimensional electron gas is obtained by spraying the surface of liquid helium with electrons. Charge neutrality is maintained by having a positively charged background. The areal density of electrons is low enough for quantum effects to be ignored. In an equilibrium distribution, the external force is zero—on any given electron the Coulomb force due to the other electrons and the charged background cancel out. The equilibrium distribution f_0 is then independent of the position coordinate and equilibrium is achieved by collisions with f_0 given by (in this chapter we will write k_B for Boltzmann constant, reserving $\mathbf{k}$ for the wavevector)

$$f_0 = n \cdot \frac{m}{2\pi\, k_B T} \cdot e^{-\frac{mv^2}{2k_B T}} \qquad \text{... (12.2.1)}$$

where $n = N/A$, N being the total number of electrons, A the area over which they are spread and the distribution, as should be checked, is normalized in two dimensions, i.e. $A \int f_0(v)\, d^2 v = N$. If a small deviation from the equilibrium distribution occurs, then the resultant force on any electron is no longer zero and this unbalanced force can check the perturbation from growing, without the effect of collisions. In what follows we will ignore the effect of collisions on the perturbation and work with the left hand side of the transport equation [Eq. (11.4.11)]. We imagine that the equilibrium distribution f_0 is changed to

$$f = f_0 + \delta f \qquad \text{... (12.2.2)}$$

The deviation δf leads to local density variations $\delta n(\mathbf{r}, t) = \int (\delta f)\, d^2 v$ and due to local density variations there will be an unbalanced force $\delta \mathbf{F}(\mathbf{r}, t)$ obtained from a potential $\delta\varphi(\mathbf{r}, t)$ which are related as

$$\delta \mathbf{F} = -(e)\, \vec{\nabla}\, \delta\varphi(\mathbf{r}, t) \qquad \text{... (12.2.3a)}$$

(e is the electronic charge)
and

$$\delta\varphi = -e \int \frac{\delta n(\mathbf{r}', t)\, d^2 r'}{|\mathbf{r} - \mathbf{r}'|} \qquad \text{... (12.2.3b)}$$

The distribution function f of Eq. (12.2.2) must satisfy the transport equation

$$\frac{\partial}{\partial t} f_0 + \frac{\partial}{\partial t} \delta f + \mathbf{v} \cdot \frac{\partial}{\partial \mathbf{r}} (f_0 + \delta f) + \frac{\delta \mathbf{F}}{m} \cdot \frac{\partial}{\partial \mathbf{v}} (f_0 + \delta f)$$

$$= \text{Collision term with } (f_0 + \delta f) \qquad \text{... (12.2.4)}$$

As stated above, we ignore the δf conribution to the collision term and f_0 gives zero for the effect of collisions. Consequently, the right hand size is zero in Eq. (12.2.4). Now $\partial f_0/\partial t = \partial f_0/\partial \mathbf{r} = 0$ and working to order δf, we obtain

$$\frac{\partial}{\partial t} \delta f + \mathbf{v} \cdot \frac{\partial}{\partial \mathbf{r}} \delta f + \frac{\delta \mathbf{F}}{m} \cdot \frac{\delta}{\delta \mathbf{v}} f_0 = 0 \qquad \text{... (12.2.5)}$$

Note that in the third term $\delta \mathbf{F}$ is already $O(\delta f)$ and hence $(\delta \mathbf{F}/m \cdot \partial/\partial \mathbf{v})\, \delta f$ is higher order, i.e. $O(\delta f)^2$, and has been dropped.

To proceed further we have to make a statement for the form of δf. If , as explained before, we expect δf to be oscillatory, then the temporal and spatial dependence of δf may be taken to be of the form

$$\delta f = A\, e^{i(\mathbf{k} \cdot \mathbf{r} - \omega t)}, \qquad \text{... (12.2.6)}$$

($i = \sqrt{-1}$ as usual)

where the velocity dependence is contained in A. The spatial dependence $e^{i\mathbf{k} \cdot \mathbf{r}}$ is assumed to be such that the wavenumber k (magnitude of the wave vector) is small which would mean a mild spatial dependence. Similarly we expect the frequency ω to be small as well and the final answer would have to be consistent with these expectations. Consistent with Eq. (12.2.6), the spatial and temporal dependence of $\delta n\,(\mathbf{r}, t)$ would be

$$\delta n = B\, e^{i\,(\mathbf{k}\cdot\mathbf{r}\,-\,\omega t)} \qquad \text{... (12.2.7)}$$

with B, a constant, obtained from A through the relation

$$B = \int A\, d^2 v \qquad \text{... (12.2.8)}$$

Inserting in Eq. (12.2.5), the forms of Eqs. (12.2.6) and (12.2.7), we get

$$iA(\mathbf{k}\cdot\mathbf{v} - \omega)\, e^{i(\mathbf{k}\cdot\mathbf{r}\,-\,\omega t)} = -\frac{1}{m}\,\delta\mathbf{F}\cdot\frac{\partial}{\partial\mathbf{v}}\left(n\,\frac{m}{2\pi k_B T}\right) e^{-\frac{mv^2}{2k_BT}}$$

$$= \frac{1}{m}\,\delta\mathbf{F}\cdot\frac{nm}{2\pi\, k_BT}\cdot\left(-\frac{m}{k_BT}\right)\mathbf{v}\; e^{-\frac{mv^2}{2k_BT}}$$

$$= \frac{f_0}{(k_BT)}\;\mathbf{v}\cdot(-e^2)\,\vec{\nabla}\int\frac{\delta n\,(\mathbf{r}', t)\, d^2 r'}{|\mathbf{r} - \mathbf{r}'|}$$

$$= -\frac{f_0\, e^2 B}{(k_BT)}\;\mathbf{v}\cdot\vec{\nabla}\int\frac{e^{i(\mathbf{k}\cdot\mathbf{r}'\,-\,\omega t)}}{|\mathbf{r} - \mathbf{r}'|}\, d^2 r'$$

$$= -\frac{f_0\, e^2 B}{(k_BT)}\;\mathbf{v}\cdot\vec{\nabla}\, e^{i(\mathbf{k}\cdot\mathbf{r}\,-\,\omega t)} \times \int\frac{e^{i\mathbf{k}\cdot(\mathbf{r}'\,-\,\mathbf{r})}\, d^2 r'}{|\mathbf{r} - \mathbf{r}'|}$$

$$= -\frac{i\, f_0\, e^2 B}{(k_BT)}(\mathbf{v}.\mathbf{k})\frac{2\pi}{k}\, \mathrm{e}^{\,i(\mathbf{k}\cdot\mathbf{r} - \omega t)}$$

(The two-dimensional Fourier transform $\int e^{i\mathbf{k}\cdot\mathbf{r}}/r\, d^2 r = 2\pi/k$) leading to

$$A = B\,\frac{e^2 f_0}{k_BT}\,\frac{2\pi}{k}\cdot\frac{\mathbf{v}\cdot\mathbf{k}}{\omega - \mathbf{k}\cdot\mathbf{v}} \qquad \text{... (12.2.9)}$$

Using Eq. (12.2.8), we obtain

$$B = \int A\, d^2 v = B\,\frac{2\pi e^2}{k}\cdot\frac{1}{k_BT}\int\frac{\mathbf{v}\cdot\mathbf{k}}{\omega - \mathbf{k}\cdot\mathbf{r}}\, d^2 v$$

$$= B\frac{2\pi e^2}{k}\cdot\frac{1}{k_BT}\int\frac{\mathbf{v}\cdot\mathbf{k}}{\omega}\cdot f_0\cdot(1+\frac{\mathbf{v}\cdot\mathbf{k}}{\omega}+\ldots)\,d^2v \qquad \ldots\ (12.2.10)$$

The last step follows on the assumption that though ω and k are both small, ω is large compared to k—for self consistency the final result must bear this out. Since $B \neq 0$, we find from Eq. (12.2.10), working to the lowest order,

$$1 = \frac{2\pi e^2}{k}\cdot\frac{k^2}{k_BT}\frac{1}{2\omega^2}2\pi\int_0^\infty\frac{nm}{2\pi k_BT}e^{-\frac{mv^2}{2k_BT}}v^3\,dv$$

$$= \frac{\pi ne^2k}{m\omega^2}\int_0^\infty x^3e^{-x^2}\,dx$$

$$= \frac{2\pi ne^2k}{m\omega^2}$$

or,

$$\omega^2 = \frac{2\pi ne^2}{m}k \qquad \ldots\ (12.2.11)$$

confirming that for small k, $\omega \propto k^{1/2}$ is indeed much larger than k. Thus, we find that our Ansatz for the form of δf is indeed consistent and there is an oscillation about the equilibrium distribution if the charged particle is disturbed from equilibrium and the frequency of oscillation is related to the wavenumber by Eq. (12.2.11). These oscillations are known as plasma oscillations.

12.3 Collisions with Fixed Scatterers

The nonlinear collision integral of Eq. (11.4.11) shall henceforth be denoted by C, i.e.

$$C = \int d^3v_1\,d\Omega\,|\mathbf{v}-\mathbf{v}_1|\,\sigma(\Omega)\left[f(\mathbf{r},\mathbf{v}_2,t)f(\mathbf{r},\mathbf{v}_3,t)-f(\mathbf{r},\mathbf{v}_1,t)f(\mathbf{r},\mathbf{v},t)\right]$$

$$= \int d^3v_1\,d\Omega\,|\mathbf{v}-\mathbf{v}_1|\,\sigma(\Omega)\left[f_2f_3-f_1f\right] \qquad \ldots\ (12.3.1)$$

where instead of the elaborate $\sigma\,(\mathbf{v}, \mathbf{v}_1 \rightarrow \mathbf{v}_2, \mathbf{v}_3)$, we use $\sigma\,(\Omega)$ since the cross-section for the elastic collision will be determined by the solid angle Ω alone, and instead of $f\,(\mathbf{r}, \mathbf{v}_i, t)$ we use the notation f_i. For the rest of this chapter, we will assume a complete absence of external forces **F**—an assumption almost always valid for uncharged particles in a force-free environment. Boltzmann transport equation now reads

$$\frac{\partial f}{\partial t} + \mathbf{v} \cdot \frac{\partial f}{\partial \mathbf{r}} = C \qquad \text{... (12.3.2)}$$

with the equilibrium distribution appropriate to a three-dimensional space set up by binary collisions as

$$f_0 = n \left(\frac{m}{2\pi\, k_B T} \right)^{3/2} e^{-\frac{mv^2}{2k_B T}} \qquad \text{... (12.3.3)}$$

A fixed density of scatterers (n_0) is assumed and particles colliding with the scatterers have the directions of their velocity changed but the magnitude is unchanged. Particles with velocity **v** when reflected from the scatterer go off with velocity $\mathbf{v}'$ ($|\mathbf{v}'| = |\mathbf{v}|$) and thus there is a depletion of particles in the volume element $d^3r\, d^3v$. Similarly, a velocity $\mathbf{v}'$ when scattered gives rise to a velocity **v** and there is addition to the number of particles in $d^3r\, d^3v$. The collision integral simplifies to

$$C = \int d\Omega\ n_0\, |\mathbf{v}|\, \sigma\,(\Omega)\ (f' - f) \qquad \text{... (12.3.4)}$$

(Note that in $f f_1$, f_1 refers to scatterers and is simply n_0, while in $f_2 f_3$, if f_2 refers to the particles with velocity $\mathbf{v}'$ and is f', then f_3 refers to the scatterers and is n_0. The relative velocity between scatterer' and particle is $|\mathbf{v}| = |\mathbf{v}'|$). If $\sigma(\Omega)$ is taken to be a constant, we have Eq. (12.3.4) further simplified to

$$\begin{aligned} C &= n_0\, |\mathbf{v}|\ \sigma \int d\Omega\ (f' - f) \\ &= \Gamma \int d\Omega\, (f' - f) \qquad \text{... (12.3.5)} \end{aligned}$$

where $\Gamma = n_0\, |\mathbf{v}|\, \sigma$ is the collision frequency. For the equilibrium distribution f_0 set up by the binary collisions, C of Eq. (12.3.5) vanishes and for the time development of perturbations to f_0, we will consider the effect of the fixed scatterers alone in this section. The algebra is much simpler than if we were dealing with equilibration by binary collision and hence the basic mathematical technique will be easier to implement.

If we produce a small deviation from the equilibrium distribution, so that

$$f = f_0 + \delta f \qquad \text{... (12.3.6)}$$

then δf satisfies

$$\frac{\partial}{\partial t}\delta f + \mathbf{v}\cdot\frac{\partial}{\partial \mathbf{r}}\delta f = \Gamma \int d\Omega\,(\delta f' - \delta f) \qquad \text{... (12.3.7)}$$

For the spatial variation of δf, we assume again a slowly varying oscillatory form (as being the smallest deviation from a uniform state) and thus write

$$\delta f(\mathbf{r}, \mathbf{v}, t) = \psi(\mathbf{v}, t)\, e^{i\,\mathbf{k}\cdot\mathbf{r}} \qquad \text{... (12.3.8)}$$

Inserting in Eq. (12.3.7), we obtain

$$\frac{\partial \psi}{\partial t} = -K\psi = -(K_0 + K_1)\psi \qquad \text{... (12.3.9)}$$

where

$$K_0\psi = \Gamma \int d\Omega\,(\psi - \psi') \qquad \text{... (12.3.10a)}$$

and

$$K_1\psi = i\,\mathbf{k}\cdot\mathbf{v}\,\psi \qquad \text{... (12.3.10b)}$$

If we can determine the eigenvalues and eigenfunctions of K (λ_n an φ_n respectively, $n = 0, 1, 2, ...$) then the solution ψ can be written as

$$\psi = \sum a_n \varphi_n e^{-\lambda_n t} \qquad \text{... (12.3.11)}$$

where a_n are constants which can be determined from the initial condition.

For the equilibrium to make sense, we must have the λ_n such that Re $\lambda_n > 0$ for all n, as otherwise there will be a runaway solution and the equilibrium will not be stable. Further, of all the λ_n, our interest is only in the lowest λ, i.e. λ_0 that determines the persistent time scale. If one waits for a time $T >> \lambda_0^{-1}$, all other terms become negligible and only the term $a_0 \varphi_0 e^{-\lambda_0 t}$ in Eq. (12.3.11) is of appreciable magnitude. To find λ_n one has to employ perturbation theory. The operator K_1 having an explicit factor of k is small and hence can be treated perturbatively once the eigenvalues of K_0 are determined. In general, we will be interested in the lowest eigenvalue of K_0. In the special case of Eq. (12.3.10a), however, all the eigenvalues and eigenfunctions of K_0 can be exactly determined.

Inspection of Eq. (12.3.10a) shows that ψ = constant is an eigenfunction of K_0 with eigenvalue 0. On the other hand, all functions φ having the property $\int \varphi \, d\Omega = 0$ are also eigenfunctions of K_0 with eigenvalue $4\pi\Gamma$

$$K_0\varphi = \Gamma \int d\Omega \, (\varphi - \varphi') = \Gamma \, \delta \int d\Omega$$

$$= 4\pi \, \Gamma\varphi \qquad \text{... (12.3.12)}$$

(note that φ' is the function that is Ω-dependent and is inegrated over). A set of functions φ having the property $\int \varphi \, d\Omega = 0$ are the spherical harmonics Y_{lm} with $l \neq 0$ ($m = -l, -(l-1) \ldots l-1, l$ for a given l and $l = 1, 2, \ldots$). For $l = 0$, Y_{00} is a constant and hence we have the following complete set of eigenfunctions for the operator K_0.

Eigenvalue	Eigenfunction
0	Y_{00}
$4\pi\Gamma$	Y_{lm} $(l \neq 0)$

As is well known, Y_{lm} forms a complete set on the space of all solid angles and thus we have solved for K_0 completely. The relevant eigenvalue is clearly $\lambda_0 = 0$ since the correction to it will be proportional to some power of k and hence small. The other eigenvalue $4\pi\Gamma$ is important only insofar as it contributes to the correction to the zero eigenvalue but for purposes of setting a time scale it is clearly irrelevant.

The next step is the evaluation of the correction to $\lambda_0 = 0$ due to the operator K_1. In first-order perturbation theory, the correction is $\langle K_1 \rangle$ which is clearly zero, i.e. $\lambda_0^{(1)} = 0$. The second-order perturbation theory contribution to λ_0 is

$$\lambda_0^{(2)} = \sum_{ex} \frac{\left(\int \varphi_{ex} K_1 \varphi_0\right)^2}{0 - \lambda_{ex}} \qquad \text{... (12.3.13)}$$

where φ_{ex} stands for all the excited staes and λ_{ex} are their respective eigen-values. In this case

$$\lambda_0^{(2)} = \sum_{l \neq 0, m} \frac{\left(\int d\Omega\, Y_{lm} K_1 Y_{00}\right)^2}{-4\pi\Gamma} \qquad \text{... (12.3.14)}$$

Now $K_1 = -i\,kv\cos\theta$ and hence $\int d\Omega\, Y_{lm}\cos\theta\, Y_{00}$ is non-zero only if $Y_{lm} = Y_{10} = \sqrt{3/4\pi}\cos\theta$ (Y_{00} being a constant, the orthogonality of Y_{lm} leads to a nonzero integral only for Y_{10}). Thus, $\int \sqrt{1/4\pi}\ \sqrt{3/4\pi}\cos^2\theta\, d\Omega = 1/\sqrt{3}$ and Eq. (12.3.14) becomes

$$\lambda_0^{(2)} = \frac{k^2 v^2}{4\pi\Gamma} \cdot \frac{1}{3} = \frac{k^2 v^2}{3\Gamma'} \qquad \text{... (12.3.15)}$$

where $\Gamma' = 4\pi\Gamma$. Hence to second order in perturbation theory, $\lambda_0 = k^2 v^2/3\Gamma'$ and similarly for the eigenvalue $4\pi\Gamma$, $\lambda_{ex} = 4\pi\Gamma + O(k^2) \approx 4\pi\Gamma$ leading to (from Eq. (12.3.11))

$$\psi = a_0\, e^{-\frac{k^2 v^2}{3\Gamma'}t} + \sum_{lm \neq 0} a_{lm} Y_{lm}\, e^{-4\pi\Gamma t} \qquad \text{... (12.3.16)}$$

The result is now transparent. If one waits long enough, only the first term is going to survive if k is small enough. To estimate how small k must be, it is clear that for the first term to survive

$$\frac{k^2 v^2}{\Gamma} << \Gamma$$

or $$k^2 << \frac{\Gamma^2}{v^2} = \frac{1}{l_{mfp}^{\,2}} \qquad \text{... (12.3.17)}$$

which implies that the wavelength must be much greater than the mean free path ($l_{\rm mfp}$). We thus see that with a fixed set of scatterers, a small deviation from equilibrium distribution slowly disappears as the particles undergo diffusive motion on collision with the scatterers.

The general strategy for attacking all problems involving deviation from equilibrium and the collision integral is now clear.

i) Linearize the collision integral in the deviation δf and assuming the spatial variation of the kind shown in Eq. (12.3.8), set up a time evolution equation of the form of Eq. (12.3.9).
ii) Find all the zero eigenvalues of K_0.
iii) Treat K_1 perturbatively and obtain the first nonvanishing contribution to the zero eigenvalue of k_0.
iv) Interpret the eigenvalues and eigenfunctions if possible

12.4 Binary Collisions

We are now ready to tackle the binary collision integral. A small deviation from the equilibrium distribution is imagined and for later convenience we write this deviation as

$$f = f_0\,(1 + \delta f) \qquad \text{... (12.4.1)}$$

The term $f_2 f_3 - f f_1$, when linearized in terms of δf becomes

$$\begin{aligned} f_2 f_3 - f f_1 &= f_{20} f_{30} - f_0 f_{10} + f_{20} f_{30}\,(\delta f_2 + \delta f_3) - f_0 f_{10}\,(\delta f + \delta f_{10}) + O(\delta f)^2) \\ &= f_0 f_{10}\,(\delta f_2 + \delta f_3 - \delta f_1 - \delta f) + O((\delta f)^2) \end{aligned} \qquad \text{... (12.4.2)}$$

where $f_0 f_{10} = f_{20} f_{30}$ for the equilibrium distribution.

The analogue of Eq. (12.3.7) is

$$\frac{\partial}{\partial t}\delta f + \mathbf{v}\cdot\frac{\partial}{\partial r}\delta f = \int d^3 v_1\, d\Omega\, |\mathbf{v} - \mathbf{v}_1|\, \sigma(\Omega)\, f_{10}\,(\delta f_2 + \delta f_3 - \delta f_1 - \delta f) \qquad \text{... (12.4.3)}$$

Now with $\delta f = \psi\,(\mathbf{p}, t)\, e^{i\mathbf{k}\cdot\mathbf{r}}$

$$\frac{\partial \psi}{\partial t} = -K\psi = -K_0\psi - K_1\psi \qquad \text{... (12.4.4)}$$

where

$$K_0\psi = \int d^3v_1\, d\Omega\, |\mathbf{v} - \mathbf{v}_1|\, \sigma(\Omega)\, f_{10}\, (\psi_2 + \psi_3 - \psi_1 - \psi) \qquad \text{... (12.4.5a)}$$

$$K_1\psi = i\,\mathbf{k}\cdot\mathbf{v}\,\psi \qquad \text{... (12.4.5b)}$$

Before proceeding any further, we need to define a scalar product of two vectors ψ and φ in the velocity space as

$$(\psi, \varphi) = \int \psi\,(\mathbf{v})\, f_0\,(v)\, \varphi\,(\mathbf{v})\, d^3v \qquad \text{... (12.4.6)}$$

Having defined the scalar product, we now address the question of zero eigenvalues of K_0, if any. This automatically settles the question of the lowest eigenvalue because K_0 cannot have any negative eigenvalues. We establish this by proving

$$(\psi_0, K_0\,\psi) \geq 0 \qquad \text{... (12.4.7)}$$

The proof is as follows:

$$(\psi, K_0\,\psi) = \int d^3v_1\, d\Omega\, |\mathbf{v} - \mathbf{v}_1|\, \sigma(\Omega)\, f_0 f_{10}\, (\psi_2 + \psi_3 - \psi_1 - \psi)\psi$$

$$= \int d^3 v_1\, d^3 v\, |\mathbf{v} - \mathbf{v}_1|\, \sigma\,(\mathbf{v}_1, \mathbf{v} \rightarrow \mathbf{v}_2, \mathbf{v}_3)\, f_0 f_{10}\, (\psi_2 + \psi_3 - \psi_1 - \psi)\, \psi \qquad \text{... (a)}$$

$$= \int d^3 v_1\, d^3v\, |\mathbf{v} - \mathbf{v}_1|\, \sigma\,(\mathbf{v}_1, \mathbf{v} \rightarrow \mathbf{v}_2, \mathbf{v}_3)\, f_0 f_{10}\, (\psi_2 + \psi_3 - \psi_1 - \psi)\, \Psi_1 \qquad \text{... (b)}$$

(by interchanging $\mathbf{v}$ and $\mathbf{v}_1$)

$$= \int d^3 v_2\, d^3 v_3\, |\mathbf{v}_2 - \mathbf{v}_3|\, \sigma\,(\mathbf{v}_2, \mathbf{v}_3 \rightarrow \mathbf{v}_1, \mathbf{v})\, f_{20} f_{30}\, (\psi + \psi_1 - \psi_2 - \psi_3)\, \psi_2$$

(by exchanging $\mathbf{v}$ and $\mathbf{v}_2$, $\mathbf{v}_1$ and $\mathbf{v}_3$) ... (c)

$$= \int d^3 v_2\, d^3 v_3\, |\mathbf{v}_2 - \mathbf{v}_3|\, \sigma\,(\mathbf{v}_2, \mathbf{v}_3 \rightarrow \mathbf{v}_1, \mathbf{v})\, f_{20} f_{30}\, (\psi + \psi_1 - \psi_2 - \psi_3)\, \psi_3$$

(by exchanging $\mathbf{v}$ and $\mathbf{v}_3$, $\mathbf{v}_1$ and $\mathbf{v}_2$) ... (d)

Noting that $f_{10} f_0 = f_{20} f_{30}$, $|\mathbf{v} - \mathbf{v}_1| = |\mathbf{v}_2 - \mathbf{v}_3|$, $d^3 v_2 \, d^3 v_3 = d^3 v \, d^3 v_1$ and $\sigma(\mathbf{v}_1, \mathbf{v} \to \mathbf{v}_2, \mathbf{v}_3) = \sigma(\mathbf{v}_2, \mathbf{v}_3 \to \mathbf{v}, \mathbf{v}_1)$ (see Chapter 11) we can add Eqs. (a)–(d) above to write

$$(\psi, K_0 \psi) = \frac{1}{4} \int d^3 v_1 \, d^3 v \, |\mathbf{v} - \mathbf{v}_1| \, \sigma(\mathbf{v}, \mathbf{v}_1 \to \mathbf{v}_2, \mathbf{v}_3) f_0 f_{10} (\psi + \psi_1 - \psi_2 - \psi_3)^2 \geq 0 \quad \text{... (12.4.8)}$$

This completes the proof.

We now list the zero eigenvalues of K_0. It is obvious from Eq. (12.4.5a) that the following wavefunctions lead to $\lambda_0 = 0$. (i) ψ = const. (ii) $\psi = p_x, p_y, p_z$, (iii) $\psi = p^2/2m$ (each of them is a collision invariant and hence $\Psi + \Psi_1 = \Psi_2 + \Psi_3$). As in our previous example, the other eigenvalues are of the order Γ—the collision frequency.

The effect of the perturbation operator K_1 on the zero eigenvalue now needs to be calculated. For this purpose we need to employ first order perturbation theory as appropriate to degenerate states since the zero eigenvalue is five-fold degenerate. The procedure for degenerate state perturbation theory is
(i) form normalized orthogonal functions out of the wavefunctions of the degenerate state.

In our case we call these wavefunctions $\varphi_1, \varphi_2, \varphi_3, \varphi_4, \varphi_5$ and write

$$\begin{aligned}
\varphi_1 &= 1 \\
\varphi_2 &= \sqrt{2/3} \, \frac{1}{k_B T} \left(\frac{1}{2} m v^2 - \frac{3}{2} k_B T \right) \\
\varphi_3 &= \sqrt{\frac{m}{k_B T}} \, v_z \\
\varphi_4 &= \sqrt{\frac{m}{k_B T}} \, v_y \\
\varphi_5 &= \sqrt{\frac{m}{k_B T}} \, v_x
\end{aligned} \quad \text{... (12.4.9)}$$

It should be verified that the above φ_i do satisfy $(\varphi_i, \varphi_j) = \delta_{ij}$. While the conserved quantities in $\varphi_2, \varphi_3, \varphi_4$ and φ_5 are clear, φ_1 can be taken to represent the number density.

(ii) the second step is to diagonalize the perturbation operator in the space of the above orthonormal set.

In this case, we have to from the matrix elements $(K_1)_{ij} = (\varphi_i, K_1 \varphi_j)$. For this purpose it helps to choose the direction of the **k** vector. Without any loss of generality we can consider the wavevector in the z-direction. The operator K_1 is ikv_z and clearly all its diagonal elements are zero. Further $(\varphi_4, K_1 \varphi_i) = (\varphi_5, K_1 \varphi_i) = 0$ for all i and hence we simply need to consider $(K_1)_{ij}$ were i and j range from 1 to 3. Straightforward calculation of the matrix elements of K_1 leads to the matrix

$$K_1 = ik\,\frac{k_BT}{m}\begin{pmatrix} 0 & 1 & 0 \\ 1 & 0 & \sqrt{2/3} \\ 0 & \sqrt{2/3} & 0 \end{pmatrix} \qquad \text{... (12.4.10)}$$

The eigenvalues of the above matrix are 0 and $\pm\, ik\sqrt{5k_BT/3m}$. Thus, the first-order perturbation theory does not lift the degeneracy altogether. For a wavevector in the z-direction the momentum wavefunctions in the x- and y-direction and a combination of the number and energy (φ_1 and φ_3) wavefunctions—a combination which can be identified with the entropy continue to be degenerate. Finally, in first-order perturbation theory the perturbation δf has the form

$$\delta f = A_0\, e^{ik\,(z\,\pm\,\sqrt{5k_BT/3m}\;t)} + \sum_{m=1}^{\infty} A_m\, e^{-\lambda_m t} \qquad \text{... (12.4.11)}$$

where the eigenvalues λ_m are of the order Γ and hence for times greater than Γ^{-1}, we have

$$\delta f \approx A_0\, e^{ik\,(z\,\pm\,\sqrt{5k_BT/3m}\;t)} \qquad \text{... (12.4.12)}$$

with three eigenvalues still zero which can acquire nonzero values only from second-order perturbation thoery and hence are expected to be of order k^2.

The result at $O(k)$ exhibited in Eq. (12.4.12) shows the propagation of a wave along the z-direction with velocity being $\sqrt{5k_BT/3m}$. It is the familiar sound wave with the velocity appropriate to a monatomic gas correctly obtained. The isentropic condition is also correctly obtained as the eigenvalue corresponding to the entropy variable continues to be zero—indicating the conservation of entropy.

Proceeding to second order in perturbation theory, we note that the threefold degenerate zero eigenvalue will be split producing eigenvalues of $O(k^2)$. Exactly as in Sec. 12.3, the contribution of second-order perturbation theory has the typical form $\alpha\, v^2 k^2/\Gamma$, where v is an average velocity and α is a number of order unity. The modes φ_4 and φ_5 produce a coefficient $\alpha\, v^2/\Gamma$ which corresponds to shear viscosity, while the combination of modes φ_1 and φ_3 that corresponds to entropy, produces a coefficient which is the thermal diffusivity. The modes corresponding to the sound wave which give rise to an imaginary eigenvalue at $O(k)$ also acquire a correction at $O(k^2)$ from second-order perturbation theory—the coefficient now corresponding to a combination of thermal and mass diffusivity. The actual calculations of the constants α are laborious, but some idea of what they could be can be had by writing the typical velocity as $v \approx l_{\text{mfp}}\, \Gamma$, whence the dissipative (diffusive) contributions $\alpha v^2/\Gamma$ are of the form $\alpha\, l_{\text{mfp}} \cdot v$. This should be compared with the elementary kinetic theory results: kinematic shear viscosity = thermal diffusivity = mass diffusivity = $1/3\, l_{\text{mfp}} \cdot v$.

12.5 H-Theorem

The foregoing discussion shows that the equilibrium distribution obtained by setting the collision term $C = 0$ in Eq. (12.3.1) is stable against small perturbations due to the collision term itself. It is the collision term which is responsible for the setting up of equilibrium and for providing the form of the equilibrium distribution. In this section, we establish that the condition $f_2 f_3 = f f_1$ that obviously sets $C = 0$ is not only sufficient but also necessary. We proceed by defining

$$H = \int f \ln f\, d^3v, \qquad \text{... (12.5.1)}$$

whence

$$\frac{dH}{dt} = \int \frac{\partial f}{\partial t}(1 + \ln f)\, d^3v \qquad \text{... (12.5.2)}$$

For $\partial f/\partial t$ to be zero it is necessarily true that dH/dt has got to be zero. In what follows we will assume an absence of external force and the distribution

f to be independent of $\mathbf{r}$, and hence the evolution will be governed by $\partial f/\partial t = C$.

Explicitly writing out the collision term we obtain

$$\frac{dH}{dt} = \int d^3v\, d^3v_1\; \sigma\left(\mathbf{v}, \mathbf{v}_1 \rightarrow \mathbf{v}_2, \mathbf{v}_3\right) d\Omega\; |\mathbf{v} - \mathbf{v}_1| \left(f_2 f_3 - f f_1\right) \times (1 + \ln f) \quad \text{... (12.5.3)}$$

The next few steps are identical to those carried out in arriving at Eq. (12.4.8) [essentially recognizing that the above integral is unaffected by interchanging $\mathbf{v}$ and $\mathbf{v}_1$, also $\mathbf{v}$ with $\mathbf{v}_2$ and $\mathbf{v}_1$ with $\mathbf{v}_3$ as well as $\mathbf{v}$ with $\mathbf{v}_3$ and $\mathbf{v}_1$ with $\mathbf{v}_2$]. We easily obtain

$$\frac{dH}{dt} = \frac{1}{4} \int d^3v\, d^3v_1\, |\mathbf{v} - \mathbf{v}_1|\; \sigma\; (\mathbf{v}, \mathbf{v}_1 \rightarrow \mathbf{v}_2, \mathbf{v}_3)\; d\Omega$$

$$\left(f_2 f_3 - f f_1\right) \left(\ln f f_1 - \ln f_2 f_3\right) \leq 0 \quad \text{... (12.5.4)}$$

The equality holds only if $f f_1 = f_2 f_3$ and hence it is a necessary condition for the equilibrium distribution.

Chapter 13

Conservation Laws and Hydrodynamics

13.1 A Theorem

In the previous chapter, we have repeatedly seen that quantities which are conserved in the elastic collision play a privileged role in the solution of the Boltzmann transport equation. We can now prove a general result: if χ is a conserved quantity in a binary elastic collision, i.e. $\chi_1 + \chi_2 = \chi_3 + \chi_4$, where subscripts 1 and 2 refer to the quantities before collision and 3 and 4 refer to the same quantities after collision, then

$$\int \left.\frac{\partial f}{\partial t}\right|_{\text{coll}} \chi \, d^3 v = 0 \qquad \text{... (13.1.1)}$$

The proof uses a technique that has been widely used in Chapter 12. We write

$$\int \left.\frac{\partial f}{\partial t}\right|_{\text{coll}} \chi \, d^3 v_1 = \int d^3 v_1 \, d^3 v_2 \, \sigma\left(\mathbf{v}_1, \mathbf{v}_2 \rightarrow \mathbf{v}_3, \mathbf{v}_4\right) d\Omega$$

$$\times \left|\mathbf{v}_1 - \mathbf{v}_2\right| \left(f_3 f_4 - f_1 f_2\right) \chi_1 \qquad \text{... (13.1.2)}$$

Now the following interchanges cannot change the integral on the right hand side:

i) interchange $\mathbf{v}_1$ and $\mathbf{v}_2$
ii) interchange $\mathbf{v}_1$ with $\mathbf{v}_3$ and $\mathbf{v}_2$ with $\mathbf{v}_4$
iii) interchange $\mathbf{v}_1$ with $\mathbf{v}_4$ and $\mathbf{v}_2$ with $\mathbf{v}_3$.

Each of these interchanges leaves $|\mathbf{v}_1 - \mathbf{v}_2|, d^3 v_1 d^3 v_2$ and $\sigma(\mathbf{v}_1, \mathbf{v}_2 \to \mathbf{v}_3) d\Omega$ unchanged and hence

$$\int \chi_1 \left.\frac{\partial f}{\partial t}\right|_{\text{coll.}} d^3 v_1 = \int d^3 v_1 d^3 v_2 \, \sigma(\mathbf{v}_1, \mathbf{v}_2 \to \mathbf{v}_3, \mathbf{v}_4) d\Omega$$

$$\times |\mathbf{v}_1 - \mathbf{v}_2| (f_3 f_4 - f_1 f_2) \chi_1$$

$$= \int d^3 v_1 d^3 v_2 \, \sigma(\mathbf{v}_1, \mathbf{v}_2 \to \mathbf{v}_3, \mathbf{v}_4) d\Omega \times |\mathbf{v}_1 - \mathbf{v}_2|$$

$$\times (f_3 f_4 - f_2 f_1) \chi_2 \quad \text{(1 and 2 interchanged)}$$

$$= \int d^3 v_1 d^3 v_2 \, \sigma(\mathbf{v}_1, \mathbf{v}_2 \to \mathbf{v}_3, \mathbf{v}_4) d\Omega \times |\mathbf{v}_1 - \mathbf{v}_2|$$

$$\times (f_1 f_2 - f_3 f_4) \chi_3 \quad \text{(1 and 3 interchanged and 2 and 4 interchanged)}$$

$$= \frac{1}{4} \int d^3 v_1 d^3 v_2 \, \sigma(\mathbf{v}_1, \mathbf{v}_2 \to \mathbf{v}_3, \mathbf{v}_4) d\Omega$$

$$\times |\mathbf{v}_1 - \mathbf{v}_2| (f_3 f_4 - f_1 f_2)(\chi_1 + \chi_2 - \chi_3 - \chi_4) = 0$$

which completes the proof.

Using the transport equation (Eq. (11.4.11) we now rewrite Eq. (13.1.1) as

$$\int \chi \left(\frac{\partial f}{\partial t} + \mathbf{v} \cdot \frac{\partial f}{\partial \mathbf{r}} + \frac{\mathbf{F}}{m} \cdot \frac{\partial f}{\partial \mathbf{v}} \right) d^3 v = 0 \qquad \text{... (13.1.3)}$$

This is the general statement of the conservation law.

13.2. Transport Equation for Conserved Quantities

For any conserved quantity χ, we want to transform the conservation law of Eq. (13.1.3) into an equation for the time dependence of the average $\langle\chi\rangle$, which in defined as

$$<\chi> = \int \chi f d^3v / \int f d^3v$$

$$= \int \chi\, f d^3v / n\,(\mathbf{r}, t) \qquad \text{... (13.2.1)}$$

where

$$n\,(\mathbf{r}, t) = \int d^3v \cdot f(\mathbf{r}, \mathbf{v}, t) \qquad \text{... (13.2.2)}$$

which is a space- and time-dependent number density.

From Eq. (13.2.1), we find

$$\int \chi f d^3v = n <\chi> = <n\chi>, \qquad \text{... (13.2.3)}$$

where in the last step we have taken n inside the angular bracket since the angular bracket averaging is over velocity, while $n\,(\mathbf{r}, t)$ has already been averaged over velocity. With the above definition we can rewrite Eq. (13.1.3) as

$$\frac{\partial}{\partial t} \int \chi f d^3v + \int \left[\frac{\partial}{\partial \mathbf{r}} \cdot (\chi f \mathbf{v}) - f \vec{\nabla} \cdot (\chi \mathbf{v})\right] d^3v +$$

$$\frac{1}{m} \int \left[\frac{\partial}{\partial \mathbf{v}} (\chi f \mathbf{F}) - f\mathbf{F} \cdot \frac{\partial \chi}{\partial \mathbf{v}} - \chi f \frac{\partial}{\partial \mathbf{v}} \cdot \mathbf{F}\right] d^3v = 0 \qquad \text{... (13.2.4)}$$

Now,

$$\int \frac{\partial}{\partial \mathbf{v}} (\chi\, \mathbf{F} f)\, d^3v = \int_S f \chi\, \mathrm{F} \,.\, d\mathrm{S}_v = 0 \qquad \text{... (13.2.5)}$$

under the very reasonable assumption that for very large velocities, the distribution function f tends to zero.

If we ignore velocity-dependent forces, then

$$\frac{\partial \mathbf{F}}{\partial \mathbf{v}} = 0 \qquad \text{... (13.2.6)}$$

and using the definition that, in general, for any quantity A,

$$<A> = \int A f d^3\, v/n \qquad \text{... (13.2.7)}$$

we can write Eq. (13.2.4) as

$$\frac{\partial}{\partial t}<n\chi> + \vec{\nabla}\cdot<n\,\chi\,\mathbf{v}> - <n\,\vec{\nabla}\cdot(\chi\,\mathbf{v})> - <n\frac{\mathbf{F}}{m}\cdot\frac{\partial\chi}{\partial\mathbf{v}}> = 0 \qquad \text{... (13.2.8)}$$

This is the form of the time development for the expectation value of a conserved quantity χ. For the binary collision there are three such conserved quanities χ.

i) a trivial constant which we can take as the mass m of the individual colliding particles

ii) the momentum $\mathbf{P} = m\mathbf{v}$ of the individual particles

iii) the thermal kinetic energy $\varepsilon = \frac{1}{2}\,m\,(v - <v>)^2$

The above form is a linear combination of conserved quantities namely, v^2, $\mathbf{v}$ and constant) and hence is itself a conserved quantity.

We take up the three cases separately.

i) If $\chi = m$, we can write Eq. (13.2.8) as

$$\frac{\partial}{\partial t}<mn> + \vec{\nabla}\cdot<(mn\,\mathbf{v})> - <n\vec{\nabla}\cdot(m\mathbf{v})> = 0 \qquad \text{... (13.2.9)}$$

the last term in Eq. (13.2.8) not contributing for obvious reasons. Noting that $mn = \rho$, the mass density, and that $\vec{\nabla}\cdot\mathbf{v} = 0$ as $\mathbf{v}$ and $\mathbf{r}$ are independent of each other, we have

$$\frac{\partial}{\partial t}<\rho> + \vec{\nabla}\cdot<(\rho\,\mathbf{v})> = 0 \qquad \text{... (13.2.10)}$$

If we write $<\mathbf{v}> = \mathbf{u}$ and note that $<\rho> = \rho$ since ρ is a function of $\mathbf{r}$ and t alone, we can write the above as the continuity equation

$$\frac{\partial\rho}{\partial t} + \vec{\nabla}\cdot(\rho\,\mathbf{u}) = 0 \qquad \text{... (13.2.11)}$$

It should be borne in mind that the distribution function f which we are using to average over $\mathbf{v}$ is not, in general, an even function of $\mathbf{v}$ and hence $<\mathbf{v}> \neq 0$. The function f is a solution of the transport equation [Eq. (11.4.11)].

ii) If $\vec{\chi} = m\mathbf{v}$, then for the ith component $\chi_i = mv_i$, we have from Eq. (13.2.8)

$$\frac{\partial}{\partial t} <mn\, v_i> + \frac{\partial}{\partial x_j} (mn\, v_i v_j) - <n \frac{\partial}{\partial x_j} (mv_i\, v_j) - <nF_j \frac{\partial v_i}{\partial v_j}> = 0$$

or

$$\frac{\partial}{\partial t} <\rho v_i> + \frac{\partial}{\partial x_j} \cdot <\rho\, v_i\, v_j) = n <F_i> \qquad \text{... (13.2.12)}$$

where use has been made of the fact that **v** and **r** are independent variables.

We now define a tensor

$$T_{ij} = \rho <(v_i - u_i)\ (v_j - u_j)>$$

$$= \rho\ <v_i\, v_j> - \rho\, u_i\, u_j\ , \qquad \text{... (13.2.13)}$$

having made use of the fact $<v_i> = u_i$. In terms of the tensor, Eq. (13.2.12) becomes

$$\frac{\partial \rho}{\partial t} u_i + \rho \frac{\partial u_i}{\partial t} + \frac{\partial}{\partial x_j} T_{ij} + \frac{\partial}{\partial x_j} (\rho\, u_i\, u_j) = \frac{<F_i\, \rho>}{m},$$

$$\left[\frac{\partial \rho}{\partial t} + \frac{\partial}{\partial x_j} (\rho u_j)\right] u_i + \rho \frac{\partial u_i}{\partial t} + \rho u_j \frac{\partial u_i}{\partial x_j} = - \frac{\partial}{\partial x_j} T_{ij}$$

or

$$\frac{\partial u_i}{\partial t} + u_j \frac{\partial u_i}{\partial x_j} = - \frac{1}{\rho} \frac{\partial}{\partial x_j} T_{ij} + \frac{1}{m} <F_i> \qquad \text{... (13.2.14)}$$

using Eq. (13.2.11) in arriving at the final form.

iii) If $\chi = m_\varepsilon = m \cdot \frac{m}{2} |\mathbf{v} - \mathbf{u}|^2$, i.e. ε is the thermal energy, then Eq. (13.2.8) becomes

$$\frac{\partial}{\partial t} <mn\varepsilon> + \vec{\nabla} \cdot <mn\, \mathbf{v}\, \varepsilon>$$

$$- <mn\, (\mathbf{v} \cdot \vec{\nabla}) \frac{m}{2} |(\mathbf{v} - \mathbf{u})|^2> = <mn\, \mathbf{F} \cdot (\mathbf{v} - \mathbf{u})> = 0$$

$$\rho \frac{\partial <\varepsilon>}{\partial t} + <\varepsilon> \frac{\partial \rho}{\partial t} + \vec{\nabla} \cdot <\rho \varepsilon \mathbf{v}> - m <\rho\, v_i\, (v_j - u_j)> \frac{\partial u_j}{\partial x_i} = 0 \qquad \text{... (13.2.15)}$$

If we define

$$\vec{\theta} = \rho\, \varepsilon\, (\mathbf{v} - \mathbf{u}) \qquad \text{... (13.2.16)}$$

and note that $\rho <v_i (v_j - u_j)> = T_{ij}$, then Eq. (13.2.15) becomes

$$<\varepsilon> \frac{\partial \rho}{\partial t} + \rho \frac{\partial}{\partial t} <\varepsilon> + \vec{\nabla} \cdot <\vec{\theta}> + \vec{\nabla} \cdot <(\rho \varepsilon \mathbf{u})> + m\, T_{ij} \frac{\partial u_j}{\partial x_i} = 0$$

or

$$\rho \frac{\partial}{\partial t} <\varepsilon> + \rho\, (\mathbf{u} \cdot \vec{\nabla}) <\varepsilon> + <\varepsilon> \left[\vec{\nabla} \cdot (\rho \mathbf{u}) + \frac{\partial \rho}{\partial t} \right] = - \vec{\nabla} <\vec{\theta}> - m\, T_{ij} \frac{\partial u_j}{\partial x_i}$$

or

$$\frac{\partial}{\partial t} <\varepsilon> + (\mathbf{u} \cdot \vec{\nabla}) <\varepsilon> = - \frac{1}{\rho} \vec{\nabla} \cdot <\vec{\theta}> - \frac{m}{2\rho} T_{ij} \left(\frac{\partial u_j}{\partial x_i} + \frac{\partial u_i}{\partial x_j} \right) \qquad \text{... (13.2.17)}$$

where the symmetry of T_{ij} has been used in the last term.

The five conservation laws are thus expressed through Eq. (13.2.11), Eq. (13.2.14) for $i = 1, 2, 3$ and Eq. (13.2.17). To make them more useful, however, we must be able to calculate T_{ij} and $<\theta_i>$.

13.3 Equations of Hydrodynamics

As stated at the end of the last section, the conservation laws become usable only if we know the distribution function f, as otherwise it is not possible to calculate the expectation values. Since solving the transport equation and obtaining f is virtually impossible, we have to make an Ansatz for it. We take it to be locally Maxwell-Boltzmann with a mean velocity $\mathbf{u}$, ie.

$$\mathbf{f}\,(\mathbf{v}, \mathbf{r}, t) = n \left(\frac{m}{2\pi k T} \right)^{3/2} e^{-\frac{m\,(\mathbf{v} - \mathbf{u})^2}{2kT}} \qquad \text{... (13.3.1)}$$

where n and T (the temperature) are both local variables depending on $\mathbf{r}$ and t. Clearly

$$<v_i> = u_i \qquad \text{... (13.3.2a)}$$

and

$$T_{ij} = \rho <(v_i - u_i)(v_j - u_j)> = \frac{kT}{m} \rho \delta_{ij} \qquad \text{... (13.3.2b)}$$

If we consider a local equation of state as

$$P = nkT = \rho \frac{kT}{m} \qquad \text{... (13.3.3)}$$

then

$$T_{ij} = P \delta_{ij} \qquad \text{... (13.3.4)}$$

and Eq. (13.2.14) becomes Euler's equation of hydrodynamics. With the above distribution $<\theta_i> = 0$ and hence Eq. (3.2.17) becomes

$$\rho \frac{\partial}{\partial t} <\varepsilon> + \rho (\mathbf{u} \cdot \vec{\nabla}) <\varepsilon> = - m P \frac{\partial u_i}{\partial x_i} \qquad \text{... (13.3.5)}$$

It should be noted that

$$<\varepsilon> = <\frac{1}{2} m (\mathbf{v} - \mathbf{u})^2> = \frac{3}{2} kT \qquad \text{... (13.3.6)}$$

and substituting in Eq. (13.3.5), we obtain

$$\frac{3}{2} k \left(\frac{\partial T}{\partial t} + (\mathbf{u} \cdot \vec{\nabla}) T \right) + \rho kT \frac{\partial u_i}{\partial x_i} = 0$$

which gives the time dependence of the local temperature as

$$\frac{\partial T}{\partial t} + (\mathbf{u} \cdot \vec{\nabla}) T = - \frac{2}{3} T (\vec{\nabla} \cdot \mathbf{u}) \qquad \text{... (13.3.7)}$$

The quantity $\partial/\partial t + \mathbf{u} \cdot \vec{\nabla}$ is the convective derivative since it is the time rate of change as seen by an observer moving with the local average velocity $\mathbf{u}$. Such an observer moves along a streamline. We note that at this order of evaluation,

$$\frac{3}{2T} \left(\frac{\partial}{\partial t} + \mathbf{u} \cdot \vec{\nabla} \right) T = - (\vec{\nabla} \cdot \mathbf{u}) \qquad \text{... (13.3.8)}$$

while from the equation of continuity

$$\frac{1}{\rho}\left(\frac{\partial}{\partial t} + \mathbf{u}\cdot\vec{\nabla}\right)\rho = -\left(\vec{\nabla}\cdot\mathbf{u}\right) \qquad \text{... (13.3.9)}$$

Subtracting Eq. (13.3.9) from Eq. (13.3.8), we get

$$\left(\frac{\partial}{\partial t} + \mathbf{u}\cdot\vec{\nabla}\right)\rho\, T^{-3/2} = 0$$

or $\rho T^{3/2}$ = constant along a streamline. ... **(13.3.10)**

It we use $P = mkT/\rho$, then $P\rho^{-5/3}$ is constant along the streamline which is the adiabatic law since for an ideal gas $\gamma = C_p/C_v = 5/3$ and we have found that PV^{γ} = constant. Thus for the streaming observer only adiabatic processes occur in a dilute gas in this approximation.

Finally, using Eq. (13.3.4.) in Eq. (13.2.1), the velocity is found to satisfy

$$\frac{\partial u_i}{\partial t} + u_j\frac{\partial u_i}{\partial x_j} = -\frac{1}{\rho}\frac{\partial P}{\partial x_i} + \frac{F_i}{m} \qquad \text{... (13.3.11)}$$

If we have steady state conditions ($\partial/\partial t = 0$) under a conservative force $F_i = -m\vec{\nabla}\varphi$, then we can write Eq. (13.3.11) as

$$\vec{\nabla}\left(\frac{1}{2}u^2 + \frac{P}{\rho} + \frac{\varphi}{m}\right) = \mathbf{u}\times(\vec{\nabla}\times\mathbf{u}) - \frac{kT}{m}\frac{\vec{\nabla}\rho}{\rho},$$

where the equation of state and the identity $(\mathbf{u}\cdot\vec{\nabla})\,\vec{u} = 1/2\,\vec{\nabla}(u^2) - \mathbf{u}\times(\vec{\nabla}\times\mathbf{u})$ have been used. Two cases can be noted.

i) if the flow is irrotational ($\vec{\nabla}\times\mathbf{u} = 0$) and the density is uniform (ρ = const)

$$\frac{u^2}{2} + \frac{P}{\rho} + \frac{\varphi}{m} = \text{constant in space} \qquad \text{... (13.3.12a)}$$

ii) if the flow is irrotational and the temperature is constant then

$$\rho = \rho_0\, e^{-\frac{1}{kT}\left(\frac{1}{2}mu^2 + \varphi\right)} \qquad \text{... (13.2.12b)}$$

What is missing from the above hydrodynamic equations are the dissipative terms. This is due to the particular form of the distribution that we have chosen. With that as the starting Ansatz, we can build up corrections to it. These corrections can be obtained to the lowest order by solving the transport equation in the relaxation time approximation. This involves writing $f = \bar{f} + \delta f$, with the collision term replaced by an expression of the form $-\delta f/\tau$, where τ is the relaxation time. In the absence of external forces, δf satisfies the relation $\partial/\partial t\, \delta f + \mathbf{v} \cdot \partial \bar{\mathbf{f}}/\partial \mathbf{r} = -\delta f/\rho$, and re-evaluation of the averages using $f = \bar{f} + \delta f$, we arrive at Navier-Stokes equation and the heat conduction equations including the shear viscosity and thermal conductivity in terms of the relaxation time τ.

Chapter 14

Boltzmann Equation for Quantum Particles

14.1 Form of the Transport Equation

We have so far discussed situations where quantum effects were not important since the De Broglie wavelength ($\lambda \sim h/\sqrt{mkT}$) was much smaller than the mean free parth (inversely proportional to the density). At low temperatures or high densities this approximation breaks down and we have to consider the quantum effects. If we consider the transport equation for classical particles [Eq. (11.4.11)] we find that the left hand side of the equation is simply the total rate of change in the particle number and hence is independent of the classical or quantum nature of the particles. The right hand side, on the other hand, depends on the collision process going on in the system and that is going to be sensitive to the nature of the particles. This is immediately obvious if the particles considered are identical and are fermions. The requirment that there can be no more than one fermion in a given state implies that in the scattering process all final states which are occupied cannot be accessed and this will change the collision integral. The case of the bosons is more subtle but certainly there will be an effect on the collision integral as we will see in what follows. Hence the effect of the quantum nature of the particles will be showing up in the collision term and it will depend on whether the particles are fermions or bosons.

Accordingly, the collision term will be altered to take into account the quantum nature of the particles. However, the Boltzmann equation that we will be writing down is sensible only in the quasi-classical approximation. It is only in this limit that it is permissible to talk about $f(\mathbf{r}, \mathbf{v}, t)$. In a strict quantum theory, uncertainty principle puts constraints on the simultaneous measurements of $\mathbf{r}$ and $\mathbf{v}$. In the semiclassical picture that we will be employing, one can interpret the $\mathbf{r}$ and $\mathbf{v}$ as quantum expectation values with the quantum fluctuations small. It is in this spirit that we keep the left hand side of Eq. (11.4.11) unaltered and proceed to alter the collision term depending on whether the particles are fermions or bosons.

4.1.1 Fermions

As we have already stated the collision term has to account for the fact that all occupied final states are inaccessible by collisions. Thus, if we consider the number of particles with velocity in the vicinity of $\mathbf{v}$, then the depletion rate due to the elastic two-body scattering process for classical particles was found in Eq. (11.4.3). The fermionic nature implies that in the scattering process where a particle with velocity $\mathbf{v}$ collides with an identical particle with velocity $\mathbf{v}_1$ producing final states with velocities $\mathbf{v}_2$ and $\mathbf{v}_3$, the scattering cross-section will be zero if there is already a particle with velocity $\mathbf{v}_2$ or $\mathbf{v}_3$ in the physical volume d^3r under consideration. To incorporate this fact, we need to multiply the expression in Eq. (11.4.3) by the factor $(1 - f_2)(1 - f_3)$ which ensures that if either of the possible final states is occupied, the cross-section vanishes. Hence the depletion rate is

$$\int d^3r\, d^3v_1\, d^3v\, \sigma(\mathbf{v}, \mathbf{v}_1 \rightarrow \mathbf{v}_2, \mathbf{v}_3)\, |\mathbf{v} - \mathbf{v}_1|\, f(\mathbf{r}, \mathbf{v}, t) f(\mathbf{r}, \mathbf{v}_1, t) \times$$

$$[(1 - f(\mathbf{r}, \mathbf{v}_3, t)] \times [(1 - f(\mathbf{r}, \mathbf{v}_2, t)]$$

Similarly, if we consider the rate of addition to the number of particles in the volume d^3r with velocities in the neighbourhood of velocity $\mathbf{v}$, the additional factor that will enter in Eq. (11.4.4) is $(1 - f)(1 - f_1)$ which ensures that the scattering cannot occur if there is already a particle with velocity $\mathbf{v}$ or $\mathbf{v}_1$. Hence the addition rate is

$$\int d^3r\, d^3r_2\, d^3v_3\, \sigma(\mathbf{v}_2, \mathbf{v}_3 \rightarrow \mathbf{v}, \mathbf{v}_1)\, |\mathbf{v}_2 - \mathbf{v}_3|\, f(\mathbf{r}, \mathbf{v}_2, t) f(\mathbf{r}, \mathbf{v}, t)$$

$$[1 - f(\mathbf{r}, \mathbf{v}, t)]\, [1 - f(\mathbf{r}, \mathbf{v}_1, t)]$$

The Boltzmann equation follows from arguments identical to those given in Chapter 11 and we have the Boltzmann transport equation for fermions as

$$\frac{\partial f}{\partial t} + \mathbf{v} \cdot \frac{\partial f}{\partial \mathbf{r}} + \frac{\mathbf{F}}{m} \cdot \frac{\partial f}{\partial \mathbf{v}} = \int d^3v_1 \, |\mathbf{v} - \mathbf{v}_1| \, \sigma(\mathbf{v}, \mathbf{v}_1 \rightarrow \mathbf{v}_2, \mathbf{v}_3) \, d\Omega$$

$$\left[f_2 f_3 (1 - f)(1 - f_1) - f f_1 (1 - f_2)(1 - f_3)\right] \quad \text{... (14.1.1)}$$

14.1.2 Bosons

For bosons there is no restriction to the number of particles that can occupy a given state. The consequence of this on a scattering process can be found in textbooks on quantum mechanics (e.g. see *Feynman Lectures*, Vol. III). The result is that in a scattering process if there are N particles already in a potential final state, then the probability of scattering into that state is enhanced by the factor $(N + 1)$. Hence when we consider the depletion rate due to the two-body scattering, the cross-section is enchanced if there are particles present with velocities $\mathbf{v}_2$ or $\mathbf{v}_3$,, the velocities of the final state in the scattering process, and the factor which expresses this is $(1 + f_2)(1 + f_3)$. Similarly, the rate of addition of particles in the phase space volume $d^3r \, d^3v$ is enchanced by the factor $(1 + f)(1 + f_1)$ since the final state in this two-body process now has particles with velocities $\mathbf{v}$ and $\mathbf{v}_1$. It should be obvious that the transport equation for bosons will consequently be

$$\frac{\partial f}{\partial t} + \mathbf{v} \cdot \frac{\partial f}{\partial \mathbf{r}} + \frac{\mathbf{F}}{m} \cdot \frac{\partial f}{\partial \mathbf{v}} = \int d^3v_1 \, |\mathbf{v} - \mathbf{v}_1| \, \sigma(\mathbf{v}, \mathbf{v}_1 \rightarrow \mathbf{v}_2, \mathbf{v}_3) \, d\Omega$$

$$\times \left[f_2 f_3 (1 + f)(1 + f_1) - f f_1 (1 + f_2)(1 + f_3)\right] \quad \text{... (14.1.2)}$$

14.2 Equilibrium Distribution

As in the case of classical particles, the equilibrium distribution is first obtained for the specific case of no external force, i.e. $\mathbf{F} = 0$. In this case the distribution f_0 is independent of $\mathbf{r}$ and since in equilibrium $\partial f_0 / \partial t = 0$, we must have the collision term vanish when f is taken to be the equilibrium distribution. In the following, we will be using the upper sign for fermions and the lower sign for bosons. In equilibrium conditions, we thus have

$$f_0 f_{10}(1 \pm f_{20})(1 \pm f_{30}) = f_{20} f_{30}(1 \pm f_0)(1 \pm f_{10})$$

or

$$\frac{f_0}{1 \pm f_0} \times \frac{f_{10}}{1 \pm f_{10}} = \frac{f_{20}}{1 \pm f_{20}} \times \frac{f_{30}}{1 \pm f_{30}}$$

or

$$\ln \frac{f_0}{1 \pm f_0} + \ln \frac{f_{10}}{1 \pm f_{10}} = \ln \frac{f_{20}}{1 \pm f_{20}} + \ln \frac{f_{30}}{1 \pm f_{30}} \qquad \text{... (14.2.1)}$$

In complete analogy with the classical case, we now argue that ln $f_0/(1 \pm f_0)$ must be a linear function of the collision invariant. Hence,

$$\ln \frac{f_0}{1 \pm f_0} = \alpha - \beta\,\varepsilon + \gamma \cdot \mathbf{v}_0$$

We can choose the reference velocity to be zero without any loss of generality and obtain

$$\ln \frac{f_0}{1 \pm f_0} = \alpha - \beta\,\varepsilon$$

or

$$f_0 = \frac{1}{e^{\varepsilon\beta - \alpha} \pm 1} \qquad \text{... (14.2.2)}$$

The constants β and α now need to be specified. To do so, we note that the classical distribution obtained in chapter 11 must be a limiting case of the quantum distribution of Eq. (14.2.2) at high temperatures and low densities. The low density condition implies $f_0 << 1$, which is achieved if we drop the constant term in the denominator of Eq. (14.2.2). This leads to

$$f_0 \approx e^{\alpha}\, e^{-\beta\varepsilon} \qquad \text{... (14.2.3)}$$

and comparing with Eq. (11.4.23), we infer that

$$\beta = (kT)^{-1} \qquad \text{... (14.2.4a)}$$

$$e^{\alpha} = n \left(\frac{m}{2\pi kT} \right)^{3/2} \quad \text{... (14.2.4b)}$$

As can be checked by referring to chapter 4, $\alpha = \mu\beta$, where μ is the chemical potential. Thus for fermions,

$$f_0 = \frac{1}{e^{(\varepsilon - \mu)/kT} + 1} \quad \text{... (14.2.5a)}$$

while for bosons

$$f_0 = \frac{1}{e^{(\varepsilon - \mu)/kT} - 1} \quad \text{... (14.2.5b)}$$

A brief comment should be made here about the interpretation of $f_0(\mathbf{v})$ in this quantum case. The classical interpretation was in terms of a density in phase space, the number of particles in the volume $d^3r\, d^3v$ being given by $f_0(\mathbf{v})\, d^3r\, d^3v$ in equilibrium. In the quantum case it is more appropriate to interpret $f_0(\mathbf{v})$ as the average number of particles in the state with velocity $\mathbf{v}$ (the average being over different realizations) and $d^3r\, d^3v$ as being the number of states available to the quantum particle in that region of phase space multiplied by the factor h^3/m^3.

For the fermions, we notice that if the limit $T \to 0$ is taken, then the distribution reduces to a theta function, with $f_0(\mathbf{v}) = 1$ if $\varepsilon = 1/2\, mv^2 < \mu$ and $f_0(\mathbf{v}) = 0$ if $\varepsilon > \mu$. All states below the energy μ, called the fermi energy ε_F, are thus fully occupied and those above the energy μ are totally empty.

$$f_0(\varepsilon) = \theta(\varepsilon_F - \varepsilon), \text{ for } T = 0 \quad \text{... (14.2.6)}$$

where the theta function is defined by $\theta(x) = 1$ if $x > 0$ and $\theta(x) = 0$ if $x < 0$. For the bosons, it should be noted that the average number of particles with $\varepsilon = 0$ is $(e^{-\mu/kT} - 1)^{-1}$ and this constrains μ to be less than or equal to zero. If we consider bosons, then the number conservation need not be valid and in that case we must have $\alpha = 0$ in Eq. (14.2.2) and consequently $\mu = 0$. This is true for bosons which can be viewed as excitations in a system characterized as a collection of oscillators (photons in a cavity, phonons arising from the vibrations of the atoms in a solid about their equilibrium position or oscillations in density in a liquid at very low temperatures, specifically helium which alone remains a liquid at temperatures where quantum effects matter).

14.3 Approach to Equilibrium (F ≠ 0 but No Collisions)

We now see how a system of fermions behaves when a small deviation from equilibrium conditions is made. We ignore the collisions and consider the weak interaction between the particles as responsible for bringing the system towards equilibrium. This is analogous to the case of plasma oscillations dealt with in Chapter 12, where for a system of charged particles, against a neutralizing background we studied deviations from equilibrium. It was the Coulomb forces which came into play on departure from equilibrium in that case that was responsible for the oscillations. We now consider an uncharged system of fermions—this is provided by ^{3}He atoms—at extremely low temperatures. There is an attractive force (weak) between ^{3}He atoms but in an equilibrium situation, however, the net force on each atom by the collection of the other atoms is zero. The equilibrium distribution is consequently given by Eq. (14.2.6). We take the weak interaction between atoms to be governed by some potential $u(r)$ and if there is a deviation δf from the equilibrium distribution, then the deviation in local number density is $\delta n(\mathbf{r}, t)$ and the potential at the point $\mathbf{r}$ will be

$$\delta\varphi = \int u(\mathbf{r} - \mathbf{r}')\ \delta n(\mathbf{r}', t)\, d^3r' \qquad \text{... (14.3.1)}$$

The potential is first order in δn and hence the force $\mathbf{F}$ is also first order in δn. The transport equation in the form $\partial f/\partial t + \mathbf{v}\cdot\partial f/\partial\mathbf{r} + F/m\cdot\partial f/d\mathbf{v} = 0$, then becomes to first order in δn,

$$\frac{\partial}{\partial t}\,\delta f + \mathbf{v}\cdot\frac{\partial}{\partial\mathbf{r}}\,\delta f - \frac{\vec{\nabla}\delta\varphi}{m}\cdot\frac{\partial f_0}{\partial\mathbf{v}} = 0 \qquad \text{... (14.3.2)}$$

Assuming that the deviation δf will be oscillatory in nature, we can make the Ansatz

$$\delta f = A\, e^{i(\omega t - \mathbf{k}\cdot\mathbf{r})} \qquad \text{... (14.3.3.a)}$$

$$\delta n = B\, e^{i(\omega t - \mathbf{k}\cdot\mathbf{r})} \qquad \text{... (14.3.3b)}$$

where B is a constant and A is a function of velocity such that

$$B = \int A(\mathbf{v})\, d^3v \qquad \text{... (14.3.4)}$$

We can write the force $\vec{\nabla}\delta\varphi$ as

$$\vec{\nabla}\,\delta\varphi = \vec{\nabla}\int u(\mathbf{r}-\mathbf{r}')\;\delta n\,(\mathbf{r}',t)\,d^3r'$$

$$= B\,e^{i\omega t}\,\vec{\nabla}\left[e^{i\mathbf{k}\cdot\mathbf{r}}\int u(\mathbf{r}-\mathbf{r}')\;e^{i\mathbf{k}\cdot(\mathbf{r}-\mathbf{r}')}\,d^3r'\right]$$

$$= B\,e^{i\omega t}\,u(k)\,(-\,i\mathbf{k})e^{-i\mathbf{k}\cdot\mathbf{r}} \qquad \text{... (14.3.5)}$$

where $u(k) = \int d^3r\;u(r)\,e^{i\mathbf{k}\cdot\mathbf{r}}$ is the Fourier transform of $u(r)$ and Eq. (14.3.2) becomes

$$A\,(i\omega - i\mathbf{k}\cdot\mathbf{v})\,e^{i(\omega t - \mathbf{k}\cdot\mathbf{v})} = -\,B\,u(k)\,i\,\mathbf{k}\cdot\frac{\partial f}{\partial \mathbf{v}}\times e^{i(\omega t - \mathbf{k}\cdot\mathbf{r})}$$

Now, $\partial f_0/\partial\mathbf{v} = -\,m\mathbf{v}\;\delta(\varepsilon_F - 1/2\,mv^2)$ and hence Eq. (14.3.6) becomes

$$A = B\,m\,u\,(k)\,\frac{\mathbf{k}\cdot\mathbf{v}}{\omega - \mathbf{k}\cdot\mathbf{v}}\,\delta\left(\varepsilon_F - \frac{1}{2}mv^2\right) \qquad \text{... (14.3.7)}$$

Insertion of the value of A from Eq. (14.3.7) in Eq. (14.3.4) yields

$$1 = m\,u(k)\int\frac{kv\cos\theta}{\omega - kv\cos\theta}\;\delta\left(\varepsilon_F - \frac{1}{2}mv^2\right)d^3v$$

$$= 2\pi\,m\,u(k)\int_{-1}^{1}\frac{\cos\theta\;d(\cos\theta)}{\dfrac{\omega}{k}\left(\dfrac{m}{2\,\varepsilon_F}\right)^2 - \cos\theta}\;\frac{2}{m}\,\frac{1}{2v_F}\cdot v_F^2$$

$$= 2\pi\,v_F\,u(k)\left[\ln\frac{\dfrac{c}{v_F}+1}{\dfrac{c}{v_F}-1} - 2\right] \qquad \text{... (14.3.8)}$$

where $1/2\,mv_F^2 = \varepsilon_F$ and $\omega = ck$. Clearly if $u(k)\to 0$, then $c\to v_F$ for Eq. (14.3.8) to make sense. For small $u(k)$, we accordingly get

$$\frac{c}{v_F} = 1 + 2\,e^{-\frac{1}{2\pi\,u(k)\,v_F}} \qquad \text{... (14.3.9)}$$

Thus, the interaction between the helium atoms allows a wave to be set up in this system of fermions and the wave velocity is given by Eq. (14.3.9) which is a remarkable result in the sense that it is non-perturbative in the strength of the interaction. The deviation of the propagation speed from the Fermi velocity v_F is small if the interaction is small, but this smallness cannot be expressed in a perturbation series. These oscillations are known as zero sound. They are very different from the usual sound (first sound) wave which, as we have seen, for the classical gas, can be propagated only in the presence of collisions. In the next section, we will investigate the effect of collisions and find the velocity of first sound at very low temperatures. However, it should be noted that as $T \to 0$, the damping of first sound increases very strongly and hence the zero sound oscillations dominate in that temperature range.

14.4 Effect of Collisions in a Fermion Gas

We now consider a situation where the external force $\mathbf{F} = 0$ and a deviation from equilibrium has to relax or propagate by the action of the collision term. The Boltzmann equation now takes the form

$$\frac{\partial f}{\partial t} + \mathbf{v} \cdot \frac{\partial f}{\partial \mathbf{r}} = \int d^3v \, d\Omega \, \sigma \, |\mathbf{v} - \mathbf{v}_1| \left[f_2 f_3 (1 - f)(1 - f_1) - f f_1 (1 - f_2)(1 - f_3) \right] \quad \text{... (14.4.1)}$$

with the equilibrium distribution

$$f_0 = \left(e^{(\varepsilon - \mu)/kT} + 1 \right)^{-1} \quad \text{... (14.4.2)}$$

The deviation from f_0 is taken to be oscillatory in space with a long wavelength (exactly as in the classical case in Chapter 13) and for convenience we take the small deviation from equilibrium to have the explicit form:

$$f = f_0 + f_0 (1 - f_0) \, e^{i \cdot \mathbf{k} \cdot \mathbf{r}} \, \psi(\mathbf{v}, t) \quad \text{... (14.4.3)}$$

where $\psi(\mathbf{v}, t)$ is to be found from the explicit solution of the linearized form of Eq. (14.4.1). The f-dependent part of the collision integral of Eq. (14.4.1) when linearized in Ψ is found by noting that:

$$f_2 f_3 (1 - f)(1 - f_1) = f_{20} \, [1 + (1 - f_{20}) \psi_2] \cdot f_{30} \, [1 + (1 - f_{30}) \psi_3] \cdot$$

$$[(1 - f_0)(1 - f_0 \Psi)] \times [(1 - f_{10})(1 - f_{10} \Psi_1)]$$

$$= f_{20} f_{30} (1 - f_0)(1 - f_{10}) [1 + (1 - f_{20}) \psi_2 + (1 - f_{30}) \psi_3 - f_0 \psi - f_{10} \psi]$$

and in a similar way

$$f_1 f (1 - f_2)(1 - f_3) = f_{10} f_0 (1 - f_{20})(1 - f_{30}) \left[1 + (1 - f_{10}) \psi + (1 - f_0) \psi_1 - f_{20} \psi_2 - f_{30} \psi_3 \right]$$

leading to

$$f_2 f_3 (1 - f)(1 - f_1) - f f_1 (1 - f_2)(1 - f_3) = f_0 f_{10} (1 - f_{20})(1 - f_{30}) \times \left[\psi_2 + \psi_3 - \psi_1 - \psi \right] \quad \text{... (14.4.4)}$$

The transport equation in terms of ψ, becomes

$$\frac{\partial \psi}{\partial t} + \mathbf{ik} \cdot \mathbf{r} \psi = \int d^3 v \, d\Omega \, \sigma \, |\mathbf{v} - \mathbf{v}_1| \, \frac{f_{10} (1 - f_{20})(1 - f_{30})}{1 - f_0} \times (\psi_2 + \psi_3 - \psi_1 - \psi) \quad \text{... (14.4.5)}$$

or

$$\frac{\partial \psi}{\partial t} = - K \psi = - K_0 \psi - K_1 \psi, \quad \text{... (14.4.6)}$$

where

$$K_0 \psi = \int d^3 v \, d\Omega \, \sigma \, |\mathbf{v} - \mathbf{v}_1| \, \frac{f_{10} (1 - f_{20})(1 - f_{30})}{1 - f_0} (\psi_2 + \psi_3 - \psi_1 - \psi) \quad \text{... (14.4.6a)}$$

$$K_1 \psi = - \mathrm{i} \, \mathbf{k} \cdot \mathbf{v} \, \psi \quad \text{... (14.4.6b)}$$

Exactly as in Chapter 12, the lowest eigenvalues of the operator K will determine the time dependence of ψ. Since k is small, this lowest eigenvalue is to be found perturbatively in K_1 and the zero eigenvalues of K_0 are the important ones provided K_0 has no negative eigenvalues. By defining the scalar product as

$$(Q, \psi) = \int Q(\mathbf{v}) f_0 (1 - f_0) \psi(\mathbf{v}) \, d^3v \qquad \text{... (14.4.7)}$$

it is clear (the reader should prove this using arguments identical to those given in Chapter 12) that

$$(\psi, K_0 \psi) \geq 0 \qquad \text{... (14.4.8)}$$

and K_0 has non-negative eigenvalues. Consequently the zero eigenvalue alone of K_0 is important and this eigenvalue is five-fold degenerate corresponding to the five collision invariants: the number of particles, the three components of momentum and the kinetic energy. It should be noted that the weighting factor $f_0(1 - f_0)$ in the definition of scalar product in Eq. (14.4.7) is of significance. The probability of a fermion occupying a state is f_0: consequently $1 - f_0$ is the probability of a state being unoccupied or as is customarily described for $\varepsilon < \varepsilon_F$ as being occupied by a hole. The product $f_0(1 - f_0)$ has to do with the probability of finding electron-hole pairs and this is natural since the collision term acts by producing such pairs—the collision process leading to the formation of an electron state as well as a hole state.

To simplify the calculation we consider the propagation at very low temperatures where the distribution f_0 is essentially the theta function $\theta(\varepsilon_F - \varepsilon)$ and hence the factor $f_0(1 - f_0)$ behaves like a slightly smeared delta function $\delta(\varepsilon - \varepsilon_F)$. For integration purposes it is convenient and accurate to take it as a delta function. If we consider this form of the weight function, then there is no difference between the wavefunction which is a constant and the wavefunction which corresponds to the kinetic energy since all energies must be considered at the Fermi surface and that effectively make the energies a constant. In other words, for $T \to 0$, the only pairs which are formed are at the Fermi surface.

We now take the propagation of the disturbance to be along the z-direction, i.e. $\mathbf{k}$ has a z-component alone. The degenerate zero-eigenvalue is now treated in first order perturbation theory. Because of the energy effectively becoming a constant, there are now only four linearly independent eigenfunctions. We need to diagonalize $K_1 = ikv_z$ in the space of these four functions. Obviously the eigenfunctions corresponding to p_x and p_y do not couple at all and hence K_1 has to be diagonalized in the space of the two functions:

i) $\varphi_1 =$ constant $= A$ (say) and ii) $\varphi_2 = Bp_z$.

Clearly $(\varphi_1, K_1 \varphi_1) = (\varphi_2, K_1 \varphi_2) = 0$

and we need to worry about the off diagonal elements alone. First, the states $\varphi_{1,2}$ need to be normalized. We take the phase space factor to be $\varepsilon^{1/2}\, d\varepsilon\, d\Omega/4\pi$ where constant factors have been dropped without any loss of generality for the purpose at hand. We have,

$$(\varphi_1, \varphi_1) = A^2 \int \varepsilon^{1/2}\, \delta(\varepsilon - \varepsilon_F)\, d\varepsilon\, \frac{d\Omega}{4\pi} = A^2\, \varepsilon_F^{1/2}$$

Hence

$$A^2 = \varepsilon_F^{-1/2}, \text{ and}$$

$$(\varphi_2, \varphi_2) = B^2 \int p_z^2\, \varepsilon^{1/2}\, \delta\, (\varepsilon - \varepsilon_F)\, d\varepsilon\, d\Omega/4\pi$$

$$= \frac{B^2}{3} \int p^2\, \varepsilon^{1/2}\, \delta\, (\varepsilon - \varepsilon_F)\, d\varepsilon$$

$$= \frac{2m\, B^2}{3} \int \varepsilon^{3/2}\, \delta\, (\varepsilon - \varepsilon_F)\, d\varepsilon$$

$$= \frac{2m}{3}\, B^2\, \varepsilon_F^{3/2},$$

leading to $B^2 = (3/2m)\, \varepsilon_F^{-3/2}$. The off diagonal matrix elements can be calculated as

$$(\varphi_1, K_1\, \varphi_2) = AB\, \frac{ik}{m} \int p_z^2\, \varepsilon^{1/2}\, \delta(\varepsilon - \varepsilon_F)\, d\varepsilon\, \frac{d\Omega}{4\pi}$$

$$= \frac{ik2AB}{3} \int \varepsilon^{3/2}\, \delta\, (\varepsilon - \varepsilon_p)\, d\varepsilon$$

$$= \frac{2}{3}\, ik\, AB\, \varepsilon_F^{3/2}$$

The matrix to be diagonalized is

$$M = \begin{pmatrix} 0 & \frac{2}{3}\, ik\, AB\, \varepsilon_F^{3/2} \\ \frac{2}{3}\, ik\, AB\, \varepsilon_F^{3/2} & 0 \end{pmatrix}$$

If the eigenvalue is λ, then it is clear that $\lambda = \pm i\omega$, where we must have

$$\omega^2 = \frac{4}{9} \varepsilon_F^3 A^2 B^2 k^2 = \frac{2}{3} \frac{\varepsilon_F}{m} k^2 = \frac{1}{3} v_F^2 k^2$$

Writing $\omega = kc$,

$$c = \pm \frac{v_F}{\sqrt{3}} \qquad \text{... (14.4.9)}$$

leading to the result that for $T \to 0$, the first sound propagates at the speed $v_F/\sqrt{3}$, which is very different from the speed of the zero sound (Eq. 14.3.9) under the same conditions.

To obtain the damping, we need to go to second order in perturbation theory and obtain the correction to the zero eigenvalue of K_0. This correction is of the form

$$\sum_{m \neq 0} \frac{<o\,|ik \cos\theta\,|\,m>^2}{E_0 - E_m} \sim k^2 \frac{M}{E_1} \qquad \text{... (14.4.10)}$$

where M is $< O\,|\cos^2\theta\,|\,O>$ and E_1 is the first nonzero eigenvalue of K_0. This eigenvalue is proportional to the number of electron-hole pairs which as $T \to 0$, is proportional to T^2 and thus the damping diverges as $T \to 0$ obliterating the sound wave in spite of its finite velocity. Thus as $T \to 0$, only the zero sound propagates.

14.5 Collision Term for Bosons

As an application of the transport equation to the case of bosons, we choose a gas of phonons which is obtained when one has density oscillations in a low temperature liquid (super fluid helium) or atomic vibrations in a solid. The Boltzmann equation is ($\mathbf{F} = 0$, no external force)

$$\frac{\partial f}{\partial t} + \mathbf{v} \cdot \frac{\partial f}{\partial \mathbf{r}} = \int d^3v_1\, d\Omega\, \sigma\, |\mathbf{v} - \mathbf{v}_1| \left[f_2 f_3 (1 + f_1)(1 + f) - f f_1 (1 + f_2)(1 + f_3) \right] \qquad \text{... (14.5.1)}$$

with the equilibrium distribution function for phonons being ($\mu = 0$)

$$f_0 = \frac{1}{e^{\varepsilon/kT} - 1} \qquad \text{... (14.5.2)}$$

We consider a deviation from equilibrium and write the general distribution f as

$$f = f_0 + f_0(1 + f_0)\, e^{i\mathbf{k}\cdot\mathbf{r}}\, \psi(\mathbf{v}, t) \qquad \text{... (14.5.3)}$$

If we linearize in ψ, then proceeding exactly as in Sec. 14.4, we obtain the dynamics of ψ as being governed by

$$\frac{\partial\psi}{\partial t} = -K\psi = -(K_0\psi + K_1\psi) \qquad \text{... (14.5.4)}$$

where

$$K_0\psi = \int d^3v_1\, d\Omega\, \sigma\, |\mathbf{v} - \mathbf{v}_1|\, \frac{f_1}{1 + f_0}(1 + f_2)(1 + f_3)(\psi_2 + \psi_3 - \psi_1 - \psi) \qquad \text{... (14.5.5a)}$$

and

$$K_1\psi = i\,\mathbf{k}\cdot\mathbf{v}\,\psi \qquad \text{... (14.5.5b)}$$

The scalar product is defined as

$$(\varphi, \psi) = \int \varphi(\mathbf{v})\, f_0(1 + f_0)\, \psi(\mathbf{v})\, d^3v, \qquad \text{... (14.5.6)}$$

which ensures $(\psi, K_0\psi) \geq 0$. Hence the eigenvalues of K_0 are equal to or greater than zero.

Once again the zero eigenvalue is the one which we are interested in. This is fourfold degenerate since number is no longer a conserved quantity. The four eigenvectors are consequently p_x, p_y, p_z and the energy ε. Once again, we consider the propagation direction to be along the z-axis, so that the perturbation operator is $K_1 = ik\,p_z/m$. First-order perturbation theory requires the diagonalization of K_1 in the four-dimensional space. Once again p_x and p_y cannot couple and hence the diagonalization is to be carried out in the p_z, ε subspace. Before proceeding further, we need to clarify what $\mathbf{p}$ and ε are for a phonon. The phonons are energy packets in a sound field and each packet carries energy $\varepsilon = \hbar\omega$, where ω is the frequency of the sound wave. If c is the speed of sound and k the wavenumber, then $\omega = ck$ and the momentum p

of the packet is given by $p = \hbar k = h\omega/c = \varepsilon/c$. Once again, dropping irrelevent constant factors, one can take the phase space differential element to be $\varepsilon^{1/2}\, d\varepsilon\, d\Omega/4\pi$ and the two functions φ_1 and φ_2 in the space of which K_1 is to be diagonalized as $\varphi_1 = Ap_z$ and $\varphi_2 = B\,\varepsilon$, with the normalization conditions

$$1 = (\varphi_1, \varphi_1) = A^2 \int p_z^2 f_0 (1 + f_0)\, \varepsilon^{1/2}\, d\varepsilon \frac{d\Omega}{4\pi}$$

$$= \frac{A^2}{3} \int p^2 f_0 (1 + f_0)\, \varepsilon^{1/2}\, d\varepsilon$$

$$= \frac{A^2}{3c^2} \int \varepsilon^{5/2} f_0 (1 + f_0)\, d\varepsilon \qquad \text{... (14.5.7)}$$

and

$$1 = (\varphi_2, \varphi_2) = B^2 \int \varepsilon^{5/2} f_0 (1 + f_0)\, d\varepsilon\, d\Omega/4\pi$$

$$= B^2 \cdot \frac{3c^2}{A^2}$$

or,

$$A^2/B^2 = 3c^2 \qquad \text{... (14.5.8)}$$

where we have made use of Eq. (14.5.7). The diagonal elements of K_1 are zero. The off-diagonal element is (note that v for the phonon is the sound speed c)

$$ik \int c \cos\theta\, A\, p_z\, B\, \varepsilon \frac{d\Omega}{4\pi} \varepsilon^{1/2}\, d\varepsilon = ikc\, AB \int P\, \varepsilon \cos^2\theta f_0 (1 + f_0)\, \varepsilon^{1/2}\, d\varepsilon \frac{d\Omega}{4\pi}$$

$$= ik \frac{AB}{3} \int \varepsilon^{5/2} f_0 (1 + f_0)\, d\varepsilon$$

$$= ik \frac{AB}{3} \cdot \frac{3c^2}{A^2} = ik\, c^2 \frac{B}{A}$$

The matrix to be diagonalized is

$$M = \begin{pmatrix} 0 & ik\, c^2 \dfrac{B}{A} \\ ik\, c^2 \dfrac{B}{A} & 0 \end{pmatrix}$$

The eigenvalues λ of M are of the form $i\omega$, where clearly

$$\omega^2 = k^2 c^2 \frac{B^2}{A^2} = k^2 \frac{c^2}{3}$$

Writing $\omega = c_2 k$, c_2 being the propagation speed of the wave in the phonon gas,

$$c_2 = \pm \frac{c}{\sqrt{3}}, \qquad \text{... (14.5.9)}$$

so that the speed of propagation of a disturbance in the phonon gas is related to the sound speed by the factor of $\sqrt{3}$. Phonons dominate in superfluid Helium, as $T \to 0$, and hence at these temperatures, the propagation of second sound, as these waves are called, occurs at a speed of $c/\sqrt{3}$. The experimental confirmation of this in the early forties was one of the spectacular successes of Landau's theory of superfluidity.

Chapter 15
Metastable States

5.1 Introduction

Our discussion in Chapter 12 has shown that if a small disturbance from equilibrium conditions is made, the system relaxes back to equilibrium or oscillates about the equilibrium position. The relaxation towards the equilibrium distribution is brought about by collisions or interparticle interactions. In every case one managed to make progress because it was permissible to linearize around the equilibrium solution. What we have not yet dealt with is a situation where the system is far from equilibrium and attempts to make a transition to the equilibrium state. Dealing with a situation far from equilibrium is analytically difficult and one needs to make some simplifying assumptions. Often, in studying systems far from equilibrium it may so happen that the system may be trapped in a state of local equilibrium. Eventually the transition to the true equilibrium will occur but the time scales involved will be very large. Such states are called metastable states.

One of the simple situations where a system can be put far from equilibrium is to consider the existence of an external force derivable from a potential. If the potential has multiple minimum, then the equilibrium distribution will be peaked in the global minimum. If, however, one considers a distribution peaked in one of the local minima, then this acts as a pseudo-equilibrium state and it takes a lot of time for the distribution to leak out of this state. This is an example of a metastable state.

In the presence of a conservative external force $\mathbf{F} = -\vec{\nabla}\varphi$, the equilibrium distribution can be written as $f_0(\mathbf{r}, \mathbf{v}) = f_0(\mathbf{v})\, g(\mathbf{r})$, where $f_0(\mathbf{v})$ is obtained as before from the vanishing of the collision term and for classical particles will be the usual Maxwell-Boltzmann distribution, while $g(\mathbf{r})$ will have to be determined from the condition that to satisfy the transport equation, [Eq. (11.4.11)], we must have

$$f_0 \mathbf{v} \cdot \vec{\nabla} g(\mathbf{r}) + (-\vec{\nabla}\varphi) \cdot \frac{\partial f_0}{\partial \mathbf{v}} g(\mathbf{r}) = 0 \qquad \text{... (15.1.1)}$$

But f_0 being the Boltzmann distribution, we have

$\dfrac{\partial f_0}{\partial \mathbf{v}} = -\dfrac{m}{kT}\mathbf{v} f_0$, so that Eq. (15.1.1) becomes

$$\frac{\vec{\nabla} g(\mathbf{r})}{g(\mathbf{r})} = \frac{m}{kT}\vec{\nabla}\varphi = 0 \qquad \text{... (15.1.2)}$$

implying

$$g(\mathbf{r}) = e^{-m\varphi/kT} \qquad \text{... (15.1.3)}$$

The equilibrium distribution is consequently given by $f_0 = Ae^{-m\left(v^2/2+\varphi\right)/kT}$ and it is clear that if φ has a minimum, then there will be concentration of the distribution there and if φ has multiple minima, then the local minima are the metastable states.

5.2 A Phenomenological Approach

One of the alternatives to using the transport equation is to use the conservation laws as one learnt in Chapter 13. The conservation law corresponding to the number of particles is the equation of continuity

$$\frac{\partial \rho}{\partial t} + \vec{\nabla} \cdot \mathbf{J} = \frac{\partial \rho}{\partial t} + \vec{\nabla} \cdot (\rho \mathbf{v}) = 0 \qquad \text{... (15.2.1)}$$

The problem is the evaluation of the current $\mathbf{J} = \rho\, \mathbf{v}$ since this requires the solution of the transport equation. It is here that a phenomenological approach is effective in the form of a connection between the current $\mathbf{J}$ and the density ρ. If there is a gradient in ρ, then there will be a diffusive flow of particles from higher to lower concentrations and this contributes $-D\vec{\nabla}\rho$ to the current where D is a constant. Because of the external force $\mathbf{F}$, there will be an Ohm's law-like contribution ($\mathbf{J} = \sigma\, \mathbf{E}$) to the current which can be written as $\gamma \mathbf{F} \rho$; thus, we have

$$\begin{aligned} \mathbf{J} &= -D\,\vec{\nabla}\,\rho + \gamma\,\mathbf{F}\rho \\ &= -D\vec{\nabla}\,\rho - \gamma\rho\,\vec{\nabla}\varphi \qquad \text{... (15.2.2)} \end{aligned}$$

The equilibrium condition obviously implies no time dependence and zero current and hence the equilibrium distribution ρ_0 is given by

$$\rho_0 = A\ e^{-(\gamma/D)\,\varphi} \qquad \text{... (15.2.3)}$$

where A is a constant. Comparison of Eqs. (15.2.3) and (15.1.2) shows that

$$\frac{\gamma}{D} = \frac{m}{kT} \qquad \text{... (15.2.4)}$$

which is a form of the fluctuation-dissipation theorem.

As an example of the calculation of the current $\mathbf{J}$ we take the special case of a one-dimensional problem involving charged particles. The particles are in a constant force field F which exists up to $x = b$.

The potential energy $\varphi(x)$ will be (see Fig. 15.1)

$$\begin{aligned} \varphi(x) &= \mathrm{F}x \qquad \text{for } 0 < x < \mathrm{b} \\ &= 0, \text{ otherwise} \end{aligned}$$

Particles incident from the left with energy less than Fb will be reflected at $x = b$. Under equilibrium conditions there is no current in the setup. We now impose an electric field qE to the right and find the current. The new potential is

$$\begin{aligned} \varphi'(x) &= Fx - q\,Ex \quad 0 < x < b \\ &= -\,q\,Ex \quad \text{otherwise} \end{aligned}$$

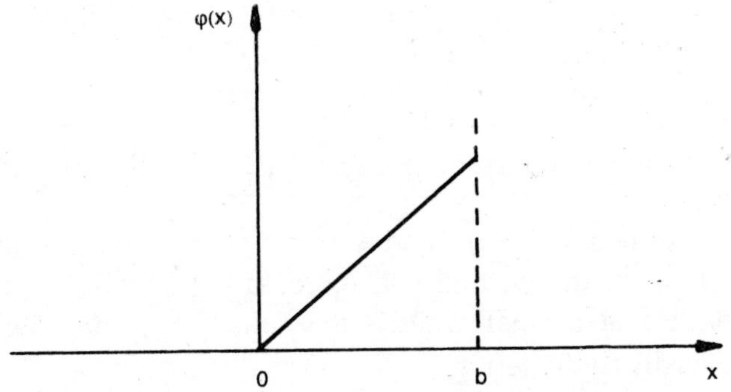

Fig. 15.1 A one-dimensional potential barrier

and the current can be written as

$$J = -D\frac{\partial \rho}{\partial x} - \frac{D}{kT}\frac{\partial \varphi'}{\partial x}\rho$$

$$= -D\left[\frac{\partial \rho}{\partial x} + \frac{1}{kT}\frac{\partial \varphi'}{\partial x}\rho\right]$$

$$= -D\,e^{-\varphi'/kT}\frac{\partial}{\partial x}(\rho\, e^{\varphi'/kT})$$

or

$$\frac{J}{D}e^{\varphi'/kT} = -\frac{\partial}{\partial x}(\rho\, e^{\varphi'/kT}) \qquad \text{... (15.2.5)}$$

We can now integrate from $x = 0$ to $x = b$, keeping in mind that J is independent of x in the steady state situation ($\partial\rho/\partial t = 0$ and thus from continuity $\partial J/\partial x = 0$) and also that ρ is constant outside the region $0 < x < b$ under the same conditions. From Eq. (15.2.5), we then obtain

$$\frac{J}{D}\int_0^b dx\; e^{(F - qE)x/kT} = -\rho\left[e^{(-qE)b/kT} - 1\right]$$

or

$$\frac{J}{D} \cdot \frac{kT}{F - qE}\left[e^{(F - qE)\,b/kT} - 1\right] = \rho\left[1 - e^{-qEb/kT}\right]$$

or

$$J \approx \frac{DF}{kT}\, e^{-Fb/kT} \rho\left(e^{qEb/kT} - 1\right) \qquad \text{... (15.2.6)}$$

where we have assumed $F >> qE$ and $e^{(F-qE)b/kT} >> 1$. We see that if the field is to the right ($E > 0$), the current can increase exponentially, but for $E < 0$, the current saturates at a small value—thus electrons can flow easily to the right and not so easily to the left

15.3 Decay of Metastable States

We now see how a metastable state will evolve towards equilibrium and over what time scales. We consider a potential $\varphi(x)$ with two minima—one the true minimum and the other a metastable state. The potential is shown in Fig. (15.2).

The minima at $x = x_a$ and x_b have the values φ_a and φ_b respectively. The intervening maximum at $x = x_0$ has the value $\varphi = \varphi_0$. The equilibrium distribution is going to be proportional to $e^{-m\varphi/kT}$ and hence the maximum population is going to be at $x = x_b$ where φ has a true minimum. We assume that a

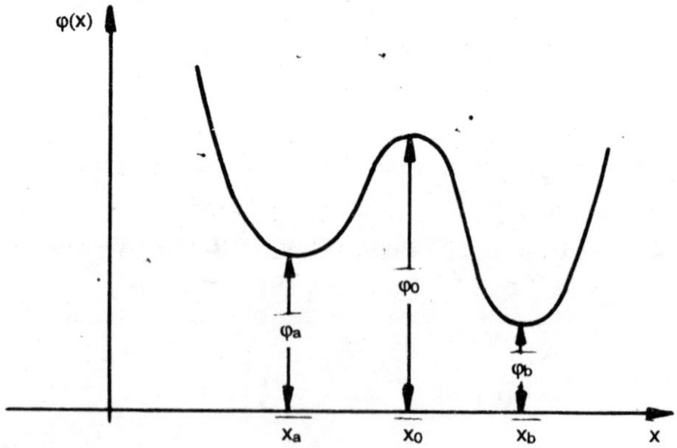

Fig. 15.2 A double-well potential. The local minimum at x_a corresponds to a metastable state.

far-from-equilibrium distribution is formed where the population is put entirely in the left hand well centred about $x = x_a$ which is a local minimum of the potential. This state will decay because of a current that will be set up in this out-of-equilibrium condition.

The current as in Sec. (15.2) will be given by

$$J = -D \frac{\partial \rho}{\partial x} - m \frac{D}{kT} \rho \frac{\partial \varphi}{\partial x}$$

$$= -D e^{-m\varphi/kT} \frac{\partial}{\partial x} (\rho e^{m\varphi/kT})$$

or

$$\frac{J}{D} e^{m\varphi/kT} = -\frac{\partial}{\partial x} (\rho e^{m\varphi/kT}) \qquad \text{... (15.3.1)}$$

We now make the assumption that the deviation of ρ from its initial value is small—this implies that the spatial derivative of J is small and to a first approximation, we can take J to be a constant. With this approximation we can integrate Eq. (15.3.1) from $x = x_a$ to $x = x_b$ and write

$$\frac{J}{D} \int_{x_a}^{x_b} dx\, e^{m\varphi/kT} = -\int_{x_a}^{x_b} \frac{\partial}{\partial x} (\rho\, e^{m\varphi/kT})\, dx$$

$$= \rho (x_a)\, e^{m\varphi_a/kT} - \rho (x_b)\, e^{m\varphi_b/kT}$$

$$= \rho_0\, e^{m\varphi_a/kT} \qquad \text{... (15.3.2)}$$

where the assumption is that not much of the distribution has leaked out yet and hence $\rho (x_b) \cong 0$ and $\rho (x_a) \cong \rho_0$, the initial density.

In attempting to integrate $e^{m\varphi/kT}$ between x_a and x_b, we approximate $\varphi (x)$ by a quadratic approximation about the maximum φ_0 at $x = x_0$. Consequently, we assume

$$\varphi (x) = \varphi_0 - \frac{1}{2} \varphi'' (x_0) (x - x_0)^2 \quad x_a \leq x \leq x_b \qquad \text{... (15.3.3)}$$

Now, $e^{-1/2\,\varphi''(x_0)\,(x_{a,b}-x_0)^2/kT} << 1$ (unless the temperature is very high) and hence we can take the integration limits of the variable $x - x_0$ as being from $-\infty$ to ∞. With this approximation Eq. (15.3.2) becomes

$$\frac{J}{D}\int_{-\infty}^{\infty} \exp\left[\frac{m\varphi_0}{kT} - \frac{m}{2}\,\frac{\varphi''(x_0)}{kT}\,(x - x_0)^2\right] d\,(x - x_0)$$

$$= \rho\,(0)\;e^{m\varphi_a/kT}$$

or

$$J = D\,\rho(0)\sqrt{\frac{m\varphi''(x_0)}{2\,kT\pi}}\;e^{-(m/kT)(\varphi_0-\varphi_a)} \qquad \text{... (15.3.4)}$$

The important factor is the exponential in Eq. (15.3.4). The term $m(\varphi_0 - \varphi_a)$ is the potential barrier that the particle at $x = x_a$ has to overcome in order to get over to the other side. At a temperature T, the typical energy available to the system to climb over this barrier is kT. This gives rise to the factor $e^{-m(\varphi_0 - \varphi_a)/kT}$ in the current.

Having found the current, we can estimate the time it takes the system to come to equilibrium. Since the current is like a velocity, the relevant time scale is proportional to the inverse of J and thus the equilibrium time τ_s is of the order of

$$\tau_s = \frac{1}{D\rho\,(0)}\sqrt{\frac{2\pi\,kT}{m\varphi''\,(x_0)}}\;e^{(m/kT)(\varphi_0-\varphi_a)} \qquad \text{... (15.3.5)}$$

The time scale is exponentially large and hence the long-lived character of metastable states. It is on this long time scale that the state far-from-equilibrium comes to equilibrium—but the Boltzmann transport equation is a cumbersome apparatus for the treatment of this equilibration. In the next two chapters, we introduce two other techniques for studying systems which are out of equilibrium.

Chapter 16
Langevin Equations

16.1 Introduction

From equations for distribution functions, we are now going to focus on equation for a single particle. However, since it is the same kind of questions that we intend to study, equilibrium distributions and approaches to equilibrium, there must be a probabilistic element to this differential equation for single particle motion. One can easily picture how such a situation may arise in practice. Imagine a particular molecule, in a gas of randomly moving molecules, being dragged with a velocity $\mathbf{v}$. As the different molecules strike this chosen molecule at random, there will be forces acting in all directions and of all possible magnitudes, so that the result of averaging this force (over a period of time say) will be zero. Because of the velocity $\mathbf{v}$ of the chosen molecules, there will, however, be a viscous drag corresponding to the larger number of collisions per unit time with molecules travelling in the opposite direction than with molecules travelling in the same direction. For a spherical molecule of radius r, this resistive force is $6\pi\eta rv$ opposite in direction to $\mathbf{v}$, where η is the coefficient of viscosity which contains information about the various collision processes. Thus the required form of the equation of motion will be

$$\dot{\mathbf{v}} = -\mathbf{f}(v) + \mathbf{g}(t) \qquad \text{... (16.1.1)}$$

where **f** (v) is the resistive force due to the surrounding and **g** (t) is the random force due to the surrounding with the average value $<g> = 0$. An equation of the above form is called a Langevin equation. It is a stochastic, as opposed to a deterministic, equation because of the random force (i.e. a force that is not definitely known) **g** (t) and arises whenever the effect of a particular particle's coupling to its environment is represented in some averaged manner. It reflects the fact that one lacks precise knowledge of the environment and hence is required to perform an averaging process over the environmental degrees of freedom. A particularly obvious example of this is provided by the case of a particle interacting with a collection of oscillators. This is presented in the next section.

16.2 An Example with Oscillators

We consider a particle with momentum p and coordinate x (for simplicity, we take the motion to be one-dimensional; additional dimensions bring no complications) interacting with a set of harmonic oscillators having momenta p_i and coordinates x_i. All masses are taken to be unity for convenience. The Hamiltonian describing the system is taken to be

$$H = \frac{p^2}{2} + \sum \frac{p_j^2}{2} + \sum \frac{1}{2} \cdot \omega_j^2 \left(q_j - \frac{\gamma_j}{\omega_j^2} x\right)^2 \qquad \text{... (16.2.1)}$$

The key point is that the coupling of the particle to the oscillators is linear. This makes the problem exactly solvable. The equations of motion for the different variables are obtained as follows:

$$\dot{p} = -\frac{\partial H}{\partial x} = -\sum_j \omega_j^2 \left(q_j - \frac{\gamma_j}{\omega_j^2} x\right) \left(-\frac{\gamma_j}{\omega_j^2}\right)$$

$$= \sum_j \gamma_j \left(q_j - \frac{\gamma_j}{\omega_j^2} x\right) \qquad \text{... (16.2.2)}$$

$$\dot{q}_j = \frac{\partial H}{\partial p_j} = p_j \qquad \text{... (16.2.3)}$$

and

$$\dot{p}_j = -\frac{\partial H}{\partial q_j} = -\omega_j^2\left(q_j - \frac{\gamma_j}{\omega_j^2}x\right) \qquad \text{... (16.2.4)}$$

From Eqs. (16.2.3) and (16.2.4), we get

$$\ddot{q}_j + \omega_j^2 q_j = \gamma_j\, x \qquad \text{... (16.2.5)}$$

The solution of Eq. (16.2.5) can be written down as

$$q_j(t) = q_j(0)\cos\omega_j t + \frac{p_j(0)}{\omega_j}\sin\omega_j t + \int_0^t \gamma_j \cdot \frac{\sin\omega_j(t-s)}{\omega_j}\, x(s)\, ds$$

$$= q_j(0)\cos\omega_j t + \frac{p_j(0)}{\omega_j}\sin\omega_j t + \gamma_j \left.\frac{\cos\omega_j(t-s)}{\omega_j^2}\, x(s)\right|_0^t$$

$$- \int_0^t \gamma_j \frac{\cos\omega_j(t-s)}{\omega_j^2}\,\dot{x}(s)\, ds$$

$$= \left[q_j(0) - \frac{\gamma_j}{\omega_j^2}x(0)\right]\cos\omega_j t + \frac{p_j(0)}{\omega_j}\sin\omega_j t$$

$$+ \frac{\gamma_j}{\omega_j^2}x(t) - \int_0^t \gamma_j \frac{\cos\omega_j(t-s)}{\omega_j^2}\, v(s)\, ds$$

Hence,

$$q_j(t) - \frac{\gamma_j}{\omega_j^2}x(t) = \left[q_j(0) - \frac{\gamma_j}{\omega_j^2}x(0)\right]\cos\omega_j t + \frac{p_j(0)}{\omega_j}\sin\omega_j t$$

$$- \int_0^t \gamma_j \frac{\cos\omega_j(t-s)}{\omega_j^2}\, v(s)\, ds \qquad \text{... (16.2.6)}$$

In the above $v = \dot{x}$ and is the same as p since the mass has been taken to be unity. Inserting in Eq. (16.2.2), the value of $q_j - \frac{\gamma_j}{\omega_j^2}x$ obtained from Eq. (16.2.6), we find

$$\dot{v} = -\sum_j \frac{\gamma_j^2}{\omega_j^2} \int_0^t \cos \omega_j (t-s)\, v(s)\, ds$$

$$+\gamma_j \left(q_j(0) - \frac{\gamma_j}{\omega_j^2} x(0)\right) \cos \omega_j t + \frac{\gamma_j}{\omega_j} p_j(0) \sin \omega_j t$$

$$= -\int K(t-s)\, v(s)\, ds + f(t) \qquad \text{... (16.2.7)}$$

where,

$$K(t-s) = \sum_j \frac{\gamma_j^2}{\omega_j^2} \cos \omega_j (t-s) \qquad \text{... (16.2.8)}$$

$$f(t) = \sum_j \gamma_j \left(q_j(0) - \frac{\gamma_j}{\omega_j^2} x(0)\right) \cos \omega_j t + \sum_j \frac{\gamma_j p_j(0)}{\omega_j} \sin \omega_j t \qquad \text{... (16.2.9)}$$

At this point, Eq. (16.2.7) is completely deterministic as expected. It ceases to be deterministic when it is not possible to specify all the $p_j(0)$ and $q_j(0)$ exactly. Now suppose that the distribution of $p_j(0)$ and $(q_j(0) - \frac{\gamma_j}{\omega_j^2} x(0))$ is specified and that this distribution is a Maxwell-Boltzman distribution at a temperature T (i.e. quadratic in the variables and each j independent of the other: it should be pointed out that such distributions are also called Gaussian). If the distribution alone is known, then only averages over this distribution makes sense and the assumed nature of the distribution implies

$$\langle p_j(0) \rangle = 0 = \langle q_j(0) - \frac{\gamma_j}{\omega_j^2} x(0) \rangle \qquad \text{... (16.2.10)}$$

$$\langle p_i(0)\, p_j(0) \rangle = k_B T\, \delta_{ij} \qquad \text{... (16.2.11)}$$

$$\langle (q_i(0) - \frac{\gamma_i}{\omega_i^2} x(0))(q_j(0) - \frac{\gamma_j}{\omega_j^2} x(0)) \rangle = \frac{k_B T}{\omega_i^2} \delta_{ij} \qquad \text{.. (16.2.12)}$$

Now, the force $f(t)$ of Eq. (16.2.9) makes sense only in terms of expectation values and Eq. (16.2.10) yields

$$<f(t)> = 0 \quad \text{... (16.2.13)}$$

For the second moment, on the other hand, we have

$$<f(t)\, f(s)> = <\sum_j \left[\gamma_j \left(q_j(0) - \frac{\gamma_j}{\omega_j^2}\, x(0)\right) \cos \omega_j t + \frac{\gamma_j}{\omega_j}\, p_j(0) \sin \omega_j t\right] \sum_i \left[\gamma_i \left(q_i(0) - \frac{\gamma_i}{\omega_j^2}\, x(0)\right) \cos \omega_i s + \frac{\gamma_i}{\omega_i} \cdot p_i(0) \sin \omega_i s\right]>$$

$$= \sum_{i,j} \gamma_i \gamma_j <\left(q_j(0) - \frac{\gamma_j}{\omega_j^2}\, x(0)\right)\left(q_i(0) - \frac{\gamma_i}{\omega_i^2}\, x(0)\right)> \cos \omega_j t \cos \omega_i s$$

$$+ \sum_{i,j} \frac{\gamma_i \gamma_j}{\omega_i \omega_j} <p_i(0)\, p_j(0)> \sin \omega_j t \sin \omega_i s$$

$$= \sum_i \frac{\gamma_i^2}{\omega_i^2} k_B T \cos \omega_i (t-s)$$

$$= k_B T\, K(t-s) \quad \text{... (16.2.14)}$$

In arriving at the above result we have made use of the fact that the cross correlation $<p_i(0)\left(q_j(0) - \frac{\gamma_j x(0)}{\omega_j^2}\right)> = 0$ for all i, j. It should be noted that due to the averaging over the initial values $q_i(0)$ and $p_i(0)$, Eq. (16.2.7) is no longer a deterministic equation. The force $f(t)$ makes sense only when its moments are specified. Before proceeding further, we make a digression on Gaussian distributions.

16.3 Gaussian Distribution and the Central Limit Theorem

A variable x is said to have a Gaussian distribution if the probability $P(x)$ of finding a value x for it is given as

$$P(x) = \frac{1}{\sqrt{2\pi\sigma^2}} e^{-x^2/2\sigma^2} \qquad \text{... (16.3.1)}$$

where we have arranged the pre-factor to ensure that the distribution is normalized, i.e. $\int_{-\infty}^{\infty} P(x)\, dx = 1$. The constant σ is called the width of the distribution. Clearly the average value $<x>$ of x is zero and the mean square average is

$$<x^2> = \int_{-\infty}^{\infty} x^2 P(x)\, dx = \frac{1}{(2\pi\sigma^2)^{1/2}} \int_{-\infty}^{\infty} x^2 e^{-x^2/2\sigma^2} dx = \sigma^2 \qquad \text{... (16.3.2)}$$

To find the average value of any power of x, we can perform the integration very easily, but it is instructive to follow a different route. This involves calculating

$$<e^{-ikx}> = \int_{-\infty}^{\infty} e^{-ikx} \frac{1}{(2\pi\sigma^2)^{1/2}} e^{-x^2/2\sigma^2} dx$$

$$= \int_{-\infty}^{\infty} \frac{1}{(2\pi\sigma^2)^{1/2}} \exp\left[-\frac{x^2}{2\sigma^2} - 2\frac{ik\sigma}{\sqrt{2}}\frac{x}{\sqrt{2}\,\sigma} - \left(\frac{ik\sigma}{\sqrt{2}}\right)^2 - \frac{k^2\sigma^2}{2}\right] dx$$

$$= \frac{e^{-k^2\sigma^2/2}}{(2\pi\sigma^2)^{1/2}} \int_{-\infty}^{\infty} \exp\left[-\frac{1}{2\sigma^2}\left(x + \frac{ik\sigma}{\sqrt{2}}\right)^2\right] dx \qquad \text{... (16.3.3)}$$

Expanding the two sides of Eq. (16.3.3)

$$< \sum_{n=0}^{\infty} \frac{(-ik)^n}{n!} x^n > = \sum \left(-\frac{k^2\sigma^2}{2}\right)^m \frac{1}{m!}$$

$$\text{or} \quad \sum_{n=0}^{\infty} \frac{(-i)^n k^n}{n!} <x^n> = \sum (-)^m \frac{k^{2m}\sigma^{2m}}{2^m} \frac{1}{m!} \qquad \text{... (16.3.4)}$$

Clearly

$$< x^{2m+1} > \; = 0 \qquad \text{... (16.3.5a)}$$

and

$$< x^{2m} > \; = \frac{(2m)!\, \sigma^{2m}}{2^m m\,!} \qquad \text{... (16.3.5b)}$$

The importance of the above derivation of $< x^n >$ is explained in the following: if we imagine that for an unknown distribution, we find the moments to be given by Eqs. (16.3.5a) and (16.3.5b), then it follows that under this distribution $< \mathrm{e}^{-ikx} > = \mathrm{e}^{-k^2\sigma^2/2}$ and now an inverse Fourier transform shows that the distribution is Gaussian. Thus the above derivation of the moments clearly shows that given the moments, we can arrive at the distribution.

If we have N independent variables $x_1, x_2, ... x_N$ and the joint probability distribution is Gaussian, then it implies that

$$P\,(x_1, x_2, \ldots x_N) = \frac{1}{(2\pi\sigma^2)^{1/2}} \exp\left(-\frac{x_1^2 + x_2^2 + \ldots + x_N^2}{2\sigma^2}\right) \qquad \text{... (16.3.6)}$$

It will be shown that for N independent random variables $x_1, x_2, \ldots x_N$ if a variable $Y = \frac{1}{\sqrt{N}}\,(x_1 + x_2 + \ldots + x_N)$ is formed then the distribution of Y is also Gaussian in the limit $N \to \infty$. The variables $\{x_i\}$ are characterized by the moments $< x_i > \; = 0$ and $< x_i^2 > \; = \sigma^2$ (the distribution of $\{x_i\}$ is not taken to be Gaussian; if it were, then the result would be trivial for all N).

Consider $< \exp^{-ikY} >$ and write it as

$$< \mathrm{e}^{-ikY} > \; = 1 - ik < Y > + \frac{(ik)^2}{2!} < Y^2 >$$

$$- \frac{(ik)^3}{3\,!} < Y^3 > + \frac{k^4}{4\,!} < Y^4 > + \ldots \qquad \text{... (16.3.7)}$$

Now $< Y > \; = \; < Y^3 > \; = 0$ since odd terms give vanishing moments for $\{ x_i \}$, while

$$< Y^2 > = \frac{1}{N} < x_1^2 + x_2^2 + \ldots + x_N^2 + 2x_1 x_2 + 2x_2 x_3 + \ldots >$$

$$= \frac{1}{N} \cdot N\sigma^2 = \sigma^2 \qquad \text{... (16.3.8)}$$

$$< Y^4 > = \frac{1}{N^2} < \sum_i x_i^4 + 6 \sum_{i>j} x_i^2 x_j^2 + 4 \sum_{j>k} x_i^2 x_j x_k > \qquad \text{... (16.3.9)}$$

For large N, we note $< \sum x_i^4 >$ is $O(N)$ while $< \sum x_i^2 x_j^2 >$ is $O(N^2)$ and hence working to $O(1)$ accuracy,

$$< Y^4 > = \frac{1}{N^2} \cdot 6 \cdot \frac{N^2}{2} \cdot \sigma^4 = 3\sigma^4$$

Thus the pattern being formed is

$$< e^{-iky} > = 1 - \frac{k^2 \sigma^2}{2} + \frac{1}{2} \cdot \frac{k^4 \sigma^4}{4} + \ldots$$

$$= e^{-k^2 \sigma^2/2} \qquad \text{... (16.3.10)}$$

Thus from our previous considerations, the distribution of Y is Gaussian as $N \to \infty$ (note that this limit is vital in the reduction of $< Y^4 >$ to $3\sigma^4$ and will be so for all higher moments) and has the width σ^2. The variable $\sum x_i$ would then have the width $N\sigma^2$. The above results constitute the central limit theorem.

To end this section, we will consider yet another Gaussian distribution—that appropriate to a random function $f(t)$. We now talk about the probability of f being a definite function $f(t)$ and express this probability as

$$P(f) = A\, e^{-\int_{-\infty}^{\infty} dt\, f(t)^2/\sigma^2} \qquad \text{... (16.3.11)}$$

where the constant A is given by

$$A^{-1} = \int D[f]\, e^{-\int_{-\infty}^{\infty} dt\, f(t)^2/\sigma^2} \qquad \text{... (16.3.12)}$$

The integration $D[f]$ is over all possible functions $f(t)$ and the definition of A assures the normalization $\int D[f]\; P(f) = 1$. Operationally the functional integral is best treated in the Fourier space by introducing the transform

$$f(t) = \frac{1}{\sqrt{2\pi}} \int_{-\infty}^{\infty} f(\omega)\; e^{i\omega t}\; d\omega \qquad \text{... (16.3.13)}$$

the requirement that $f(t)$ is real implying $f(-\omega) = f^*(\omega)$. A given set of $f(\omega)$ leads to a given function $f(t)$. The $f(\omega)$ are simply numbers and if we scan every possible value of $f(\omega)$ for every value of ω, then the space of all functions of $f(t)$ has been taken into account. Accordingly,

$$D[f] = \prod_{\omega} df(\omega)$$

$$\int_{-\infty}^{\infty} f(t)^2\, dt = \frac{1}{2\pi} \int_{-\infty}^{\infty} f(\omega_1)\; e^{i\omega_1 t}\; d\omega_1\; f(\omega_2)\; e^{i\omega_2 t}\; d\omega_2\; dt$$

$$= \int_{-\infty}^{\infty} f(\omega_1)\, f(\omega_2)\; d\omega_1\, d\omega_2 \int_{-\infty}^{\infty} \frac{dt}{2\pi}\; e^{i(\omega_1 + \omega_2)t}$$

$$= \int_{-\infty}^{\infty} f(\omega_1)\, f(\omega_2)\; d\omega_1\; d\omega_2\; \delta(\omega_1 + \omega_2)$$

$$= \int_{-\infty}^{\infty} f(\omega_1)\, f(-\omega_1)\; d\omega_1$$

$$= \int_{-\infty}^{\infty} |f(\omega)|^2\, d\omega = \sum_{\omega} |f(\omega)|^2$$

The distribution can now be thought of as

$$P(f) = A\; e^{-\sum_{\omega} |f(\omega)|^2/\sigma^2} \qquad \text{... (16.3.14)}$$

with

$$A^{-1} = \int \prod_{\omega} df(\omega)\ e^{-\sum_{\omega} |f(\omega)|^2/\sigma^2} \qquad \text{... (16.3.15)}$$

One can now express the various odd moments as

$$\langle f(\omega) \rangle = \langle f(\omega_1)\, f(\omega_2)\, f(\omega_3) \rangle = \ldots$$

$$= \langle f(\omega_1)\, f(\omega_2)\ \ldots\ f(\omega_{2m+1}) \rangle = 0 \qquad \text{... (16.3.16)}$$

The even moments are expressed as

$$\langle f(\omega_1)\, f(\omega_2) \rangle = 0 \text{ if } \omega_1 \neq -\omega_2,$$

while for $\omega_1 = -\omega_2$, from the properties of the Gaussian, we have

$\langle f(\omega_1)\, f(-\omega_1) \rangle = \sigma^2$, leading to the result

$$\langle f(\omega_1)\, f(\omega_2) \rangle = \sigma^2\, \delta_{\omega_1, -\omega_2} \qquad \text{... (16.3.17)}$$

It should be clear that if we consider the fourth moment, then,

$$\langle f(\omega_1)\, f(\omega_2)\, f(\omega_3)\, f(\omega_4) \rangle = \sigma^4 \Big[\delta_{\omega_1, -\omega_2}\, \delta_{\omega_3, -\omega_4} + \delta_{\omega_1, -\omega_3}\, \delta_{\omega_2, -\omega_4} + \delta_{\omega_1, -\omega_4}\, \delta_{\omega_2, -\omega_3} \Big]$$

and obviously the fourth moment is zero if $\omega_1 + \omega_2 + \omega_3 + \omega_4 \neq 0$. For an arbitrary even order moment

$$\langle f(\omega_1)\, f(\omega_2)\ \ldots\ f(\omega_{2n}) \rangle = \sigma^{2n} \Big[\sum_{\text{all pairwise breakup}} \delta_{\omega_1, -\omega_2} \times \delta_{\omega_3, -\omega_4}\, \delta_{\omega_5, -\omega_6} \ldots \delta_{\omega_{2n-1}, -\omega_{2n}} \Big] \qquad \text{... (16.3.18)}$$

The number of pairs is $\dfrac{(2n)!}{n!\,2^n}$.

Finally, we need to discuss the correlations of the original function $f(t)$. Clearly

$$\langle f(t) \rangle = 0 \qquad \text{... (16.3.19)}$$

$$\langle f(t_1) f(t_2) \rangle = \Big\langle \frac{1}{2\pi} \int d\omega_1\, f(\omega_1)\, e^{i\omega_1 t_1}\, d\omega_2\, f(\omega_2)\, e^{i\omega_2 t_2} \Big\rangle$$

$$= \frac{1}{2\pi} \int d\omega_1\, d\omega_2\, e^{i(\omega_1 t_1 + \omega_2 t_2)} < f(\omega_1)\, f(\omega_2) >$$

$$= \frac{\sigma^2}{2\pi} \int d\omega_1\, d\omega_2\, e^{i(\omega_1 t_1 + \omega_2 t_2)}\, \delta_{\omega_1, -\omega_2}$$

$$= \frac{\sigma^2}{2\pi} \int d\omega_1\, e^{i\omega_1 (t_1 - t_2)}$$

$$= \sigma^2\, \delta(t_1 - t_2) \qquad \ldots (16.3.20)$$

where we have made use of Eq. (16.3.17). It should now be obvious that

$$< f(t_1)\, f(t_2)\, \ldots\, f(t_{2n+1}) > = 0 \qquad \ldots (16.3.21)$$

and

$$< f(t_1)\, f(t_2) \ldots f(t_{2n}) > = \sigma^{2n} \Big[\delta(t_1 - t_2)\, \delta(t_3 - t_4) \ldots$$

$$\delta(t_{2n-1} - t_{2n}) + \text{all such pairwise breakups}\Big] \qquad \ldots (16.3.22)$$

We note that if a random function f (t) is characterized by moments of the form shown in Eqs. (16.3.21) and (16.3.22), then the distribution will be given by Eq. (16.3.11).

16.4 Langevin Equation

We now return to Eq. (16.2.7) with the $f(t)$ stochastic and the second moment given by Eq. (16.2.14). It is clear that all odd moments vanish because of the Gaussian nature of the distribution of p_i (0) and $\bar{q}_i(0) = q_i\,.(0) - \frac{\gamma_i}{\omega_i^2} x(0)$. Thus

$$< f(t_1)\, f(t_2)\, \ldots\, f(t_{2n+1}) > = 0 \qquad \ldots (16.4.1)$$

To find the even moments, we write

$$< f(t_1)\, f(t_2) \ldots f(t_{2n}) > = < \sum_{j_1} \Big[\gamma_{j_1} \bar{q}_{j_1}(0) \cos\omega_{j_1} t_1$$

$$+ \frac{\gamma_{j_1}}{\omega_{j_1}} p_{j_1}(0) \sin\omega_{j_1} t_1\Big] \sum_{j_2} \gamma_{j_2} \bar{q}_{j_2}(0) \cos\omega_{j_2} t_2$$

$$+\frac{\gamma_{j_2}}{\omega_{j_2}} p_{j_2}(0) \sin \omega_{j_2} t_2 \Big] \ldots \sum_{j_{2n}} \Big[\gamma_{j_{2n}} \bar{q}_{j_{2n}}(0) \cos \omega_{j_{2n}} t_{2n}$$

$$+\frac{\gamma_{j_{2n}}}{\omega_{j_{2n}}} p_{j_{2n}}(0) \sin \omega_{j_{2n}} t_{2n} \Big]$$

and now using the property of Gaussian distribution for an even string of $\{\bar{q}_j(0)\}$, $\{p_j(0)\}$ or the combination of $(2n-2m)$ number of $\{\bar{q}_j(0)\}$ and $2m$ number of $\{p_j(0)\}$, we have

$$< f(t_1) f(t_2) \ldots f(t_n) > = (kT)^n \Big[K(t_2 - t_1)$$

$$\times K(t_4 - t_3) \ldots K(t_{2n} - t_{2n-1}) + \text{all such pairwise breakups} \Big] \quad \ldots \textbf{(16.4.2)}$$

Recalling the definition of $K(t-s)$, we note that it is crucially dependent on what form of frequency weighting we employ. Thus

$$K(t-s) = \sum_j \frac{\gamma_j^2}{\omega_j^2} \cos \omega_j (t-s)$$

$$= \int_{-\infty}^{\infty} \frac{\gamma^2}{\omega^2} \cos \omega (t-s) \, P(\omega) \, d\omega \quad [P(\omega) \text{ is the weighting factor}]$$

$$= \Gamma \int_0^{\infty} \cos \omega (t-s) \, d\omega \quad \left(\text{choosing } \frac{\gamma^2}{\omega^2} P(\omega) = \Gamma \right)$$

$$= \Gamma \underset{\omega \to \infty}{\text{Lt}} \frac{\sin \omega (t-s)}{t-s}$$

$$= \Gamma \delta (t-s) \quad \ldots \textbf{(16.4.3)}$$

The equation of motion for v, that we had obtained in Eq. (16.2.7), now becomes

$$\dot{v} = -\Gamma v + f(t) \quad \ldots \textbf{(16.4.4)}$$

where $f(t)$ is a random function with the moments,

$$< f(t) > = 0 \quad \ldots \textbf{(16.4.5a)}$$

$$< f(t_1)\, f(t_2) > \; = D\,\delta\,(t_2 - t_1) \qquad \text{... (16.4.5b)}$$

where $D/\Gamma = kT$. For an arbitrary number of factors, f, we have from Eqs. (16.4.2) and (16.4.3),

$$< f(t_1)\, f(t_2)\; \ldots\; f(t_{2n+1}) > \; = 0 \qquad \text{... (16.4.6a)}$$

$$< f(t_1)\, f(t_2)\; \ldots\; f(t_{2n}) > \; = \; D^n \Big[\delta\,(t_2 - t_1)\,\delta\,(t_4 - t_3)$$

$$\delta(t_6 - t_5)\; \ldots\; \delta\,(t_{2n} - t_{2n-1}) + \text{all possible pairwise breakups}\Big] \qquad \text{... (16.4.6b)}$$

The equation of motion given in Eq. (16.4.4) is a linear Langevin equation with white noise. The linearity comes from the fact that the dissipative term is linear in v (i.e. $-\Gamma v$) — a nonlinear function of v at this stage would make it a nonlinear Langevin equation. The characteristic of the stochastic force $f(t)$ given in Eqs. (16.4.6a) and (16.4.6b) makes it a white noise.

The steps leading to Eq. (16.4.3) show that all frequencies contribute equally when one is dealing with white noise. For a different frequency weighting the kernel $K(t - s)$ has "memory" i.e. is non-zero for $t \neq s$ and we have a 'coloured noise'. The relation between the strength of the correlation of f in Eq. (16.4.5b) and the strength of the dissipation (Γ) in Eq. (16.4.4) [namely $D = \Gamma\,(kT)$] is called the fluctuation dissipation theorem. It should be clear when $f(t)$ is a white noise term with the moments given by Eqs. (16.4.6a) and (16.4.6b), the probability distribution associated with f is

$$P[f] = A\, e^{-\int_{-\infty}^{\infty} f(t)^2\, dt/D} \qquad \text{... (16.4.7)}$$

where

$$A^{-1} = \int D[f]\; e^{-\int_{-\infty}^{\infty} f(t)^2\, dt/D}$$

16.5 Langevin Equation and Equilibrium

We are going to deal almost exclusively with the linear Langevin equation with a white noise force term, i.e. the equation

$$\dot{v} = -\Gamma v + f(t) \qquad \text{... (16.5.1)}$$

with the statistics of $f(t)$ as specified. The advantage of the linear equation is that it can be easily solved. We start with an initial velocity $v(0)$ and see what the Langevin equation tells us. The solution of Eq. (16.5.1) can be written as

$$v(t) = e^{-\Gamma t} \int_0^t f(t') \, e^{\Gamma t'} \, dt' + v(0) \, e^{-\Gamma t} \qquad \text{... (16.5.2)}$$

Since $f(t)$ is stochastic, so will the $v(t)$ in Eq. (16.5.2) be stochastic and only the moments of $v(t)$ will make sense.

Clearly $< v(t) > = v(0) \; e^{-\Gamma t}$ and for the second moment, we have

$$< v(t_1) \, v(t_2) > = < \left[e^{-\Gamma t_1} \int_0^{t_1} f(t') \, e^{\Gamma t'} \, dt' + v(0) \, e^{-\Gamma t_1} \right]$$

$$\times \left[e^{-\Gamma t_2} \int_0^{t_2} f(t'') \, e^{-\Gamma t''} \, dt'' + v(0) \, e^{-\Gamma t_2} \right] >$$

$$= e^{-\Gamma (t_1 + t_2)} \int_0^{t_1} dt' \int_0^{t_2} dt'' \; e^{\Gamma (t' + t'')} < f(t') \, f(t'') >$$

$$+ \, v(0)^2 \; e^{-\Gamma (t_1 + t_2)}$$

$$= D e^{-\Gamma (t_1 + t_2)} \int_0^{t_1} dt' \int_0^{t_2} dt'' \; e^{\Gamma (t' + t'')} \delta (t' - t'')$$

$$+ \, v(0)^2 \; e^{-\Gamma (t_1 + t_2)} \qquad \text{... (16.5.3)}$$

In the integral on the right hand side of Eq. (16.5.3) if $t_1 > t_2$, then the integration over t' needs to carried out first and the integral become $e^{-\Gamma (t_1 + t_2)} \int_0^{t_2} dt'' \; e^{2\Gamma t''} = (e^{-\Gamma (t_1 - t_2)} - e^{-\Gamma (t_1 + t_2)}) / 2\Gamma$. On the other hand, if $t_2 > t_1$, then the integral over t'' needs to be done first and we have the result as $(e^{-\Gamma (t_2 - t_1)} - e^{-\Gamma (t_1 + t_2)}) / 2\,\Gamma$.

Clearly, then

$$< v(t_1)\, v(t_2) > = \frac{D}{2\Gamma}\; e^{-\Gamma |t_1 - t_2|} + \left[v(0)^2 - \frac{D}{2\Gamma} \right] e^{-\Gamma (t_1 + t_2)} \qquad \text{... (16.5.4)}$$

For the equal time correlation function $< v(t)^2 >$, we have

$$\underset{t \to \infty}{\text{Lt}} < v(t)^2 > = \frac{D}{2\Gamma} \qquad \text{... (16.5.5)}$$

If we want to determine $\underset{t \to \infty}{\text{Lt}} < v(t)^n >$ it is not necessary to consider $v(0)$, since the term involving $v(0)$ will not survive in that limit. From the statistics of $f(t)$ it is obvious that $\underset{t \to \infty}{\text{Lt}} < v(t)^{2m+1} > = 0$, while the relevant part of the even order correlation is

$$< v(t)^{2n} > = < \left[e^{-\Gamma t} \int_0^t f(t_1)\; e^{\Gamma t_1}\, dt_1 \right]$$

$$\times \left[e^{-\Gamma t} \int_0^t f(t_2)\; e^{+\Gamma t_2}\, dt_2 \right] \cdots \left[e^{-\Gamma t} \int_0^t f(t_{2n})\; e^{\Gamma t_{2n}}\, dt_{2n} \right] >$$

$$= e^{-2n\Gamma t} \int_0^t dt_1 \int_0^t dt_2 \cdots \int_0^t dt_{2n} < f(t_1)\, f(t_2) \,..\, f(t_{2n}) >$$

$$\times\; e^{\Gamma (t_1 + t_2 \,\ldots\, t_{2n})} \qquad \text{... (16.5.6)}$$

At this point we use the statistics of $f(t)$ and note that it breaks up into a chain of pairs in $(2n)!/n!2^n$ ways and thus the time-independent part of the right hand side of Eq. (16.5.6) is extracted as

$$< v^{2n}(t) > = \frac{D^n}{(2\Gamma)^n} \; \frac{(2n)!}{n!2^n} + \text{terms which involve powers of } e^{-2\Gamma t}$$

$$\underset{t \to \infty}{\text{Lt}} < v^{2n}(t) > = \left(\frac{D}{2\Gamma} \right)^n \frac{(2n)!}{2^n\; n!} \qquad \text{... (16.5.7)}$$

Since $v(t)$ is a random variable, there is a probability distribution $P\{v(t)\}$ associated with it and as $t \to \infty$, this probability distribution tends to the equilibrium value with the moments specified as in Eq. (16.5.7). The odd moments are zero. The equilibrium probability distribution is then given by

$$P_{eq}(v) = \left(\frac{2\Gamma}{2\pi D}\right)^{1/2} e^{-\frac{v^2}{2D}\cdot 2\Gamma}$$

$$= \left(\frac{1}{\sqrt{\pi kT}}\right) e^{-\left(v^2/kT\right)} \qquad \text{... (16.5.8)}$$

which is a Gaussian distribution. The important result is that the Langevin dynamics is another way of describing the approach to equilibrium. If we begin with an initial distribution of v which is not according to equilibrium, then as $t \to \infty$, the equilibrium distribution is attained. If, on the other hand, we begin with an equilibrium distribution, then in Eq. (16.5.4), $v(0)^2 = \dfrac{D}{2\Gamma}$ and the dynamics governed by the Langevin equation corresponds to decay of correlations.

The above results about the establishment of an equilibrium distribution are not restricted to the linear Langevin equation. If we have a Langevin equation with the structure

$$\dot{v} = -\Gamma \frac{\partial F}{\partial v} + f(t) \qquad \text{(16.5.9)}$$

where F is an arbitrary function of v and $f(t)$ is the random noise with $< f(t) > = 0$ and $< f(t_1)\, f(t_2) > = \varepsilon\, \Gamma\, \delta(t_1 - t_2)$, then the equilibrium distribution is proportional to $e^{-F(v)/\varepsilon}$. This is most conveniently demonstrated by setting up an equation for the probability distribution $P\{v(t)\}$. This will be done in the next chapter.

16.6 Brownian Motion

The heavily damped motion of a particle in a colloidal solution is generally referred to as Brownian motion. Observations showed that for a tagged particle the displacement directions were random and hence on an average the displacement was zero, but if one considered the square of the displacement, then in a time t, the square of the displacement was proportional to t. The theory of Brownian motion was worked out by Einstein, who assumed that the motion of the tagged particle was i) highly damped, ii) subject to a random force by the other molecules. This led to a Langevin equation of the form

$$\dot{v} = -\Gamma v + f(t) \qquad \text{... (16.6.1)}$$

where

$$< f(t) > = 0 \text{ and } < f(t_1)\, f(t_2) > = D\,\Gamma\,\delta\,(t_1 - t_2).$$

The solution of Eq. (16.6.1) is

$$v\,(t) = e^{-\Gamma t} \int_0^t e^{\Gamma t'}\, f(t')\, dt' + v\,(t)\; e^{-\Gamma t} \qquad \text{... (16.6.2)}$$

The displacement Δx in time t is clearly

$$\Delta x = \int_0^t v\,(t_1)\, dt_1$$

the expectation value of which goes to zero as $t \to \infty$. The mean square displacement on the other hand is

$$< (\Delta x)^2 > = < \int_0^t v\,(t_1)\; dt_1 \int_0^t v\,(t_2)\; dt_2 >$$

$$= < \int_0^t \left[e^{-\Gamma t_1} \int_0^{t_1} e^{\Gamma t'} f(t')\, dt' + v(0)\, e^{-\Gamma t_1} \right] dt_1$$

$$\times \int_0^t \left[e^{-\Gamma t_2} \int_0^{t_2} e^{\Gamma t''} f(t'')\, dt'' + v(0)\, e^{-\Gamma t_2} \right] dt_2 >$$

$$= \langle v(0)^2 \rangle \int_0^t dt_1 \int_0^t dt_2 \ e^{-\Gamma(t_1+t_2)}$$

$$+ \int_0^t dt_1 \ e^{-\Gamma t_1} \int_0^t dt_2 \ e^{-\Gamma t_2} \int_0^{t_1} dt' \int_0^{t_2} dt'' \langle f(t') \, f(t'') \rangle \, e^{\Gamma(t'+}$$

$$= \langle v(0)^2 \rangle \left(\frac{1-e^{-\Gamma t}}{\Gamma} \right)^2 + \int_0^t dt_1 \int_0^t dt_2 \ e^{-\Gamma(t_1+t_2)}$$

$$\times \left[D\Gamma \int_0^{t_1} dt' \ e^{2\Gamma t'} + \text{similar term if } t_1 < t_2 \right]$$

$$= \langle v(0)^2 \rangle \left(\frac{1-e^{-\Gamma t}}{\Gamma} \right)^2 + D \int_0^t dt_1 \int_0^t dt_2 \left[e^{-\Gamma(t_2-t_1)} - e^{-\Gamma(t_1+t_2)} \right]$$

$$= \left[\langle v(0)^2 \rangle - D \right] \left(\frac{1-e^{-\Gamma t}}{\Gamma} \right)^2 + D \int_0^t dt_1 \int_0^t dt_2 \ e^{-\Gamma(t_2-t_1)\,\theta(t_2-t_1)}$$

... **(16.6.3)**

In the integral on the right hand side of Eq. (16.6.3), we have the constraint $t_2 > t_1$ and hence we fix t_2 at a given value and integrate t_1 from 0 *to* t_2 and then integrate t_2 from 0 to t. Thus,

$$\langle (\Delta x)^2 \rangle = \left[\langle v(0)^2 \rangle - D \right] \left(\frac{1-e^{-\Gamma t}}{\Gamma} \right)^2 + D \int_0^t dt_2 \int_0^{t_2} dt_1 \times e^{-\Gamma(t_2-t_1)}$$

$$= \left[\langle v(0)^2 \rangle - D \right] \left(\frac{1-e^{-\Gamma t}}{\Gamma} \right)^2 + \frac{D}{\Gamma} t - \frac{D}{\Gamma^2} (1-e^{-\Gamma t})$$

On the right hand side of the above equation if $t \to \infty$, $e^{-\Gamma t} \to 0$ and the linear term in t dominates, so that for large t

$$< (\Delta x)^2 > = \frac{D}{\Gamma} t \qquad \text{... (16.6.4)}$$

The result has been obtained for a one-dimensional case, but there is nothing in the above manipulations which is special to $D = 1$. The reader can easily verify that we could have started with $\dot{v}_i = - \Gamma v_i + f_i$ $(i = 1, 2, 3$ for $D = 3)$, with $< f_i > = 0$ and $< f_i (t_1) f_j (t_2) > = D \delta (t_1 - t_2) \delta_{ij}$ and obtained $< \sum_{i=1}^{3} \Delta x_i \Delta x_i >$ as $\frac{3D}{\Gamma} \cdot t$. The coefficient of t is known as the diffusion coefficient.

Chapter 17
Fokker-Planck Equations

1.7.1 Introduction

This chapter will consider distributions, equilibrium distributions and the approach to equilibrium from yet another point of view. The dynamics that will be studied is the dynamics of the probability distribution that is associated with Langevin equation. As we have found in the last chapter, the Langevin equation refers to the stochastic dynamics of a variable v (v can actually be anything)

$$\dot{v} = -\frac{\partial S}{\partial v} + f(t) \qquad \text{... (17.1.1)}$$

where S is an arbitrary function of v and $f(t)$ is the noise term. The variable $v(t)$ is characterized by its distribution $P\{v(t)\}$ and the Fokker-Planck equation is an equation of motion for the distribution P. A stationary solution for the equation of motion will yield the equilibrium distribution $P_{eq.}$ Finding the equilibrium distribution and studying the approach to equilibrium is one of the major problems associated with Fokker-Planck equations which we will take up in detail. For an arbitrary initial condition, the solution of Fokker-Planck equation can be written down in general, which can be very useful in tackling several out-of-equilibrium situations. Before providing one of the many standard derivations of Fokker-Planck equations in the next section we wish to point out that the Boltzmann transport equation is also an equation for a probability distribution and it stands to reason that we should be able to construct the Fokker-Planck equation from the transport equation. A good clear cut connection, however, is lacking although some literature does exist.

17.2 Fokker-Planck Equation: Derivation

Our starting point will be the Langevin equation for a random variable as written down in Eq. (17.1.1). The statistics of the random force $f(t)$ is specified as

$$\langle f(t)\rangle = 0 \qquad \text{... (17.2.1)}$$

$$\langle f(t)\, f(t')\rangle = 2\varepsilon\delta\,(t-t') \qquad \text{... (17.2.2)}$$

and all the higher moments as appropriate to a Gaussian distribution for $f(t)$. The probability $P\, f(t)$ of getting a particular $f(t)$ is given by

$$P\,\{f(t)\} = A\, e^{-\int f(t)^2\, dt/2\varepsilon} \qquad \text{... (17.2.3)}$$

where

$$A^{-1} = \int D[f]\; e^{-\int f(t)^2\, dt/2\varepsilon} \qquad \text{... (17.2.4)}$$

All averaging that we will have occasion to introduce would mean an averaging over the distribution $P\,(f\,(t))$, i.e.

$$\langle \,...\, \rangle = \int D[f]\, P\{f(t)\}\;(...) \qquad \text{... (17.2.5)}$$

It is easiest to proceed by considering time as a discrete variable and take a sequence $t_0, t_1, t_2, \ldots t_n, t_{n+1}, \ldots$, such that $t_{n+1} - t_n = \delta t$ is an infinitesimal. Writing $v_n = v\,(t_n)$, we can write the discretized Langevin equation as

$$v_{n+1} = v_n - g \qquad \text{... (17.2.6)}$$

where

$$g = \frac{\partial S}{\partial v}\,\delta t - \bar{f}(t)\sqrt{\delta t} \qquad \text{... (17.2.7)}$$

with the definition

$$\bar{f}(t) = f(t)\sqrt{\delta t} \qquad \text{... (17.2.8)}$$

so that the correlation of $\bar{f}(t)$ is

$$<\bar{f}(t)\bar{f}(t')> = 2\,\varepsilon\,\delta_{tt'} \qquad \text{... (17.2.9)}$$

$\delta_{tt'}$ being the Kroenecker delta.

Our aim is to describe the dynamics that takes one from $P\{v(t)\}$ to $P\{v(t + \mathrm{d}t)\}$, the probability distribution at time $t + dt$. In terms of the discretized variable we want to go from $P(v_n, t_n)$ *to* $P(v_{n+1}, t_{n+1})$. The connection is

$$P(v_{n+1}, t_{n+1}) = <\int P(v_n, t_n)\;\delta(v_{n+1} - v_n + g)\,dv_n> \qquad \text{... (17.2.10)}$$

Note that the integration over the δ-function is not enough although that does take one from step n to step $n+1$. This step is dependent on g which involves the noise term and hence to get the final answer we must average over the different realizations of the random force. But g is small as shown in Eq. (17.2.7) and we can expand in it as

$$\delta(v_{n+1} - v_n + g) = \delta(v_{n+1} - v_n) + g\,\delta'(v_{n+1} - v_n) + \frac{g^2}{2!}\,\delta''(v_{n+1} - v_n) + O(dt^{3/2}) \qquad \text{... (17.2.11)}$$

Inserting this expansion in Eq. (17.2.10) we find, correct to $O(\delta t)$,

$$P(v_{n+1}, t_{n+1}) = <\int P(v_n, t_n)\,\delta(v_{n+1} - v_n)dv_n + \int P(v_n, t_n)\,g\delta'(v_{n+1} - v_n)\,dv_n$$

$$+ \int P(v_n, t_n)\frac{g^2}{2}\,\delta''(v_{n+1} - v_n)dv_n> \qquad \text{... (17.2.12)}$$

Now,

$$\int P(v_n, t_n)\,\delta'(v_{n+1} - v_n)dv_n = P(v_{n+1}, t_n) \qquad \text{... (17.2.13)}$$

$$\int P(v_n, t_n)\,g\,\delta'(v_{n+1} - v_n)\,dv_n = -P(v_n, t_n)\,g\,\delta(v_{n+1} - v_n)$$

$$+ \int \frac{d}{dv_n} \{gP(v_n, t_n)\} \delta(v_{n+1} - v_n) dv_n$$

$$= + \frac{d}{dv_{n+1}} \cdot (g\, P(v_{n+1}, t_n)) \qquad \text{... (17.2.14)}$$

(we have assumed that $P \to 0$ at the end points of the integration.)

$$\int P(v_n, t_n) \frac{g^2}{2} \delta''(v_{n+1} - v_n) dv_n = -P(v_n, t_n) \frac{g^2}{2} \delta'(v_{n+1} - v_n)$$

$$+ \int \frac{d}{dv_n} (\frac{g^2}{2} P(v_n, t_n))\, \delta'(v_{n+1} - v_n)\, dv_n$$

$$= \int \frac{d}{dv_n} \left(\frac{g^2}{2} P(v_n, t_n) \right) \delta'(v_{n+1} - v_n) dv_n$$

$$= - \frac{d}{dv_n} \left\{ \frac{g^2}{2} P(v_n, t_n) \right\} \delta(v_{n+1} - v_n)$$

$$+ \int \frac{d^2}{dv_n^2} \left(\frac{g^2}{2} P(v_n, t_n) \right) \delta(v_{n+1} - v_n) dv_n$$

$$= \frac{d^2}{dv_n^2} \left\{ \frac{g^2}{2} P(v_{n+1}, t_n) \right\} \qquad \text{... (17.2.15)}$$

(P as well as its derivative have been assumed to vanish at the end points.)

Using Eqs. (17.2.13) – (17.2.15) in Eq. (17.2.12), we arrive at

$$P(v_{n+1}, t_{n+1}) = P(v_{n+1}, t_n) + < \frac{d}{dv_{n+1}} (g\, P(v_{n+1}, t_n))>$$

$$+ < \frac{d^2}{dv_{n+1}^2} \left(\frac{g^2}{2} P(v_{n+1}, t_n) \right) > \qquad \text{... (17.2.16)}$$

For the averages, we have to $O(\delta t)$

$$<g> = < \frac{\partial S}{\partial v} > \delta t - <\bar{f}> \sqrt{\delta t}$$

$$= \frac{\partial S}{\partial v} \cdot \delta t$$

while

$$<g^2> = <\bar{f}^2> \delta t$$
$$= 2\varepsilon\, \delta t$$

Using the above averages in Eq. (17.2.16)

$$P(v_{n+1}, t_{n+1}) - P(v_{n+1}, t_n) = \frac{d}{dv_{n+1}} (P \frac{\partial S}{\partial v}) \delta t + \varepsilon \frac{d^2 P}{dv_{n+1}^{\;2}} \delta t \qquad \text{... (17.2.17)}$$

Reverting to the continuum variables by taking the limit of $\delta t \to 0$, we have

$$\frac{\partial P}{\partial t} = \frac{\partial}{\partial v}\left(P \frac{\partial S}{\partial v}\right) + \varepsilon \frac{\partial^2 P}{\partial v^2} \qquad \text{... (17.2.18)}$$

This is the well known Fokker-Planck equation for the probability distribution $P\{v(t)\}$ for a random variable $v(t)$.

What if we start with N variables $\{v_\alpha\}$ ($\alpha = 1,2, \ldots\ldots N$) which satisfy Langevin equations of the form

$$\dot{v}_\alpha = -\frac{\partial S}{\partial v_\alpha} + f_\alpha \qquad \text{... (17.2.19)}$$

where $S = S(\{v_\alpha\})$ and the random forces f_α have the correlation

$$<f_\alpha(t_2)\, f_\beta(t_1)> = 2\varepsilon_\alpha \delta\,(t_2 - t_1)\, \delta_{\alpha\beta} \qquad \text{... (17.2.20)}$$

If $P[\{v_\alpha(t)\}]$ is the probability of the set $\{v_\alpha\}$ acquiring the value $\{v_\alpha(t)\}$ at time t, then the Fokker-Planck equation for P is

$$\frac{\partial P}{\partial t} = \frac{\partial}{\partial v_\alpha}\left\{P \frac{\partial S}{\partial v_\alpha}\right\} + \varepsilon_\alpha \frac{\partial^2 P}{\partial v_\alpha \partial v_\alpha} \qquad \text{... (17.2.21)}$$

where a repeated index means that it has to be summed over.

To end this subsection, we return to Eq. (17.2.18) and find the equilibrium distribution by setting $\frac{\partial P_{eq}}{\partial t} = 0$. This leads to

$$\frac{\partial}{\partial v}\left(P_{\text{eq}}\frac{\partial S}{\partial v}\right) + \varepsilon\frac{\partial}{\partial v}\left\{\frac{\partial P_{\text{eq}}}{\partial v}\right\} = 0$$

or

$$\frac{\partial P_{\text{eq}}}{\partial v} + \frac{P_{\text{eq}}}{\varepsilon}\frac{\partial S}{\partial v} = \text{constant}$$

The constant can be set equal to zero by evaluating the expression for very large values of v, where both P and $\frac{\partial P}{\partial v}$ have to vanish. Thus,

$$\frac{1}{P_{\text{eq}}}\frac{\partial P_{\text{eq}}}{\partial v} = -\frac{\partial}{\partial v}\left(\frac{S}{\varepsilon}\right)$$

or

$$P_{\text{eq}} = P_0\, e^{-S/\varepsilon} \qquad \textbf{... (17.2.22)}$$

the constant P_0 can be found by the normalization condition $\int P_{\text{eq}}(v)dv = 1$. We have thus found the equilibrium distribution to be proportional to $e^{-S/\varepsilon}$, as we had said would happen for an arbitrary $S(v)$ in the previous chapter. It now remains to be shown that an arbitrary distribution must reach the equilibrium distribution as $t \to \infty$. This will be dealt with in the next section.

17.3 The General Solution

The standard technique of writing down a formal solution to the Fokker-Planck equation [Eq. (7.2.18)] is to write it in the form of a Schroedinger equation by carrying out the transformation

$$P = \sqrt{P_{\text{eq}}}\cdot\psi = e^{-\frac{S}{2\varepsilon}}\,\psi \qquad \textbf{... (17.3.1)}$$

Differentiation yields

$$\frac{\partial}{\partial v}\left(P\frac{\partial S}{\partial v}\right) = \frac{\partial}{\partial v}\,(\mathrm{e}^{-\frac{S}{2\varepsilon}}\,\psi S')$$

$$= (-\frac{S'^2}{2\,\varepsilon}\,\psi + \psi' S' + \psi S'')\,\mathrm{e}^{-\frac{S}{2\varepsilon}} \qquad \text{... (17.3.2)}$$

where the prime denotes derivative with respect to v. Again

$$\frac{\partial P}{\partial v} = \frac{\partial}{\partial v}\left(\psi \mathrm{e}^{-S/2\varepsilon}\right) = \left(\psi' - \frac{S'\psi}{2\varepsilon}\right)\mathrm{e}^{-S/2\varepsilon}$$

and

$$\frac{\partial^2 P}{\partial v^2} = \frac{\partial}{\partial v}\left(\left(\psi' - \frac{S'\psi}{2\varepsilon}\right)\mathrm{e}^{-S/2\varepsilon}\right)$$

$$= \left(\psi'' - \frac{S''\psi}{2\varepsilon} - \frac{\psi' S'}{\varepsilon} + \frac{S'^2\Psi}{4\varepsilon^2}\right)\mathrm{e}^{-\frac{S}{2\varepsilon}} \qquad \text{... (17.3.3)}$$

Inserting Eqs. (17.3.2) and (17.3.3) in Eq. (17.2.18), we get

$$\frac{\partial \psi}{\partial t} = \varepsilon\,\psi'' + \left(\frac{S''}{2} - \frac{S'^2}{4\varepsilon}\right)\psi \qquad \text{... (17.3.4)}$$

Writing the stationary state solutions of this equations as $\psi = e^{-\lambda t}\varphi(v)$, we find the eigenvalue equation satisfied by $\varphi(v)$ as

$$-\varepsilon\,\varphi'' + \left(\frac{S'^2}{4\varepsilon} - \frac{S''}{2}\right)\varphi = \lambda\varphi \qquad \text{... (17.3.5)}$$

If we define the operator

$$A = \frac{\partial}{\partial v} + \frac{S'}{2\varepsilon} \qquad \text{... (17.3.6)}$$

then the adjoint operator is

$$A^{+} = -\frac{\partial}{\partial v} + \frac{S'}{2\varepsilon} \qquad \text{... (17.3.7)}$$

and

$$A^{+}A\varphi = \left(-\frac{\partial}{\partial v}+\frac{S'}{2\varepsilon}\right)\left(\frac{\partial}{\partial v}+\frac{S'}{2\varepsilon}\right)\varphi$$

$$= -\frac{\partial^2\varphi}{\partial v^2}-\frac{S''}{2\varepsilon}\varphi-\frac{S'\varphi'}{2\varepsilon}+\frac{S'\varphi'}{2\varepsilon}+\frac{S'^2\varphi}{4\varepsilon^2}$$

$$= \left(-\frac{\partial^2}{\partial v^2}+\frac{S'^2}{4\varepsilon^2}-\frac{S''}{2\varepsilon}\right)\varphi \quad \text{... (17.3.8)}$$

Consequently, we can write Eq. (17.3.5) as

$$\varepsilon A^{+} A \varphi = \lambda \varphi \quad \text{... (17.3.9)}$$

The spectrum of $\varepsilon A^{+}A$ is non-negative.

The zero eigenvalue of $\varepsilon A^{+}A$ is clearly obtained for $A\varphi_0 = 0$ which means that the eigenfunction φ_0 corresponding to the zero eigenvalue is obtained from

$$\frac{\partial\varphi_0}{\partial v}+\frac{S}{2\varepsilon}\varphi_0 = 0$$

or

$$\varphi_0 = B\, e^{-S/2\varepsilon} \quad \text{... (17.3.10)}$$

where B is a constant.

The remaining eigenvalues of $\varepsilon A^{+}A$ are written in order of increasing magnitude as $\lambda_1, \lambda_2, \lambda_3, \ldots$, with eigenfunctions $\varphi_1, \varphi_2, \varphi_3, \ldots$ Since the operator $\varepsilon A^{+}A$ is hermitian, the eigenfunctions of it form a complete set and hence an arbitrary solution of Eq. (17.3.4) can be written as

$$\psi(v,t) = a_0\, e^{-S/2\varepsilon} + \sum_{n=1}^{\infty} a_n\, e^{-\lambda_n t}\, Q_n \quad \text{... (17.3.11)}$$

and the probability distribution $P(v,t)$ is obtained from Eq. (17.3.1) as

$$P(v,t) = A_0\, e^{-S/\varepsilon} + \sum_{n=1}^{\infty} a_n\, e^{-\lambda_n t}\, \varphi_n\, e^{-S/2\varepsilon} \quad \text{... (17.3.12)}$$

Since all λ_n $(n = 1, 2 \ldots) > 0$, as $t \to \infty$, we have

$$\underset{t \to \infty}{\text{Lt}} \; P(v, t) = A_0 \, e^{-S/\varepsilon} = P_{eq} \quad \text{... (17.3.13)}$$

Thus we have demonstrated that as $t \to \infty$, the equilibrium distribution is indeed obtained. If the initial distribution is prescribed as $P(v, t = 0) = \psi_0 \, e^{-S/2\varepsilon}$, then from Eq. (17.3.11)

$$\psi_0 = \sum_{n=0}^{\infty} a_n \, \varphi_n$$

so that the a_n are prescribed as,

$$a_n = (\varphi_n, \psi_0) \quad \text{... (17.3.14)}$$

and Eq. (17.3.12) is indeed a complete solution to the problem.

As the final stage of approach to equilibrium is reached, the time scale for reaching equilibrium is set by the smallest eigenvalue λ_1 in the expansion of Eq. (17.3.12). For a variety of problems, this eigenvalue is thus of great importance. Being an excited state eigenvalue it is difficult to obtain it variationally. So we describe a transformation which for the one-dimensonal problem casts the eigenvalue λ_1 as the ground state eigenvalue of a Hamiltonian that we shall construct and thence it can be variationally obtained.

To proceed we define the operator

$$Q = \begin{pmatrix} 0 & 0 \\ A & 0 \end{pmatrix} \quad \text{... (17.3.15)}$$

The adjoint is

$$Q^{+} = \begin{pmatrix} 0 & A^{+} \\ 0 & 0 \end{pmatrix} \quad \text{... (17.3.16)}$$

The operator Q is nilpotent with $Q^2 = 0 = Q^{+2}$. Now, consider the operator

$$\begin{aligned} H_s &= Q\,Q^{+} + Q^{+}\,Q \\ &= \begin{pmatrix} A^{+}A & 0 \\ 0 & AA^{+} \end{pmatrix} = \begin{pmatrix} H_{+} & 0 \\ 0 & H_{-} \end{pmatrix} \end{aligned} \quad \text{... (17.3.17)}$$

Our concern is with the operator H_{+}, the eigenvalues and eigenfunctions of which can be written as

$$H_+ \varphi_n = \frac{\lambda_n}{\varepsilon} \varphi_n \quad \dots (17.3.18)$$

We note that the column vector $\begin{pmatrix} \varphi_n \\ 0 \end{pmatrix}$ is an eigenvector of H_s with eigenvalue λ_n/ε since

$$H_s \begin{pmatrix} \varphi_n \\ 0 \end{pmatrix} = \begin{pmatrix} H_+ & 0 \\ 0 & H_- \end{pmatrix} \begin{pmatrix} \varphi_n \\ 0 \end{pmatrix} = \begin{pmatrix} H_+ \varphi_n \\ 0 \end{pmatrix}$$

$$= \frac{\lambda_n}{\varepsilon} \begin{pmatrix} \varphi_n \\ 0 \end{pmatrix} \quad \dots (17.3.19)$$

Further, the operator Q commutes with H_s,

$$\begin{aligned} [Q, QQ^+ + Q^+Q] &= Q[Q, Q^+] + [Q, Q^+]Q \\ &= Q\begin{pmatrix} -A^+A & 0 \\ 0 & AA^+ \end{pmatrix} + \begin{pmatrix} -A^+A & 0 \\ 0 & AA^+ \end{pmatrix} Q \\ &= \begin{pmatrix} 0 & 0 \\ -A & 0 \end{pmatrix} \begin{pmatrix} -A^+A & 0 \\ 0 & AA^+ \end{pmatrix} + \begin{pmatrix} -A^+A & 0 \\ 0 & AA^+ \end{pmatrix} \begin{pmatrix} 0 & 0 \\ A & 0 \end{pmatrix} \\ &= \begin{pmatrix} 0 & 0 \\ -AA^+A & 0 \end{pmatrix} + \begin{pmatrix} 0 & 0 \\ AA^+A & 0 \end{pmatrix} \quad \dots (17.3.20) \\ &= 0. \end{aligned}$$

Hence $Q\begin{pmatrix} \varphi_n \\ 0 \end{pmatrix}$ is also an eigenvector of H_s with eigenvalue λ_n/ε, except when $\lambda_n = \lambda_0 = 0$, since in that case $Q\varphi_0 = 0$.

To each λ_n of H_+, there can be only one φ_n in this one-dimensional problem and hence for each λ_n of H_s, we have a double degeneracy—an eigenvector $\begin{pmatrix} \varphi_n \\ 0 \end{pmatrix}$ and another $Q\begin{pmatrix} \varphi_n \\ 0 \end{pmatrix}$.

The structure of

$$Q\begin{pmatrix} \varphi_n \\ 0 \end{pmatrix} \text{ is } \begin{pmatrix} 0 & 0 \\ A & 0 \end{pmatrix} \begin{pmatrix} \varphi_n \\ 0 \end{pmatrix} = \begin{pmatrix} 0 \\ A\varphi_n \end{pmatrix} = \begin{pmatrix} 0 \\ \psi_n \end{pmatrix}$$

But $\begin{pmatrix} 0 \\ \psi_n \end{pmatrix}$ is an eigenvector of H_s with eigenvalue λ_n/ε and consequently

$$\lambda_n \begin{pmatrix} 0 \\ \psi_n \end{pmatrix} = H_s \begin{pmatrix} 0 \\ \psi_n \end{pmatrix} = \begin{pmatrix} H_+ & 0 \\ 0 & H_- \end{pmatrix} \begin{pmatrix} 0 \\ \psi_n \end{pmatrix}$$

$$= \begin{pmatrix} 0 \\ H_- \psi_n \end{pmatrix} \qquad \text{... (17.3.21}$$

Thus

$$H_- \psi_n = \frac{\lambda_n}{\varepsilon} \psi_n \qquad \text{... (17.3.22)}$$

The functions ψ_n are the eigenfunctions of the operator H_- with eigenvalues λ_n/ε. The lowest eigenvalue is λ_1/ε (the above arguments do not hold for $\lambda_0 = 0$ as explained earlier). A variational calculation on the Hamiltonian H_- for the ground state yields λ_1. Thus, we have the following result:

For the one-dimensional Hamiltonian εA^+A, with the nondegenerate eigenvalues in increasing order $0, \lambda_1, \lambda_2, ...,$ the eigenvalues $\lambda_1, \lambda_2, \lambda_3, \ldots$ are the nondegenerate eigenvalues of the partner Hamiltonian εAA^+. The eigenvalue λ_1 being the ground state energy of εAA^+, can be obtained variationally.

17.4 Metastable State and the Lowest Eigenvalue

We consider the function $S(v)$ to be of the form shown in Fig 5.2. Instead of using the notation v for the variable, we will shift to x and thus consider $S(x)$ to be a function with two minima at $x = x_1$ and $x = x_2$ and a maximum at $x = x_m (x_1 < x_m < x_2)$. The values of $S(x)$ at $x = x_1$, x_m and x_2 are S_1, S_m' and S_2 respectively. The function is shown by the solid curve in Fig. 17.1.

This is the kind of potential where a metastable state is formed—an initial distribution peaked around $x = x_1$ is such a metastable state and as has been generally demonstrated for the Fokker-Planck equation, eventually this state must reach the equilibrium distribution where $P_0 e^{-S(x)}/\varepsilon$, where P_0 is a normalizing constant. The approach to equilibrium will be governed by the first excited state eigenvalue λ_1 of the operator $-\varepsilon\frac{d^2}{dx^2} + \frac{S'^2}{4\varepsilon} - \frac{S''}{2}$ (we note that λ_1 will be

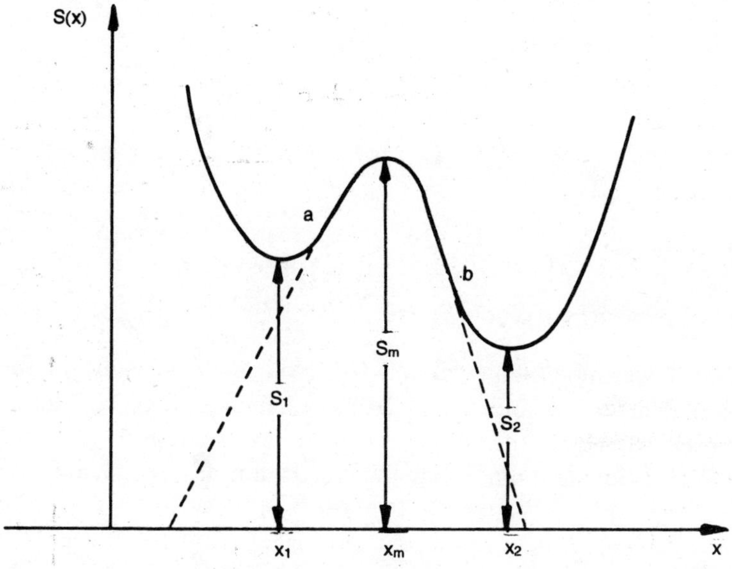

Fig. 17.1 ***S (x)* is the potential exhibiting a metastable state. The point $x = a$ and $x = b$ and the dashed lines are relevant to the variational trial function of Eqs. (17.4.2) – (17.4.4).**

exponentially small ε since qualitative reasoning shows that the potential $\frac{S'^2}{4\varepsilon} - \frac{S''}{2}$ has three minima and the tunneling probabilities between the different minima brings the ground state and the first excited state exponentially close. Since $\lambda_0 = 0$, we have λ_1 exponentially small). As explained in the previous section λ_1 can be found as the ground state energy of the partner Hamiltonian $\overline{H}$, given by

$$\overline{H} = -\varepsilon \frac{d^2}{dx^2} + \frac{S'^2}{4\varepsilon} + \frac{S''}{2} \quad \text{... (17.4.1)}$$

We now have to find a good trial wavefunction that will guarantee a small eigenvalue. First note that if $\psi = e^{S/2\varepsilon}$, then $\overline{H}\,\psi$ is identically zero, but clearly ψ is not an admissible eigenfunction since it cannot be normalized. However, it can serve as a guide in choosing the trial function.

We break up the region $-\infty \le x \le \infty$ into three distinct regions

i) $-\infty \le x \le a$

ii) $a \leq x \leq b$
iii) $b \leq x \leq \infty$
and assume the general form of the trial function ψ_g as $e^{f(x)/2\varepsilon}$, where we choose

$$\text{Region (ii) } f(x) = S(x) \qquad \text{... (17.4.2)}$$

$$\text{Region (i) } f(x) = S(a) - \alpha(a - x) \qquad \text{... (17.4.3)}$$

$$\text{Region (iii) } f(x) = S(b) - \beta(x - b) \qquad \text{... (17.4.4)}$$

The variational parameters are 'a' and 'b'. Note that the value of the function $f(x)$ is matched at $x = a$ and $x = b$ from either side. The constants α and β will be chosen to match the slopes as well. In the region $a \leq x \leq b$, we approximate $S(x)$ by a parabola about the maximum at $x = x_m$, i.e.

$$S(x) = S_m - \frac{1}{2} S''(x_m)(x - x_m)^2 \; ; a \leq x \leq b \qquad \text{... (17.4.5)}$$

Consequently the slopes at $x = a$, b, of $S_{(x)}$ are

$$- S''(x_m)(a - x_m), \; - S''(x_m)(b - x_m)$$

Hence, we identify

$$\alpha = - S''(x_m)(a - x_m) \qquad \text{... (17.4.6)}$$

$$\beta = S''(x_m)(b - x_m) \qquad \text{... (17.4.7)}$$

The values $S(a)$ and $S(b)$ can be obtained by expanding $S(x)$ about x_1 and x_2 respectively in a parabolic approximation as

$$S(a) = S_1 + \frac{1}{2} S''(x_1)(a - x_1)^2 \qquad \text{... (17.4.10)}$$

$$S(b) = S_2 + \frac{1}{2} S''(x_2)(b - x_2^2) \qquad \text{... (17.4.9)}$$

We thus obtain the following approximation for $f(x)$.

$$\text{Region (i), } f(x) = S_1 + \frac{1}{2} S''(x_1)(a - x_1)^2 + S''(x_m)(a - x_m)(a - x)$$

Region (ii), $f(x) = S_m - \frac{1}{2} S''(x_m)(x - x_m)^2$

Region (iii), $f(x) = S_2 + \frac{1}{2} S''(x_2)(b - x^2)^2 - S''(x_m)(b - x_m)(x - b)$

The variational estimate for λ_1 is now

$$\lambda_1 = \int_{-\infty}^{\infty} e^{\frac{f(x)}{2\varepsilon}} \overline{H} e^{\frac{f(x)}{2\varepsilon}} dx \Big/ \int_{-\infty}^{\infty} e^{\frac{f(x)}{\varepsilon}} dx \quad \text{... (17.4.10)}$$

The integral can be evaluated for $\varepsilon << 1$, by assuming that the range $a \le x \le b$ for integration purposes can be extended from $-\infty$ to ∞. The resulting integral λ_1 is a function of a and b and these variational parameters are found by setting.

$$\frac{\partial \lambda_1}{\partial a} = \frac{\partial \lambda_1}{\partial b} = 0 \quad \text{... (17.4.11a)}$$

leading to

$$a = \sqrt{\frac{3\varepsilon}{|u''(a)|}}, \quad b = \sqrt{\frac{3\varepsilon}{|u''(b)|}} \quad \text{... (17.4.11b)}$$

The eigenvalue λ_1 is found to be

$$\lambda_1 = \frac{e^{3/2}}{6\sqrt{6\pi}} [e^{-[S_m - S(a)]/\varepsilon} + e^{-[S_m - S(b)]/\varepsilon}] \quad \text{... (17.4.12)}$$

The characteristic time $\tau_s = \lambda_1^{-1}$ and is known as Kramer's time. It can be obtained from a WKB calculation as well. In that case the pre-factor of λ_1 is $1/2\pi$. The advantage of Eq. (17.4.12) is that, if being variational, this λ_1 is known to be a strict upper limit on the actual eigenvalue.

Chapter 18
Fokker-Planck Equation: Examples

18.1 Introduction

In this chapter we will discuss the application of Fokker Planck equation to two different kinds of situations. One is the approach to equilibrium of a non-equilibrium initial state. We prepare the system which has an equilibrium distribution localized in space (localized meaning that the probability of its extending out to infinity is zero) in a far from equilibrium condition and then study the time evolution of the probability distribution of the state. This sort of time development occurs in a wide variety of physical phenomena. The one which we will deal with explicitly in Sec. 18.2 will be the sudden switching on of a laser. Similar problems arise when the temperature of a system undergoing phase transformation is suddenly lowered from that corresponding to a one-phase region to that corresponding to a two-phase situation.

A different kind of problem is one where the potential is periodic. In this case the particle does not settle down but can jump from one minimum to another due to the noise and its mean square displacement is expected to go on increasing with time. This increase for the one-dimensional potential can be written as

$$<|x(t) - x(0)|^2> = 2Dt \qquad \text{... (18.1.1)}$$

where D is the diffusion constant. A related quantity called the mobility can be defined by applying an additional force F (independent of spatial position) which will make the particle diffuse preferentially in the direction of F. For small F, the mobility defined is by the relation

$$\langle v \rangle = \mu F \quad \text{... (18.1.2)}$$

A few examples of particles moving in a periodic potential are

i) a pendulum oscillating in a viscous fluid. The equation of motion in terms of the angle θ is

$$m\, l\, \ddot{\theta} + m\, k\, l\, \dot{\theta} + mg \sin\theta = lf(t) \quad \text{...(18.1.3)}$$

where l is the length of the pendulum. The second term on the left is the viscous damping produced by the fluid and the torque on the right is the effect of the fluctuating forces produced by the fluid molecules.

ii) Superionic conductor: Here some ions are highly mobile in a nearly fixed ion lattice. A typical example is AgI, where the lattice consists of iodide ions and the silver ions Ag^+ are highly mobile. The damping and the fluctuating force are provided by the lattice vibrations.

iii) Josephson tunneling junctions: Here two superconductors are separated by a thin oxide layer. It is the phase difference between the wave functions of the Cooper pairs on either side that satisfies an equation similar to Eq. (18.1.3).

18.2 Passage to Equilibrium of a Nonequilibrium State

We consider the sudden switching on of a lasser, which can be considered as a far from equilibrium phenomenon. The laser switches on as a pump parameter μ is varied and for $\mu > \mu_c$, the zero intensity state in the cavity is unstable and the intensity begins to grow and saturates at a value determined by μ. The dynamics can be described by the phenomenological equation

$$\frac{\partial n}{\partial t} = (\mu - \mu_c)\, n - \beta n^3 + f(t) \quad \text{... (18.2.1)}$$

The noise term $f(t)$ comes from the fact that the atomic levels involved in the lasing action are not sharp but broadened due to various effects and the averaging over the broadening gives noise. It should also be noted that n is not strictly a scalar. The above equation, if derived from a microscopic starting point, would involve the electric field $\mathbf{E} = \text{Re}\,\mathbf{E}(t)\,e^{i\omega t}$. The vector $\mathbf{E}(t)$ would involve a magnitude and phase for each of its components. Each component (E_x, E_y or E_z) would thus be characterized by two functions E_1 and E_2 (the real and imaginary parts) and we would actually have two equations

$$\dot{E}_1 = (\mu - \mu_c)\,E_1 - \beta\,(E_1^2 + E_2^2)E_1 + f_1(t) \quad \text{... (18.2.2a)}$$

$$\dot{E}_2 = (\mu - \mu_c)\,E_2 - \beta\,(E_1^2 + E_2^2)\,E_2 + f_2\,(t) \quad \text{... (18.2.2b)}$$

with

$$\langle f_1 f_2 \rangle = 0 \quad \text{... (18.2.3)}$$

$$\langle f_1(t_1)\,f_1(t_2) \rangle = \langle f_2(t_1) f_2(t_2) \rangle = 2\varepsilon\delta\,(t_1 - t_2) \quad \text{... (18.2.4)}$$

We now recall the Fokker-Planck equation for N variables. The Langevin equations are taken to be

$$\dot{q}_\alpha = -\frac{\partial S}{\partial q_\alpha} + f_\alpha \quad \text{... (18.2.5)}$$

where

$$S = S\,[q_\alpha] \quad \text{... (18.2.6)}$$

and

$$\langle f_\alpha(t_2) f_\beta(t_1) \rangle = 2\varepsilon\delta\,(t_2 - t_1)\,\delta_{\alpha\beta} \quad \text{... (18.2.7)}$$

The corresponding Fokker-Planck equation for $P(\{q_\alpha(t)\})$, the probability of the set $\{q_\alpha(t)\}$ acquiring the value $\{q_\alpha(t)\}$ at time t is given by

$$\frac{\partial P}{\partial t} = \frac{\partial}{\partial q_\alpha}\left(P\frac{\partial S}{\partial q_\alpha}\right) + \varepsilon\frac{\partial^2 P}{\partial q_\alpha\,\partial q_\alpha} \quad \text{... (18.2.8)}$$

Hence the Fokker-Planck equation for the problem of Eq. (18.2.2) with $E_1 = q_1$ and $E_2 = q_2$ is

$$\frac{\partial P}{\partial t} = \vec{\nabla} \cdot [P\,\mathbf{q}\,(\alpha - \beta\,q^2)] + \varepsilon\,\nabla^2 P, \quad \text{... (18.2.9)}$$

where $q^2 = q_1^2 + q_2^2$, $\mathbf{q} = (q_1, q_2)$, $\nabla^2 = \frac{\partial^2}{\partial q_1^2} + \frac{\partial^2}{\partial q_2^2}$

Since P will not depend on the phase of $\mathbf{q}$ its functional dependence is on the magnitude of $\mathbf{q}$ alone and Eq. (18.2.9) becomes

$$\frac{\partial P}{\partial t} = \frac{1}{q}\frac{\partial}{\partial q}[-Pq^2(1-q^2)] + \frac{\varepsilon}{q}\frac{\partial}{\partial q}\left(q\frac{\partial P}{\partial q}\right) \qquad \text{... (18.2.10)}$$

where we have set $\alpha = \beta = 1$ without any loss of generality.

The above equation needs to be supplemented by a boundary condition at $t = 0$. We set this as $P(q, t = 0) = \delta(q)$. This physically corresponds to suddenly setting the pump parameter at a value $\mu > \mu_c$ from a value below μ_c. This implies that at $t = 0$, there are no photons in the cavity (a non-equilibrium state), while the equilibrium has to correspond to a definite intensity of radiation, namely $P \alpha\, e^{-S/\varepsilon}$, where $S = -\frac{q^2}{2} + \frac{q^4}{4}$. The approach to equilibrium has three different regimes which we will now consider separately.

18.2.1 Diffusive Regime

The probability, concentrated near $q = 0$, begins to spread out and the appropriate approximation is to use $S = -q^2/2$ for the initial growth period. The Fokker-Planck equation can be solved exactly for the quadratic potential. We illustrate this for one-dimensional situation where $S = -x^2/2$. This leads to the following form for the operator on the left hand side of Eq. 17.3.5 (calling the operator L)

$$L = -\varepsilon\frac{\partial^2}{\partial x^2} + \frac{1}{4\varepsilon}x^2 + \frac{1}{2} \qquad \text{... (18.2.11)}$$

The eigenvalues λ_n with the eigenfunctions φ_n satisfy

$$\left(-\varepsilon\frac{\partial^2}{\partial x^2} + \frac{1}{4\varepsilon}x^2\right)\varphi_n = \left(\lambda_n - \frac{1}{2}\right)\varphi_n \qquad \text{... (18.2.12)}$$

Clearly

$$\lambda_n = n + 1 \qquad \text{... (18.2.14)}$$

with

$$\varphi_n = N_n \, e^{-x^2/\varepsilon} \, H_n\left(\frac{x}{2\varepsilon}\right) \qquad \text{... (18.2.14)}$$

where H_n is the Hermite Polynomial of order n and N_n is the normalization constant.

Thus,

$$P(x, t) = \sum C_n \, e^{-(n+1)t} \, N_n \, e^{-x^2/\varepsilon} \, H_n\left(\frac{x}{2\varepsilon}\right) \qquad \text{... (18.2.15)}$$

where the C_n are obtained from the fact that at $t = 0$, $P(x, t) = \delta(x)$ and hence $C_n = H_n(0)$. We now obtain

$$P(x, t) = \sum e^{-(n+1)t} N_n \, e^{-x^2/\varepsilon} H_n(0) \, H_n\left(\frac{x}{2\varepsilon}\right)$$

$$= \frac{1}{\left[2\pi\varepsilon\left(e^{2t}-1\right)\right]^{1/2}} \, e^{-x^2/2\varepsilon\left(e^{2t}-1\right)} \qquad \text{... (18.2.16)}$$

For the D-dimensional oscillator potential x^2 becomes the r^2 where r is the radius vector. For the two-dimensional vector q of our case

$$P(\underline{q}, t) = \frac{1}{[2\pi\varepsilon\,(e^{2t}-1)]^{1/2}} \, e^{-q^2/2\varepsilon\,(e^{2t}-1)} \quad \text{(for short time)} \qquad \text{... (18.2.17)}$$

The short-time approximation means that the inverted oscillator part of the potential $V(q) = -q^2/2 + q^2/4$ is being experienced. As expected for the inverted potential, the probability distribution spreads out in time and as is obvious from Eq. (18.2.12), the expectation value $\langle q^2 \rangle$ increases as $2\varepsilon(e^{2t} - 1) \sim 4\varepsilon t$ for short times. The validity of the approximation of dropping q^4 is going to be reasonable for times such that $\langle q^4 \rangle < 1$, which implies $2\varepsilon e^{2t} < 1$ which is for times $t < t_0 \sim 0\,(\ln \frac{1}{\varepsilon})$. For $t \gtrsim t_0$ it is no longer permissible to drop the q^4 term of the potential. However, yet another simplification sets in.

18.2.2 "Scaling" or "Sliding" Regime

When q becomes somewhat larger, we can no longer ignore the q^4 term but now the effect of noise is minimal while the system slides down the potential. In this limit the ε-containing terms in Eq. (18.2.10) can be dropped and we have

$$\frac{\partial P}{\partial t} = \frac{1}{q}\frac{\partial}{\partial q}\,[-Pq\,(1-q^2)]$$

$$= -\left[(q\,(1-q^2)\frac{\partial P}{\partial q} + 2\,(1-2q^2)\,P\right] \qquad \dots \textbf{(18.2.18)}$$

Substituting $\tau = e^t$ and $q^2 = y$, we find

$$\frac{\tau}{2}\frac{\partial P}{\partial \tau} = -\left[y\,(1-y)\,\frac{\partial P}{\partial y} + (1-2y)\,P\right] \qquad \dots \textbf{(18.2.19)}$$

This has the same form as the Callan-Symanzik equation and hence the solution can be written down as

$$P(y, \tau) = \frac{1}{y(1-y)}\,\Phi\left(\ln \tau^2\frac{(1-y)}{y}\right) \qquad \dots \textbf{(18.2.20)}$$

where Φ is any function. The functional form of Φ is determined by matching between regimes (i) and (ii). In regime (i), the solution has the form $P \sim e^{-q^2/\varepsilon\,(e^{2t}-1)}$, which is valid for $\langle q^2\rangle < 1$. To match this form with the solution given in Eq. (18.2.21), we require $\Phi\,[\ln e^{2t}\frac{(1-q^2)}{q^2}] = e^{-q^2/\varepsilon(1-q^2)\,(e^{2t}-1)}$, which is of course $e^{-q^2/\varepsilon e^{2t}(1-q^2)}$ for $e^{2t} >> 1$, i.e. corresponding to the regime (ii). Taking care of the normalization, we have for the combined regimes (i) and (ii)

$$P\,(q, t) = \frac{1}{q^2(1-q^2)}\,e^{-q^2/\varepsilon(1-q^2)\,(e^{2t}-1)} \qquad \dots \textbf{(18.2.21)}$$

The time development of the probability is shown in the sequence of figures in Fig. 18.1.

The solution given in Eq. (18.2.21) does show that the probability acquires peaks near $q^2 \doteq 1$, but it does not spread beyond $q^2 = 1$ and this cannot give the true equilibrium distribution as $t \to \infty$. The limit of Eq. (18.2.21) as $t \to \infty$ is the accumulation of the probability at $q^2 = 1$. The true equilibrium is obviously proportional to $e^{\left(\frac{q^2}{2}-\frac{q^4}{4}\right)/\varepsilon}$ and hence there is a third regime where the true equilibrium sets in.

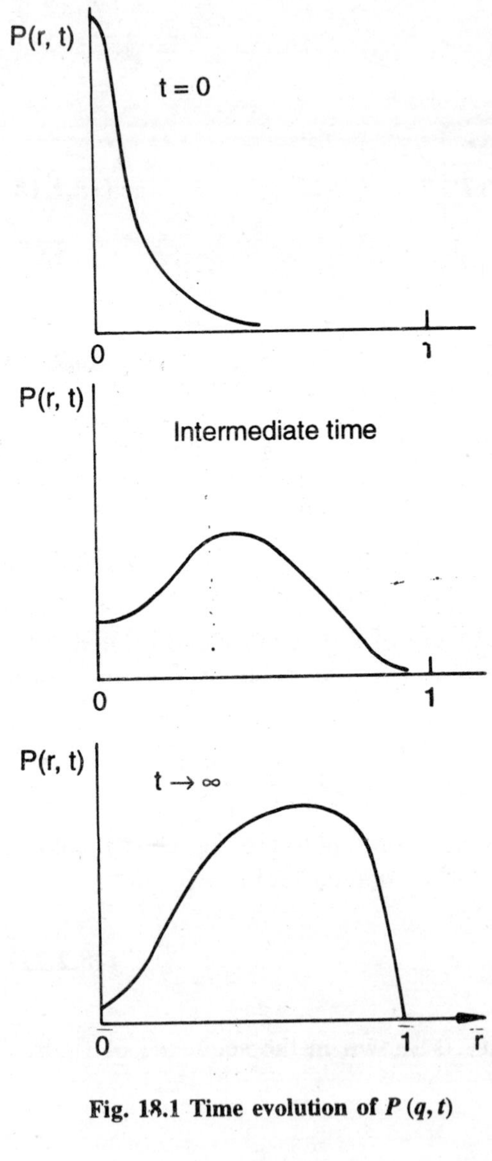

Fig. 18.1 Time evolution of $P(q, t)$

18.2.3 Kramers Regime

In this regime which sets in when $q<^2> \sim 1$, i.e. for time scales greater than that obtained from $\varepsilon < 1 - q^2 > e^{2t} \sim 1$, the noise is important once more and causes occasional transition between the two wells. The solution for $P\ (q,\ t)$ in this final regime has the form

$$P(q, t) = N\left[e^{\frac{q^2}{2\varepsilon}-\frac{q^4}{4\varepsilon}} + a_1\ \varphi(q)\, e^{-\lambda_1 t}\right] \qquad \text{... (18.2.21)}$$

where N is a normalization constant, $\Phi(q)$ is the first excited state wavefunctions in the equivalent Schröedinger equation corresponding to the potential $q^2/2 + q^4/4$, a_1 is an expansion coefficient in a complete set expansion and λ_1 is the eigenvalue for the first excited state. Equation (18.2.21) is valid for $t \to \infty$, when in the complete let expansions, the λ_1-term alone survives. The general arguments of the previous chapter tell us that $\lambda_1 = \Delta S/\varepsilon$, where ΔS is the height of the potential barrier. In this case $\Delta S = 1/2$.

18.3 Diffusion in a Periodic Potential

As written down in Eq. (18.1.3), the form of the equation of motion for a periodic potential $V(x)$ and an externally applied constant force F is

$$\ddot{x} + k\dot{x} = F - V'(x) + f(t) \qquad \text{... (18.3.1)}$$

where

$$< f(t) f(t') > = 2\Gamma\, \delta(t - t') \qquad \text{... (18.3.2)}$$

and

$$V(x) = V(x + 2\pi) \qquad \text{... (18.3.3)}$$

We will deal with the simplest situation, namely the high friction limit and consequently will drop the $\ddot{x}$ term in Eq. (18.3.1) and work with the Langevin equation which acquires the form

$$\dot{x} = F - V'(x) + f(t) \qquad \text{... (18.3.4)}$$

where the constant k has been absorbed in a choice of time scale. This will lead to a Fokker-Planck equation for the probability distribution $P(x, t)$:

$$\frac{\partial P}{\partial t} = -\frac{\partial}{\partial x}[(F - V')P] + \varepsilon \frac{\partial^2 P}{\partial x^2} \qquad \text{... (18.3.5)}$$

where 2ε is the coefficient of the delta function of the fluctuating force when the re-scaled time variable is used. The stationary solution P_0 is clearly given by $\left(\frac{\partial P_0}{\partial t} = 0\right)$

$$(F - V)\, P_0 - \varepsilon \frac{\partial P_0}{\partial x} = \text{const} = S \qquad \text{... (18.3.6)}$$

leading to

$$P_0(x) = e^{-w(x)/\varepsilon} \left[N - \frac{S}{\varepsilon} \int_0^x e^{W(x')/\varepsilon}\, dx' \right]$$

where

$$W(x) = V(x) - Fx \qquad \text{... (18.3.8)}$$

is the total potential.

We now show that if $P_0(x)$ is bounded for large x, then P_0 is periodic with $P_0(x) = P_0(x + 2\pi)$. The proof proceeds by constructing the integral

$$\int_0^{2\pi n + x} e^{W(x')/\varepsilon}\, dx' = \int_0^{2\pi} e^{W(x')/\varepsilon}\, dx' + \ldots + \int_{2\pi(n-1)}^{2\pi n} e^{W(x')/\varepsilon} dx'$$

$$+ \int_{2\pi n}^{2\pi n + x} e^{W(x')/\varepsilon}\, dx'$$

Now,

$$\begin{aligned} W(x + 2n\pi) &= V(x + 2n\pi) - (x + 2n\pi)\, F \\ &= V(x) - Fx - 2n\pi F \\ &= W(x) - 2\pi n F \end{aligned}$$

and thus the above integral becomes

$$\int_0^{2\pi n + x} e^{W(x')/\varepsilon}\, dx = I + I\, e^{-2\pi F/\varepsilon} + \ldots + I\, e^{-2\pi(n-1)F/\varepsilon}$$

$$+ \int_0^x e^{W(x')/\varepsilon}\, dx'\, e^{-2\pi n F/\varepsilon}$$

$$= I \frac{1 - e^{2\pi n F/\varepsilon}}{1 - e^{-2\pi F/\varepsilon}} + e^{-2\pi n F/\varepsilon} \int_0^x e^{W(x')/\varepsilon} dx' \qquad \text{... (18.3.9)}$$

where

$$I = \int_0^{2\pi} e^{W(x')/\varepsilon} dx' \qquad \text{... (18.3.10)}$$

Thus,

$$P_0(x + 2n\pi) = e^{-w(x)/\varepsilon} \left[N - \frac{S}{\varepsilon} \int_0^{2n\pi + x} e^{W(x')/\varepsilon} dx' \right] e^{2\pi n F/\varepsilon}$$

$$= e^{-W(x)/\varepsilon} e^{2\pi n F/\varepsilon} \left[N - \frac{S}{\varepsilon} I \frac{1}{1 - e^{-2\pi F/\varepsilon}} \right]$$

$$+ \frac{e^{-W(x)/\varepsilon}}{1 - e^{-2\pi F/\varepsilon}} \frac{SI}{\varepsilon} - \frac{S}{\varepsilon} e^{-W(x)/\varepsilon} \int_0^x e^{W(x')/\varepsilon} dx' \qquad \text{... (18.3.11)}$$

where we have used Eq. (18.3.9). If we now require that $P_0(x + 2n\pi)$ is bounded for large n ($n > 0$ if $F > 0$ and $n < 0$ if $F < 0$), then first term in Eq. (18.3.11) needs to be removed and hence we require that

$$N = \frac{S}{\varepsilon} I \cdot \frac{1}{1 - e^{-2\pi F/F}} \qquad \text{...(18.3.12)}$$

Now using this in Eq. (18.3.11) in the non-vanishing part, we have (see Eq. (18.3.7)

$$P_0(x + 2n\pi) = P_0(x) \qquad \text{... (18.3.13)}$$

Because of the periodicity, the probability is normalized in the periodicity interval and we have

$$\int_0^{2\pi} P_0(x)\, dx = 1 \qquad \text{... (18.3.14)}$$

The mean drift velocity in the stationary state can be found from the Langevin equation given in Eq. (18.3.4) as

$$
\begin{aligned}
<\dot{x}> &= < F - V'(x) + f(t) > \\
&= < F - V'(x) > \\
&= \int_0^{2\pi} (F - V'(x))\, P_0(x) dx \\
&= \int_0^{2\pi} \left(S + \varepsilon \frac{\partial P_0}{\partial x} \right) dx \\
&= 2\pi S \qquad \text{... (18.3.15)}
\end{aligned}
$$

where we have used Eq. (18.3.6). The integration constants N and S are obtained from Eqs. (18.3.12) and (18.3.14). Elimination of N between these two equations, and taking the limit of $F \to 0$, yields the mobility μ as

$$
\mu = \underset{F \to 0}{\text{Lt}} \cdot \frac{<v>}{F} = 2\pi \frac{2\pi}{\int_0^{2\pi} e^{V(x)/\varepsilon}\, dx} \cdot \frac{2\pi}{\int_0^{2\pi} e^{-V(x)/\varepsilon}\, dx} \qquad \text{... (18.3.16)}
$$

Chapter 19
Approach to Equilibrium: Quantum Systems

In the previous chapter our concern has been with the approach to thermodynamic equilibrium for classical systems. In this chapter we will focus on the quantum case and provide two possible equations of motion that can describe the passage to equilibrium—one is the Pauli master equation and the other a quantum Langevin equation.

19.1 Pauli Master Equation

This requires the description of a quantum system in terms of the density matrix. Usually a quantum state is specified by giving the state vector $|\psi(t)>$ at any time t. The state vector can be expanded in terms of a complete set of states $|u_n>$ for the system as

$$|\psi(t)> = \sum c_n(t)|u_n> \quad \text{... (19.1.1)}$$

In terms of the numbers c_n the average value of any operator A can be written as

$$<A> = <\psi|A|\psi> = \sum c_n^* c_m A_{nm} \quad \text{... (19.1.2)}$$

where

$$A_{nm} = \int u_n^* A\, u_m\, dx \qquad \text{... (19.1.3)}$$

The description of a state in terms of the density matrix involves the specification of the numbers $c_n^* c_m$, i.e. the density operator is given by

$$\rho = |\psi> <\psi| \qquad \text{... (19.1.4)}$$

The matrix element ρ_{mn} is given by

$$\rho_{mn} = <u_m|\psi> < \psi| u_n >$$

$$= c_m^* c_n \qquad \text{... (19.1.5)}$$

Clearly the diagonal element ρ_{mm} is the probability of the systems being in the *m*th state. Note that specification of ρ gives the expectation value of any operator *A* as

$$<A> = \text{Tr}\ (\rho A) \qquad \text{... (19.1.6)}$$

and the equation of motion for ρ is clearly

$$\frac{d\rho}{dt} = \frac{i}{\hbar}[\rho, H] \qquad \text{... (19.1.7)}$$

Let us now write the Hamiltonian operator as

$$H = H_0 + \lambda V \qquad \text{... (19.1.8)}$$

where H_0 is the unperturbed Hamiltonian (the non-interacting particles or quasi-particles part) and λV is the perturbation whose strength is measured by λ. The eigenvalues and eigenfunctions of H_0 can be found as

$$H_0\, |u_n> = E_n|u_n> \qquad \text{... (19.1.9)}$$

We use the $|u_n>$ as our basis functions and calculate the matrix elements ρ_{mn} in this basis. The matrix elements of *H* are

$$H_{mn} = E_m\, \delta_{mn} + \lambda V_{mn} \qquad \text{... (19.1.10)}$$

where without any loss of generality we have let $V_{nn} = 0$.

If $V = 0$, $\dot{\rho}_{mm} = 0$ as should be obvious form Eq. (19.1.7). The central question is what is the probability of finding the particle in the *m*th state, i.e.

ρ_{mm} as a function of time when V is not equal to 0. The answer provided by Pauli is

$$\frac{\partial \rho_{mm}}{\partial t} \approx \left\{ \sum_n (W_{mn}\, \rho_{nn}(t) - W_{nm}\, \rho_{mm}(t) \right\} \qquad \text{... (19.1.11)}$$

where the transition rates are given by Fermi's "Golden Rule" as

$$W_{mn} = 2\,\pi\,\lambda^2\, |V_{mn}|^2\, \delta\,(E_m - E_n) \qquad \text{... (19.1.12)}$$

The derivation of Pauli master equation as follows:

$\Delta\rho_{mm}(t)$ = net increase of occupation of the state $|u_n>$ during an interval Δt

$$= \text{gain in } |u_m> - \text{ loss in } |u_n> \qquad \text{... (19.1.13)}$$

The gain can come only from states other than $|u_m>$:

Gain in $|u_m> = \sum\limits_{n \neq m}$ [prob. of transition from n to m during Δt] $\times$ [occupation probability of $|u_n>$]

$$= \sum_{n \neq m} (W_{mn}\, \Delta t)\, \rho_{nn} \qquad \text{... (19.1.14)}$$

The loss can only be to other states:

$$\text{loss of } |u_m> = \sum_{n\mu} (W_{nm}\, \Delta t)\, \rho_{mm} \qquad \text{... (19.1.15)}$$

Now, writing

$$\frac{\partial \rho_{mm}}{\partial t} = \frac{\Delta \rho_{mm}}{\Delta t} \qquad \text{... (19.1.16)}$$

we get the master equation of Eq. (19.1.11), provided there exists a time-independent transition rate W_{mn} for all m and n with $m \neq n$.

First-order perturbation theory ensures the existence of W_{mn} under certain circumstances. To explore the constraints, we write the wavefunction of the time-dependent Schrödinger equation

$$i\hbar \frac{\partial|\psi>}{\partial t} = H_0|\psi> + \lambda V|\psi> \qquad \text{... (19.1.17)}$$

$$\text{as } |\psi> = \sum c_n(t)\, e^{-iE_n t/\hbar}|n> \qquad \text{... (19.1.18)}$$

where $c_n(0) = 1$ if $n = m$ and 0 otherwise. So the system is initially in the mth state and now $c_n(t)$ will give the transition amplitude from the mth to the nth state. Inserting the solution of Eq. (19.1.18) into Eq. (19.1.17),

$$\dot{c}_n = \sum_j c_j(t) < u_n|V|u_j > e^{-i(E_j - E_n)t/\hbar}$$

$$\approx \lambda < u_n|V|u_m > e^{-i(E_m - E_n)t/\hbar}$$

leading to

$$c_n = \lambda V_{mn} \int_0^t e^{-iE_{mn}t'/\hbar}\, dt'$$

$$= \lambda V_{mn}\hbar \left\{ \frac{1 - e^{-iE_{mn}t/\hbar}}{E_{mn}} \right\}$$

$$= \lambda V_{mn} \frac{1 - e^{-i\omega_{mn}t}}{\omega_{mn}}$$

$$|c_n|^2 = 4\lambda^2 |V_{mn}|^2 \frac{\sin^2 \frac{\omega_{mn}t}{2}}{\omega_{mn}^2}, \qquad \text{... (19.1.19)}$$

which is the probability of making a transition from the state m to the state n in a time t.

$$\text{Now } \underset{t\to\infty}{\text{Lt}} \frac{\sin^2 \frac{\omega_{mn}t}{2}}{\omega_{mn}^2} = \frac{\pi}{2} t\, \delta(\omega_{mn}) \qquad \text{... (19.1.20)}$$

and hence the W_{mn} of Eq. (19.1.12) follows. We note:

i) t is never really infinite and hence the delta function is never really exact

ii) The true energy levels of the system differ from E_n by terms of $O(\lambda)^2$. Hence $\omega_{mn} \sim O(\lambda^2)$ and $\omega_{mn}\, t \sim O(\lambda^2 t)$.

If we restrict overselves to be within the first peak of $\dfrac{\sin^2(\frac{\omega_{mn} t}{2})}{\omega_{mm}^2}$ then $\omega_{mn}\, t < 0\ 2\pi)$ and hence the constraint

$$\lambda^2 t < O(1) \qquad \text{... (19.1.21)}$$

Thus for $\lambda << 1$ 1, we can indeed think in terms of t becoming very big.

A more rigorous derivation and one that holds for larger values of λ was given by Van Hove.

iii) Note that if the levels are very densely spaced, then one can define a density of states and the total loss from state m to state n has to be obtained by integrating over the bunch of states near n (all of them satisfy the approximate energy conservation and this leads to the factor

$$\int \frac{\sin^2 \frac{\omega_{mn}t}{2}}{\omega_{mn}^2} \times \rho(\omega_{mn})\, d\omega_{mn}$$

$$\approx \rho(\omega_{mn}) \int \frac{\sin^2 \omega_{mn}t}{\omega_{mn}^2}\, d\omega_{mn}$$

$$\propto \rho(\omega_{mn})t$$

yielding once again a time-independent transition rate.

The master equation written down in Eq. (19.1.11) does not have the invariance under $t \to -t$, which was possessed by the Schrödinger equation on which we based our considerations. This equation describes the irreversible behaviour of a system. If we call $P_n = \rho_{nn}$, the probability of being in the nth state, then the master equation can be written as

$$\frac{dp_m}{dt} = \sum_{n \neq m} (W_{nm} P_n - W_{mn} P_m) \qquad \text{... (19.1.22)}$$

For an isolated system in equilibrium $P_m = 0$. Also by our construction $W_{mn} = W_{nm}$ and hence in equilibrium $P_m = 0$ yields as a sufficient condition

$$P_n = P_m \quad \text{... (19.1.23)}$$

which is the principle of equal a priori probability. If $W_{mn} \neq W_{nm}$, then the sufficient condition for vanishing of P_m is

$$W_{mn} P_m = W_{nm} P_n \quad \text{... (19.1.24)}$$

We now describe the situation where $W_{mn} \neq W_{nm}$.

Consider a system S_l in contact with a large system S_0 forming the total system $S = S_1 + S_0$. The total Hamiltonian can be written as

$$H = H_1 + H_0 + H_{\text{int}} \quad \text{... (19.1.25)}$$

where H_1 is the Hamiltonian of S_1, H_0 that of S_0 and H_{int} is a weak interaction between S_0 and S_1 which is responsible for the equilibrium between S_0 and S_1. The system S_0 (heat reservoir) is so large that irrespective of what S_1 does, S_0 is always in thermal equilibrium characterized by a temperature T. We now want the transition probability W_{mn} for the system S_1. To find W_{mn} we proceed as follows: we consider a combined state where S_1 is in state m and S_0 is in a state m' and as a transition to state n is made for S_1, S_0 ends up in the state n'. The transition probability for this is denoted by $W^0\,(mm' \to nn')$. To find W_{mn}, we have to evaluate

$$W_{mn} = \sum_{m',n'} P'_{m'}\, W^0\,(mm' \to nn') \quad \text{... (19.1.26)}$$

where for a given m' of S_0, we first sum over the possible final states labelled by n' and then sum over m'. The probability of finding S_0 in the initial state m' is $P'_{m'}$. The fact that S_0 is in equilibrium at temperature T means that it is characterized by a Maxwell-Boltzmann distribution, i.e.

$$P'_{m'} \propto e^{-\beta E_{m'}} \;(\beta^{-1} = kT) \text{ and hence}$$

$$W_{mn} = C \sum_{m',n'} e^{-\beta E_{m'}}\, W^0\,(mm' \to nn')$$

Similarly

$$W_{nm} = \sum_{m',n'} P'_{n'}\, W^0\,(nn' \to mm')$$

$$= c \sum_{m', n'} e^{-\beta E_{n}'} W^0 (nn' \to mm') \quad \text{(19.1.27)}$$

Now note that $E_m + E_m' \approx E_n + E_n'$ since the transitions are energy conserving and

$$W^0 (nn' \to mm') = W^0 (mm' \to nn') \quad \text{... (19.1.28)}$$

and hence Eq. (19.1.27) can be written as

$$W_{nm} = c \sum e^{-\beta E_{m}' - \beta(E_m - E_n)} W^0 (mm' \to nn')$$

$$= e^{-\beta(E_m - E_n)} W_{mn} \quad \text{... (19.1.29)}$$

As the system reaches equilibrium, we have

$W_{mn} P_m \to W_{nm} P_n$ and hence

$$\frac{P_m}{P_n} \to \frac{e^{-\beta E_m}}{e^{-\beta E_n}} \quad \text{... (19.1.30)}$$

which is to say that, in equilibrium, the Maxwell-Boltzmann distribution is achieved.

19.2 Quantum Langevin Equation

We now show how a Langevin equation can be obtained for the type of situation envisaged towards the end of the previous section.

Our system S_1 is a single particle (a Brownian particle) and the system S_0 is a system of heat bath particles, all of them connected to the Brownian particle by a spring, thus bringing in the interaction part of the Hamiltonian in Eq. (19.1.25)

$$H = \frac{p^2}{2m} + V(x) + \sum_j \frac{p_j^2}{2mj} + \frac{1}{2} k_j (q_j - x)^2 \quad \text{... (19.2.1)}$$

Here x and p are the coordinate and momentum operators of the Brownian particle and the set (q_j, p_j) is the set of coordinates and momentum operators for the heat bath particles. The mass of the jth bath particle is m_j and k_j is the spring constant of the spring connecting it to the Brownian particle. The potential $V(x)$ is the external potential for the Brownian particle. To this Hamiltonian must be added the commutation relations

$$[x, p] = i\hbar \ , \ [q_j, p_k] = i\hbar \ \delta_{jk} \quad \text{... (19.2.2)}$$

The equation of motion for any Heisenberg operators O is

$$\dot{O} = \frac{1}{i\hbar}[O, H] \quad \text{... (19.2.3)}$$

and hence

$$\dot{x} = \frac{p}{m} \quad \text{... (19.2.4a)}$$

$$\dot{p} = -V'(x) + \sum_j k_j (q_j - x) \quad \text{... (19.2.4b)}$$

$$\dot{q}_j = p_j/m_j \quad \text{... (19.2.4c)}$$

$$\dot{p}_j = -k_j (q_j - x) \quad \text{... (19.2.4d)}$$

The solution of Eqs. (19.2.4c) and (19.2.4d) gives

$$q_j(t) = q_j(O)\cos\ \omega_j t + \frac{p_j(O)}{m_j w_j}\ \sin\ \omega_j t + x(t) - x(O)\cos\omega_j t$$
$$-\int_0^t dt' \cos\omega_j (t - t')\, x(t') \quad \text{... (19.2.5)}$$

where $\omega_j = (k_j/m_j)^{1/2}$ Inserting the value of q_j from Eq. (19.2.5) in Eq. (19.2.4b) we easily obtain

$$m\ddot{x} + \int_0^t dt'\, B(t - t')\, x(t') + V'(x) + B(t)\, x(o) = F(t) \quad \text{... (19.2.6)}$$

where

$$F(t) = \sum_j [q_j(0)\, k_j \cos\omega_j t + p_j(0)\, \omega_j\ \sin\ \omega_j t] \quad \text{... (19.2.7)}$$

and

$$B(t) = \sum_j k_j \cos \omega_j t \qquad \text{... (19.2.8)}$$

Note that the initial variables of the heat bath occur in the force term in Eq. (19.2.6) Apart from that it is an equation in $x(t)$.

Now, we introduce the statistical average over the initial variables of the heat bath. To do so we assume that at $t = 0$, the oscillators are canonically distributed with respect to the free oscillator Hamiltonian

$$H_0 = \sum \left(\frac{p_j^2}{2m_j} + \frac{1}{2} k_j q_j^2 \right\} \qquad \text{... (19.2.9)}$$

For an arbitrary operator O, we introduce the expectation value,

$$<O> = \frac{\text{Tr } O \, e^{-H_o/kT}}{\text{Tr } e^{-H_o/kT}} \qquad \text{... (19.2.10)}$$

where the trace is with respect to the bath particles. It is straightforward to check

$$< q_j(o) \, q_k(o) > = \frac{<p_j(o) \, p_k(o)>}{\omega_j^2 \, m_j^2} = \delta_{jk} \frac{\hbar \coth\left(\frac{\hbar\omega_j}{2kT}\right)}{2m_j\omega_j}$$

$$<q_j(o) \, p_k(o)> = -<p_k(o) \, q_j(o)> = \frac{1}{2} \, i\hbar \;\; \delta_{jk} \qquad \text{... (19.2.11)}$$

In addition, there is the Gaussian property that the expectation value of an odd number of factors of q_j and p_j vanishes and the expectation value of an even number of factors is the sum of the products of the pairwise expectations with the order of the factors preserved. These properties lead to the results:

$$\frac{1}{2} < F(t) \, F(t') + F(t') \, F(t)> = \sum \frac{1}{2} k_j h \; \omega_j \left(\coth \frac{\hbar \, \omega_j}{2kT} \right) \cos \omega_j \; (t - t') \qquad \text{... (19.2.12)}$$

and

$$[F(t), F(t')] = -i\hbar \sum_j k_j \omega_j \sin \omega_j (t - t')$$

We now turn to $B(t)$ in Eq. (19.2.8) and assuming an infinite number of oscillators in the bath with a continuous distribution of frequency, obtain

$$B(t) = \int_0^\infty N(\omega)\, k(\omega) \cos \omega t \quad \text{... (19.2.13)}$$

Choosing $N(\omega)\, k(\omega) \approx$ constant, $B(t)\ \alpha\ \delta(t)$ and hence at all finite times, $B(t)$ can be dropped, so that Eq. (19.2.6) becomes

$$m\ddot{x} + \int_0^t dt'\, B(t - t')\, x(t') + V'(x) = F(t) \quad \text{... (19.2.14)}$$

with $F(t)$ a random operator prescribed through the correlation in Eq. (19.2.12). We have thus obtained the quantum Langevin equation. Eq (19.2.14).

The quantum Langevin equation has been obtained in a very special situation. Can the structure of Eq. (19.2.14) be universal? The best one can assert is that if there is a universal description, then it must be of the form of Eq. (19.2.14).

The last question is: Does one have the approach to the correct equilibrium state in this Langevin equation? This means that we would have to demonstrate for the stationary solution $x(t)$ of the Langevin equations, that

$$\underset{t \to \infty}{\text{Lt}} \langle x(t)^N \rangle = \frac{\sum_m e^{-E_m/kT} \int (\psi_m x^N \psi_n)\, dx}{\sum_n e^{-E_n/kT}}$$

where ψ_n and E_n are the eigenfunctions and eigenvalues of the Schrödinger equation

$$\left[\frac{-\hbar^2}{2m}\frac{d^2}{dx^2} + V(x)\right]\psi_n = E_n\, \psi_n \quad \text{... (19.2.15)}$$

This has not been achieved so far.

Chapter 20
Fluctuation Dissipation Theorem

In this chapter, we present the general derivation by Callen and Welton of the fluctuation dissipation theorem and apply it to have yet another look at the Brownian motion. We begin with definitions.

A system is said to be dissipative if it is capable of absorbing energy when subjected to a time-periodic perturbation (e.g. electrical resistor absorbing energy from an impressed sinusoidal voltage). The system is linear if the power dissipated is proportional to the square of the magnitude of the perturbation. For the electrical case, power = (voltage)$^2 \dfrac{R}{|Z|^2}$, Z being the complex number $Z = Z_1 + iZ_2$ with $i = \sqrt{-1}$, $Z_1 = R$, the resistance and Z_2 is the inductive and capacitative part. The applied perturbation will be treated quantum mechanically and the power dissipated related to the matrix element of the perturbation operator.

The Hamiltonian is,

$$H = H_0 + V(t)\,Q \qquad \text{... (20.1)}$$

where Q is a function of coordinates and momenta and $V(t)$ is a function of time alone. For a sinusoidal variation $V(t) = V_0 \sin \omega t$. We write the eigenvalues and eigenfunctions of H_0 as $\{E_n\}$ and $\{\psi_n\}$ with

$$H_0\,\psi_n = E_n\,\psi_n \qquad \text{... (20.2)}$$

and expand the wavefunction $\psi(t)$ at any time as

$$\psi(t) = \sum_k c_k(t)\,\psi_k \qquad \text{... (20.3)}$$

and proceeding exactly as in the previous chapter, obtain (to the lowest order) for the transition probability of a system initially in the state ψ_n the quantity

$$\frac{\pi}{2} V_0^2 \hbar^{-1} \left[|<E_n + \hbar\omega | Q | E_n>|^2 \rho(E_n + \hbar\omega) \right.$$

$$\left. + |<(E_n - \hbar\omega) | Q | E_n>|^2 \rho(E_n - \hbar\omega) \right] \qquad \text{... (20.4)}$$

where $<E_n + \hbar\omega|Q|E_n>$ is the matrix element of Q between unperturbed states whose energies are E_n and $E_n + \hbar\omega$ and $\rho(E)$ is the density in energy of the quantum states between E and $E + dE$.

Each transition from E_n to $E_n + \hbar\omega$ involves an absorption of energy $\hbar\omega$ and each transition from E_n to $E_n - \hbar\omega$ involves a release of energy $\hbar\omega$ and hence the rate of absorption of energy by a system initially in the state n is

$$\frac{\pi}{2} V_0^2\omega \left[|<E_n + \hbar\omega|Q|E_n>|^2 \rho(E_n + \hbar\omega) \right.$$

$$\left. - |<E_n - \hbar\omega)|Q|E_n>|^2 \rho(E_n - \hbar\omega) \right].$$

Now the thermodynamic probability of a state with energy E_n is the Boltzmann factor $e^{-E_n/kT}$ (in general we will denote this probability by $f(E_n)$) and the power dissipated is then

$$\text{Power} = \omega \frac{\pi}{2} V_0^2 \sum_n \left[|n <E_n + \hbar\omega|Q|E_n>|^2 \rho(E_n + \hbar\omega) \right.$$

$$\left. - |<E_n - \hbar\omega|Q|E_n>|^2 \rho(E_n - \hbar\omega) \right] f(E_n)$$

$$= \frac{\pi}{2} V_0^2 \omega \int_0^\infty \rho(E) f(E)\, dE \left\{ |<E + \hbar\omega|Q|E>|^2 \rho(E + \hbar\omega) \right.$$

$$- |<E - \hbar\ \omega|Q\ |E>|^2\ \rho(E - \hbar\ \omega)\ \Big\} \quad \text{... (20.5)}$$

where we have converted the sum to an integral by noting

$$\sum_n () \to \int \rho(E)\ ()\ dE$$

We now define the impedance Z as the ratio of V to the response $\dot{Q}$ and thus $V = Z(\omega)\ \dot{Q}$ ... **(20.6)**

The instantaneous power is $V\ \dot{Q}\ R/|Z|$ and the average power absorbed at frequency ω is

$$\text{Power} = \frac{1}{2} V_0^2\ R(\omega)/|\ Z(\omega)|^2 \quad \text{... (20.7)}$$

Comparison with Eq. (20.5) leads to

$$\frac{R}{|Z|^2} = \pi\omega \int_0^\infty \rho(E)\ f(E)\ \{|< E + \hbar\ \omega\ |Q|E>|^2 \rho(E + \hbar\ \omega)$$

$$- |(E - \hbar\ \omega\ |Q|\ E>|^2\ \rho(E - \hbar\ \omega)] \quad \text{... (20.8)}$$

We now remove the applied force V and leave the system in thermal equilibrium. In this isolated condition there can be spontaneous fluctuations of $\dot{Q}$ with a mean square value $<\dot{Q}^2>$, which can be calculated as follows:

Suppose the system is in the nth eigenstate. Clearly $<E_n|\dot{Q}|E_n> = 0$ since $\dot{Q} = \frac{1}{i\hbar}[Q, H_0]$.

The mean square fluctuation of $\dot{Q}$ in the nth state is

$$<E_n|\dot{Q}^2|E_n> = \sum_m <E_n|\dot{Q}|E_m><E_m|\dot{Q}|E_n>$$

$$= \frac{1}{\hbar^2} \sum <E_n\ |H_0Q - QH_0|\ E_m> <E_m|\ H_0Q - QH_0\ |\ E_n>$$

$$= \frac{1}{\hbar^2} \sum_m (E_n - E_m)^2\ |<E_m\ |Q\ |\ E_n>|^2 \quad \text{... (20.9)}$$

Introducing a frequency $\omega = |E_n - E_m|/\hbar$, the sum over m can be split into two parts $E_n < E_m$, and $E_n > E_m$) and changing the summation to an integration, we have

$$<E_n|\dot{Q}^2|E_n> = \hbar \int_0^\infty d\omega\, \omega^2 \{|<E_n + \hbar\omega|Q|E_n>|^2 \rho(E_n + \hbar\omega)$$

$$+|<E_n + \hbar\omega|Q|E_n>|^2 \rho(E_n - \hbar\omega)\} \quad \text{... (20.10)}$$

The actually observed fluctuation is obtained by multiplying the fluctuations in the nth state by the thermodynamic weight factor (E_n) and summing. We get

$$<\dot{Q}^2> = \hbar \int_0^\infty \omega^2 d\omega \int_0^\infty \rho(E) f(E) \{|<E + \hbar\omega|Q|E>|^2 \rho(E + \hbar\omega)$$

$$+ |<E - \hbar\omega|Q|E>|^2 \rho(E - \hbar\omega)\} \quad \text{... (20.11)}$$

The "voltage" fluctuation is consequently

$$<V^2> = \int_0^\infty |Z|^2 \omega^2 d\omega \int_0^\infty p(E) f(E)$$

$$\times\{|<E + \hbar\omega|Q|E>|^2 \rho(E + \hbar\omega)$$

$$+ |<E - \hbar\omega|Q|E>|^2 \rho(E - \hbar\omega)\} dE \quad \text{... (20.12)}$$

Using Eqs. (20.8) and (20.12), we now need to relate $R/|Z|^2$ with $<V^2>$

To do so, we need to relate the two integrals $C_\pm$, where

$$C_\pm = \int_0^\infty f(E) \{ |<E - \hbar\omega|Q|E>|^2 \rho(E - \hbar\omega)$$

$$\pm |<E - \hbar\omega|Q|E>|^2 \rho(E - \hbar\omega)\} \rho(E)\, dE \quad \text{... (20.13)}$$

Working with second part of the integral in Eq. (20.13) we note that $< E - \hbar\,\omega | Q | E > = 0$ if $E < \hbar\omega$ and hence one can write

$$\int_0^\infty f(E)\,|<E-\hbar\;\omega\,|Q|\,E>|^2\,\rho(E-\hbar\;\omega)\,\rho(E)\,dE$$

$$= \int_{\hbar\,\omega}^\infty f(E)\,|< E-\hbar\;\omega|Q|E>|^2\,\rho\,(E-\hbar\;\omega)\,\rho(E)dE$$

$$= \int_0^\infty f(E+\hbar\;\omega)\,|<E-|Q|E+\hbar\;\omega>|^2\,\rho(E)\,\rho(E-\hbar\;\omega)\,dE$$

Using the above result in the $C_\pm$ of Eq. (20.13), we get

$$C_\pm = \int_0^\infty |<E+\hbar\;\omega\,|Q|E>|^2\,\rho(E+\hbar\;\omega)\,\rho(E)\,f(E)$$

$$\times\left(1 \pm \frac{f(E+\hbar\;\omega)}{f(E)}\right)dE$$

$$= \left(1 \pm e^{-\hbar\;\omega/k_BT}\right)\int_0^\infty |<E+\hbar\;\omega|Q|E>|^2\,\rho(E+\omega)\,\rho(E)\,f(E)\,dE$$

where we have used the fact that $f(E) \propto e^{-E/k_BT}$.

We return to Eq. (20.12) to write

$$<V^2> = \int_0^\infty |Z|^2\,\hbar\;\omega^2\,d\omega\,C_+ = \int_0^\infty |Z|^2\,\hbar\;\omega^2\,C_-\,\frac{1+e^{-\hbar\;\omega/kT}}{1-e^{-\hbar\;\omega/kT}}$$

$$= \int_0^\infty |Z|^2\,\hbar\;\omega^2\,\frac{R}{|Z|^2}\,\frac{1}{\pi\omega}\,\frac{1+e^{-\hbar\;\omega/kT}}{1-e^{-\hbar\;\omega/kT}}\,d\omega\cdot$$

$$= \int R(\omega)\,\frac{\hbar\;\omega}{\pi}\left(\frac{e^{\hbar\;\omega/kT}+1}{e^{\hbar\;\omega/kT}-1}\right)d\omega$$

$$= \int R(\omega) \frac{\hbar\,\omega}{\pi} \left(1 + \frac{2}{e^{\hbar\,\omega/kT} - 1} \right) d\omega$$

$$= \frac{2}{\pi} \int R(\omega)\, E(\omega)\, d\omega \qquad \text{... (20.15)}$$

where

$$E(\omega) = \frac{1}{2}\,\hbar\,\omega \left(1 + \frac{2}{e^{\hbar\,\omega/k_BT} - 1} \right) \qquad \text{... (20.16)}$$

The fluctuation dissipation theorem is given by Eqs. (20.15) and (20.16). At high temperatures,

$$E(\omega) \approx \frac{1}{2}\, k_BT$$

and we get

$$<V^2> \approx \frac{2k_BT}{\pi} \int R(\omega)\, d\omega \qquad \text{... (20.17)}$$

Turning to the question of a particle moving through a fluid with velocity v, we know that the drag force (friction) exerted on the particle is

$$F = -\eta\, v \qquad \text{... (20.18)}$$

where η is the viscosity. According to the fluctuation dissipation theorem for $k_BT >> \hbar\,\omega$, the corresponding fluctuating force has the mean square value

$$<F_x^2> = <F_y^2> = <F_z^2> = \frac{2kT}{\pi}\, \eta \int d\omega \qquad \text{... (20.19)}$$

This is consistent with the fluctuating force written down in the Langevin equation of Chapter 16.

Chapter 21

Phase Transitions

21.1 Introduction

We came across two models exhibiting second-order phase transitions—transitions which occur without requiring any latent heat—in Chapter 5. One was the model of real gases given by Van der Waals equation where in the P–V diagram there were coexisting regions (i.e. coexisting liquid and gas phases) for sufficiently low temperatures but at a critical temperature $T = T_c$, the coexisting region shrunk to zero and above T_c, there was a single phase (gaseous) region alone. Setting $P = P_c$ and $V = V_c$, i.e. the pressure and volume corresponding to the point at which the coexisting region shrinks to zero, if one lowers the temperature from above T_c to below T_c, then a transition occurs from a one-phase region to a two-phase region—this transition requires no latent heat. Based on Van der Waals equation, properties near the critical point were examined in Chapter 5. The result was an example of what is known as a mean field theory. We found characteristically universal critical behaviour—the universality was so strong that the results were independent of the dimensionality of space. Starting with Van der Waals equation, one cannot but help get the mean field behaviour since the equation itself can be considered an equation of state in the mean field approximation. A more interesting behaviour is found from the Ising model where we consider classical spins at each site of a regular lattice. The spins can point up or down relative to an external field, i.e. they can take the values ± 1. At the site $\boldsymbol{n}$ ($\boldsymbol{n}$ is a D-dimensional vector), the spin is denoted by $S(\mathbf{n})$ and the energy of interaction between two spins $S(\mathbf{n})$ and $S(\mathbf{m} + \mathbf{n})$ at two different sites is given by $-J(\mathbf{n})\, S(\mathbf{m})\, S(\mathbf{m}+\mathbf{n})$. The main task is the evaluation of the partition function

$$Z = \sum_{\{S(\mathbf{m})\}} e^{\beta \sum_{m,n} J(\mathbf{n})\, S(\mathbf{m})\, S(\mathbf{m} + \mathbf{n})} = \sum_{\{S(\mathbf{m})\}} e^{A} \quad \text{... (21.1.1)}$$

where the action A is given by $\beta \sum_{m,n} J(\mathbf{n})\, S(\mathbf{m})\, S(\mathbf{m} + \mathbf{n})$. If the interaction is of the nearest neighbour variety, then

$$J(\mathbf{n}) = \sum_{\mathbf{a}} J(\mathbf{a})\, \delta_{\mathbf{n},\, \mathbf{n} + \mathbf{a}}\,,$$ where $\mathbf{a}$ is the lattice spacing.

There is a sum over $\mathbf{a}$ since the coupling can be in any one of the D-directions available in the D-dimensional lattice. In different spatial dimensions, the action A can consequently be written as

$D = 1$,

$$A = \beta J \sum_{i} S(i)\, S(i+1) \quad \text{... (21.1.2)}$$

(i labels the one-dimensional lattice sites)

$D = 2$,

$$A = \beta \sum_{i,j} \left(J_1\, S(i,j)\, S(i+1,j) + J_2 S(i,j)\, S(i,j+1)\right) \quad \text{... (21.1.3)}$$

Note that in two dimensions, we require two integers i and j to label the spatial sites.

$D = 3$

$$A = \beta \sum_{i,j,k} \left[J_1\, S(i,j,k)\, S(i+1,j,k) + J_2 S(i,j,k)\, S(i,j+1,k) + J_3\, S(i,j,k)\, S(i,j,k+1)\right] \quad \text{... (21.1.4)}$$

and so on.

In Chapter 5, we have already dealt with the situation in $D = 1$ by constructing the transfer matrix and obtaining the different thermodynamic quantities. We now give a somewhat different demonstration of the fact that there can be no long-range order in the one-dimensional model at any finite temperature. To do so, we imagine a completely ordered state at some $T \neq 0$ and investigate the effect of flipping some of the spins. The flipping which costs the least energy is shown in Fig. (21.1)

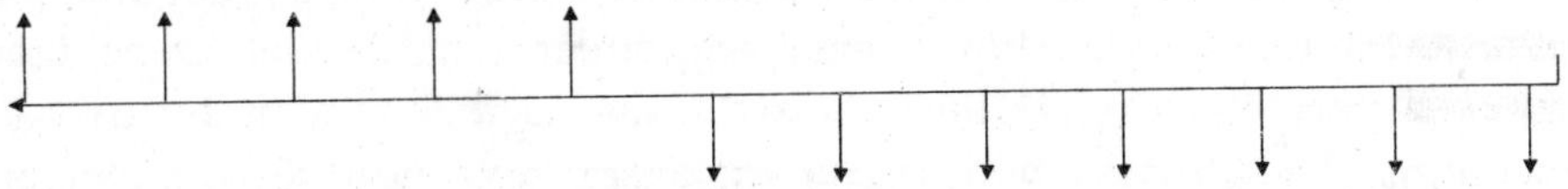

Fig. 21.1

The change ΔE in energy caused by the flipping shown in Fig. (21.1) is clearly $2J$. However, whether the flipped state is thermodynamically favourable or not is determined by the change ΔF in the free energy which is given by $\Delta F = \Delta E - T\Delta S$. To estimate ΔS, we note that the length of the overturned spin can be anything between the lattice spacing a and the total length L of the system. Consequently the change in entropy ΔS is of the order of ln L and thus

$$\Delta F = \Delta E - T\Delta S = 2J - T \ln L < 0 \qquad \text{... (21.1.5)}$$

for all $T \neq 0$. Hence it is thermodynamically favourable to over-turn the spins and consequently the system cannot order except at $T = 0$.

Since the above model does not show any ordered behaviour, it is natural to ask whether the one dimensional Ising model can have an ordered state at $T \neq 0$ if the interaction $J(n)$ is long ranged. This would mean that the action of Eq. (21.1.2) now has the form

$$A = \beta \sum_{i,n} J(n)\, S(i)\, S(i+n) \qquad \text{... (21.1.6)}$$

and $J(n) \propto n^{-\alpha}$. If we now flip the spins starting from an ordered configuration as shown in Fig. (21.1), then the energy expended would be

$$\Delta E \propto \sum_{i,j} \frac{1}{|i-j|^{\alpha}} \propto \int \frac{dx\, dy}{|x-y|^{\alpha}} \qquad \text{... (21.1.7)}$$

If we want the energy barrier to overcome the disordering tendency of the entropy term then ΔE must be at least of $O(\ln L)$. Clearly the integral in Eq. (21.1.7) is $O(\ln L)$ if $\alpha = 2$. For $\alpha > 2, \Delta E$ is a finite number and hence the entropy term will always dominate. Consequently long-range interactions falling off faster than n^{-2} cannot cause the Ising model to have an ordered state at $T \neq 0$. On the other hand, interactions which are slower than n^{-2} give a ΔE larger than $O(\ln L)$ and hence will cause a transition to an ordered state for $T \neq 0$.

We now turn to the two-dimensional model and try to construct an argument similar to the one for the short ranged one-dimensional model. Once again we imagine a perfectly aligned system and cause a defect in it by overturning some spins. The flipping with minimum energy expenditure is shown in Fig. (21.2).

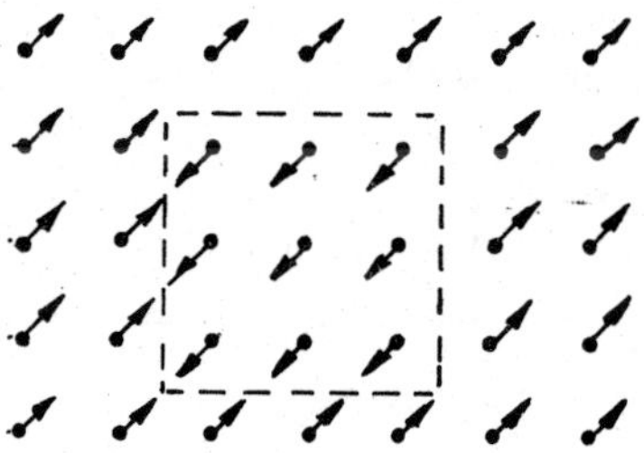

Fig. 21.2 Formation of a region of upturned spins

The flipping shown in Fig. (21.2) which creates the smallest possible domain of overturned spin requires an amount of energy $\Delta E = 8J$. Keeping this energy the same, we cannot have any larger domains. It is easy to see that if larger domains of overturned spins are created then a larger amount of energy is expended. This is the major difference from the one-dimensional case. There we could create arbitrarily long regions of overturned spins without extra expenditure of energy. The change in entropy is consequently of $O(\ln d)$ where d is the domain size. Since ΔE and ΔS are of the same order it is possible to have the ordered state favoured at low enough temperatures. Hence it is possible for the nearest neighbour Ising model to exhibit a phase transition in two dimensions. The above argument is not rigorous—it is to be viewed as a plausibility argument. Needless to say that two-dimensional models with long-range interactions will also exhibit phase transitions at a finite temperature.

21.2 Transfer Matrix for the Two-dimensional Ising Model

The nearest neighbour two-dimensional Ising model is described by the action of Eq. (21.1.3). Defining $\beta_1 = \beta J_1$ and $\beta_2 = \beta J_2$, our task is to calculate the partition function

$$Z = \sum_{\{S(i,j)\}} e^{\sum_{ij}[\beta_1 S(i,j)S(i+1,j) + \beta_2 S(i,j)S(i,j+1)]} \qquad \text{... (21.2.1)}$$

We take j to label the sites along a row and i to label the different rows Fig. 21.3). The transfer matrix that we are seeking will be from a given row to the next row. For the one-dimensional problem in Chapter 5, the transfer matrix took us from one site to the next. In this two-dimensional case it is from one row to the next and we can easily imagine that for the three-dimensional case it will be from one plane to the next.

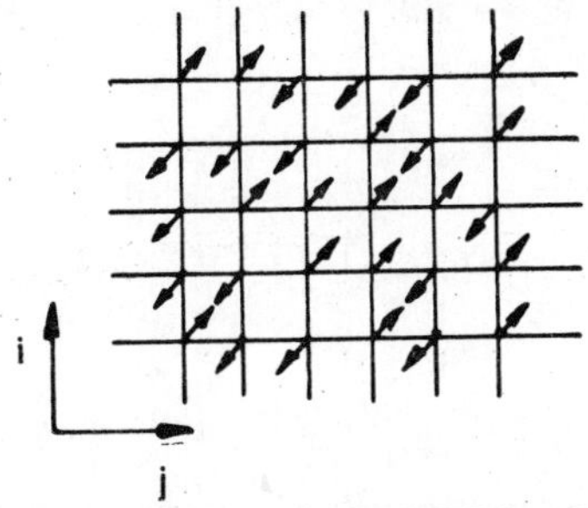

Fig. 21.3 Two-dimensional Ising lattice.

In what follows we will not try to find the transfer matrix for the general case of arbitrary β_1 and β_2. Instead, we will concentrate on a limit which allows the transfer matrix operator to have the form $e^{-H\tau}$, which becomes $1 - H\tau$ in the limit $\tau << 1$. The couplings β_1 and β_2 will have to be adjusted to obtain this limit. Since we expect the model to have a phase transition, one would find a relation between β_1 and β_2 which would give the transition point. At the transition point, we expect long-range order, which would imply an infinite correlation length and we would expect all dependence on lattice spacings to vanish. The latter may be taken as the signal for the transition point. Thermodynamic quantities, if calculated, would show a non-analytic behaviour at the transition point. We will not be able to calculate the spontaneous magnetization, however, since that would require the presence of an external field which would have to be set equal to zero after the derivative of the free energy with respect to the field is taken.

We start with the restricted goal of finding the transfer matrix by noting that apart from constants, the action of Eq. (21.2.1) can be written as

$$A = -\sum_{i,j} \left[\beta_1 S(i,j) S(i+1,J) + \beta_2 S(i,j) S(i,j+1)\right]$$

$$= \sum_{i,j} \left[\frac{\beta_1}{2} [S(i,j) - S(i+1,j)]^2 - \beta_2 S(i,j) S(i,j+1)\right] \quad \ldots \textbf{(21.2.2)}$$

we now focus on a particular i and write

$$A = \sum_i L(i, i+1) \quad \ldots \textbf{(21.2.3)}$$

where

$$L(i, i+1) = \sum_j \frac{\beta_1}{2} [S(i,j) - S(i+1,j)]^2 - \frac{\beta_2}{2} [S(i,j) S(i,j+1) +$$

$$+ S(i+1,j) S(i+1,j+1)] \quad \ldots \textbf{(21.2.4)}$$

Introducing the notation

$$S(i,j) = \sigma_3(j), \; S(i+1,j) = s_3(j) \quad \ldots \textbf{(21.2.5)}$$

we write

$$L(i, i+1) = \sum_j \left\{\frac{\beta_1}{2} [\sigma_3(j) - s_3(j)]^2 - \frac{\beta_2}{2} [\sigma_3(j) \sigma_3(j+1)\right.$$

$$\left. + s_3(j) s_3(j+1)\right\} \quad \ldots \textbf{(21.2.6)}$$

If there are N sites on each row, then there are 2^N spin configurations and the transfer matrix will be a $2^N \times 2^N$ matrix. The elements of the transfer matrix can be labelled in terms of the number of spin flips between the two rows. The diagonal elements are the ones with no spin flips. The various off diagonal elements can be characterized by one flip, two flips, etc. The matrix elements of $L(i, i+1)$ are clearly

$$L(i, i+1)|_{0 \text{ flips}} = -\beta_2 \sum_j \sigma_3(i) \sigma_3(j+1) \quad \ldots \textbf{(21.2.7a)}$$

($\sigma_3 = s_3$ for no flips)

$$L(i, i+1)|_{1\ \text{flip}} = 2\beta_1 - \frac{\beta_2}{2}\sum_j [\sigma_3(j)\,\sigma_3(j+1) + s_3(j)s_3(j+1)]. \quad \textbf{(21.2.7b)}$$

(the j-value at which the flip occurs leads to $\sigma_3(j) - s_3(j) = \pm 2$.)

$$L(i,\ i+1)|_{n\ \text{flips}} = 2n\beta_1 - \frac{\beta_2}{2}\sum_j [\sigma_3(j)\,\sigma_3(j+1) + s_3(j)s_3(j+1)]. \quad \textbf{(21.2.7c)}$$

Since $T(i,\ i+1) = e^{-L(i,\, i+1)}$, the corresponding matrix elements for T are

$$T(i,\ i{=}1)|_{0\ \text{flip}} = e^{\beta_2 \sum_j \sigma_3(j)\,\sigma_3(j+1)} \quad \textbf{... (21.2.8a)}$$

$$T(i,\ i=1)|_{1\ \text{flip}} = e^{-2\beta_1}\ e^{\left[\sum_j \sigma_3(j)\,\sigma_3(j+1) + \sum_j s_3(j)s_3(j+1)\right]\frac{\beta_2}{2}} \quad \textbf{... (21.2.8b)}$$

.

.

.

$$T(i, i+1)|_{n\ \text{flips}} = e^{-2n\beta_1}\ e^{\frac{\beta_2}{2}\left[\sum_j \sigma_3\sigma_3 + \sum s_3 s_3\right]} \quad \textbf{... (20.2.8c)}$$

At this point, we bring in the simplification of writing $T = e^{-H\tau} \sim 1 - H\tau$, which would lead to the following relations between matrix elements

$$T|_{0\ \text{flip}} = 1 - \tau\, H|_{0\ \text{flip}} \quad \textbf{... (21.2.9a)}$$

$$T|_{1\ \text{flip}} = -\ \tau\, H|_{1\ \text{flip}} \quad \textbf{... (21.2.9b)}$$

$$T|_{n\ \text{flip}} = -\tau\, H|_{n\ \text{flips}} \quad \textbf{... (21.2.9c)}$$

[Note that the operator 1 cannot have matrix elements between states with flipped spins and hence does not appear in Eqs. (21.2.9b) and (21.2.9c)].

Comparing Eqs. (21.2.8a) and (21.2.9a), we conclude that for the form of the latter to hold, we must have $\beta_2 = O(\tau)$ in the former. Similarly comparison of eqs. (21.2.8b) and (21.2.9b) leads to the conclusion that $e^{-2\beta_1} = O(\tau)$ and thus we can write

$$\beta_2 = \lambda\, e^{-2\beta_1} \quad \textbf{... (21.2.10)}$$

where λ is a number independent of τ. If we now choose

$$e^{-2\beta_1} = \tau \qquad \text{... (21.2.11a)}$$

then it follows that

$$\beta_2 = \lambda\tau \qquad \text{... (21.2.11b)}$$

If we now compare Eqs. (21.2.8c) and (21.2.9c), we find that the matrix element of H for n flips must be of the order τ^{n-1} and hence for $n \geq 2$ the matrix elements are negligible in the limit $\tau \to 0$. Thus, in this limit the only relevant matrix elements of H are those with zero flips (diagonal elements) and those with a single flip. The matrix elements are

$$H|_{0\text{ flip}} = -\lambda \sum_k \sigma_3\ (j)\, \sigma_3\ (j+1) \qquad \text{... (21.2.12a)}$$

$$H|_{1\text{ flip}} = -1 \qquad \text{... (21.2.12b)}$$

Clearly the representation of H can be in terms of a 2×2 matrix and the obvious candidate for the spin flipping matrix is the Pauli matrix $\sigma_1 = \begin{pmatrix} 0 & 1 \\ 1 & 0 \end{pmatrix}$. As for the diagonal elements, we can use the Pauli matrix $\sigma_3 = \begin{pmatrix} 1 & 0 \\ 0 & -1 \end{pmatrix}$ at each site j and thus obtain the Hamiltonian operator

$$H = -\left[\sum_j \sigma_1(j) + \lambda \sum_j \sigma_3(j)\, \sigma_3(j+1)\right] \qquad \text{... (21.2.13)}$$

which will reproduce correctly all the matrix elements of Eqs. (21.2.12a) and (21.2.12b).

The eigenvalue spectrum of the transfer matrix is thus to be obtained from the spectrum of the quantum mechanical Hamiltonian of Eq. (21.2.13). It should be noted that the quantum Hamiltonian is not a single particle one but a many-particle Hamiltonian for spins sitting on a one-dimensional lattice. To bring out explicitly the fermionic character of the above Hamiltonian it is convenient to consider an equivalent Hamiltonian $\overline{H}$, defined as

$$\overline{H} = -\left[\sum_n \sigma_3(n) + \lambda \sum_n \sigma_1(n)\, \sigma_1(n+1)\right] \qquad \text{... (21.2.14)}$$

which is obtained from H by the orthogonal transformation that takes $\sigma_1 \to \sigma_3$ and $\sigma_3 \to \sigma_1$.

We begin by noting that the Pauli matrices for raising and lowering spins can be constructed as

$$\sigma^{\pm}(n) = \frac{1}{2}[\sigma_1(n) \pm i\,\sigma_2(n)] \qquad \text{... (21.2.15)}$$

The matrices $\sigma^{\pm}(n)$ anticommute on the same site and $\sigma^{+2} = \sigma^{-2} = 2$. In this they behave like fermions. But true fermion operators have anticommutation properties on separate lattice sites as well. With this in mind, we construct the following operators for a lattice where n runs from $-N$ to N in steps of unity:

$$C(n) = \prod_{j=-N}^{n-1} \left[e^{i\pi\sigma^+(j)\sigma^-(j)}\right] \sigma^-(n) \qquad \text{... (21.2.16a)}$$

and

$$C^+(n) = \sigma^+(n) \prod_{j=-N}^{n-1} \left[e^{-i\pi\sigma^+(j)\sigma^-(j)}\right] \qquad \text{... (21.2.16b)}$$

Making use of the results

$$\sigma^-(n)\,\sigma^+(n) = \frac{1}{2}\,[1 - \sigma_3(n)] \qquad \text{... (21.2.17a)}$$

$$\sigma^+(n)\,\sigma^-(n) = \frac{1}{2}\,[1 + \sigma_3(n)] \qquad \text{... (21.2.17b)}$$

$$e^{i\pi\sigma_3/2} = i\sigma_3 \qquad \text{... (21.2.17c)}$$

we find

$$C(n) = \prod_{j=-N}^{n-1} [-\sigma_3(j)]\,\sigma^-(n) \qquad \text{... (21.2.18a)}$$

$$C^+(n) = \sigma^+(n) \prod_{j=-N}^{n-1} [-\sigma_3(j)] \qquad \text{... (21.2.18b)}$$

and then it is straightforward to verify that

$$\{C(n), C^+(m)\} = \sigma_{mn} \qquad \text{... (21.2.19a)}$$

$$\{C(n), C(m)\} = 0 \qquad \text{... (21.2.19b)}$$

where $\{AB\} = AB \pm BA$ is the anticommutator.

The $C(n)$ and $C^+(n)$ are thus true fermionic operators and we next have to express the Hamiltonian of Eq. (21.2.14) in terms of them. From Eq. (21.2.17b)

$$\sigma_3(n) = 2\sigma^+(n)\ \sigma^-(n) - 1 \qquad \text{... (21.2.20)}$$

and thence using Eqs. (21.2.18a) and (21.2.18b), we have $\sigma^+(n)\,\sigma^-,$ $(n) = C^+(n)\,C(n)$ so that

$$\sigma_3(n) = 2C^+(n)\,C(n) - 1 \qquad \text{... (21.2.21)}$$

The coupling term is tackled by writing

$$\begin{aligned}\sigma_1(n)\,\sigma_1(n+1) &= [\sigma^+(n) + \sigma^-(n)]\,[\sigma^+(n+1) + \sigma^-(n+1)] \\ &= \sigma^+(n)\sigma^+(n+1) + \sigma^+(n)\,\sigma^-(n+1) + \sigma^-(n)\,\sigma^+(n+1) \\ &\quad + \sigma^-(n)\,\sigma^-(n+1)\end{aligned} \qquad \text{... (21.2.22)}$$

Now,

$$\begin{aligned}C^+(n)\,C(n+1) &= \sigma^+(n)\,[-\sigma_3(n)]\sigma^-(n+1) \\ &= \sigma^+(n)\,\sigma^-(n+1)\end{aligned} \qquad \text{... (21.2.23)}$$

since $\sigma^+(n)\,\sigma_3(n) = -\sigma^+(n)$. Similarly, one finds

$$\begin{aligned}C(n)\,C^+(n+1) &= -\sigma^-(n)\,\sigma^+(n+1) \\ C^+(n)\,C^+(n+1) &= \sigma^+(n)\,\sigma^+(n+1) \\ C(n)\,C(n+1) &= -\sigma^-(n)\,\sigma^-(n+1),\end{aligned} \qquad \text{... (21.2.24)}$$

leading to (dropping constant terms)

$$\overline{H} = -2\sum_n C^+(n)\,C(n) - \lambda\sum_n [C^+(n) - C(n)]\,[C^+(n+1) + C(n+1)] \qquad \text{... (21.2.25)}$$

Since $\overline{H}$ is quadratic, the model will be exactly solvable. It is convenient to work in momentum space by defining the operators

$$a_k = \left(\frac{1}{2N+1}\right)^{1/2} \sum_{n=-N}^{N} e^{ikn}\, C(n) \qquad \text{... (21.2.26)}$$

where,

$k = 0, \pm \dfrac{2\pi}{2N+1}, \pm \dfrac{4\pi}{2N+1}, \ldots \dfrac{2\pi N}{2N+1}$. We have in choosing k made a commitment to a boundary condition linking a_{-N} to a_N which, however, is not vital for our present purpose. The operators a_k have the correct fermionic anticommutation relations, namely,

$$\{a_k^{+}, a_k\} = \delta_{k,k} \qquad \text{... (21.2.27)}$$

$$\{a_k, a_k\} = \{a_k^{+}, a_k^{+}\} = 0$$

Inverting Eq. (21.2.26), we can write

$$C(n) = \left(\frac{1}{2N+1}\right)^{1/2} \sum_{k} e^{ikn}\, a_k \qquad \text{... (21.2.28)}$$

and now it is possible to write each term of $\overline{H}$ in terms of a_k and a_k^{+} to obtain,

$$\overline{H} = -2\sum_{k} (1+\lambda \cos k)\, a_k^{+} a_k - \lambda \sum_{k} (e^{-ik}\, a_k^{+} a_{-k}^{+} - e^{ik}\, a_k a_{-k}) \qquad \text{... (21.2.29)}$$

if we restrict the sum over k to only $k > 0$ by explicitly writing out the terms with $k < 0$ and carrying out the transformation $k = -k$ for them, we find

$$\overline{H} = -2 \sum_{k>0} (1+\lambda \cos k)\,(a_k^{+} a_k + a_{-k}^{+} a_{-k})$$

$$+\, 2i\,\lambda \sum_{k>0} \sin k\,(a_k^{+} a_{-k}^{+} + a_k a_{-k}) \qquad \text{... (21.2.30)}$$

The diagonalization of $\overline{H}$ is done by a canonical transformation (Bogoliubov-Valatin transformation used in superconductivity). The idea is to obtain a set of operators η_k such that $\overline{H}$ is written in terms of them as

$$\overline{H} = \sum_k \Lambda_k \eta_k^+ \eta_k + \text{constant} \qquad \text{... (21.2.31)}$$

We define

$$\eta_k = u_k a_k + i v_k a_{-k}^+ \qquad \text{... (21.2.32)}$$

$$\eta_k^+ = u_k a_k^+ - i v_k a_{-k} ,$$

where u_k and v_k depend only on the magnitude of k and are real quantities. The requirement that the transformation is canonical, i.e.

$$\{\eta_{k'}, \eta_k^+\} = \delta_{k'k} \qquad \text{... (21.2.33)}$$

$$\{\eta_{k'}, \eta_k\} = \{\eta_{k'}^+, \eta_k^+\} = 0$$

leads immediately to

$$u_k^2 + v_k^2 = 1 \qquad \text{... (21.2.34)}$$

Demanding that $\overline{H}$ have the form of Eq. (21.2.31), requires

$$4(1 + \lambda \cos k) u_k v_k + 2\lambda \sin k (u_k^2 - v_k^2) = 0 \qquad \text{... (21.2.35)}$$

Straightforward algebraical manipulation now yields

$$\Lambda_k = 2\sqrt{1 + 2\lambda \cos k + \lambda^2} \qquad \text{... (21.2.36)}$$

for the energy spectrum of the particles whose creation operator is η_k^+. The minimum value of the excitation energy is obtained for $k = \pm \pi$ and is given by

$$\Lambda_\pi = 2 |1 - \lambda| \qquad \text{... (21.2.37)}$$

At this point it is instructive to introduce dimensions of the various quantities. In the above, the dimensionless wavenumber is k. In the vicinity of $k = \pi$, we introduce the dimensional wavenumber p (also the momentum if $\hbar = 1$) and write

$$k = \pi + p.a \qquad \text{... (21.2.38)}$$

where a is a length equal to the lattice spacing. We are interested in the range $p.a << 1$. The dimensional energy ($\hbar = c = 1$) will similarly be defined as $E_k = \Lambda_k/2a$ and we have from Eq. (21.2.36) for $pa < 1$

$$E_p = \frac{\Lambda_{\pi + pa}}{2a} = \frac{1}{a}\sqrt{1 + 2\lambda \cos(\pi + pa) + \lambda^2}$$

$$= \frac{1}{a}\sqrt{1 - 2\lambda \cos pa + \lambda^2}$$

$$\cong \frac{1}{a}\sqrt{(1-\lambda)^2 + p^2a^2}$$

$$= \sqrt{\frac{(1-\lambda)^2}{a^2} + p^2} \quad \text{... (21.2.39)}$$

The energy spectrum of the η-particles is the relativistic energy of a particle of momentum p. The mass of the particle depends on the lattice spacing and thus the spectrum is dependent on the lattice variable a except at $\lambda = 1$, when we have the energy spectrum given by $E_p = |p|$ independent of the lattice spacing. In accordance with our discussion at the beginning of this section the critical point (the phase transition point) is then $\lambda = 1$. To see how this compares with the exact result obtained by Kramers and Wannier, we note that $\lambda = 1$ implies [see Eqs. (21.2.11a) and (21.2.11b)

$$\beta_2 = e^{-2\beta_1} \quad \text{... (21.2.40a)}$$

The exact result for the transition curve is

$$\sin h\,(2\beta_2)\,\sin h\,(2\beta_1) = 1 \quad \text{... (21.2.40b)}$$

If $\beta_1 \to \infty$ and $\beta_2 \to 0$ (the limits considered by us for $\tau \to 0$), then $\sin h\, 2\beta_1 \to \frac{1}{2}e^{2\beta_1}$ and $\sin h\, 2\beta_2 \to 2\beta_2$ and Eq. (21.2.40b) reduce to Eq. (20.2.40a). Thus, the identification of $\lambda = 1$ as the transition point is indeed correct.

Although, we have identified the transition point, we have not yet found the largest eigenvalue of the transfer matrix. To do so, it is best to return to Eq. (21.2.30) and work near $\lambda = 1$ and $k \cong \pi$. We will write $k = \pi + K$, where $K << 1$ and replace $\sin k = \sin(\pi + K) = -\sin K$ by $-K$. We will work to the lowest (in this case the first) order in $\lambda - 1$ and K. Consequently, the Hamiltonian of Eq. (21.2.30) becomes

$$\overline{H} = 2(\lambda - 1) \sum_{K>0} (a_k^+ a_k + a_{-k}^+ a_{-k}) \qquad \text{... (21.2.41)}$$

$$-2i \sum_{K>0} K (a_k^+ a_{-k}^+ + a_k a_{-k})$$

Introducing a fermion field $\varphi(x)$ [instead of the fermionic variables $C(n)$ on the lattice], by the relation

$$\varphi(x) = \frac{1}{\sqrt{L}} \sum_K a_k e^{-iKx} \qquad \text{... (21.2.42)}$$

$$\int_0^L \varphi(x) \nabla\varphi(x)\, dx = -\sum_K iK a_k a_{-k} = -2i \sum_{K>0} K a_k a_{-k}$$

and to within additive constants the Hamiltonian of Eq. (21.2.41) can be written as

$$\overline{H} = \int dx\, H(x) \qquad \text{... (21.2.43)}$$

where

$$H(x) = (\varphi \nabla \varphi - \varphi^+ \nabla \varphi^+) + 2(\lambda - 1)\, \varphi^+ \varphi \qquad \text{... (21.2.44)}$$

The Hamiltonian density of the Dirac equation for a particle of mass m in one dimension is

$$H_D(x) = -i\psi^+ \alpha \nabla\psi + m\psi^+ \beta\psi \qquad \text{... (21.2.45)}$$

where the Dirac matrices α and β need to satisfy

$$\alpha\beta + \beta\alpha = 0 \qquad \text{... (21.2.46)}$$

$$\alpha^2 = \beta^2 = 1$$

which can be satisfied in a 2×2 representation by the Pauli matrices giving

$$\alpha = \sigma_2 = \begin{pmatrix} 0 & -i \\ i & 0 \end{pmatrix}$$

$$\beta = \sigma_3 = \begin{pmatrix} 1 & 0 \\ 0 & -1 \end{pmatrix} \qquad \text{... (21.2.47)}$$

making the wavefunction ψ a two-component object. The similarity between Eqs. (21.2.44) and (21.2.45) is obvious but they cannot be identical unless we introduce a two-component object involving the $\varphi's$. To do so we imagine a second Ising lattice superposed on the first but not interacting with it. The fields on the two lattices are $\varphi_1(x)$ and $\varphi_2(x)$ and we define the connection between ψ and φ by writing

$$\psi = \begin{pmatrix} \varphi_1 + i\varphi_2 \\ \varphi_1 + \varphi_2 \end{pmatrix} \qquad \text{... (21.2.48)}$$

where the fields φ_1 and φ_2 are supposed to anticommute, so that

$$\{\varphi'_i(x), \varphi_j^+(x')\} = \delta_{ij}\, \delta(x - x') \qquad \text{... (21.2.49)}$$

$$\{\varphi_i(x), \varphi_j(x')\} = 0$$

The equivalence is now complete since

$$-i\psi^+ \alpha \nabla \psi = \psi^+ \begin{pmatrix} 0 & -1 \\ 1 & 0 \end{pmatrix} \psi$$

$$= (\varphi_1 \nabla \varphi_1 - \varphi_1^+ \nabla \varphi_1^+) + (\varphi_2 \nabla \varphi_2 - \varphi_2^+ \nabla \varphi_2^+$$

$$+ i \nabla (\varphi_1 \varphi_2 + \varphi_2^+ \varphi_2^+ \varphi_1^+)$$

The last term is a total derivative and cannot contribute to $\overline{H}$ on integration. Similarly, the mass term is

$$\psi^+ \beta \psi = \psi \begin{pmatrix} 1 & 0 \\ 0 & -1 \end{pmatrix} \psi = \left[\varphi_1^+, \varphi_1\right] + \left[\varphi_2^+, \varphi_2\right]$$

$$+ i\{\varphi_1^+, \varphi_2\} - i\{\varphi_2^+, \varphi_1\}$$

$$= 2\varphi_1^+ \varphi_1 + 2\varphi_2^+ \varphi_2 - 2\delta(0),$$

since the anticommutators vanish and the commutators have been evaluated with Eq. (21.2.49). The delta function is to be smeared out over a lattice spacing and hence not dangerous. Discarding it as another constant, we have the identification

$$\int H_D(x)\, dx = \int [H(\varphi_1) + H(\varphi_2)]dx$$

where the mass m of the Dirac equation is given by $m = (\lambda - 1)$. The ground state energy is found by filling up all the negative energy single-particle states of the Dirac equation and this determines the largest eigenvalue of the transfer matrix. The ground state energy is clearly

$$E_G = -\sum_k (m^2 + k^2)^{1/2} = \frac{L}{\pi}\int_0^{k_c} dk\, \sqrt{m^2 + k^2}$$

$$-\frac{L}{\pi}\int_0^{k_c} dk\, \sqrt{(\lambda - 1)^2 + k^2} \qquad \text{... (21.2.50)}$$

where the cutoff has to come since our approximation has to break down at large values of k.

To make the connection with thermodynamics clear, we need to bring in the temperature variable. To do so we note that $\lambda = e^{2\beta_1}\beta_2$ is a function of temperature, i.e. $\lambda = f(T)$ since $\beta_1 = \beta J_1$ and $\beta_2 = \beta J_2$ are determined entirely by $\beta = (kT)^{-1}$ for a given J_1 and J_2 and the $\lambda - 1$ defines the critical temperature T_c such that $f(T_c) = 1$. Hence $\lambda - 1 = f(T) - f(T_c) = A_1(T - T_c) + A_2(T - T_c)^2 + \ldots$ in a Taylor expansion. If $\lambda \cong 1$, we can retain the first term alone and thus $\lambda - 1 = A_1(T - T_c)$ in our approximation, where A_1 is a constant of order unity. The largest eigenvalue of the transfer matrix e^{E_G} will dominate in the thermodynamic limit when the lattice size becomes very large and consequently the partition function $Z = e^{-2E_GL}$ ($2L$ for the two non-interacting lattices) and the free energy F per unit length is (in units of kT)

$$F = \frac{1}{2L}\ln Z = -E_G$$

$$= -\frac{L}{\pi}\int_0^{k_c} \sqrt{A_1^2(T - T_c)^2 + k^2}\, dk \qquad \text{... (21.2.51)}$$

The entropy is obtained as $S = -\dfrac{\partial F}{\partial T}$ and thus

$$S = A_1(T-T_c)\frac{L}{\pi}\int_0^{k_c} \frac{dk}{\sqrt{A_1^2(T-T_c)^2+k^2}}$$

A further differentiation yields the specific heat as

$$C = T\frac{\partial S}{\partial T} = TA_1\frac{L}{\pi}\left[\int_0^{k_c}\frac{dk}{\sqrt{A_1^2(T-T_c)^2+k^2}} - A_1(T-T_c)^2\int_0^{k_c}\frac{dk}{[A_1^2(T-T_c)^2+k^2]^{3/2}}\right] \quad \text{... (21.2.52)}$$

We want the asymptotic behaviour of C as $T \to T_c$.

Note that both the integrals in Eq. (21.2.52) diverge at $k = 0$ if $T = T_c$. However the second integral is multiplied by $(T - T_c)^2$ which cancels the $(T-T_c)^2$ divergence of the integral. The critical behaviour is contained in the first term from which we find for $T \cong T_c$

$$C \sim C_0 \ln (T - T_c)^{-1} \quad \text{... (21.2.53)}$$

showing that the specific heat is singular at the transition point and that the singularity is logarithmic—a result first established by Onsager on the basis of his complete solution.

21.3 Planar Model and Heisenberg Model

The discussion of the Ising model shows that it does not order at any finite temperature for $D = 1$, but does so for $D = 2$. The question we can ask is whether the model defined on lattices with D between 1 and 2 will show a finite temperature transition or not. To answer this question, we note that in any dimension D, if we begin with ordered spins and then overturn a few to form a domain of 'wrong' spins, then the size of the domain will be given by L^{D-1}, where L is the typical length scale in the problem. If $2J$ is the amount of energy required to turn a spin, then the energy cost of having a disordered domain is of the order JL^{D-1}. For $D > 1$, it costs a large amount of energy to create a large domain of disorder—in particular, the larger the domain of wrong spins, more the energy expenditure to create it (it is only in $D = 1$ that arbitrarily large domains can be produced with no extra cost in energy). The entropy change can at most be) $O(\ln L)$ and hence for $D > 1$, the energy change dominates, making it possible for the Ising model to exhibit a finite temperature phase transition for all $D > 1$. The case $D = 1$ is accordingly called the lower critical dimension.

The Ising spins can only point up or down. However, magnetic moments in general are vectors and hence we can, in general, consider $S(\mathbf{m})$ to be a vector $S(\mathbf{m})$at the lattice point **m**. The energy of the interaction (nearest neighbour interaction and isotropy being assumed) is accordingly

$$E = -J \sum_{\mathbf{m}} \mathbf{S}(\mathbf{m}) \cdot \mathbf{S}(\mathbf{m} + \mathbf{1}) \qquad \text{... (21.3.1)}$$

(ferromagnetic if $J > 0$) and antiferromagnetic if $J < 0$).

If **S** is a three-dimensional vector, we have the Heisenberg model and if **S** is a two-dimensional vector, then we have the planar model. The most important feature of the vector spin models is the alteration in the lower critical dimension from the $D = 1$ for the Ising case.

The alteration in the lower critical dimension is best seen by considering domains of upturned spin at any finite temperature assuming that at $T = 0$ the spins are all aligned. If the upturned spin domain is of size L, then we have to consider the boundary between the spins pointing one way and the spins pointing the other way. This boundary is not sharp since the spins can be canted and the transition from up-spins to down-spins occurs in a region of size of order L. In a D-dimensional space, the sharp interface for the Ising model implies an energy cost of $O(L^{D-1})$. The blurred interface with an extension of $O(L)$ will cost a lower amount of energy— namely of $O(L^{D-2})$ since the turning of spins is over a domain of length L and in that length L, the energy cost is negligible. Thus the effective dimension for calculating the energy change is reduced by one. Thus to form a macroscopic domain of disordered spin, the energy cost is $O(L^{D-2})$. The entropy gained is $O(\ln L)$ and consequently the free energy change on disordering is

$$\Delta F = O(L^{D-2}) - T\, O(\ln L) \qquad \text{... (21.3.2)}$$

Clearly for $D < 2$, it is not possible to have a positive ΔF at any finite temperature and hence the disordering will lower the free energy. Consequently, there is no long-range order at any nonzero T for $D < 2$. The lower critical dimension is thus $D = 2$. To investigate the situation at $D = 2$, one needs a more sophisticated argument. We simply mention here that it was shown by Mermin and Wagner and Hohenberg that at any nonzero T, the planar model does not order in $D = 2$. However, there is a particular kind of ordering for the planar model in $D = 2$, which was found by Kosterlitz and Thouless.

21.4 Statistical Mechanics and Field Theory

In Sec. 21.2, we found that the two-dimensional Ising model was equivalent to a many-body quantum problem in one spatial dimension which is a field theory in 1 + 1 dimensions, i.e. one spatial dimension and time. Had we started with the one-dimensional Ising model, the equivalent quantum Hamiltonian would be a single-particle Hamiltonian, i.e. the one-dimensional statistical mechanics problem is equivalent to a problem in quantum mechanics. We anticipate from the above the general result that a $(D + 1)$-dimensional statistical mechanical problem would be having a quantum field theory counterpart in D spatial dimensions and time.

To illustrate the above, we consider a self-coupled scalar field in D spatial dimensions with the Lagrangian

$$L = \int d^D x \left[\frac{1}{2} \left(\frac{\partial \varphi}{\partial t} \right)^2 - \frac{1}{2} (\vec{\nabla} \varphi)^2 - \frac{1}{2} m^2 \varphi^2 - \frac{\lambda}{4} \varphi^4 \right] \qquad \text{... (21.4.1)}$$

A formal expression for a transition amplitude from time t_i to t_f can be written down as

$$Z = \int D[\varphi]\, e^{\,i \int_{t_i}^{t_f} L\, dt/\hbar} \qquad \text{... (21.4.2)}$$

We now consider the imaginary time variable

$$t' = i\, t \qquad \text{... (21.4.3)}$$

and introducing a lattice spacing a in the spatial direction and a spacing τ in the temporal direction t', find for the action

$$S = \mathrm{i}\int L\,dt = -\int d^D x\,dt'\left[\frac{1}{2}\left(\frac{\partial\varphi}{\partial t'}\right)^2 + \frac{1}{2}(\vec{\nabla}\varphi)^2 + \frac{m^2}{2}\varphi^2 + \frac{\lambda}{4}\varphi^4\right]$$

$$= -\sum_{\mathbf{n}}\left\{\frac{a^D}{2\tau}[\Delta_0\,\varphi(\mathbf{n})]^2 + \frac{\tau}{2}a^{D-2}\sum_{l}[\Delta_l\,\varphi(\mathbf{n})]^2 + \frac{\tau}{2}a^D m^2\varphi^2(\mathbf{n}) + \frac{\tau}{4}\lambda\, a^D\varphi^4(\mathbf{n})\right\}$$

$$= -\,S_E \qquad \text{... (21.4.4)}$$

where S_E is called the Euclidean action and

$$\Delta_\mu\,\varphi(\mathbf{n}) = \varphi(\mathbf{n}+\mu) - \varphi(\mathbf{n}) \qquad \text{... (21.4.5)}$$

while n labels the lattice points (n has D + 1as is components), $\mu = 0$ is the 'temporal direction' and $\mu = l\ (l = 1, 2, \ldots D)$ stands for the spatial directions.
Defining

$$K_\tau = \frac{a^D}{2\tau},\ K = \frac{\tau}{2}a^{D-2},\ b = \frac{\tau}{2}a^D m^2 \qquad \text{... (21.4.6)}$$

$$u = \frac{\tau}{4}\lambda\, a^D$$

the Euclidean action can be written as

$$S_E = \sum_n\left[K_\tau(\Delta_0\,\varphi)^2 + K_l\,\Sigma(\Delta_l\,\varphi)^2 + b\,\varphi(n)^2 + u\,\varphi(n)^4\right] \qquad \text{... (21.4.7)}$$

and the generating functional Z *of the field theory is now*

$$Z = \prod_n\int_{-\infty}^{\infty} d\varphi(n)\ e^{-S/\hbar} \qquad \text{... (21.4.8)}$$

where the functional integration over the φ-field has been given the meaning of integrating each site variable $\varphi(n)$ over all possible values. From Eq. (21.4.8), it is clear that Z *is the partition function of a statistical mechanical system where the variable is* $\varphi(n)$ *defined on lattice points in a* $(D + 1)$-dimensional lattice and thus a field theory in D spatial dimensions and time leads to a statistical mechanical problem in $(D + 1)$ dimensions. Note that $\hbar$ of the quantum problem plays the role of kT (temperature) in the statistical mechanical problem.

For the statistical mechanical problem, we can construct the transfer matrix to propagate the field $\varphi(n)$ at one time slice to the field $\varphi'(n)$ on the next time slice. The n in the argument are the D "spatial" variables—the different time values on the two successive slices is taken care of by φ and φ'. We can define the transfer matrix by (we set $\hbar = 1$).

$$< \varphi'|T|\varphi > = e^{-\sum_{\mathbf{n}} \{K_\tau [\varphi'(\mathbf{n}) - \varphi(\mathbf{n})]^2 + \frac{1}{2} K [\sum_l (\Delta_l \varphi'^2) + \sum_l (\Delta_l \varphi^2] + \frac{b}{2} [\varphi'^2(\mathbf{n}) + \varphi^2(\mathbf{n})] + \frac{u}{2} [\varphi'(\mathbf{n})^4 + \varphi(\mathbf{n})^4]\}} \quad \text{... (21.4.9)}$$

and obtain the partition function as

$$Z = \text{Tr } T^{N+1} \quad \text{... (21.4.10)}$$

by introducing complete sets of $|\varphi>$ states at every time slicing, imposing periodic boundary conditions and summing over all possible initial conditions. We get an operator expression for T by introducing at every spatial point $\mathbf{l}$, the conjugate field π by the relation

$$[\varphi(\mathbf{l}), \pi(\mathbf{l}')] = i\, \delta_{\mathbf{l},\mathbf{l}'} \quad \text{... (21.4.11)}$$

which readily yields

$$T = e^{-\frac{1}{2}\sum_n \{K \sum_l [\Delta_l \varphi']^2 + b\, \varphi'(n)^2 + u\varphi'(n)^4\}} \times e^{-\sum_n \frac{1}{4k_\tau} \pi^2(n)} \times e^{-\frac{1}{2}\sum_n \{K \sum_l (\Delta_l \varphi)^2 + b\varphi(n)^2 + u\varphi(n)^4\}} \quad \text{... (21.4.12)}$$

We can formally write $T = e^{-H_s\tau}$, although H_s is not very revealing, except where $\tau \to 0$ and H_s is the canonical Hamiltonian of the original field theory.

The operator T is Hermitian and hence its eigenvalues will be real and be written as $e^{-E_i\tau}$. If the corresponding eigenfunction is $|\, i>$, we can write a spectral decomposition of T as

$$T = \sum_i |i> e^{-E_i\tau} <i| \qquad \text{... (21.4.13)}$$

knowing that the set $\{|i>\}$ is complete. To get the partition function we need T raised to a large power N and clearly

$$T^N = \sum_i |i> e^{-NE_i\tau} <i| \qquad \text{... (21.4.14)}$$

As $N \to \infty$, the sum will be dominated by the lowest value of E_i if it is unique and hence

$$\underset{N\to\infty}{\text{Lt}} \; T^N \to |0> e^{-N_\tau E_0} <0|$$

$$= |0> e^{-E_0 L_\tau} <0| \qquad \text{... (21.4.15)}$$

where L_τ is the total length in the temporal direction. The partition function is thus

$$Z = e^{-E_0 L_\tau} = e^{-\omega_0 V L_\tau} \qquad \text{... (21.4.16)}$$

where we have written $E_0 = \omega_0 V$ keeping in mind the extensive nature of E_0 (V is the spatial volume). On the other hand the thermodynamic identity relates Z to the *free energy* F as

$$Z = e^{-F} = e^{-fVL_\tau} \qquad \text{... (21.4.17)}$$

where we have taken the extensive nature of F in $(D + 1)$ dimensions into account. Comparing Eqs. (21.4.17) and (21.4.16)

$$f = \omega_0 \qquad \text{... (21.4.18)}$$

This yields the first connection between the statistical mechanical problem and the field theory. Further connections are revealed by considering the propagator of the field theory defined as ($t > 0$)

$$\Delta(t, \mathbf{x}) = < 0 | \varphi(t, \mathbf{x}) \varphi(0, \mathbf{0}) | 0 > \qquad \text{... (21.4.19)}$$

where $|0>$ *is the Heisenberg picture ground state of the Hamiltonian* H_s and $\varphi(x, t)$ is the field operator in the Heisenberg picture. The corresponding correlation function in the statistical mechanical problem is

$$\Gamma(n) = \frac{1}{Z}\int \prod_{n'_0,\, \mathbf{n}'} d\varphi(n'_0, \mathbf{n}')\, \varphi(\mathbf{n}, \mathbf{n})\varphi(0, 0)\, e^{-S} \qquad \text{... (21.4.20)}$$

where n_0 is the temporal part of n and $\mathbf{n}$ the spatial part.

Writing the Heisenberg operator $\varphi(x, t)$ in terms of the Schrödinger operator as $\varphi(x, t) = e^{iH_s t}\, \varphi(x)\, e^{-iH_s t}$, we have

$$\Delta(t, \mathbf{x}) = \langle 0 \mid \varphi(x)\, e^{-iH_s t}\, \varphi(0) \mid 0\rangle\, e^{iE_0 t} \qquad \text{... (21.4.21)}$$

and manipulations of Eq. (21.4.20) in terms of the transfer matrix shows the correspondence

$$\Gamma(n_0, \mathbf{n}) = \Delta(-in_0\tau, \mathbf{n}) \qquad \text{... (21.4.22)}$$

Finally, if one is not at $T = T_c$ (the critical point), then the statistical mechanical correlation function falls of exponentially, characterized by a definite correlation length. In particular, we will have

$$\Gamma(n_0, 0) \sim e^{-n_0/\xi} \qquad \text{... (21.4.23)}$$

where ξ is the correlation length. Now from Eq. (21.4.21), we have

$$\Delta(-in_0\tau, 0) = \langle 0 \mid \varphi(0)\, e^{-H_s n_0 \tau}\, \varphi(0) \mid 0\rangle e^{E_0 n_0 \tau}$$

$$= \sum_l |\langle 0| \varphi(0) \mid l\rangle|^2\; e^{(E_0 - E_l) n_0 \tau}$$

$$\cong |\langle 0| \varphi(0) | 1\rangle|^2\, e^{-(E_1 - E_0)\, n_0 \tau} \qquad \text{... (21.4.24)}$$

if n_0 is very large (we have simply taken the sum to be dominated by the largest term and in the ordered sequence of eigenvalues, E_1 is closest to E_0).

Comparing with Eq. (21.4.23), ξ is simply $(E_1 - E_0)^{-1}$. But this energy difference is the mass of the lightest particle at zero momentum and thus the reciprocal of the correlation length is the mass gap of the field theory. The complete correspondence is as follows:

Statistical Mechanics	*Field Theory*
Free energy density	Vacuum energy density
Correlation function	Propagator
Inverse correlation length	Mass gap

Turning now to our analysis of the 2-D Ising model in Sec. 21.2, we note that the energy spectrum was found to be $E_p = \sqrt{\frac{(1-\lambda)^2}{a^2} + p^2}$ and hence the lightest particle at zero momentum (i.e. $p = 0$) has a mass $m = (1-\lambda)/a$. Thus at $\lambda = 1$, the mass gap vanishes yielding an infinite correlation length, implying that $\lambda = 1$ is indeed the critical point of the theory—a fact which we had established then by arguing the absence of lattice effects at $\lambda = 1$.

21.5 Kac-Hubbard-Stratanovich Transformation

The Kac-Hubbard-Stratanovich transformation is in general a procedure for converting one model into another. It is particularly useful in critical phenomena where it is employed for converting a model defined on a lattice to a field theoretic model. Here, we will consider the simplest model which is the Ising model on a D-dimensional lattice. The Ising spins at each lattice site can take the values ± 1 only and the partition function is given by

$$Z_N(J) = \sum_{\{s_m\}} e^{\Sigma J_{ij}\, s_i\, s_j} \qquad \text{... (21.5.1)}$$

where s_i is the spin at site i (the site is a point on a D-dimensional lattice in general and requires D coordinates for specification) and J_{ij} is the interaction with the property that $J_{ij} = J_{ji}$ and $J_{ii} = 0$ (we are using units where $kT = 1$). The total number of spins is N.

We will demonstrate that Eq. (21.5.1) can be rewritten as

$$Z_N(J) = e^{f_0(J)} \int_{-\infty}^{\infty} \prod_{i=1}^{N} d\,\bar{s}_i\; e^{-\Sigma_{i,j} Q_{ij}\, \bar{s}_i\, \bar{s}_j}\; e^{-\Sigma W(\bar{s}_i)} \qquad \text{... (21.5.2)}$$

where $f_0\,(J)$ is a smooth analytic function of J. The integration is over each continuous spin variable $\bar{s}_i$ of which there are as many as there were original sites. The limits for each integration are $-\infty$ and $+\infty$. The weighting function $W(s_i)$ will be

$$-W(s_i) = \ln\cosh s_i - \frac{1}{2}\, Q_{ii}\, s_i^2 = -\frac{1}{2}\,(Q_{ii}-1)\, s_i^2 - \frac{1}{12}\, s_i^4 + \ldots \qquad \text{... (21.5.3)}$$

where Q_{ii} will be defined as we go along.

We note that

$$\frac{1}{2}(s_i + s_j)^2 = \frac{1}{2}s_i^2 + \frac{1}{2}s_j^2 + \frac{1}{2}\cdot 2s_i s_j = 1 + s_i s_j \geq 0$$

and that interactions can be rewritten as

$$J_{ij}\, s_i s_j = (P_0 I + P_{ij})\, s_i s_j$$

$$= (-P_0\, \delta_{ij} + P_{ij})\, s_i s_j$$

(the diagonal elements of p_{ij} are p_0 to ensure $j_{ii} = 0$)

and hence

$$\sum J_{ij}\, s_i s_j = -\frac{NP_0}{2} + \frac{1}{2}\sum_{i=1}^{N}\sum_{j=1}^{N} P_{ij}\, s_i s_j \qquad \text{... (21.5.4)}$$

If $[s_1, ..., s_N]$ is the row vector s^T and the transposed column vector is s, then the partition function becomes

$$Z_N = e^{-\frac{NP_0}{2}} \sum_{\{S_m\}} e^{\frac{1}{2}s^T Ps} \qquad \text{... (21.5.5)}$$

At this point consider a quadratic form in N continuous variables Y_i given by

$$Q(Y_i) = \sum_{i=1}^{N}\sum_{j=1}^{N} Q_{ij}\, Y_i Y_j \qquad \text{... (21.5.6)}$$

where Q_{ij} is a symmetric, positive definite matrix and can be diagonalized by a proper orthogonal transformation O. In terms of new variables x_i

$$x_i = \sum_j O_{ij}\, Y_j \qquad \text{... (21.5.7)}$$

the quadratic form is

$$Q(y) = \sum_{j=1}^{N} \lambda_j x_j^2 \qquad \text{... (21.5.8)}$$

where λ_j are the real positive eigenvalues of Q. It follows that the determinant of Q is

$$\text{Det } |Q| = \prod_{j=1}^{N} \lambda_j \qquad \text{... (21.5.9)}$$

The partition function-like expression

$$I(Q) = \int_{-\infty}^{\infty} d^N \, Y \, e^{-\frac{1}{2} Y^t QY} \qquad \text{... (21.5.10)}$$

can be evaluated by making the change from y to x, noting that the Jacobian of the transformation is 1 and the resulting integrals are Gaussian, yielding

$$I(Q) = \prod_{j=1}^{N} \sqrt{\frac{2\pi}{\lambda_j}} = \frac{(2\pi)^{N/2}}{\sqrt{\text{Det } |Q|}} \qquad \text{... (21.5.11)}$$

A shift in the variable y is now made according to

$$Y = \bar{s} + Q^{-1} \, s \qquad \text{... (21.5.12)}$$

The $\bar{s}$ are the continuous variables, while the s represent fixed shift parameters. The transpose of Eq. (21.5.12) is given by

$$Y^T = \bar{s}^T + s^T \, Q^{-1} \qquad \text{... (21.5.13)}$$

and the quadratic form of Eq. (21.5.6) becomes

$$\begin{aligned} Y^T QY &= ((\bar{s}^T + s^T Q^{-1}) \, Q \, (\bar{s} + Q^{-1} \, s) \\ &= \bar{s}^T \, Q \bar{s} + \bar{s}^T \bar{s} + s^T \bar{s} + s^T \, Q^{-1} s \\ &= s^{-T} \, Q \bar{s} + 2 \sum_{i=1}^{N} s_i \bar{s}_i + s^T \, Q^{-1} s \end{aligned}$$

leading to

$$I(Q) = e^{-\frac{1}{2} s^T P s} \int_{-\infty}^{\infty} d^N \bar{s} \, e^{-\frac{1}{2} \bar{s}^T Q \bar{s}} \; e^{-\sum_{i=1}^{N} s_i \bar{s}_i}$$

where $P = Q^{-1} = (P_0 I + J)^{-1}$. Combining with Eq. (21.5.11),

$$e^{-\frac{1}{2}s^T Ps}\int_{-\infty}^{\infty} d^N\bar{s}\; e^{-\frac{1}{2}\bar{s}^T Q\bar{s} - \sum_{i=1}^{N} s_i\bar{s}_i} = \frac{(2\pi)^{N/2}}{\sqrt{\det |Q|}} = (2\pi)^{N/2}\sqrt{\mathrm{Det}\,|P|} \qquad \text{... (21.5.14)}$$

Using the form of $e^{+\frac{1}{2}s^T Ps}$ obtained from Eq. (21.5.14) the partition function of Eq. (21.5.5) becomes

$$Z_N = \frac{e^{-NP_0/2}}{(2\pi)^{N/2}\sqrt{\mathrm{Det}\,|P|}} \sum_{\{ss\}} \int_{-\infty}^{\infty} d^N\bar{s}\; e^{-\frac{1}{2}\bar{s}^T Q\bar{s} - \sum_{i=1}^{N} s_i\bar{s}_i\}}$$

$$= \frac{e^{-NP_0/2}}{(2\pi)^{N/2}\sqrt{\mathrm{Det}\,|P|}} \int_{-\infty}^{\infty} d^N\bar{s}\; e^{-\frac{1}{2}\bar{s}^T Q\bar{s}}\; e^{+\ln\cosh \bar{s}_i}$$

$$= \frac{e^{-NP_0/2}}{(2\pi)^{N/2}\sqrt{\mathrm{Det}\,|P|}} \int_{-\infty}^{\infty} d^N\bar{s}\; e^{-\sum_{(i,j)}^{N} Q_{ij}\bar{s}_i\bar{s}_j - \sum_i \left(1/2\; Q_{ii} s_i^2 - \ln\cosh s_i\right)} \qquad \text{... (21.5.15)}$$

We have thus reduced the partition function to the form of Eq. (21.5.2) and we have effectively transformed the discrete sum over spins to an integration over continuous variables.

21.6 Critical Phenomena: Scaling Laws

In this Section, we will use the terminology of the the magnetic transition to introduce the phenomenology. At high temperatures the individual magnetic moments are disordered and point in random directions and the entropy contribution to the free energy dominates. As the temperature is lowered, the short-range interaction, which favours alignment of the individual magnetic moments, begins to dominate and the equilibrium state of the system is one in which there is overall magnetization—this spontaneous magnetization grows as the temperature is lowered. The temperature at which the spontaneous magnetization first appears is the critical temperature T_c. The growth of the magnetization M, as the temperature is lowered is continuous (Fig. 21.4) and in the immediate vicinity of the critical temperature is characterized by a relation of the form

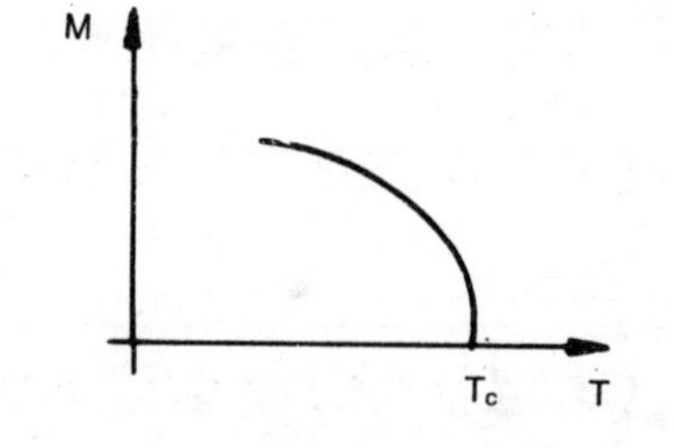

Fig. 21.4 Magnetization near T_C

$$M \propto |T_c - T|^{\beta} \qquad \dots (21.6.1)$$

A relation of this form is called a scaling law and the number β is called a critical exponent. In Chapter 5, we had encountered such a behaviour in both the van der Waals gas and the approximate solutions of the Ising model in arbitrary dimensions.

As the transition is approached from above, the response of the system becomes stronger and stronger to an infinitesimal external magnetic field H. This response is characterized by the isothermal susceptibility $X = \partial M/\partial H$ which grows as the critical point is approached (Fig 21.5). Close to $T = T_c$, whether above or below T_c, the fluctuations in the magnetization are extremely strong and the large susceptibility is characterized by

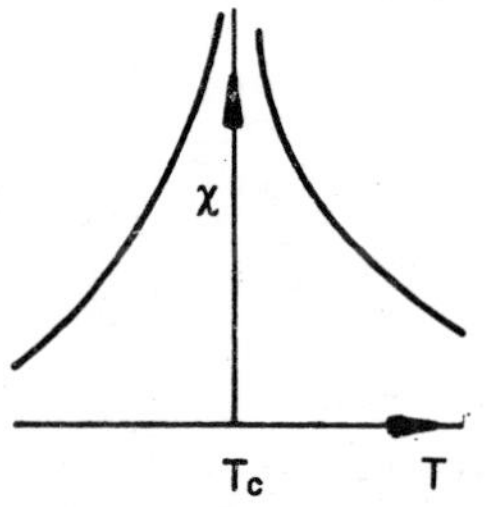

Fig. 21.5 The divergence of the susceptibility χ at T_C

$$\chi \propto |T - T_c|^{-\gamma} \qquad \dots (21.6.2)$$

What lies behind this large susceptibility is the long range correlation that develops in the system as T approaches T_c. To consider the correlation one has to consider the local magnetization function $\varphi(\mathbf{x})$ which is the average over the individual magnetic moment in a small volume surrounding the point $\mathbf{x}$ (this volume is macroscopic with regard to atomic dimensions but microscopic with regard to sample dimensions, so that $\varphi(\mathbf{x})$ is a continuous function). We consider the correlation function $G(x_1, x_2)$ defined as

$$G(\mathbf{x}_1, \mathbf{x}_2) = \langle \varphi(\mathbf{x}_1)\, \varphi(\mathbf{x}_2) \rangle \qquad \dots (21.6.3)$$

where the angular bracket denotes an average over the equilibrium distribution of the field $\varphi(x)$. translational invariance implies that $G(\mathbf{x}_1, \mathbf{x}_2)$ is a function of $\mathbf{x}_1 - \mathbf{x}_2 = \mathbf{x}_{12}$.

For $T > T_c$, the correlation function dies off exponentially, being characterized by an asymptotic long distance form $G(\mathbf{x})_{12} \sim e^{-|x_{12}|/\xi} / |x_{12}|^{D-2+\eta}$, where

ξ is the correlation length and D is the dimensionality of space. As the temperature approaches T_c, the correlation length ξ increases, approaching infinity according to the relation

$$\xi \propto |T - T_c|^{-\nu} \qquad \text{... (21.6.4)}$$

At T_c, when ξ is infinite, the correlation function decays according to a power law—this implies long-range behaviour, and is characterized by

$$G(\mathbf{x}_{12}) = \frac{1}{|x_{12}|^{D-2+\eta}} \qquad \text{... (21.6.5)}$$

where the exponent η is the anomalous dimension exponent. As we shall establish later, there is a fluctuation dissipation theorem relating the susceptibility to an integral over the correlation function and the long-range nature of $G(\mathbf{x}_{12})$ leads to the divergent susceptibility.

The magnetization M at $T = T_c$ grows if an external magnetic field H is applied and thus the growth law is characterized by the relation

$$M \sim H^{1/\delta} \qquad \text{... (21.6.6)}$$

Another characteristic quality that develops a singularity at the transition point (Fig 21.6) is the specific heat, C, which diverges as

$$C \sim |T - T_c|^{-\alpha} \qquad \text{... (21.6.7)}$$

The first astounding thing about the exponents α, β, δ, ν and η is that they are universal. This implies that they do not depend on the nature of the material considered. We must clarify, of course, that they do depend on whether the magnet considered is Ising-like (the magnetization field $\varphi(x)$ is scalar or Heisenberg-like ($\varphi(x)$ is a vector). Also, the exponents would differ for the superfluid transition for which $\varphi(x)$ is a complex quantity. The exponents differ with differing spatial dimensionality as well, i.e. whether $D = 3$ or $D = 2$ is being considered. If we consider the

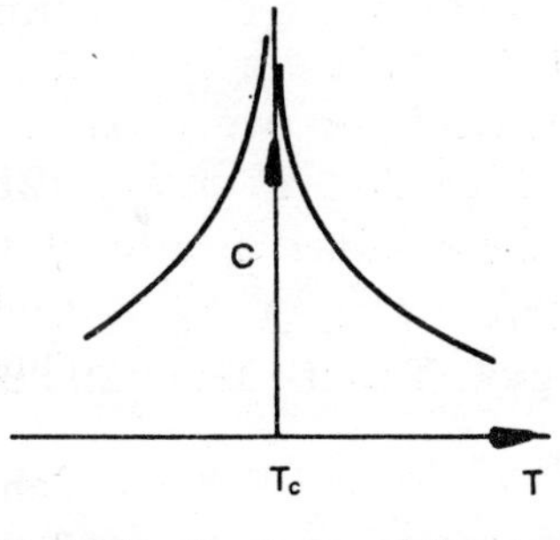

Fig. 21.6 The divegence of the specific heat C at T_c

dimensionality n of the field $\varphi(\mathbf{x})$, then $n = 1$, if $\varphi(\mathbf{x})$ is a scalar; $n = 2$, if $\varphi(\mathbf{x})$ is complex and $n = 3$ if $\varphi(\mathbf{x})$ is a vector. Universality means that the exponents introduced above are functions of n and D above.

The second remarkable fact is that the exponents are not all independent but connected by the scaling relations

$$\alpha + 2\beta + \gamma = 2 \quad \text{... (21.6.8a)}$$

$$(2-\eta)\,\nu = \gamma \quad \text{... (21.6.8b)}$$

$$\beta(1 + \delta) + \alpha = 2 \quad \text{... (21.6.8c)}$$

$$2 - \alpha = D\nu \quad \text{... (21.6.8d)}$$

Considering the correlation function $G(x_{12})$ at a temperature different from T_c (i.e. ξ is finite), we note that it has to be a function of two variables x_{12} and ξ. The $\xi = \infty$ form of Eq. (21.1.5) immediately suggests that $G(x_{12}, \xi)$ is a homogeneous function of the two variables, so that we can write

$$G(x_{12}, \xi) = x_{12}^{-(D-2+\eta)}\, g(x_{12}/\xi) \quad \text{... (21.6.9)}$$

where the function $g(x)$ has the property that $g(0)$ is a nonzero constant. From the experimental point of view, a more convenient quantity is the Fourier transform $G(k, \xi)$ $G(x_{12}, \xi)$, since it can be directly measured by neutron scattering. The Fourier transform is defined as

$$G(k, \xi) = \frac{1}{(2\pi)^{1/2}} \int d^D x_{12}\, e^{i\mathbf{k}\cdot\mathbf{x}_{12}}\, G\left(\frac{x_{12}}{\xi}\right)$$

$$= k^{-2+\eta}\, \bar{g}(k\xi) \quad \text{... (21.6.10)}$$

The function $\bar{g}(x)$ has the property that $\bar{g}(x) \propto x$ if $x \ll 1$ and $\bar{g}(x) \rightarrow$ constant if $x \rightarrow \infty$. The correlation function $G(k, \xi)$ is a function of two variables the scattering momentum transfer k and the correlation length ξ. In an experiment, both k and ξ can be varied and hence $G(k, \xi)$ can be obtained as a function of both variables. The prediction of Eq. (21.6.10) is that if $G(k, \xi)\,.\,k^{2-\eta}$ is plotted against $k\xi$, the data would collapse on a single plot. This is demonstrated in Fig. (21.7).

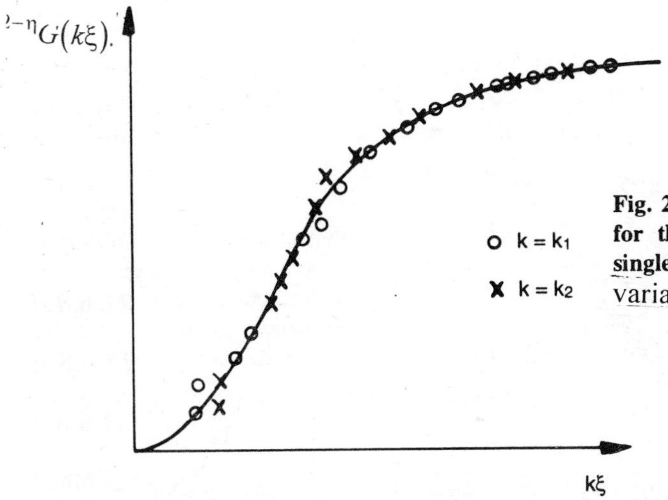

Fig. 21.7 Collapse of different curves for the correlation function onto a single plot when the correct scaling variable is used.

The equation of state for a magnet is a relation between M, H and T. Near the critical point it can be cast as a relation between M, H and ξ. From Eq. (21.1.1) and (21.1.6), we note that

$$M \propto \xi^{-\beta/\nu} \text{ if } H = 0 \qquad \text{... (21.6.11)}$$

and

$$M \propto H^{1/\delta} \text{ if } \xi = \infty$$

The consequence of Eq. (21.6.11) is that the equation for M must be a homogeneous function of $\xi^{-\beta/\nu}$ and $H^{1/\delta}$ and thus we can write

$$M = \xi^{-\beta/\nu}\, m(H\,\xi^{\beta\delta/\nu})$$

$$= t^{\beta}\, m\left(\frac{H}{t^{\beta\delta}}\right) \qquad \text{... (21.6.12)}$$

where $t = (T_c - T)/T_c$. The function $m(x)$ has the property that $m\,(0)$ goes to a constant and $m(x) \sim x^{1/\delta}$ when $x \to \infty$. The magnetization M can be measured as a function of two variables t and H and the consequence of Eq. (21.6.12) is that if $Mt^{-\beta}$ is plotted against $H/t^{\beta\delta}$, the data should lie on a single curve. This is illustrated in Fig. 21.8.

The immediate task of a theory of critical phenomena is to establish universality, prove the scaling laws [Eqs. (21.6.8a) – (21.6.8d)] and determine values of the critical exponents as functions of n and D. The tremendous difficulty in doing any of these is the long range nature of the correlations. This is a

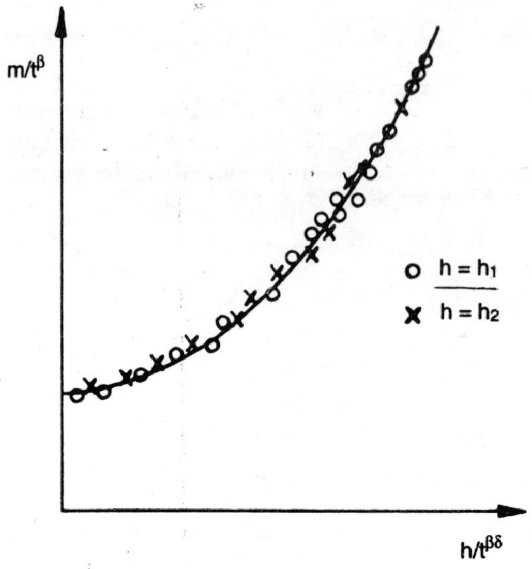

Fig. 21.8 Collapse of different curves for the magnetization on a single curve when the appropriate scaling variables are used.

quintessential many-body problem where perturbation theory refuses to work because of strong correlations. The successful handling of the difficulty required the advent of the renormalization group. However, before discussing that, we will explain the first attempt at solving the problem—the Landau theory—in the next chapter. The amplitudes of individual quantities like magnetizations, susceptibility, specific heat etc. are not universal but certain combinations of amplitudes are. A proper theory needs to identify and calculate the universal amplitude ratios.

Chapter 22

Landau Theory and Related Models

22.1 Landau Theory

In any statistical mechanics problem the central concept is that of a partition function. All thermodynamic quantities follow once the partition function is known. Accordingly, one needs to construct the partition function for a system near the critical point. The description of the system is in terms of a field. Near the critical point the local value of the magnetization (the generic term is the order parameter) is the important quantity. There are large fluctuations in the system as it tries to settle down to a state with a spontaneous magnetization and thus the local magnetization is the field variable. The individual magnetic moments when averaged over a small volume about the field point **x** gives the magnetization (order parameter field $\varphi_\alpha(\mathbf{x})$ ($\alpha = 1, 2 \ldots, n$) and the energy of the system will be a function of φ. From the standard definitions of the partition function

$$Z = \sum_{\text{all possible } \varphi} w(\varphi)\, e^{-E(\varphi)/kT} \qquad \ldots \textbf{(22.1.1)}$$

where k is the Boltzmann constant and T is the temperature. The factor w is the degeneracy—it is the number of microscopic states associated with a given macroscopic description in terms of $\varphi(\mathbf{x})$. The entropy $S(\varphi) = k \ln w(\varphi)$ and thus

$$Z = \sum_{\text{all } \varphi} e^{-[E(\varphi) - TS(\varphi)]/kT}$$

$$= \sum_{\text{all } \varphi} e^{-F(\varphi)/kT} \quad \text{... (22.1.2)}$$

where $F(\varphi)$ is the free energy. It is convenient to work in units in which $kT = 1$ and we have the result

$$Z = \sum_{\text{all } \varphi} e^{-F(\varphi)} \quad \text{... (22.1.3)}$$

It is important to realize that it is the free energy functional which enters the calculation of the partition function and to remember that the thermodynamic equilibrium state of a system is the one which minimizes the thermodynamic free energy. The construction of $F(\varphi)$ follows from general principles

i) Only even functions of φ are admissible.
ii) Since long range properties are of interest the lowest admissible derivative will survive
iii) For an isotropic system there will be $O(n)$ symmetry.

We thus write

$$F(\varphi) = \int d^D x \left[\frac{a}{2} \sum_{\alpha} {\varphi_\alpha}^2 + \frac{1}{2} \sum_{\alpha} (\vec{\nabla} \varphi_\alpha)^2 + \frac{b}{4} \left(\sum {\varphi_\alpha}^2 \right)^2 + \ldots \right] \quad \text{... (22.1.4)}$$

The simplest evaluation of the partition function follows by evaluating the minimum of $F(\varphi)$ and writing $Z \cong e^{-F_{\min}}$. In trying to do so, we note that φ_0 must be independent of space—the spatial dependence always adds to the free energy. Writing $M = \sum_{\alpha} \varphi_\alpha^2$ we now need to minimize

$$F = V \left[\frac{a}{2} M^2 + \frac{b}{4} M^4 + \cdots \right] \quad \text{... (22.1.5)}$$

where V is the total volume of the system. If the above is to be a description of the phase transition, then the minimization of F should lead to $M = 0$ for $T > T_c$ and $M \neq 0$ for $T < T_c$. Clearly, the minima of Eq. (10.2.5) are obtained for

$$M^2 = 0, \ -a/b \qquad \text{... (22.1.6)}$$

and the corresponding $F_{\min}$ are

$$F_{\min} = 0, \ -V \cdot \frac{a^2}{4b} \qquad \text{... (22.1.7)}$$

It is immediately obvious that the theory will describe critical phenomena if we choose

$$a = a_0\,(T - T_c) \qquad \text{... (22.1.8)}$$

$$b = \text{constant}$$

For $T > T_c$, F is minimized for $M = 0$, while for $T < T_c$, F is minimized for

$$M = \left[\frac{a_0}{b}\,(T_c - T)\right]^{1/2} \qquad \text{... (22.1.9)}$$

Comparison with Eq. (21.1.1) shows that

$$\beta = 1/2 \qquad \text{... (22.1.10)}$$

Since the specific heat C is

$$-T\,\frac{\partial^2 F}{\partial T^2}$$

we find that $C = 0$ for $T > T_c$ and $C = a_0^2\, T_c/2b$ for $T < T_c$. Thus the specific heat has a finite discontinuity at $T = T_c$, but no singularity. Hence, from Eq. (21.6.7),

$$\alpha = 0 \qquad \text{... (22.1.11)}$$

To calculate the susceptibility, we need to include an external magnetic field H and the quantity to be minimized is

$$F(M, H) = V\left[\frac{a}{2}\,M^2 + \frac{b}{4}\,M^4 - MH\right] \qquad \text{... (22.1.12)}$$

This leads to

$$aM + bM^3 - H = 0 \qquad \text{... (22.1.13)}$$

The susceptibility follows as

$$\chi = \left.\frac{\partial M}{\partial H}\right|_{M\to 0} = \frac{1}{a} = \frac{1}{a_0}(T - T_c)^{-1} \qquad \text{... (22.1.14)}$$

for $T > T_c$. For $T < T_c$, we need to consider fluctuations about $M^2 = -a/b$, and we find

$$(a + 3bM^2)\ \delta M = \delta H$$

or

$$\chi = \frac{\delta M}{\delta H} = \frac{1}{a + 3bM^2} = -\frac{1}{2a}$$

$$= \frac{1}{2a_0}\ (T_c - T)^{-1} \qquad \text{... (22.1.15)}$$

Once again $\chi \propto |T - T_c|^{-1}$ and hence

$$\gamma = 1 \qquad \text{... (22.1.16)}$$

Setting $T = T_c$ or $a = 0$ in Eq. (21.6.13), we find

$$\delta = 3 \qquad \text{... (22.1.17)}$$

which completes the Landau theory of phase transition. Note that the scaling relations (21.1.8a) and (21.1.8c) are satisfied. There is universality—in fact superuniversality in that α, β, γ and δ are independent of everything including n and D.

The Landau theory is important as the first nontrivial theory of critical phenomena—it is obviously not quantitatively correct—the problem lies with the ignoring of fluctuations. However, the task of the theory is now clear. One has to evaluate the partition function of Eq. (22.1.1) with the free energy functional of Eq. (22.1.4). It is more conventional to write $a = m^2$ and $b/4 = \lambda/4!$ and then one is faced with the functional integral

$$Z(H) = \int D[\varphi]\ e^{-F(\varphi)} \qquad \text{... (22.1.18)}$$

where

$$F(\varphi) = \int d^D x \left[\frac{m^2}{2} \sum_{\alpha=1}^{n} \varphi_\alpha{}^2 + \frac{1}{2} \sum_{\alpha} (\vec{\nabla} \varphi_\alpha)^2 + \frac{\lambda}{4!} \left(\sum \varphi_\alpha{}^2 \right)^2 - H\varphi_1 \right] \qquad \text{... (22.1.19)}$$

in the presence of an external field H in the one-direction.

We end this section by establishing some thermodynamic identities. First the magnetization M which is defined as the average of $\varphi_1\ (x)$ is related to $Z\ (H)$ by

$$M = \underset{H \to 0}{\text{Lt}} < \int \varphi_1\ (x)\ dx > = \underset{H \to 0}{\text{Lt}} \frac{1}{2} \int D[\varphi] \int \varphi_1\ (x)\ e^{-F(\varphi)}\ d^D x$$

$$= a \underset{H \to 0}{\text{Lt}} \frac{1}{Z} \frac{\partial Z}{\partial H} \qquad \text{... (22.1.20)}$$

The susceptibility is

$$\chi = \frac{\partial M}{\partial H} = - \left(\frac{1}{Z} \frac{\partial Z}{\partial H} \right)^2 + \frac{1}{Z} \frac{\partial^2 Z}{\partial H^2}$$

$$= \int <\varphi_1\ (x)\ \varphi_1(y)> d^D x\ d^D y - \left[\int <\varphi_1(x)\ > d^D x \right]^2$$

$$= \int <\varphi_1(x)\ \varphi_1(y)>_c\ d^D y \qquad \text{... (22.1.21)}$$

where the subscript 'c' denotes the connected part of the correlation function. Translational invariance implies $G(x, y) = <\varphi_1(x)\ \varphi_1(y)>_c\ = G(x - y)$ and thus

$$\chi = V \int d^D\ x_{12} < \varphi_1\ (\mathbf{x}_1)\ \varphi_2\ (\mathbf{x}_2) >_c \qquad \text{... (22.1.22)}$$

This is the static fluctuation dissipation theorem.

The specific heat is found to be

$$C = -T\frac{\partial^2 F}{\partial T^2} = +T\frac{\partial^2}{\partial T^2}(\ln Z)$$

$$= T\left[\frac{1}{Z}\frac{\partial^2 Z}{\partial T^2} - \left(\frac{1}{Z}\frac{\partial Z}{\partial T}\right)^2\right]$$

$$= a_0^2\, T \int d^D x\, d^D y < \sum_{\alpha=1}^{n} \varphi_\alpha^2(x) \sum_{\beta=1}^{n} \varphi_\beta^2(\gamma) >_c \qquad \text{... (22.1.23)}$$

where the subscript 'c' means that the factored part $< \sum_{\alpha=1}^{n} \varphi_x^2(x) >^2$ has been substracted. Thus the critical problem in the field theory of phase transition is the calculation of various correlation functions.

22.2 The Gaussian Model

The Landau theory leaves out fluctuations altogether. An exactly solvable model that includes the effect of fluctuations but only up to the quadratic order is called the Gaussian model. The quartic term in the free energy functional of Eq. (22.1.4) is now dropped and the partition function in the Gaussian approximation is Z_G, given by

$$Z_G = \int D[\varphi] \exp\left\{-\int d^D x\left[\frac{a}{2}\sum_{\alpha=1}^{n}\varphi_\alpha^2 + \frac{1}{2}\sum_{\alpha=1}^{n}(\vec{\nabla}\varphi_\alpha)^2\right]\right\} \qquad \text{... (22.2.1)}$$

The standard method of evaluating the functional integral is to work in terms of the Fourier components of $\varphi_\alpha(x)$. The functional integral now becomes

$$Z_G = \int \prod_{\mathbf{k},\alpha} d\varphi_\alpha(\mathbf{k}) \exp\left\{\sum_{\mathbf{k},\alpha} \frac{1}{2}(a + k^2)\,|\,\varphi_\alpha(k)|^2\right\}$$

$$= \prod_{\mathbf{k},\alpha} (a + k^2)^{-1/2} \qquad \text{... (22.2.2)}$$

The thermodynamic free energy is given by

$$F = -kT \ln Z_G = \frac{1}{2} kT \sum_{k,\alpha} \ln (a + k^2)$$

$$= \frac{nkT}{2} \sum_k \ln (a + k^2) \qquad \ldots \text{(22.2.3)}$$

The specific heat is to obtained by differentiating twice with respect to T. This really means differentiation with respect to a, since that is the quantity which changes most as T is varied near T_c (our interest is primarily in the behaviour close to T_c and hence, apart from where it appears in the term a, *the temperature can be set equal to* T_c everywhere else). We then find

$$C = n\, C_0 \sum_k \frac{1}{(k^2 + a)^2} = n\, C_0 \int \frac{d^D k}{(2\pi)^D} \cdot \frac{2}{(k^2 + a)^2}$$

$$\propto a^{(D-4)/2}$$

$$\propto (T - T_c)^{-(4-D)/2} \qquad \ldots \text{(22.2.4)}$$

leading to

$$\alpha = \frac{4-D}{2} \qquad \ldots \text{(22.2.5)}$$

The correlation function $G(k, a)$ is easily evaluated as

$$G(k, a) = \langle \varphi_\alpha (k)\ \varphi_\alpha(-k) \rangle$$

$$= \frac{1}{Z_a} \int \prod_{\mathbf{P},\beta} d\, \varphi_\beta (\mathbf{p})\ \varphi_\alpha(k)\ \varphi_\alpha(-k)\ \exp \left\{ \sum_{-\mathbf{p},\alpha} \frac{1}{2}\ (a + p^2) \right\}$$

$$= \frac{1}{k^2 + a} \qquad \ldots \text{(22.2.6)}$$

At $a = 0$, $G(k) \sim k^{-2}$ and thus $\eta = 0$. The susceptibility is obtained from the $k = 0$ value of $G(k, a)$ and thus

$$\chi (a) = G(k = 0, a) = \frac{1}{a} \propto (T - T_c)^{-1}$$

leading to $\gamma = 1$. The Fourier transform of Eq. (22.2.6) clearly shows that $a^{1/2}$ is the inverse correlation length. Thus $\xi^{-1} \propto (T - T_c)^{1/2}$, leading to $\nu = 1/2$. The scaling laws of Eqs. (21.6.8b) and (21.6.8d) are thus seen to be satisfied. The Gaussian model is not defined for $T < T_c$ and hence we cannot make any comments about the value of β in this model.

22.3 The Spherical Limit

We will now discuss a model where the quartic term is included but in a manner that still makes the model exactly solvable. We begin by observing that in Eq. (22.1.4), the first two terms on the right hand side are of order n, but the third term is of order n^2 and hence in the limit $n \to \infty$, the model would make no sense unless we require the coupling constant b to behave as n^{-1}. Accordingly we write $b = \frac{u}{n}$ and rewrite Eq. (22.1.4) as

$$F(\varphi) = \int d^D x \left[\frac{a}{2}\sum \varphi\alpha^2 + \frac{1}{2}\sum (\vec{\nabla}\varphi\alpha)^2 + \frac{u}{4n}(\sum \varphi\alpha^2)^2\right]$$

$$= \int d^D x \left[\left(\frac{a}{2} + \frac{u}{4} <\varphi_\alpha^2>\right)\sum \varphi_\alpha^2 + \frac{1}{2}\sum (\vec{\nabla}\varphi_\alpha)^2\right.$$

$$\left. + \frac{u}{4n}\left(\sum \varphi_\alpha^2 - n <\varphi_\alpha^2>\right)\sum \varphi_\alpha^2\right] \quad \text{... (22.3.1)}$$

In the above $<\varphi_\alpha^2>$ stands for the average value of $\int \varphi_\alpha^2\, dx$. On the right hand side of the above equation, the first two terms are clearly of order n, but the third term is $O(1)$ and hence in the limit $n \to \infty$, we can drop the last term and have a model defined by the free energy

$$F(\varphi) = \int d^D x \left[\frac{a'}{2}\sum \varphi_\alpha^2 + \frac{1}{2}\sum (\vec{\nabla}\varphi_\alpha)^2\right] \quad \text{... (22.3.2)}$$

where

$$a' = a + \frac{u}{2} <\varphi_\alpha^2> \quad \text{... (22.3.3)}$$

The new critical point is obtained when a' goes to zero. From Eq. (22.3.2), it is clear that

$$<\varphi_\alpha(k)\,\varphi_\alpha(-k)> \;=\; \frac{1}{k^2+a'} \qquad \text{... (22.3.4)}$$

(just as in the case of the Gaussian model of the previous Section) and hence

$$<\varphi_\alpha^2> \;=\; \int \frac{d^Dk}{(2\pi)^D}\,\frac{1}{k^2+a'} \qquad \text{... (22.3.5)}$$

With $a = a_0(T - T_c)$, Eq. (22.3.3) becomes

$$a' = a_0(T - T_c) + \frac{u}{2}\int \frac{d^Dk}{(2\pi)^D}\cdot\frac{1}{k^2+a'} \qquad \text{... (22.3.6)}$$

The new transition temperature is T_c', where $a' = 0$. Thus,

$$0 = a_0(T_c' - T_c) + \frac{u}{2}\int \frac{d^Dk}{(2\pi)^D}\cdot\frac{1}{k^2} \qquad \text{... (22.3.7)}$$

Subtracting Eq. (22.3.7) from Eq. (22.3.6).

$$a' = a_0\,(T - T'_c) + \frac{u}{2}\int \frac{d^Dk}{(2\pi)^D}\left[\frac{1}{k^2+a'} - \frac{1}{k^2}\right]$$

or

$$a_0\,(T - T'_c) = a' + \frac{ua'}{2}\int \frac{d^Dk}{(2\pi)^D}\,\frac{1}{k^2(k^2+a')}$$

$$= a' + Aa'^{\frac{D}{2}-1} \qquad \text{... (22.3.8)}$$

where A is a constant independent of a'. Since we are interested in small values of a', the right hand side of Eq. (22.3.8) is dominated by $a'^{\frac{D}{2}-1}$ if $4 > D > 2$. and by a' if $D > 4$ Hence,

$$a' \propto (T - T'_c)^{\frac{2}{D}-2} \text{ if } 2 < D < 4$$

$$a' \propto T - T'_c \text{ if } D > 4 \qquad \text{... (22.3.9)}$$

The above manipulations reveal a vital fact—the fluctuation contribution from the quartic term completely dominates the quadratic term for $D < 4$. This makes perturbative calculations around the Gaussian model meaningless for $D < 4$.

Just as in the Gaussian model of the previous Sections, the correlation length is given by

$$\xi = a^{-1/2}\begin{cases} \propto (T-T_c)^{-1/D-2} & for\ 2 < D < 4 \\ \propto (T-T_c)^{-1/2} & for\ D > 4 \end{cases} \quad \text{... (22.3.10)}$$

leading to

$$\nu = \frac{1}{D-2} \qquad 2 < D < 4$$

$$\nu = 1/2 \qquad D > 4 \quad \text{... (22.3.11)}$$

The anomalous dimension index η is clearly zero and $\gamma = 2\nu$.

The calculation of specific heat is more difficult. This can be appreciated from Eq. (22.1.23). Using only the quadratic part of the free energy [Eq. (22.3.2)], we can easily evaluate the specific heat. To do this, we write Eq. (22.1.23) as

$$C = a_0^2\, T \int \frac{d^D p}{(2\pi)^D}\frac{d^D q}{(2\pi)^D} < \sum_{\alpha=1}^{n} \varphi_\alpha(q)\, \varphi_\alpha(-q) \sum_{\beta=1}^{n} \varphi_\beta(p)\, \varphi_\beta(-p)>_c$$

$$= n\, a_0^2\, T \int \frac{d^D P}{(2\pi)^D} [< \varphi(p)\ \varphi(-p) >]^2$$

$$\propto n\, C_0 \int \frac{d^D P}{(2\pi)^D} \left(\frac{1}{p^2 + a'}\right)^2$$

$$\propto n\, C_0\, a'^{(D/2)-2} \quad \text{... (22.3.12)}$$

At this point it would appear on using Eq. (22.3.10) that $\alpha = \dfrac{4-D}{D-2}$ for $4 > D > 2$. If we check the scaling law $D\nu = 2 - \alpha$ at this point, we would find that it is not satisfied. Thus, the evaluation of the specific heat is a more complicated problem for $n \to \infty$.

To understand the source of the above difficulty, we imagine performing the average in Eq. (22.1.23) by using the full free energy functional of Eq. (22.3.1). This can be done by expanding the part $\exp\left\{\frac{u}{4n}\int d^Dx\,(\sum \varphi_\alpha^2 - n > \varphi_\alpha^2 >^2) \sum \varphi_\alpha^2\right\}$ in a power series.

The first term in such an expansion requires the average

$$\frac{u}{4n} a_0^2 \Gamma \int d^Dx\, d^Dx\, d^Dy\, d^Dz < \sum \varphi_\alpha^2(x) \sum \varphi_\alpha^2(z) (\sum \varphi_\alpha^2(z) - n < \varphi_\alpha^2 >)$$

$$\sum \varphi_\alpha^2(y) >$$

These being Gaussian averages, a little thought will convince the reader that there is a part of the average which is proportional to n^2 and hence the entire term is proportional to n, which is just as big as the zeroth order term found in Eq. (22.3.12). The a'-dependence of the term is $(a'^{D/2-2^2})$. If we now imagine considering the higher order terms, a trend will be seen to emerge. At the second order the combinatorial factor yields n^3 while the coupling constant is n^{-2}, giving once again a part proportional to n. The a' dependance is $(a'^{(D/2)-2})^3$. The result is a geometric series

$$C \propto nD_0\, a^{D/2-2}\left[1 - \frac{u}{4}\, a^{D/2-2} + \left(\frac{u}{4}\, a^{D/2-2}\right)^2 \cdots\right]$$

$$= nC_0\, a'^{D/2-2}\left(1 + \frac{u}{4}\, a'^{D/2-2}\right)^{-1}$$

$$= \frac{4n\, C_0/u}{1 + \frac{u}{u}\, a'^{2-(D/2)}}$$

$$\cong 4n\, C_0/n\left(1 - \frac{4}{u}\, a'^{2-(D/2)} + \ldots\right) \qquad \ldots \textbf{(22.3.13)}$$

Using the T - T_c dependence of a; from Eq. (22.3.10), we find

$$\alpha = \frac{D-4}{D-2} \text{ for } 4 > D > 2 \qquad \ldots \textbf{(22.3.14)}$$

The hyperscaling law $D\nu = 2 - \alpha$ is now satisfied. The $n \to \infty$ limit is known as the spherical model.

Chapter 23
Introduction to the Renormalization Group

23.1 Introduction

The problem with critical phenomena lies in the very large number of degrees of freedom which makes perturbation theory intractable. As is customary, we will call the free energy F of Eq. (22.1.4), the Hamiltonian H from now and we will see that the constant b has to have a special value for perturbation theory to work, implying that there is a special Hamiltonian for the perturbation theory to be effective. The meaning of that special Hamiltonian will be made clear in this chapter.

To get a preliminary feel for the techniques involved, we follow the early work of Kadanoff and consider a two-dimensional Ising model. The Hamiltonian for this model is $H=-J\Sigma\, S_{i,j}\, S_{i+1,j} + \Sigma\, S_{i,j}\, S_{i,j+1}$ and as long range order sets in near the critical point, the spins at different sites get correlated and we can imagine (see Fig. 23.1) four spins of the four vertices of a square (a block) acting conjointly to form an effective single spin $\widetilde{S}$. The original spins S have the value $S = \pm 1$ and the new spins are assumed to be properly rescaled to have the value ± 1 once more. The block spins (a block of spins giving a single spin) are assumed to have nearest neighbour interaction like the original spins and the new Hamiltonian is

$$H' = -{}^{1}J_1 \Sigma \left(\widetilde{S}_{i,j}\ \widetilde{S}_{i+1,j} + \widetilde{S}_{i,j}\ \widetilde{S}_{i,j+1}\right)$$

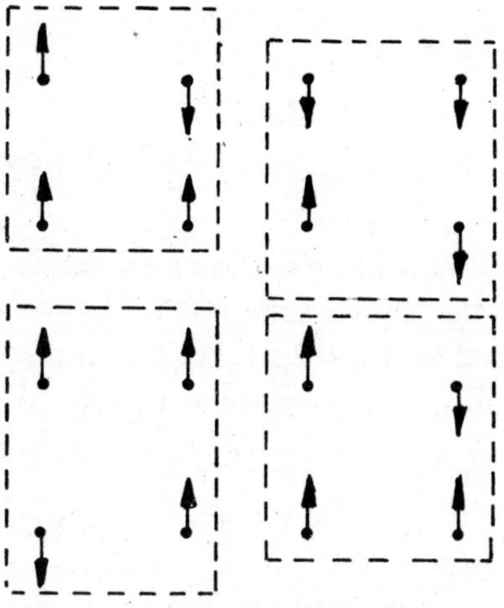

Fig. 23.1

where J_1 is the new interaction strength between the block spins. We expect J_1 to be some analytic function of J

$$J_1 = f(J) \qquad \text{... (23.1.1)}$$

We now consider what happens to the correlation length as we construct the block spins. The block spins are twice as far removed as the original spins and hence the correlation length (measured in units of lattice spacing) is halved with the formation of the block spins, i.e.

$$\xi\{f(J)\} = \frac{1}{2}\,\xi\,(J) \qquad \text{... (23.1.2)}$$

Now imagine doing this block spin construction repeatedly—a J_1 generates a J_2 and so on and eventually it is possible that J does not change any more—we have then reached a fixed point of the transformation (23.1.1)—and the corresponding value J_c of J satisfies

$$J_c = f(J_c) \qquad \text{... (23.1.3)}$$

At $J = J_c$, $\xi(J_c) = 1/2\,\xi\,(J_c)$, which implies that ξ is infinite at $J = J_c$. Hence $J = J_c$ corresponds to the critical point.

To work near the critical point, we examine $f(J)$ near $J = J_c$ and expand as a Taylor series

$$f(J) = f(J_c) + (J - J_c)\,f'(J_c) + \ldots$$

$$= f(J_c) + \lambda\,(J - J_c) + \ldots$$

$$f(J) - f(J_c) = \lambda(J - J_c) + O(J - J_c)^2 \qquad \text{... (23.1.4)}$$

Assuming that near the critical point J_c

$$\xi(J) \propto (J - J_c)^{-\nu} \qquad \text{... (23.1.5)}$$

we find

$$\frac{\xi(f(J))}{\xi(J)} = \left[\frac{(f(J)-f(J_c))}{(J-J_c)}\right]^{-\nu} = \lambda^{-\nu} \quad \text{... (23.1.6)}$$

Hence the critical exponent ν can be found if λ is known. To know λ, we must know $f(J)$, which is the central feature of the above scheme.

The philosophy of the approach is summarized in Eq. (23.1.2). The transformation reduces the correlation length. Repeating the process n times, we would have

$$\xi(J_n) = 2^{-n}\,\xi(J) \quad \text{... (23.1.7)}$$

One could always choose an n such that $\xi(J_n)$ is small and conventional perturbation theory can be employed to calculate it.

23.2 The Basic Operations of the Renormalization Group

The considerations of the previous section will have to be more carefully implemented on the Hamiltonian for the n-component field, which in momentum space has the form [Eq. (22.1.19) with the magnetic field equal to zero]

$$H = \frac{1}{2}\sum_{\substack{\alpha=1\\ k<\Lambda}}^{n} \varphi_\alpha(\mathbf{k})\,\varphi_\alpha(-\mathbf{k})\,(m^2+k^2)$$

$$+\frac{\lambda}{4!}\sum_{\alpha,\beta,\mathbf{p}_i} \varphi_\alpha(\mathbf{p}_1)\,\varphi_\alpha(\mathbf{p}_2)\,\varphi_\beta(\mathbf{p}_3)\,\varphi_\beta(\mathbf{p}_4)\,\delta^{(D)}\left(\sum \mathbf{p}_i\right) \quad \text{... (23.2.1)}$$

The momentum range is restricted to $k < \Lambda$, where Λ is a large momentum which corresponds to the lattice spacing in the discrete version of the continuum theory. As discussed in detail already, the aim is to calculate the partition function Z and the n-point correlation functions

$$Z = \sum_{[\varphi]} e^{-H(\varphi)} \quad \text{... (23.2.2)}$$

$$S_c^{(2)}(k) = \langle \varphi(k)\, \varphi(-k) \rangle = \frac{1}{Z} \sum_{[\varphi]} \varphi(k)\, \varphi(-k)\, e^{-H(\varphi)} \qquad \text{... (23.2.3)}$$

and so on.

The lesson of the previous section was that the degrees of freedom have to be reduced by increasing the length scale—in this context it implies that we must remove the Fourier components with momentum between Λ and Λ/b where $b > 1$. With the degrees of freedom reduced, one now scales the momentum by a factor b, so that the cutoff is restored to its original value Λ. The field φ is rescaled as well (corresponding to the spin always being ± 1 in the previous section) so that the gradient term in the Hamiltonian is unaltered. These three steps –

i) Integrating out the degrees of freedom
ii) Rescaling momenta
iii) Rescaling the fields,

constitute the renormalization group.

If we call the field rescaling factor ζ_b, then the process of the renormalization group is embodied in the operation

$$\int \left[\prod_{\alpha=1}^{n} \prod_{\Lambda/b < k' < \Lambda} D\, \varphi_\alpha(k')\, e^{-H} \right]_{\varphi_\alpha(k) \to \zeta_b \varphi_\alpha(bk)}$$

The result is a new exponential factor $e^{-H'}$. If we call H' the transformed Hamiltonian $R_b H$, then

$$e^{-R_b H} = \int \left[\prod_{\alpha=1}^{n} \prod_{\Lambda/b < k' < \Lambda} D\, \varphi_\alpha(k')\, e^{-H} \right]_{\varphi_\alpha(k) \to \zeta_b \varphi_\alpha(b k)} \qquad \text{... (23.2.4)}$$

If we carry out two successive renormalization group transformations, then $\varphi_\alpha(k)$ goes to $\zeta_{b'} \zeta_b \varphi_\alpha(b'bk)$. The group property requires that $R_{b'} R_b = R_{b'b}$ and hence

$$\zeta_{bb'} = \zeta_{b'} \zeta_b \qquad \text{... (23.2.5)}$$

This is achieved by the choice

$$\zeta_b = b^y \qquad \text{... (23.2.6)}$$

We will write y as

$$y = 1 - \frac{\eta}{2} \qquad \text{... (23.2.7)}$$

and later show that η is indeed the anomalous dimension η of Chapter 21.

We now need to know how the partition function and the correlation function transform under the action of R_b.

It should be clear from Eqs. (23.2.2) and (23.2.4) that the partition function is unchanged under the transformation. The correlation function $G_c^{(2)}(k, H)$ transforms from its definition [Eq. (23.2.2)] according to

$$\begin{aligned} G_c^{(2)}(k, H) &= [\zeta_b]^2\, G_c^{(2)}(bk, R_bH) \\ &= b^{2-\eta}\, G_c^{(2)}(bk, R_bH) \end{aligned} \qquad \text{... (23.2.8)}$$

23.3 Fixed Points and Scaling Laws

We now ask what happens as we repeat the group operation of the previous section. Each step generates a new Hamiltonian and after l steps

$$H_l = (R_b)^l\, H$$

A possibility after repeated applications is the emergence of a fixed point Hamiltonian H^*, which does not change on application of R_b, i.e.

$$R_bH^* = H^* \qquad \text{... (23.3.1)}$$

The existence of a fixed point cannot be guaranteed from any general considerations.

What changes as the renormalization group operation is performed? It is the coupling constants in the Hamiltonian (in principle, we imagine the Hamiltonian to contain all possible terms consistent with the symmetry) which we denote by K_α that change under the actions of the renormalization group and the fact that H has become a fixed point Hamiltonian implies that the constants have acquired their fixed point values K_α^* (note that m^2, $\lambda^4/4!$ ... etc. of our Hamiltonian (Eq. 23.2.1) constitute the set K_α').

If we linearize the transformation $(K'_\alpha) = R_b\,(K_\alpha)$ near the fixed point, we get

$$K'_\alpha = K_\alpha^* + \left.\frac{\partial K'_\alpha}{\partial K_\beta}\right|_{\{K^*\}} \delta K_\beta$$

or

$$\delta K'_\alpha = T_{\alpha\beta}\, \delta K_\beta \qquad \text{... (23.3.2)}$$

where

$$\delta K_\alpha = K_\alpha - K_\alpha^* \qquad \text{... (23.3.3)}$$

and

$$T_{\alpha\beta} = \left.\frac{\partial K'_\alpha}{\partial K_\beta}\right|_{\{K^*\}} \qquad \text{... (23.3.4)}$$

If the eigenvalues of matrix T_{ab} are $\{\lambda_{(i)}\}$ and the eigenvectors $\Psi_{(i)}$, then

$$\sum_\alpha T_{\alpha\beta}\, \Psi_{(i)\alpha} = \lambda_i \Psi_\beta \qquad \text{... (23.3.5)}$$

We now construct the fields $u_{(i)}$ out of the set δK_α by using the linear transformation

$$u_i = \sum_\alpha \Psi_{(i)\alpha}\, \delta K_\alpha \qquad \text{... (23.3.6)}$$

The fields $\{u_i\}$ transform as

$$u'_i = \sum_{\alpha} \Psi_{(i)\alpha} \delta K'_{\alpha}$$

$$= \sum_{\alpha, \beta} \Psi_{(i)\alpha} T_{\alpha\beta} \delta K_{\beta}$$

$$= \sum_{\beta} \delta K_{\beta} \sum_{\alpha} T_{\alpha\beta} \Psi_{(i)\alpha}$$

$$= \lambda_{(i)} \sum_{\beta} \Psi_{(i)\beta} \delta K_{\beta}$$

$$= \lambda_i u_i \qquad \text{... (23.3.7)}$$

The $\{u_i\}$ transform, under the action of R by a multiplicative factor and are known as scaling fields.

We note that there are three possibilities:

i) $\lambda_i > 1$, the corresponding scaling variables grow under the action of the renormalization group. These are the *relevant* variables.

ii) $\lambda_i < 1$, the corresponding scaling variables shrink and ultimately vanish under the action of the renormalization group. These are the *irrelevant* variables.

iii) $\lambda_i = 1$, the corresponding scaling variables are called *marginal*.

The correlation length transforms under the action of the group as

$$\xi\,(R_b H) = \xi\,(H)/b \qquad \text{... (23.3.8)}$$

At criticality ξ is infinite and does not transform under the group operation. The fixed point Hamiltonian does not transform under the group operation and in that sense the fixed point Hamiltonian is critical. The general Hamiltonian can be written as

$$H = H^* + \sum_i O_i u_i \qquad \text{... (23.3.9)}$$

where O_i are the operators corresponding to the variables u_i (which, in turn, are determined by K_α). It is immediately clear that $(R_b)^l\, H \rightarrow H^*$ as $l \rightarrow \infty$, if the set $\{i\}$ constitute the irrelevant set and the relevant variables u_i are held equal to zero. Hence at criticality, the relevant fields must be equal to zero. Turning now to ordinary critical phenomena, we note that for criticality the external magnetic field h must be equal to zero and the temperature T must be equal to T_c. If we construct the variable $t = (T - T_c)/T_c$, then $t = 0$ and $h = 0$ at criticality. Hence there must be two relevant variables, t and H, to describe the usual critical point. All other variables must be irrelevant. We anticipate the absence of marginal operations since the existence of a marginal variable u_m would imply that

$$H = H^* + c_m\, O_m\, u_m$$

would remain invariant for all values of the constant c_m and this would generate a line of critical points. The eigenvalues λ corresponding to the scaling fields t and h are denoted by λ_t and λ_h respectively and we denote the set of irrelevant variables eigenvalues by λ_1, λ_2, ...

The free energy per unit volume is determined by the scaling variables t, h, λ_1, λ_2, ... since ln Z does not change under the renormalization group transformation, the free energy per unit volume

$$F = \frac{1}{V} \ln Z$$

has the transformation

$$F(x) = b^{-D} F(x') \qquad \text{... (23.3.10)}$$

i.e.

$$\begin{aligned} F(t, h, u_1, u_2, ...) &= b^{-D} F(\lambda_t t, \lambda_h h, \lambda_1 u_1, \lambda_2 u_2, ...) \\ &= b^{-nD} J(\lambda_t^n t, \lambda_h^n h, \lambda_1^n u_1, \lambda_2^n u_2, ...) \end{aligned} \qquad \text{... (23.3.11)}$$

For sufficiently large n, we can choose $\lambda_t^n t = 1$, so that

$$\begin{aligned} F(t, h, u_1, u_2, \ldots) &= \left(b^{-D}\right)^{\left(\frac{\ln(1/t)}{\ln \lambda_t}\right)} \times \\ &\times F\left(1, \lambda_h^{\left(\frac{\ln(1/t)}{\ln \lambda_t}\right)} h, \lambda_l^{\left(\frac{\ln(1/t)}{\ln \lambda_t}\right)} u_t, \ldots\right) \\ &= t^{D/y_t} F\left(1, \frac{h}{t^{y_h/y_t}}, \frac{u_1}{t^{y_1/y_t}}, \ldots\right) \end{aligned} \qquad \text{... (23.3.12)}$$

where

$$y_t = \ln \lambda_t / \ln b > 0$$

$$y_h = \ln \lambda_h / \ln b > 0$$

while,

$$y_1 = \ln \lambda_1 / \ln b, \; y_2 = \ln \lambda_2 / \ln b, \; \ldots \text{ all } < 0$$

For sufficiently small t (i.e. close to the critical point. $u_1/t^{y_1/y_t} \to 0$ and we have

$$F(t, h) = t^{D/y_t} \, F(h/t^{y_h/y_t}) \quad \ldots \textbf{(23.3.13)}$$

Since the specific heat behaves as $t^{-\alpha}$, the free energy should behave as $t^{2-\alpha}$ for $h = 0$ and that leads to

$$2 - \alpha = \frac{D}{y_t} \quad \ldots \textbf{(23.3.14)}$$

The magnetization $M = \left.\frac{\partial F}{\partial h}\right|_{h=0} \propto t^{\beta}$ and we have from Eq. (23.3.13)

$$\beta = \frac{D - y_h}{y_t} \quad \ldots \textbf{(23.3.15)}$$

The susceptibility $\chi = \frac{\partial M}{\partial h} \propto t^{-\gamma}$ and differentiating the free energy twice

$$\gamma = \frac{2y_h - D}{y_t} \quad \ldots \textbf{(23.3.16)}$$

From the above relations we readily see that

$$\gamma + 2\beta = \frac{D}{y_t} = 2 - \alpha$$

or

$$\gamma + 2\beta + \alpha = 2$$

which establishes the scaling law [Eq. (21.6.8a)].

Differentiating Eq. (23.3.13)

$$M = t^{D - y_h/y_t} F'\left(\frac{h}{t^{y_h/y_t}}\right) \qquad \text{... (23.3.17)}$$

As $t \to 0$, $h/t^{y_h/y_t} \to \infty$ and to recover the behaviour $M \sim H^{1/\delta}$, the asymptotic behaviour of $F'(x)$ for large x has to be $x^{1/\delta}$, with δ the chosen to remove the t-dependence. This yields,

$$y_h/y_t \cdot \frac{1}{\delta} = \frac{D - y_h}{y_t}$$

$$\frac{1}{\delta} = \frac{D - y_h}{y_h} \qquad \text{... (23.3.18)}$$

leading to

$$\beta(1 + \delta) = \beta\delta\left(1 + \frac{1}{\delta}\right) = \frac{D - y_h}{y_t} \cdot \frac{y_h}{D - y_h} \cdot \frac{D}{y_h}$$

$$= \frac{D}{y_t} = 2 - \alpha$$

or

$$\alpha + \beta(1 + \delta) = 2,$$

which is another scaling law [Eq. (21.6.8c)].

The correlation length ξ transforms as

$$\xi(t) = \frac{1}{b} \xi(\lambda_t \cdot t) \qquad \text{... (23.3.19)}$$

under the renormalization group transformation. Repeated iterations lead to

$$\xi(t) = \frac{1}{b} \xi(\lambda_t \cdot t) = \ldots = \frac{1}{b^n} \xi(\lambda_t^n \cdot t)$$

Choosing $\lambda_t^n \cdot t = 1$, we have

$$\xi(t) = t^{-1/y_t} \xi(1) \qquad \text{... (23.3.20)}$$

identifying the exponent ν as

$$\nu = 1/y_t \,. \qquad \text{... (23.3.21)}$$

From Eq. (23.3.14) we establish the scaling law

$$D\nu = 2 - \alpha$$

We now turn to the transformation of the correlation function [Eq. (23.2.8)]

$$G_c^{(2)}\,(k,\, t) = b^{2-\eta}\; G_c^{(2)}\,(bk,\; \lambda_t\, t) \qquad \text{... (23.3.22)}$$

At $T = T_c$ $(t = 0)$, it is clear that

$$G_c^{(2)}\,(k) \;\propto\; k^{-2+\eta}$$

and thus η is indeed the anomalous dimension exponent. On the other hand at $k = 0$, repeated application of Eq. (23.2.22) implies $G_c^{(2)} = (b^n)^{2-\eta}\; G_c^{(2)}\,(\lambda_t^n\, t) \;\propto\; t^{-(2-\eta)/y_t}$ as we choose $\lambda_t^n \,.\, t = 1$. Since $G_c^{(2)}\,(k = 0)$ is the susceptibility, we have

$$\gamma = (2 - \eta)\,\frac{1}{y_t} = (2 - \eta)\;\nu$$

which is another scaling law. Thus, using general principle of the renormalization group transformation, we have succeeded in establishing all the scaling laws introduced in Chapter 21.

23.4 Gaussian Model

In the Hamiltonian of Eq. (23.2.1) we set $\lambda = 0$ and, on carrying out the renormalization group transformation, readily find

$$R_b\; H = \sum_{k < \Lambda/b} (m^2 + k^2)\;\varphi\,(k)\;\varphi\,(-k)$$

$$= \sum_{k' < \Lambda} \zeta_b^2\;(m^2 + k'^2/b^2)\;\varphi\,(k')\;\varphi\,(-k') \qquad \text{... (23.4.1)}$$

The requirement of fixing ζ_b is to keep the coefficient of the k^2 term unchanged (i.e. always equal to 1). Thus from Eq. (23.4.1), $\zeta_b' = b$ and the mass transforms as

$$m'^2 = b^2 m^2 \qquad \text{... (23.4.2)}$$

with $\zeta_b = b^{1-\eta/2}$, we have $\eta = 0$. From Eq. (23.4.2) it is immediately clear that $b^2 = \lambda_t$ since $m^2 \propto t$.

This leads to the exponent ν as

$$\nu = \frac{1}{y_t} = \frac{\ln b}{\ln b^2} = \frac{1}{2}$$

The specific heat exponent α follows from

$$2 - \alpha = \frac{D}{y_t} = \frac{D}{2}$$

or $\alpha = \varepsilon/2$, where $\varepsilon = 4 - D$.

The susceptibility exponent γ follows from

$$\gamma = (2 - \eta)\, \nu = 2\nu = 1.$$

The results are in accordance with those of Chapter 22.

23.5 Recursion Relations for the φ^4 Theory

In this section, we demonstrate how the renormalization group works in momentum space by working with the scalar field Landau-Ginzburg model which has the Hamiltonian [Eq. (23.2.1) with $n = 1$]

$$H = \int d^D x \left[\frac{r}{2}\varphi^2 + (\nabla\phi)^2 + \frac{\lambda}{4!}\varphi^2\right]$$

$$= \sum_k^{\Lambda}\left(\frac{r}{2} + \frac{k^2}{2}\right)\varphi(\mathbf{k})\,\varphi(-\mathbf{k}) + \frac{\lambda}{4!}\sum_{\mathbf{p}_1,\mathbf{p}_2,\mathbf{p}_3}\varphi(\mathbf{p}_1)\,\varphi(\mathbf{p}_2)$$

$$\times\,\varphi(\mathbf{p}_3)\,\varphi(-\mathbf{p}_1, -\mathbf{p}_2, -\mathbf{p}_3) \qquad \text{... (23.5.1)}$$

To implement the renormalization group process for the evaluation of the partition function

$$Z = \int D\varphi \ e^{-H}$$

we need to

i) integrate over the Fourier components $\varphi(k)$ which have k lying between Λ/b and Λ where b is a number greater than unity. This reduces the momentum cutoff to Λ/b.

ii) rescale the momenta by b so that the cutoff is restored to the previous value Λ.

iii) rescale the fields $\varphi(x)$, so that the renormalization group correctly focusses on the problem at hand—e.g. gives the correct form for the two point correlation function at $T = T_c$.

We proceed by splitting each field $\varphi(x)$ as

$$\varphi(x) = \frac{1}{(2\pi)^D}\int^{\Lambda/b} \varphi(\mathbf{k})\, e^{i\mathbf{k}\cdot\mathbf{x}}\, d^Dk + \frac{1}{(2\pi)^D}\int_{\Lambda/b}^{\Lambda} \varphi(\mathbf{k})\, e^{i\mathbf{k}\cdot\mathbf{x}}\, d^Dk$$

$$= \frac{1}{(2\pi)^D}\int \varphi_< \,.\, e^{ik\cdot x}\, d^Dk + \frac{1}{(2\pi)^D}\int \varphi_> \cdot e^{ik\cdot x}\, d^Dk \qquad \text{... (23.5.2)}$$

where $\varphi_<$ refers to the Fourier components with wavenumber less than Λ/b, where $\varphi_>$ refers to the components lying between Λ/b and Λ. The Hamiltonian can now be written as

$$H = \frac{1}{2}\sum_{k<\Lambda/b}(r + k^2)\,\varphi_<(k)\,\varphi_<(-k) + \frac{1}{2}\sum_{\Lambda/b<k<\Lambda}(r + k^2)\,\varphi_>(k)\,\varphi_>(-k)$$

$$+ \frac{\lambda}{4!}\sum \varphi_<(\mathbf{p}_1)\,\varphi_<(\mathbf{p}_2)\,\varphi_<(\mathbf{p}_3)\,\varphi_<(-\mathbf{p}_1 - \mathbf{p}_2 - \mathbf{p}_3)$$

$$+ 4\cdot\frac{\lambda}{4!}\sum \varphi_<(\mathbf{p}_1)\,\varphi_<(\mathbf{p}_2)\,\varphi_<(\mathbf{p}_3)\,\varphi_>(-\mathbf{p}_2 - \mathbf{p}_2 - \mathbf{p}_3)$$

$$+ 6\cdot\frac{\lambda}{4!}\sum \varphi_<(\mathbf{p}_1)\,\varphi_<(\mathbf{p}_2)\,\varphi_>(\mathbf{p}_3)\,\varphi_>(-\mathbf{p}_1 - \mathbf{p}_2 - \mathbf{p}_3)$$

$$+ 4 \cdot \frac{\lambda}{4!} \sum \varphi_>(\mathbf{p}_1)\, \varphi_>(\mathbf{p}_2)\, \varphi_>(\mathbf{p}_3)\, \varphi_<(-\mathbf{p}_1 - \mathbf{p}_2 - \mathbf{p}_3)$$

$$+ \frac{\lambda}{4!} \sum \varphi_>(\mathbf{p}_1)\, \varphi_>(\mathbf{p}_2)\, \varphi_>(\mathbf{p}_3)\, \varphi_>(-\mathbf{p}_1 - \mathbf{p}_2 - \mathbf{p}_3)$$

$$= H^< + H_2^> + H_4^> \qquad \text{... (23.5.3)}$$

where $H^<$ is that part of the Hamiltonian which involves only the fields $\varphi_<$, $H_2^>$ is the quadratic part involving the fields $\varphi_>$ and $H_4^>$ is the part of the Hamiltonian involving quartic terms at least one of which is $\varphi_>$. We now need to evaluate

$$Z = \int D\varphi_<\, D\varphi_>\, e^{-(H^< + H_2^> + H_4^>)}$$

$$= \int D\varphi_<\, e^{-H^<} \int D\varphi_>\, e^{-(H_2^> + H_4^>)} \qquad \text{... (23.5.4)}$$

The evaluation of the functional integration over $\varphi_>$ leads to a function of $\varphi_<$ which we write as $e^{-\overline{H}}$. Noting that H_4 is $O(\lambda)$, we can expand the exponential in powers of $H_4^>$ and obtain

$$e^{-\overline{H}} = \int D\varphi_>\, e^{-H_2^>} \left[1 - H_4^> + \frac{1}{2} H_4^{>2} + \dots\right]$$

$$= \left(\int D\varphi_>\, e^{-H_2^>}\right) \left[1 - \langle H_4^> \rangle + \frac{1}{2} \langle H_4^{>2} \rangle + \dots\right] \qquad \text{... (23.5.5)}$$

where

$$\langle \dots \rangle = \int D\varphi_>\, (\dots)\, e^{-H_2^>} \Big/ \int D\varphi_>\, e^{-H_2^>} \qquad \text{... (23.5.6)}$$

We can then express $\overline{H}$ as

$$\overline{H} = -\ln \left[\int D\varphi_>\, e^{-H_2^>}\right] \left[1 - \langle H_4^> \rangle + \frac{1}{2} \langle H_4^{>2} \rangle + \dots\right]$$

$$= \langle H_4^> \rangle - \frac{1}{2} \left\{\langle H_4^{>2} \rangle - \langle H_4^> \rangle^2\right\} + \text{constant} + O(\lambda^3) \qquad \text{... (23.5.7)}$$

We want exactly those parts of $\overline{H}$ which are functions of φ and hence ln $[\int D\varphi_> e^{-H_2^>}]$ which is a constant is not important for our purpose (note that it would be important for an exact calculation of the free energy). At $O(\lambda)$, we need to extract the $\varphi_<$-dependent part of $<H_4>$. This is easily seen to be [from Eq. (23.5.3)]

$$<H_4^> > = 6 \cdot \frac{\lambda}{4!} < \sum \varphi_<(\mathbf{p}_1)\, \varphi_<(\mathbf{p}_2)\, \varphi_>(\mathbf{p}_3)\, \varphi_>(-\mathbf{p}_1 - \mathbf{p}_2 - \mathbf{p}_3) >$$

$$= 6 \cdot \frac{\lambda}{4!} < \sum_{\mathbf{p}_1} \varphi_<(\mathbf{p}_1)\, \varphi_<(\mathbf{p}_1)\, \varphi_>(\mathbf{p}_2)\, \varphi_>(-\mathbf{p}_2) >$$

$$= 6 \cdot \frac{\lambda}{4!} < \sum_{\mathbf{p}_1} \varphi_<(\mathbf{p}_1)\, \varphi_<(\mathbf{p}_1) \int_{\Lambda/b}^{\Lambda} \frac{dp_2}{(2\pi)^D} \frac{1}{p_2^2 + m^2} >$$

$$= 6 \cdot \frac{\lambda}{4!} \left[\int_{\Lambda/b}^{\Lambda} \frac{d^D p_2}{(2\pi)^D} \frac{1}{p_2^2 + m^2}\right] \sum_{\mathbf{p}_1} \varphi_<(\mathbf{p}_1)\, \varphi_<(-\mathbf{p}_1) \qquad \ldots \textbf{(23.5.8)}$$

This term then modifies (i.e. renormalizes) the quadratic part of the Hamiltonian H. The quartic part of H is unaffected at this order.

If we now turn to the term $\frac{1}{2}\left[<H_4^{>2} > - < H_4^> >^2\right] = \frac{1}{2} < H_4^{>2}>_c$, we note that it produces contributions to the quadratic and quartic parts of H. Since we are interested in the lowest order calculations, we ignore at present the contributions to the quadratic part and focus on the quartic contribution instead. Thus, the quartic contribution to H from $< H_4^{>2} >_c / 2$ is

$$\frac{1}{2}\left(6 \times \frac{\lambda}{4!}\right)^2 \Big[\times < \sum \varphi_<(\mathbf{p}_1)\, \varphi_<(\mathbf{p}_2)\, \varphi_>(\mathbf{p}_3)\, \varphi_>(-\mathbf{p}_1 - \mathbf{p}_2 - \mathbf{p}_3) \Big|$$

$$\times \sum \varphi_<(\mathbf{q}_1)\, \varphi_<(\mathbf{q}_2)\, \varphi_>(\mathbf{q}_3)\, \varphi_>(-\mathbf{q}_1 - \mathbf{q}_2 - \mathbf{q}_3) >_c$$

$$= \frac{1}{2}\left(6 \times \frac{\lambda}{4!}\right)^2 \sum \varphi_<(\mathbf{p}_1)\, \varphi_<(\mathbf{p}_2)\, \varphi_<(\mathbf{q}_1)\, \varphi_<(\mathbf{q}_2)$$

$$\times < \sum \varphi_>(\mathbf{p}_3)\ \varphi_>(-\mathbf{p}_1 - \mathbf{p}_2 - \mathbf{p}_3)\ \varphi_>(\mathbf{q}_3)\ \varphi_>(-\mathbf{q}_3 - \mathbf{q}_1 - \mathbf{q}_2)>_c \qquad \text{... (23.5.9)}$$

The above step shows the first complication in the renormalization group procedure. The "renormalized" vertex produced in Eq. (23.5.9) is momentum dependent. The coefficient of $\varphi_<(\mathbf{p}_1)\ \varphi_<(\mathbf{p}_2)\ \varphi_<(\mathbf{q}_1)\ \varphi_<(\mathbf{q}_2)$ in Eq. (23.6.9) is a function of $\mathbf{p}_1, \mathbf{q}_1, \mathbf{p}_2, \mathbf{q}_2$, with the constraint $\mathbf{p}_1 + \mathbf{q}_1 + \mathbf{p}_2 + \mathbf{q}_2 = 0$.

We now expand the momentum-dependent vertex about the zero momentum value and recognize the fact that the renormalization group produces entirely new couplings of the form $p^2\varphi^4$, ... etc. Consequently the right hand side of Eq. (23.5.9) becomes

$$\frac{1}{2}\left(6 \times \frac{\lambda}{4!}\right)^2 2\sum \varphi_<(\mathbf{p}_1)\ \varphi_<(\mathbf{p}_2)\ \varphi_<(\mathbf{q}_1)\ \varphi_<(\mathbf{q}_2)$$

$$= \int \frac{d^D p_3}{(2\pi)^D} \frac{1}{p_3^2 + m^2} \int \frac{d^D q_3}{(2\pi)^D} \frac{1}{q_3^2 + m^2} \times (2\pi)^D\, \delta^D\, (\mathbf{p}_3 + \mathbf{q}_3)$$

Thus the relevant contribution to $\overline{H}$ to $O(\lambda^2)$ is

$$\overline{H} = 6\,\frac{\lambda}{4!} \int \frac{d^D p}{(2\pi)^D} \frac{1}{p^2 + m^2} \sum \varphi_<(\mathbf{p}_1)\ \varphi_<(-\mathbf{p}_1)$$

$$-\left(\frac{6\lambda}{4!}\right)^2 \int \frac{d^D p}{(2\pi)^D} \left(\frac{1}{p^2 + m^2}\right)^2 \sum \varphi_<\ \varphi_<\ \varphi_<\ \varphi_< \qquad \text{... (23.5.10)}$$

and $H^< + \overline{H}$ becomes

$$H^< + \overline{H} = \sum \frac{\lambda}{4!}\left(\frac{m^2}{2} + \frac{k^2}{2} + 6\,\frac{\lambda}{4!} \int \frac{d^D p}{(2\pi)^D} \frac{1}{p^2 + m^2}\right) \varphi_<\ \varphi_<$$

$$+ \sum \left[\frac{\lambda}{4!} - \left[\left(\frac{6\lambda}{4!}\right)^2 \int \frac{d^D p}{(2\pi)^D} \left(\frac{1}{p^2 + m^2}\right)^2\right]\right] \varphi_<\ \varphi_<\ \varphi_<\ \varphi_< \qquad \text{... (23.5.11)}$$

We now rescale k to $k' = bk$, so that the cutoff on k' is restored to Λ. Simultaneously the $\varphi(x)$ are scaled by C which scales $\varphi(k)$ by b^D/C, *i.e.*

$\varphi_r(k) = \varphi(k')\, b^D/C$. Having scaled distances and fields, we find the renormalized Hamiltonian H_R as

$$H_R = \sum \frac{b^{2D}}{C^2}\left(\frac{m^2}{2} + \frac{k'^2/b^2}{2} + \frac{6\lambda}{4!}\int \frac{d^Dp}{(2\pi)^D}\frac{1}{p^2+m^2}\right)\varphi_<(k')\,\varphi_<(-k') +$$

$$+\sum \frac{b^{4D}}{C^4}\left\{\frac{\lambda}{4!} - \left(\frac{6\lambda}{4!}\right)^2 \int \frac{d^Dp}{(2\pi)^D}\left[\frac{1}{p^2+m^2}\right]^2\right\}\varphi_<\,\varphi_<\,\varphi_<\,\varphi_<$$

$$= \sum\left(\frac{\overline{m}^2}{2} + \frac{ak'^2}{2}\right)\varphi(k')\,\varphi(-k') +$$

$$\sum \frac{\overline{\lambda}}{4!}\varphi(k'_1)\,\varphi(k'_2)\,\varphi(k'_3)\,\varphi(-k'_1 - k'_2 - k'_3) \qquad \text{... (23.5.12)}$$

where

$$\overline{m}^2 = \frac{b^D}{C^2}\left(m^2 + \frac{12\lambda}{4!}\int \frac{d^Dp}{(2\pi)^D}\frac{1}{p^2+m^2}\right) \qquad \text{... (23.5.13)}$$

$$a = \frac{b^{D-2}}{C^2}\,(1 + O(u^2)) \qquad \text{... (23.5.14)}$$

$$\overline{\lambda} = \frac{b^D\cdot\lambda}{C^4}\left[1 - \frac{36\lambda}{4!}\int \frac{d^Dp}{(2\pi)^D}\left(\frac{1}{p^2+m^2}\right)^2\right] \qquad \text{... (23.5.15)}$$

The appropriate renormalization group requires that $a = 1$, which forces $C^2 = b^{D-2}$ (we are free to choose an overall scale factor and this is done by requiring $a = 1$). The recursion relation for the coupling constant λ now becomes

$$\overline{\lambda} = b^{4-D}\,\lambda\left[1 - \frac{3\lambda}{2}\int \frac{d^Dp}{(2\pi)^D}\left(\frac{1}{p^2+m^2}\right)^2\right] \qquad \text{... (23.5.16)}$$

To the lowest order in λ, we see $\overline{\lambda} = b^{\varepsilon}\,\lambda$ and $\overline{m}^2 = b^2 m^2$ reflecting the canonical dimensions of λ and m^2. The renormalization group to $0\,(\varepsilon)$ can now be set up by expanding Eq. (23.5.16) as

$$\bar{\lambda} = \lambda(1 + \varepsilon \ln b)\left[1 - \frac{3}{2}\lambda \int \frac{d^D p}{(2\pi)^D}\left(\frac{1}{p^2 + m^2}\right)^2\right]$$

$$= \lambda\left[1 + \varepsilon \ln b - \frac{3}{2}\lambda \int \frac{d^D p}{(2\pi)^4}\left(\frac{1}{p^2 + m^2}\right)^2\right] \quad \text{... (23.5.17)}$$

where the last line follows from the expectation that the non-zero fixed point in λ is going to be of $O(\varepsilon)$. In the critical region m^2 is small and thus evaluating the integral in $D = 4$,

$$\int \frac{d^D p}{(2\pi)^D}\left(\frac{1}{p^2 + m^2}\right)^2 = \frac{2\pi^2}{(2\pi)^4}\int_{\Lambda/b}^{\Lambda} \frac{p^3 dp}{p^4} = \frac{1}{8\pi^2}\ln b$$

leading to

$$\bar{\lambda} = \lambda + \lambda\varepsilon \ln b - \frac{3}{2}\frac{\lambda^2}{8\pi^2}\ln b \quad \text{... (23.5.18)}$$

Turning to Eq. (23.5.13), we can evaluate (again in $D = 4$)

$$\int \frac{d^D p}{(2\pi)^D}\left(\frac{1}{p^2 + m^2}\right) = \frac{1}{8\pi^2}\int \frac{p^3\, dp}{p^2 + m^2}$$

$$= \frac{1}{16\pi^2}\left[\Lambda^2 - \frac{\Lambda^2}{b^2}\right] - \frac{m^2}{8\pi^2}\ln b$$

and write

$$\overline{m^2} = b^2\left\{m^2 + \frac{\lambda}{2}\left[\frac{\Lambda^2}{16\pi^2}\left(1 - \frac{1}{b^2}\right) - \frac{m^2}{8\pi^2}\ln b\right]\right\} \quad \text{... (23.5.19)}$$

It helps to consider b close to unity so that the recursion relations acquire a differential character. We put $b = e^{\delta l} \cong 1 + \delta l$ and to $O(\delta l)$ obtains

$$\overline{m^2} = (1 + 2\delta l)\left\{m^2 + \frac{\lambda}{2}\left[\frac{\Lambda^2}{16\pi^2}2\delta l - \frac{m^2}{8\pi^2}\delta l\right]\right\}$$

$$= m^2 + \delta l \left[2m^2 - \frac{2m^2}{16\pi^2} - \frac{\lambda \Lambda^2}{16\pi^2} \right]$$

or

$$\frac{dm^2}{dl} = \left(2 - \frac{\lambda}{16\pi^2} \right) m^2 - \frac{\lambda \Lambda^2}{16\pi^2} \qquad \text{... (23.5.20)}$$

while

$$\bar{\lambda} = \lambda + \left(\lambda \varepsilon - \frac{3}{2} \frac{\lambda^2}{8\pi^2} \right) \delta l$$

$$\frac{d\lambda}{dl} = \lambda \left(\varepsilon - \frac{3}{2} \frac{\lambda}{8\pi^2} \right) \qquad \text{... (23.5.21)}$$

Eqs. (23.5.20) and (23.5.21) constitute the renormalization group flow equations to the lowest order.

As explained in the previous section, once the flow has been obtained, one looks for the fixed points and the eigenvalues of the linearized transformation about the fixed point. The fixed points and the associated eigenvalues are:

i) $\lambda = 0, m^2 = 0$—the Gaussian fixed point with eigenvalues b^2 and b^{ε}(note that the eigenvalues refer to the flow equations in the form $\overline{m^2} = f(m^2, \lambda)$ and $\bar{\lambda} = g(m^2, \lambda)$). Clearly for $\varepsilon < 0$ $(D > 4)$, one eigenvalue is smaller than unity (the "λ eigenvalue") and hence the fixed point is "stable" in the sense that it can describe critical behaviour (one eigenvalue greater than unity and the others smaller). In dimensions greater than four, the quartic term is not involved in the determination of critical behaviour. The Gaussian model results are consequently exact.

ii) $\lambda = \dfrac{16\pi^2}{3} \varepsilon, m^2 = \dfrac{\Lambda \varepsilon/3}{2 - \varepsilon/3}$ the non-trivial fixed point with eigenvalues $b^{2-\varepsilon/3}$ and $b^{-\varepsilon}$. For $\varepsilon > 0$ (i.e. $D < 4$), the eigenvalue in the λ-direction is now smaller than unity while the other eigenvalue is greater than unity. Consequently, it is this fixed point which now describes the critical behaviour. It is unstable in one direction (corresponding to the temperature variable) and stable in the λ-direction. The "relevant" or "thermal" eigenvalue, $b^{2-\varepsilon/3}$, determines the exponent ν as [see Eq. (23.3.21)]

$$\nu = \frac{1}{2-\varepsilon/3} = \frac{1}{2} + \frac{\varepsilon}{12} + O(\varepsilon^2) \quad \text{... (23.5.22)}$$

At this order the anomalous dimensions exponent $\eta = 0$ and thus

$$\gamma = 2\nu = 1 + \frac{\varepsilon}{6} + O(\varepsilon^2)$$

Hyperscaling leads to the specific heat index α as

$$\alpha = 2 - D\left(\frac{1}{2} + \frac{\varepsilon}{12}\right) = \frac{\varepsilon}{6} + O(\varepsilon^2)$$

The exponent β follows from the scaling relation $\alpha + 2\beta + \gamma = 2$ as

$$2\beta = 2 - 1 - \frac{\varepsilon}{6} - \frac{\varepsilon}{6} = 1 - \frac{1}{3}\varepsilon$$

or

$$\beta = \frac{1}{2} - \frac{\varepsilon}{6} + O(\varepsilon^2)$$

It should be noted that η can also be obtained by working to $O(\lambda^2)$. This involves the extraction of the terms quadratic in $\varphi_< \varphi_<$ from the $O(\lambda^2)$ part in Eq. (23.5.9). This makes the structure of Eq. (23.5.14)

$$a = \frac{b^{D-2}}{C^2}\left[1 + 48\left(\frac{\lambda}{4!}\right)^2 \frac{3}{2}\frac{1}{2} \ln b + \ldots\right] \quad \text{... (23.5.23)}$$

The requirement of setting $a = 1$ now establishes η when we substitute the fixed point value of λ obtained from Eq. (23.5.21). The reader should go through the straightforward algebra of obtaining Eq. (23.5.23) and verify that at this order the anomalous dimension η works out to be $\varepsilon^2/54$.

Chapter 24

Dynamic Critical Phenomena

24.1 Introduction

In the last few chapters we have been studying equilibrium critical phenomena where the order parameter field exhibits no time dependence. The characteristic feature has been the establishment of a growing correlation length as the transition point is approached, leading to a characteristic divergence at the critical temperature T_c. Our aim in this chapter will be to study small deviations from equilibrium and focus on the time scale over which the system returns to equilibrium. The main characteristic of the time scale will be the fact that it will have to become longer and longer as the critical point is approached. This is easily understood since there would be larger and larger correlated regions as the critical point is neared and the dynamics would involve the time dependence of large correlated patches and processes would have to slow down. This is the critical slowing down which will manifest itself in a characteristic divergence of the time scale.

The first question that we ask is what sort of dynamical equations should we have. We have already seen that approach to equilibrium is quite easily described in terms of Langevin equations. The order parameter field $\varphi(x)$ has its dynamics specified by (we take φ to be scalar, the generalization to a n-component φ is straightforward):

$$\frac{\partial\varphi}{\partial t}(x,t) = -\Gamma\frac{\delta F}{\delta\varphi} + \eta \qquad \text{... (24.1.1)}$$

where F is the equilibrium free energy and η is a random noise obeying the fluctuation-dissipation theorem, namely,

$$<\eta(x_2,t_2)\,\eta(x_1,t_1)> \;=\; 2\Gamma\,(k_BT)\,\delta\,(t_2-t_1)\delta\,(x_2-x_1\,).$$

The dynamics governed by Eq. (24.1.1) will ensure that an initially out-of-equilibrium situation will tend to the equilibrium distribution $P(\{\varphi\}) \propto e^{-F/k_BT}$ as $t \to \infty$. It should be realized that an equation of the form of Eq. (24.1.1) is true for the long-distance behaviour of $\varphi\,(x, t)$, which in Fourier space implies the small wave number components of $\varphi\,(x, t)$. The short-distance behaviour is averaged over and that produces the thermal noise in the Langevin equation that we have written down. We should also point out that the relaxational form of Eq. (24.1.1) is valid for order parameter fields which are not conserved. If $\varphi\,(x)$ is a conserved field, i.e. $\frac{\delta}{\delta t}\int\varphi\,(x,t)\,d^Dx = 0$, then the Langevin equation would have to be modifed to

$$\frac{\partial\varphi}{\partial t} = \Gamma\,\nabla^2\,\frac{\delta F}{\partial\varphi} + \eta \qquad \text{... (24.1.2)}$$

with the noise η having the correlation function

$$<\eta\,(x_2,\,t_2)\,\eta\,(x_1,\,t_1)> = -\,2\Gamma(k_BT)\,\nabla^2\delta\,(x_2-x_1)\,\delta\,(t_2-t_1)$$

It is convenient for calculational purposes to write Eqs. (24.1.1) and (24.1.2) in Fourier space as:

Non-conserved order-parameter

$$\frac{\partial\varphi}{\partial t}(k,t) = -\Gamma\frac{\delta F}{\delta\varphi(-k)} + \eta\,(k,t) \qquad \text{... (24.1.3)}$$

$$<\eta\,(k_2.\,t_2)\,\eta\,(k_1,\,t_1)> \;=\; 2(k_BT)\,\Gamma\,\delta\,(\mathbf{k}_1\,+\,\mathbf{k}_2)\,\delta\,(t_2-t_1)$$

Conserved order-parameter

$$\frac{\partial\varphi}{\partial t}(k,t) = -\,\Gamma\,k^2\frac{\partial F}{\partial\varphi(-k)} + \eta\,(k,t) \qquad \text{... (24.1.4)}$$

$$< \eta (k_2, t_2)\, \eta (k_1, t_1) > \; = \; 2 (k_B T)\, \Gamma\, k^2\, \delta (\mathbf{k}_1 + \mathbf{k}_2)\, \delta (t_2 - t_1)$$

We are now in a position to exhibit the critical slowing down. We begin with a Gaussian model for F, which reads

$$F = \sum_k \frac{1}{2} (a + k^2)\, \varphi (k)\, \varphi (-k) \qquad \text{... (24.1.5)}$$

where $a = a_0 (T - T_c)$, T_c being the critical temperature.

With this form of F, Eq. (24.1.3) becomes

$$\frac{\partial \varphi (k, t)}{\partial t} = - \Gamma\, (a + k^2)\, \varphi (k, t) + \eta (k, t) \qquad \text{... (24.1.6)}$$

with the solution

$$\varphi (k, t) = e^{-\Gamma (a + k^2) t} \int_0^t e^{\Gamma (a + k^2) t'}\, \eta\, (k, t')\, dt'$$

$$+ \; \varphi (k, 0)\, e^{-\Gamma (a + k^2) t} \qquad \text{... (24.1.7)}$$

The time-dependent correlation function $C(k, t_2 - t_1)$ is found to be

$$C(k, t_2 - t_1) = < \varphi (k, t_2)\, \varphi (-k, t_1)$$

$$= < \varphi (k, 0)\, \varphi (-k, 0)\, e^{-\Gamma (a + k^2)(t_1 + t_2)} > + \; e^{-\Gamma (a + k^2)(t_1 + t_2)} \int_0^{t_2} dt'' \int_0^{t_1} dt'$$

$$\times e^{\Gamma (a+k^2)(t'+t'')} < \eta \left(k, t'\right) \eta \left(-k, t''\right) >$$

$$= C(k,0) e^{-\Gamma (a+k^2)(t_1 + t_2)} + 2\Gamma\, k_B T \left[\frac{e^{-\Gamma (a+k^2)|t_2 - t_1|}}{2\Gamma (a+k^2)} - \frac{e^{-\Gamma (a + k^2)(t_1 + t_2)}}{2\Gamma (a + k^2)} \right]$$

$$= \left[C(k, 0) - \frac{k_B T}{a + k^2} \right] e^{-\Gamma(a + k^2)(t_1 + t_2)} + \frac{k_B T}{a + k^2} e^{-\Gamma(a + k^2)|t_2 - t_1|} \qquad \text{... (24.1.8)}$$

Now $C(k, 0)$ is the equal-time correlation function and if we take that to be the equilibrium correlation function, i.e. $C(k, 0) = k_B T/(a + k^2)$, then the time-dependent correlations decay as

$$C(k, t) = C(k, 0) \, e^{-\Gamma(a + k^2)t} \qquad \text{... (24.1.9)}$$

which is to say that if the equilibrium is disturbed, then a relaxation towards the equilibrium occurs over a time scale $\tau = [\Gamma(a + k^2)]^{-1}$. The characteristic frequency $\omega = \tau^{-1}$ and is given by

$$\omega = \Gamma(a + k^2) = \Gamma(k^2 + \xi^{-2}) \qquad \text{... (24.1.10)}$$

where we have made use of the fact that for the Gaussian model the correlation length ξ is given by $\xi = a^{-1/2}$. We can write Eq. (24.1.10) in the form

$$\omega = k^z f(k\xi) \qquad \text{... (24.1.11)}$$

where $z = 2$ and $f(x) = \Gamma(1 + {}^1/x^2)$. The phenomenon of critical slowing down is contained in Eq. (24.1.11). As $\xi \to \infty$ at fixed k (i.e. $T \to T_c$), we note that $\omega \sim k^z$, which implies that the relaxation rate is slower for longer wavelengths. Alternatively, as $k \to 0$, we must have $f(x) \sim x^{-z}$, so that $\omega \sim \xi^{-z}$, and larger the correlation length ξ, smaller is the relaxation rate.

If we have a conserved order-parameter, then an identical analysis shows that

$$\omega = \Gamma k^2 (k^2 + \xi^{-2}) = \Gamma k^4 \left(1 + \frac{1}{k^2 \xi^2} \right) \qquad \text{... (24.1.12)}$$

In this case once again ω has the structure of Eq. (24.1.11), but one must remember that for the function $f(x)$, the asymptotic behaviour for $x << 1$ has to be $f(x) \sim x^{-z+2}$ so that the relaxation rate in the regime $\xi >> k^{-1}$ has the form

$$\omega \sim k^2 \xi^{2-z} \qquad \text{... (24.1.13)}$$

the explicit factor of k^2 showing the existence of the conservation law.

In our discussion so far, we have made the tacit assumption that the coefficient Γ (Onsager coefficient) is not affected by the critical slowing down. In the Gaussian model and the relaxational dynamics that we have considered this is quite obviously true. However for real systems:

i) F will not have the Gaussian form

ii) the dynamics will not be purely relaxational

and consequently the behaviour of Γ will have to be re-examined. As we shall see in the following sections, the effect of nonlinear terms in an equation such as Eq. (24.1.1) is to make the Onsager coefficient dependent on wavenumber and temperature. The Gaussian free energy and purely relaxational dynamics case can be viewed as a mean field theory for critical dynamics.

24.2 Time-Dependent Ginzburg-Landau Model

In this Section we are going to investigate the effect of having the full Ginzburg-Landau free energy, i.e. instead of the quadratic F of Eq. (24.1.5), we have

$$F = \sum_k \frac{1}{2}(a + k^2)\, \varphi(\mathbf{k})\, \varphi(-\mathbf{k}) + \frac{u}{4} \sum_{\mathbf{k}_1, \mathbf{k}_2, \mathbf{k}_3} \varphi(\mathbf{k}_1)\, \varphi(\mathbf{k}_2)\, \varphi(\mathbf{k}_3)\, \varphi(-\mathbf{k}_1 - \mathbf{k}_2 - \mathbf{k}_3) \qquad \text{... (24.2.1)}$$

leading to the equation of motion (non-conserved φ)

$$\frac{\partial \varphi(k)}{\partial t} = -\Gamma\, \{(a + k^2)\, \varphi(k) + u \sum_{\mathbf{k}_1, \mathbf{k}_2} \varphi(\mathbf{k}_1)\, \varphi(\mathbf{k}_2)\, \varphi(\mathbf{k} - \mathbf{k}_1 - \mathbf{k}_2)\} + \eta \qquad \text{... (24.2.2)}$$

In dealing with the static critical phenomenon, we have seen that the effect of the quartic term in F can be properly handled only within the context of the renormalization group. We will not demonstrate here that should a perturbative calculation be carried out in u to calculate the correction to the relaxation rate, that perturbation series will diverge as $\xi \to \infty$, but will directly explain the working of the renormalization group in this dynamical context. As with the static case, the first move is to divide the full wavenumber range $0 < k < \Lambda$ (some arbitrary high momentum cutoff) into $k_<$ which runs from 0 to Λ/b and $k_>$, which contains momenta between Λ/b and Λ, with $b > 1$. Since the low momentum behaviour is what is of interest, we proceed to remove by averaging over η the high wavenumber part from the equation of motion and try to find an equation of motion of the form of Eq. (24.2.2), but restricted to the momentum range $0 < k < \Lambda/b$. To do this, we split $\varphi(x)$ as

$$\varphi(x) = \varphi_<(x) + \varphi_>(x) \qquad \text{... (24.2.3)}$$

where

$$\varphi_<(x) = \frac{1}{(2\pi)^D}\int_0^{\Lambda/b} d^D k\ \varphi_<(k)\ e^{i\mathbf{k}\cdot\mathbf{x}}$$

and $$\varphi_>(x) = \frac{1}{(2\pi)^D}\int_{\Lambda/b}^{\Lambda} d^D k\ \varphi_>(k)\ e^{i\mathbf{k}\cdot\mathbf{x}}$$

Note that $\varphi_<(k)$ implies that the argument k lies in the range $0 < k < \Lambda/b$ and $\varphi_>(k)$ implies that the argument k lies in the range $\Lambda/b < k < \Lambda$. It should also be noted that the Fourier transform of $\varphi_<(x)$ is $\varphi_<(k)$ and that of $\varphi_>(x)$ is $\varphi_>(k)$. With the above observations it is straightforward to establish

$$\begin{aligned}\dot{\varphi}_<(k) &= -\Gamma(a + k^2)\,\varphi_<(k) - u\Gamma\sum_{k_1, k_2}\{\varphi_<(\mathbf{k}_1)\,\varphi_<(\mathbf{k}_2)\,\varphi_<(\mathbf{k}-\mathbf{k}_1-\mathbf{k}_2)\\ &\quad + 3\,\varphi_<(\mathbf{k}_1)\,\varphi_<(\mathbf{k}_2)\,\varphi_>(\mathbf{k}-\mathbf{k}_1-\mathbf{k}_2) + 3\,\varphi_>(\mathbf{k}_1)\,\varphi_>(\mathbf{k}_2)\,\varphi_<(\mathbf{k}-\mathbf{k}_1-\mathbf{k}_2)\\ &\quad + \varphi_>(\mathbf{k}_1)\,\varphi_>(\mathbf{k}_2)\,\varphi_>(\mathbf{k}-\mathbf{k}_1-\mathbf{k}_2)] + \eta_<(k)\end{aligned} \qquad \text{... (24.2.4)}$$

and

$$\begin{aligned}\dot{\varphi}_>(k) &= -\Gamma(a + k^2)\,\varphi_>(k) - u\Gamma\sum_{\mathbf{k}_1, \mathbf{k}_2}[\varphi_<(\mathbf{k}_1)\,\varphi_<(\mathbf{k}_2)\,\varphi_<(\mathbf{k}-\mathbf{k}_1-\mathbf{k}_2)\\ &\quad + 3\varphi_<(\mathbf{k}_1)\,\varphi_<(\mathbf{k}_2)\,\varphi_>(\mathbf{k}-\mathbf{k}_1-\mathbf{k}_2) + 3\varphi_>(\mathbf{k}_1)\,\varphi_>(\mathbf{k}_2)\,\varphi_<(\mathbf{k}-\mathbf{k}_1-\mathbf{k}_2)\\ &\quad + \varphi_>(\mathbf{k}_1)\,\varphi_>(\mathbf{k}_2)\,\varphi_>(\mathbf{k}-\mathbf{k}_1-\mathbf{k}_2)] + \eta_>(k)\end{aligned} \qquad \text{... (24.2.5)}$$

Having written out equations of motion for $\varphi_<$ and $\varphi_>$, we need to solve for $\varphi_>(k)$ from Eq. (24.2.5) in terms of $\varphi_<$ and insert that in Eq. (24.2.4) to get an equation of motion for $\varphi_<$ where $\varphi_>$ does not appear. To solve Eq. (24.2.5), we proceed perturbatively in the coupling constant u and write

$$\varphi_>(k) = \varphi_>^{(0)}(k) + u\,\varphi_>^{(1)}(k) + 0(u^2) \quad \text{... (24.2.6)}$$

Now, insert this expansion in Eq. (24.2.5), match the terms of equal powers of u on either side and arrive at

$$\dot{\varphi}_>^{(0)}(k) = -\Gamma(a + k^2)\,\varphi_>^{(0)}(k) + \eta_>(k) \quad \text{... (24.2.7)}$$

$$\dot{\varphi}_>^{(1)}(k) = -\Gamma(a+k^2)\,\varphi^{(1)}(k) - \Gamma\sum_{\mathbf{k}_1,\mathbf{k}_2}[\varphi_<^{(0)}(\mathbf{k}_1)\,\varphi_<^{(0)}(\mathbf{k}_2)\,\varphi_<^{(0)}(\mathbf{k}-\mathbf{k}_1-\mathbf{k}_2)$$

$$+ 3\varphi_<^{(0)}(\mathbf{k}_1)\,\varphi_<^{(0)}(\mathbf{k}_2)\,\varphi_>^{(0)}(\mathbf{k}-\mathbf{k}_1-\mathbf{k}_2)$$

$$+ 3\varphi_<^{(0)}(\mathbf{k}_1)\,\varphi_>^{(0)}(\mathbf{k}_2)\,\varphi_>^{(0)}(\mathbf{k}-\mathbf{k}_1-\mathbf{k}_2)$$

$$+ \varphi_>^{(0)}(\mathbf{k}_1)\,\varphi_>^{(0)}(\mathbf{k}_2)\,\varphi_>^{(0)}(\mathbf{k}-\mathbf{k}_1-\mathbf{k}_2)] \quad \text{... (24.2.8)}$$

The solution to Eq. (24.2.7) can be written as

$$\varphi_>^{(0)}(k, t) = \int_{-\infty}^{t} G(k, t-t_1)\,\eta(k, t_1)\,dt \quad \text{... (24.2.9)}$$

where

$$G(K, t-t_1) = e^{-\Gamma(a+k^2)(t-t_1)} \quad \text{... (24.2.10)}$$

The solution to Eq. (24.2.8) can be written as

$$\varphi_>^{(1)}(k, t) = -\Gamma\sum_{\mathbf{k}_1,k_2}[\int_{-\infty}^{t} G(k, t-t_2)\,[\varphi_<^{(0)}(\mathbf{k}_1, t_2)\,\varphi_<^{(0)}(\mathbf{k}_2, t_2)\,\varphi_<^{(0)}(\mathbf{k}-\mathbf{k}_1-\mathbf{k}_2, t_2)$$

$$+ 3\varphi_<^{(0)}(\mathbf{k}_1, t_2)\,\varphi_<^{(0)}(\mathbf{k}_2, t_2)\,\varphi_>^{(0)}(\mathbf{k}-\mathbf{k}_1-\mathbf{k}_2, t_2)$$

$$+ 3\varphi_<^{(0)}(\mathbf{k}_1, t_2)\,\varphi_>^{(0)}(\mathbf{k}_2, t_2)\,\varphi_>^{(0)}(\mathbf{k}-\mathbf{k}_1-\mathbf{k}_2, t_2)$$

$$+ \varphi_>^{(0)}(\mathbf{k}_1, t_2)\,\varphi_>^{(0)}(\mathbf{k}_2, t_2)\,\varphi_>^{(0)}(\mathbf{k}-\mathbf{k}_1-\mathbf{k}_2, t_2)]\,dt_2 \quad \text{... (24.2.11)}$$

with $\varphi_>^{(0)}(k, t)$ given by Eq. (24.2.9) and $G(k, t)$ by Eq. (24.2.10). We now return to Eq. (24.2.4) and introduce the expansion of Eq. (24.2.6) to write

$$\begin{aligned}
\varphi_<(k, t) = & -\Gamma(a+k^2)\,\varphi_<(k, t) - u\sum_{\mathbf{k}_1;\,\mathbf{k}_2}\varphi_<(\mathbf{k}_1, t)\,\varphi_<(\mathbf{k}_2, t)\,\varphi_<(\mathbf{k}-\mathbf{k}_1-\mathbf{k}_2, t)\\
& - u\Gamma\sum_{\mathbf{k}_1,\,\mathbf{k}_2}[3\varphi_<(\mathbf{k}_1, t)\,\varphi_<(\mathbf{k}_2, t)\,\varphi_>^{(0)}(\mathbf{k}-\mathbf{k}_1-\mathbf{k}_2, t)\\
& + 3\varphi_>^{(0)}(\mathbf{k}_1, t)\,\varphi_>^{(0)}(\mathbf{k}_2, t)\,\varphi_<(\mathbf{k}-\mathbf{k}_1-\mathbf{k}_2, t)\\
& + \varphi_>^{(0)}(\mathbf{k}_1, t)\,\varphi_>^{(0)}(\mathbf{k}_2, t)\,\varphi_>^{(0)}(\mathbf{k}-\mathbf{k}_1-\mathbf{k}_2, t)]\\
& - u^2\Gamma\sum_{\mathbf{k}_1,\,\mathbf{k}_2}[3\varphi_<(\mathbf{k}_1, t)\,\varphi_<(\mathbf{k}_2, t)\,\varphi_>^{(1)}(\mathbf{k}-\mathbf{k}_1-\mathbf{k}_2, t)\\
& + 3\varphi_>^{(1)}(\mathbf{k}_1, t)\,\varphi_>^{(0)}(\mathbf{k}_2, t)\,\varphi_<(\mathbf{k}-\mathbf{k}_1-\mathbf{k}_2, t)\\
& + 3\varphi_>^{(0)}(\mathbf{k}_1, t)\,\varphi_>^{(1)}(\mathbf{k}_2, t)\,\varphi_<(\mathbf{k}-\mathbf{k}_1-\mathbf{k}_2, t)\\
& + \{\varphi_>^{(1)}(\mathbf{k}_1, t)\,\varphi_>^{(0)}(\mathbf{k}_2, t)\,\varphi_>^{(0)}(\mathbf{k}-\mathbf{k}_1-\mathbf{k}_2, t)\\
& + 2\ \text{permutations}\} + O(u^3) + \eta_<(t) \qquad \ldots\ \mathbf{(24.2.12)}
\end{aligned}$$

We now imagine inserting Eq. (24.2.11) in the above in all those places where we have a $\varphi_>^{(1)}$ and averaging over the noise. (Note that the averaging over η implies using the time-dependent correlation function of $\varphi^{(0)}$ which, with $\varphi^{(0)}$ given by Eq. (24.2.9), will be obtained from Eq. (24.1.9)). We will henceforth use units where $k_BT = 1$. We recognize the emergence of three types of terms after this exercise,

i) terms linear in $\varphi_<$

ii) terms cubic in $\varphi_<$

iii) terms quintic in $\varphi_<$

The terms quintic in $\varphi_<$ should turn out to be irrelevant in the renormalization group sense. The terms cubic in $\varphi_<$ will serve to renormalize the coupling constant u. For the work of this section, we will imagine the flow equation of u to be known and work at the fixed point value (in an ε-expansion) of u. We are now left with the linear term. Going through the details of the linear term, we notice that these are of two types. One which has the structure

$\varphi_<(k, t) \int K_0\,(t-t')dt'$ (these we call local) and the other which has the structure $\int K(t-t')\,\varphi_<(k, t')dt'$ (these we call non-local). The local terms can be thought of as renormalizing the bare mass a. The non-local terms, on the other hand, can only be thought of as modifying the relaxation rate Γ. We now return to Eq. (24.2.12) and keeping only the linear terms, we have

$$\frac{\partial \varphi_<(k)}{\partial t} = -\,\Gamma\,(a + \Delta a + k^2)\,\varphi_<(k,\, t) + 3u^2\Gamma^2 \Bigg\{ \sum_{\Lambda/b < \mathbf{p},\, \mathbf{q} < \Lambda}$$

$$\int^{t} G\,(\mathbf{p},\, t - t')\; C\,(\mathbf{q},\, t - t')\; C\,(\mathbf{k} - \mathbf{p} - \mathbf{q},\, t - t')\; \varphi_<(\mathbf{k},\, t')\; dt'$$

$$+\; 2 \text{ permutations}\Bigg\} + \eta_<(k,\, t) \qquad \text{... (24.2.13)}$$

where Δa is the static correction discussed above and $C\,(k,\ t-t')$ is the correlation function given by

$$C\,(k,\ t - t') = \langle\, \varphi_>(k,\ t)\ \varphi_>(-k,\ t') \rangle$$

$$= (k^2 + a)^{-1}\ e^{-\Gamma(k^2 + a)\,|t - t'|} \qquad \text{... (24.2.14)}$$

To proceed further, we write Eq. (24.2.13) in Fourier space (i.e. using frequency variables) with

$$\varphi\,(k,\ t) = \frac{1}{2\pi} \int_{-\infty}^{\infty} \varphi\,(k,\ \omega)\ e^{-i\,\omega t}\ d\omega \qquad \text{... (24.2.15)}$$

and find

$$\left(-\frac{i\omega}{\Gamma} + a + \Delta a + k^2 + K\,(k,\ \omega) \right) \varphi_<(k,\ \omega) = \eta_<(k,\ \omega) \qquad \text{... (24.2.16)}$$

where

$$K\,(k,\,\omega) = -\,3u^2\,\Gamma \iint \frac{d^Dp}{(2\pi)^D}\,\frac{d^Dq}{(2\pi)^D}\,\frac{3a + p^2 + q^2 + (\mathbf{k} - \mathbf{p} - \mathbf{q})^2}{(p^2 + a)\,(q^2 + a)} \times$$

$$\times\ \frac{1}{[(\mathbf{k} - \mathbf{p} - \mathbf{q})^2 + a]\,[-\,i\omega + \Gamma\,\{3a + (\mathbf{k} - \mathbf{p} - \mathbf{q})^2 + p^2 + q^2\}]}$$

$$\text{... (24.2.17)}$$

Our task is to find an effective Γ from the above expression and we do this by expanding

$$K(k,\ \omega) = K(k,\ 0) + (-\,i\omega)\,\frac{\partial K}{\partial(-\,i\omega)}\Big|_{\omega=0} + \ldots$$

Inserting this expanded form of $K(k, \omega)$ in Eq. (24.2.16), we have

$$\left[-\,i\omega\left(\frac{1}{\Gamma} + \frac{\partial K}{\partial\,(-\,i\omega)}\Big|_{\omega=0}\right) + a + \Delta a + k^2 + K\ (k,\ 0) + \ldots\right]\varphi_< = \eta_< \qquad \ldots\ \textbf{(24.2.18)}$$

and are able to identify the new relaxation rate Γ' as

$$\frac{1}{\Gamma'} = \frac{1}{\Gamma} + \frac{\partial K}{\partial\,(-\,i\omega)}\Big|_{\omega=0} \qquad \ldots\ \textbf{(24.2.19)}$$

Clearly Γ' is dependent on k and the correlation length a^{-1}. We work at the critical point and set $a = 0$ to find the momentum-dependent Γ satisfy the recursion relation

$$\frac{1}{\Gamma'} = \frac{1}{\Gamma} + \frac{3u^2}{\Gamma}\iint \frac{d^Dp}{(2\pi)^D}\,\frac{d^Dq}{(2\pi)^D}\cdot\frac{1}{p^2}\cdot\frac{1}{q^2}\cdot\frac{1}{(\mathbf{k} - \mathbf{p} - \mathbf{q})^2}\left[\frac{1}{p^2 + q^2 + (\mathbf{k} - \mathbf{p} - \mathbf{q})^2}\right] \qquad \ldots\ \textbf{(24.2.20)}$$

It should be remembered that the integration over the momenta $\mathbf{p}$ and $\mathbf{q}$ cover the range between Λ/b and Λ only. Having found the effective relaxation rate under an elimination of the high wavenumbers, we now have to rescale the spatial and temporal variables and the field, so that the new free energy has the same form as the old one. The spatial rescaling is $x \to x' = x/b$ so that the momenta are rescaled according to $k \to k' = bk$. The field rescales as $\varphi \to \varphi' = b^a\,\varphi$ and the frequency will be rescaled according to the yet to be found dynamic scaling exponent as $\omega \to \omega' = b^z\,\omega$. These scalings will alter Γ according to its naive dimension obtained by power counting and the way to find the dimension of Γ is to note that the relaxation rate which is a frequency is Γ/χ, where χ is the static susceptibility. Consequently, the dimension of Γ is that of (frequency) $\times$ (susceptibility) and since the susceptibility at T_c

has the behaviour $k^{-2+\eta}$, it is obvious that Γ will be scaled by the factor $b^{z-2+\eta}$. Taken together with Eq. (24.2.20), this discussion implies that under the action of the renormalization group, the Onsager coefficient transforms as (from the nth to the $(n+1)^{\text{th}}$ step)

$$\frac{1}{\Gamma_{n+1}} = \frac{b^{2-z-\eta}}{\Gamma_n}\left[1 + 3u^2 \iint_{\Lambda/b < p,\, q < \Lambda} \frac{d^Dp}{(2\pi)^D}\frac{d^Dq}{(2\pi)^D}\frac{1}{p^2}\frac{1}{q^2}\frac{1}{(\mathbf{k}-\mathbf{p}-\mathbf{q})^2}\frac{1}{p^2+q^2+(\mathbf{k}-\mathbf{p}-\mathbf{q})^2}\right] \quad \text{... (24.2.21)}$$

At this point, we will make use of the ε-expansion.

This is natural since it should be clear from the integral on the right hand side that it is finite for $D > 4$ and hence there can be no effect of it on the exponent. Thus the upper critical dimension is four (as in the static case) and an expansion in $\varepsilon = 4 - D$ will be possible. The first term of such an expansion will involve using u to $O(\varepsilon)$ [note that we are assuming that we are at the fixed point so far as the statics is concerned and thus the ε-expansion for the fixed point value of u can be used] and evaluating the integral in $D = 4$ where it will yield the usual $\ln b$ behaviour. Evaluating the integral, one obtains

$$\frac{1}{\Gamma_{n+1}} = \frac{b^{2-z-n}}{\Gamma_n}\left[1 + u^2\, C_4^2\, 9 \ln\frac{4}{3} \ln b + \ldots\right] \quad \text{... (24.2.22)}$$

where

$C_4 = \dfrac{1}{8\pi^2}$. Using $u = \dfrac{\lambda}{6} = \dfrac{\varepsilon}{9C_4}$ from Eqs (23.5.1) and (23.5.21), we find

$$\frac{1}{\Gamma_{n+1}} = \frac{b^{2-z-\eta}}{\Gamma_n}\left[1 + \frac{\varepsilon^2}{9}\ln\frac{4}{3}\ln b\right]$$

$$= \frac{b^{2-z-\eta}}{\Gamma_b}\left[1 + 6\eta \ln\frac{4}{3}\ln b\right] \quad \text{... (24.2.23)}$$

having noted that to $O(\varepsilon^2)$, $\eta = \varepsilon^2/54$. Expanding $b^{2-z-\eta}$ as $1 + (2 - z - \eta)\ln b +$ higher orders, an expansion which is valid for $b \cong 1$, we can write Eq. (24.2.23) as

$$\frac{1}{\Gamma_{n+1}} = \frac{1}{\Gamma_n}\left[1 + \left(2 - z - \eta + 6\eta \ln \frac{4}{3}\right) \ln b\right] \quad \text{... (24.2.24)}$$

The condition for a fixed point $\Gamma_{n+1} = \Gamma_n$ is clearly

$$z = 2 + \left(6 \ln \frac{4}{3} - 1\right) \eta + O(\varepsilon^3) \quad \text{... (24.2.25)}$$

Thus the non-linear terms cause a change in the mean field value $z = 2$ by causing the Onsager coefficient to be k-dependent. Had the order parameter followed a conservation law, we would have proceeded in a similar fashion and obtained the analogue of Eq. (24.2.21) as

$$\frac{1}{\Gamma_{n+1}} = \frac{b^{4-z-\eta}}{\Gamma_n} [1 + u^2 I(k, b)] \quad \text{... (24.2.26)}$$

However, in this case, the integral $I(k, b)$ would turn out to be finite (one should write it down explicitly to verify it) and the condition for arriving at the fixed point would be

$$z = 4 - \eta \quad \text{... (24.2.27)}$$

correct to all order in ε.

The important thing to note is that the effect of the nonlinear terms is small. The change produced in the dynamic exponent due to these dissipative nonlinear couplings is of $O(\eta)$. The anomalous dimension exponent being of $O(10^{-2})$, we have a very small effect. In the next section we will drop these dissipative non-linearities and concentrate on the reversible terms that have to be introduced into the equations of motion for completeness. Their effects will be more spectacular.

24.3 Nonlinear Models with Reversible Terms

In this *Section* we are going to include the reversible couplings that necessarily enter the dynamics of the order parameter field. Sometimes this coupling may be to a secondary field whose dynamics has got to be considered as well (e.g. if we are dealing with the liquid-vapour critical point or the consolute point of a binary liquid, the order parameter for which is the density or the concentration respectively, then the dynamics of the order parameter is necessarily coupled to that of the velocity field). Here, however, we will deal with ferromagnets where the complication of a secondary field is absent.

To write the equation of motion for the order parameter, we first recognize the fact that we have a three-dimensional order parameter (magnetization is a vector ($\vec{\varphi}$ (**x**, t)) and the non-relaxational dynamics comes from the fact that at a given point in space one has a local magnetic field **h** (x) set up by the magnetization vectors at all the other points. A magnetic field exerts a torque on the magnetization $\vec{\varphi}$ and the equation of motion is consequently a precession of the vector $\vec{\varphi}$ given by

$$\frac{\partial \vec{\varphi}}{\partial t} = g_0 \, \vec{\varphi} \times \mathbf{h} \qquad \text{... (24.3.1)}$$

where g_0 is some coupling constant that is required to keep the dimensions correct. Now, the local field **h** (x) can be expanded as a series in $\vec{\varphi}$ and its derivatives and consequently, we can expand

$$\begin{aligned} \mathbf{h}(x) &= \lambda_0 \, \vec{\varphi}(x) \, [1 + \lambda_1 \, \vec{\varphi} \cdot \vec{\varphi} + \ldots] \\ &\quad + \mu_0 \, (\nabla^2 \, \vec{\varphi}(x)) \, [1 + \mu_1 \, \vec{\varphi} \cdot \vec{\varphi} + \ldots] \\ &\quad + \text{higher derivatives} \end{aligned} \qquad \text{... (24.3.2)}$$

Clearly the first term on the right hand side does not contribute and near the critical point, where we are interested in the long-distance behaviour, the higher derivatives can be ignored. The lowest nonvanishing term in the equation of motion comes from the second term on the right hand side of Eq. (24.3.2) and we have the Landau-Lifshitz equation

$$\frac{\partial \vec{\varphi}}{\partial t} = g_0 \, \mu_0 \, \vec{\varphi} \times \nabla^2 \, \vec{\varphi} = g \, \vec{\varphi} \times \nabla^2 \, \vec{\varphi} \qquad \text{... (24.3.3)}$$

It is more convenient to write the above equation in Fourier space and in component form. If φ_i is the ith component of $\vec{\varphi}$ and φ_i (**k**) the Fourier component at wavenumber k, then Eq. (24.3.3) can be written as

$$\dot{\varphi}_i(\mathbf{k}) = - g \, \varepsilon_{ijk} \sum_{\mathbf{p}+\mathbf{q}=\mathbf{k}} \varphi_j(\mathbf{p}) \, q^2 \, \varphi_k(\mathbf{q}) \qquad \text{... (24.3.4)}$$

where ε_{ijk} is the usual completely antisymmetric Levi-Civita symbol. The above equation can be put in a more usable form by the following manipulations

$$\dot{\varphi}_i(\mathbf{k}) = -g\,\varepsilon_{ijk} \sum_{\mathbf{p}+\mathbf{q}=\mathbf{k}} \varphi_j(\mathbf{p})\, q^2\, \varphi_k(\mathbf{q})$$

$$= -g\,\varepsilon_{ikj} \sum_{\mathbf{p}+\mathbf{q}=\mathbf{k}} \varphi_k(\mathbf{p})\, q^2\, \varphi_j(\mathbf{q})$$

(interchanging j and k)

$$= -g\,\varepsilon_{ikj} \sum_{\mathbf{p}+\mathbf{q}=\mathbf{k}} \varphi_k(\mathbf{q})\, p^2\, \varphi_j(\mathbf{p})$$

(interchanging ($\mathbf{p}$ and $\mathbf{q}$)

$$= \frac{g}{2}\,\varepsilon_{ijk} \sum_{\mathbf{p}+\mathbf{q}=\mathbf{k}} (p^2 - q^2)\, \varphi_j(\mathbf{p})\, \varphi_k(\mathbf{q})$$

If we now add the relaxational part to the above equation, we have the complete equation of motion

$$\dot{\varphi}_i(\mathbf{k}) = \frac{g}{2}\,\varepsilon_{ijk} \sum_{\mathbf{p}+\mathbf{q}=\mathbf{k}} (p^2 - q^2)\, \varphi_j(\mathbf{p})\, \varphi_k(\mathbf{q})$$

$$- \Gamma_0\, k^2 \frac{\partial F}{\partial \varphi_i(-k)} + \eta \qquad \text{... (24.3.5)}$$

Note that the relaxational term must have the explicit k^2 to accommodate the conservation law associated with the total magnetization.

We now begin an investigation of the effect of the reversible term by exploring the dynamics in the ordered phase. The field $\vec{\varphi}$ has an expectation value which is the magnetization m of the system. We take the mangetization to be along the z-direction i.e. $\langle\vec{\varphi}\rangle = \varphi_3\,\hat{k} = m\hat{k}$. The quantity m is temperature-dependent and vanishes as T approaches T_c according to $|T - T_c|^{\beta}$. In Eq. (24.3.5), φ_3 has only one Fourier component, namely m at $k = 0$. With this in mind, we can find the equation of motion for the transverse fields φ_1 and φ_2 provided we know the free energy. It should be noted that below the transition, the transverse fields are massless and hence to quadratic order $\frac{\partial F}{\partial \varphi_i(-k)} = k^2\,\varphi_i(k)$, $i = 1, 2$. For what follows the noise is not important and consequently, we have from Eq. (24.3.5),

$$\dot{\varphi}_1 (k) = g\, m\, k^2\, \varphi_2 (k) - \Gamma_0\, k^4\, \varphi_1 (k) \qquad \text{... (24.3.6)}$$

$$\dot{\varphi}_2 (k) = -\, g\, m\, k^2\, \varphi_1 (k) - \Gamma_0\, k^4\, \varphi_2 (k)$$

Writing $Z = \varphi_1 + i\, \varphi_2$, we easily see that

$$\dot{Z} = -\, g\, m\, k^2\, Z - \Gamma_0\, k^4\, Z \qquad \text{... (24.3.7)}$$

The relaxation rate $\Omega(k)$ associated with Z is

$$\Omega (k) = i\, g \cdot m\, k^2 + \Gamma_0\, k^4 \qquad \text{... (24.3.8)}$$

which shows that there is propagating mode which has the frequency $\omega_s = g \cdot mk^2$ and which is damped out at the rate $\Gamma_0\, k^4$. This damped wave in the transverse mode dynamics is the spin wave of the Heisenberg magnet, which does not allow the system to order below $D = 2$. We note that because of the temperature dependence of m, the frequency $\Omega(k)$ of Eq. (24.3.8) depends on both wavenumber and the correlation length. We recall that

$$m \propto t^{\beta} \propto (t^{\nu})^{\beta/\nu} \propto \xi^{-\beta/\nu} \qquad \text{... (24.3.9)}$$

Now, $D\nu = 2 - \alpha = \gamma + 2\,\beta = (2 - \eta)\,\nu + 2\beta$ from the scaling laws and thus (D is the dimensionality of space)

$$\frac{\beta}{\nu} = \frac{D}{2} - 1 - \frac{\eta}{2} \qquad \text{... (24.3.10)}$$

We can ignore η which is very small and consequently arrive at

$$M \propto \xi^{-D/2+1} = \kappa^{D/2-1} \qquad \text{... (24.3.11)}$$

where as usual $\kappa = \xi^{-1}$. The spin wave frequency, consequently, has the behaviour

$$\omega_s = g \cdot mk^2 \propto gk^2\, \kappa^{(D/2)-1} \qquad \text{... (24.3.12)}$$

The above considerations are valid in the so-called hydrodynamic regime where $k \ll \xi^{-1}$. For a conserved order parameter system the scaling behaviour in this region, according to Eq. (24.1.13), is $\omega \sim k^2\, \xi^{2-z} = k^2\, \kappa^{z-2}$, which identifies the dynamic exponent z in this case as

$$z = 1 + \frac{D}{2} \qquad \text{... (24.3.13)}$$

We return to Eq. (24.3.8) and now investigate the damping term. The scaling behaviour of the frequency implies that it is a homogeneous function of k and κ. This statement should be valid for the complex frequency of Eq. (24.3.8) and hence the total k and κ power-count of the damping part must add up to $1 + D/2$. This can be achieved only if Γ_0 is renormalized to a k and κ-dependent Onsager coefficient and the spin wave damping ω_d is written as

$$\omega_d = \Gamma\,(k,\ \kappa)\ k^4 \qquad \text{... (24.3.14)}$$

with

$$\Gamma\,(k,\ \kappa) \sim k^{D/2-3} \quad k >> \kappa$$

$$\Gamma\,(k,\ \kappa) \sim \kappa^{D/2-3} \quad k << \kappa \qquad \text{... (24.3.15)}$$

or

$$\Gamma\,(k,\ \kappa) = k^{D/2-3}\ \gamma\left(\frac{\kappa}{k}\right) \qquad \text{... (24.3.16)}$$

Thus for $D < 6$, the Onsager coefficient diverges at the critical point as $k \to 0$. It should be noted that for $D > 6$, there will be no critical behaviour in Γ and the relaxation rate will have $z = 4$ as happens for a conserved order parameter in the mean field limit. For the dynamics, we can then view $D = 6$ as the upper critical dimension for this problem. A renormalization group treatment is in accordance with the above statements. It should also be pointed out that in problems involving dynamics, the upper critical dimension is determined by the nature of the reversible coupling and hence differs from one class of system to another, i.e. from ferromagnets to antiferromagnets to binary liquids and so on. The universality classes are consequently more restrictive. All ferromagnets behave the same way and so do all binary liquids but the upper critical dimension and the value of the dynamic exponent z are very different. The static exponents for the ferromagnet and binary liquid are different as well but they can be obtained from the same expression by setting different values of the order parameter dimensionality n ($n = 3$ for ferromagnets and $n = 1$ for binary liquids). In the treatment of dynamics the z-values are fundamentally different.

Our considerations so far have been restricted to $T < T_c$. We now want to go above the critical point. The spin waves which one has below T_c, become purely diffusive modes once one is above T_c. At $T = T_c$, the spin wave velocity disappears for $k \rightarrow 0$ and for $T > T_c$, we have a completely damped mode. Consequently, only the relaxational part of Eq. (24.3.8) will survive and that damping has the form given by Eq. (24.3.14). However, there is yet another difference as we pass from below T_c to above it. The spin wave susceptibility was k^{-2}. The susceptibility for $T > T_c$ corresponds to massive excitations and hence one has $(k^2 + \kappa^2)^{-1}$ for the susceptibility above the transition. Thus the spin wave diffusion above is given by Eq. (24.3.16), but the relaxation rate has to be written as

$$\omega_d = k^2 \chi^{-1} (k, \kappa) \Gamma (k, \kappa) = k^2 (k^2 + \kappa^2) k^{D/2-3} \gamma\left(\frac{\kappa}{k}\right)$$

$$= k^{1+(D/2)} \left(1 + \frac{\kappa^2}{k^2}\right) \gamma (\kappa/k) \qquad \text{... (24.3.17)}$$

Neutron scattering experiments close to and above T_c in various ferromagnets have confirmed the above form for the relaxation rate in $D = 3$.

We conclude this section with an intuitive discussion of the transport coefficient for binary liquids. Let us consider a region of size of the correlation length ξ. The fluctuating order parameter for the transition is Ψ, the concentration difference of the two kinds of atoms in the binary liquid. Let the average of Ψ over this region be $\overline{\Psi}$. The average of $\overline{\Psi}^2$ is by the static fluctuation dissipation theorem proportional to the susceptibility per unit volume. Thus

$$< \overline{\Psi}^2 > \sim \chi/\xi^D \sim \xi^{2-D-n} \qquad \text{... (24.3.18)}$$

Let us imagine an external force $\vec{\mathbf{F}}$ on one type of atoms and $-\vec{\mathbf{F}}$ on the other type, which keeps the total force zero. However, on the region of dimension ξ, there is going to be a net force of magnitude

$$\mathbf{f}_{\text{applied}} \sim \overline{\Psi} \xi^D \mathbf{F} \qquad \text{... (24.3.19)}$$

This force has to be balanced by the viscous drag, which leads to

$$\overline{\Psi} \xi^D \mathbf{F} = \overline{\eta} \mathbf{v} \xi^{D-2} \qquad \text{... (24.3.20)}$$

where $\overline{\eta}$ is the viscous coefficient.

The current is given by

$$\mathbf{j} = \overline{\Psi}\ \mathbf{v} = \frac{\overline{\Psi}^2\ \xi^2}{\overline{\eta}}\ \mathbf{F} \qquad \text{... (24.3.21)}$$

But by definition $\mathbf{j} = \lambda\,\mathbf{F}$, where λ is the diffusivity and hence

$$\lambda \propto \frac{\overline{\Psi}^2\ \xi^2}{\overline{\eta}} \sim \frac{\xi^{4-D-\eta}}{\overline{\eta}} \qquad \text{... (23.3.22)}$$

if we assume that $\overline{\eta}$ remains finite at the critical point, we find that in $D = 3\ (\eta \cong 0)$,

$$\lambda \propto \xi \qquad \text{... (24.3.23)}$$

and hence the diffusion coefficient diverges as the critical point is approached. It should be noticed from Eq. (24.3.22) that for $D > 4$, this divergent behaviour is absent and thus the upper critical dimension for the binary liquid is four in contrast to the ferromagnets of the previous Section where we found the upper critical dimension to be six.

24.4 Spinodal Decomposition

In this last section we will deal with two situations where the dynamics is not confined to near-equilibrium states, but involves states which are very far from equilibrium. To set up the problem, we consider the coexistence curve of Fig. 24.1, for a binary liquid. At any temperature $T < T_i$, there are two concentrations c_A and c_B in equilibrium. The dashed curve on which the susceptibility vanishes is called the spinodal line. States inside the spinodal curve, correspond to negative values of susceptibility, are unstable. The states between the coexistence curve and the spinodal curve are called metastable states. The problem that we will be interested in is one where at $t = 0$, the system in equilibrium in the one-phase region is *suddenly* cooled to a temperature in the two-phase region through the critical point. This puts the system in a nonequilibrium state (unstable) at $t = 0$. It is in a homogeneous state whereas the equilibrium state is one in which phase separation has taken place and the two phases coexist. The time evolution of the initial state will have to be towards this equilibrium state.

This is the problem of spinodal decomposition and the time development will take place in three stages.

In the very early stages the long wavelength fluctuations will begin to grow exponentially and a length scale will develop—a scale that will determine that the fluctuations with wavelengths greater than that scale will be growing. The exponential growth of the early stages is of course an artifact of a linearized equation of motion. The non-linear terms take over in the second stage of development and correctly provide the description of the growth of the fluctuations till the typical domain size is somewhat larger than the correlation length. For the final stage of development, the domain size grows as $t^{1/3}$ where t is the time and this is known as the Lifshitz-Slyozov law.

A variant of the above process is one where the system is put onto the metastable state by an initial supersaturation. This is achieved by taking an initial equilibrium state (point on the coexistence curve) and cooling it without allowing for change in concentration. This puts the system in a metastable state. The state will decay due to the formation of localized finite amplitude fluctuations. The formation of these droplets is called nucleation.

We being by writing down the equation of motion for the system which, as is customary in such problems, is a binary alloy consisting of two types of atoms—atoms A and atoms B. The homogeneous phase situation is an even mixture of equal amounts of A and B. Below T_c, the phases separate and the coexistence curve of Fig. 24.1 results. The order parameter field is $\varphi(\mathbf{r})$ which is the local difference of the number of A and B type atoms and the equation of motion is the dynamics of Section 24.1—the one where the order parameter obeys a conservation law, i.e.

$$\frac{\partial \varphi}{\partial t} = F\,\nabla^2 \frac{\delta F}{\delta \varphi} + \eta \qquad \text{... (24.4.1a)}$$

with

$$\langle \eta(\mathbf{x}_2, t_2)\,\eta(\mathbf{x}_1, t_1)\rangle = -2\Lambda\,\nabla^2\,\delta(\mathbf{x}_1 - \mathbf{x}_2)\,\delta(t_2 - t_1) \qquad \text{... (24.4.1b)}$$

$$F = \int d^D x \left[-\frac{a}{2}\varphi^2 + \frac{1}{2}(\vec{\nabla}\varphi)^2 + \frac{u}{4}\varphi^4 \right]. \qquad \text{... (24.4.1c)}$$

The important difference with the equation of motion of Section 24.1 is in the sign of the quadratic term in F. We have $a = a_0(T_c - T)$ and the negative sign of the coefficient of the quadratic term is because the time development occurs

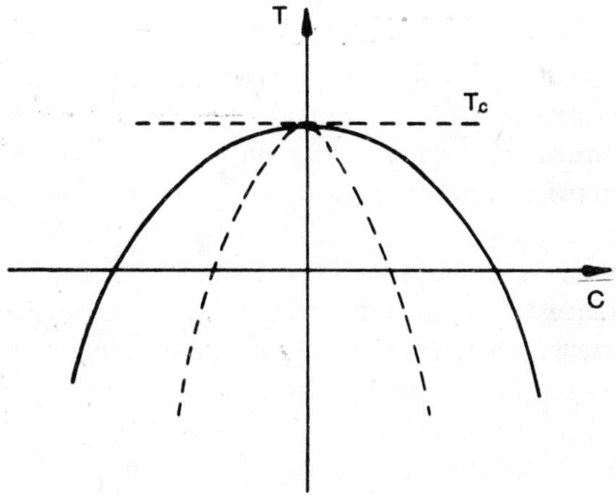

Fig. 24.1 The phase diagram in concentration temperature plane

at a temperature below T_c. In Fourier space and dropping nonlinear (i.e. the cubic) terms, we can write Eq. (24.4.1a) as

$$\frac{\partial \varphi (k)}{\partial t} = - \Gamma k^2 (-a + k^2) \varphi (k) + \eta (k) \quad \text{... (24.4.2)}$$

If we define the structure function $S(k, t)$ as

$$S(k, t) = < \varphi (k, t) \varphi (-k, t) >, \quad \text{... (24.4.3)}$$

then it is straightforward to show that Eq. (24.4.2) leads to

$$S(k, t) = S(k, 0) \, e^{-2\omega (k)t} \quad \text{... (24.4.4)}$$

where

$$\omega (k) = \Gamma k^2 (-a + k^2) \quad \text{... (24.4.5)}$$

Clearly, the structure factor grows if k^2 is smaller than a. Hence all fluctuation with wavenumber smaller than a will be growing. The growth rate will be maximum for a wavenumber k_c which satisfies $d\omega/dk = 0$, where

$$k_c^2 = a/2 \quad \text{... (24.4.6)}$$

Although the above linear theory shows the growth of long wavelength fluctuations, the effect of nonlinear terms is extremely important and real experiments or numerical simulations hardly ever show the behaviour shown in Eq. (24.4.4).

As $t \to \infty$, one has clear-cut domains formed and in the final approach to equilibrium, this problem is like a nucleation problem where one has droplets of both A and B phases but one of the phases (say B) is in minority which causes the departure from equilibrium. The B droplets will now grow on an average and the dynamics is specified by a diffusion equation describing the diffusion of the B-atoms through A and the continuity equation for the droplet distribution function $f(R, t)$. Appropriate approximations lead to the Lifshitz-Slyozov growth law. A renormalization group argument due to Bray does reproduce the Lifshitz-Slyozov law and is one of the very few exact results in this subject.

In a brief discussion of nucleation, we will consider a D-dimensional lattice of N Ising spins, with a small external field H pointing in the positive z-direction. Most of the spins in the system will be pointing in the + z direction. There will be a few clusters (droplets) of down spins, which will be well separated and can be treated as a gas of non-interacting droplets. The number of l-spin clusters at a temperature T will be given by the Boltzmann distribution

$$\eta_l = N\, e^{-\beta\varepsilon_l} \qquad \text{... (24.4.7)}$$

where ε_l is the free energy for the formation of the cluster. It is taken to consist of two parts—one proportional to H which is a bulk effect and determined by the total volume (i.e. total number of spins in the cluster) and the other a surface tension term, proportional to the surface area of the droplet. The first is proportional to l and the second to $l^{(D-1)/D}$. It is conventional to write the energy as

$$\varepsilon_l = 2Hl + \sigma\, l^{(D-1)/D} \qquad \text{... (24.4.8)}$$

We now imagine turning the magnetic field to make it point in the negative z-direction. We can now have a metastable state—the field has turned but the system not having responded fast enough still retains some positive magnetism, i.e. there are more clusters of the 'wrong' variety (spin up while the mangetic field is down). We now have a situation where in applying Eq. (24.4.8), H is negative (spins and field in different directions) and l can have a large range of values. It is clear that the surface and bulk terms are now in competition and there will be a minimum in n_l when the energy expression has a maximum, i.e. when

$$2|H| = \frac{D-1}{D}\, \sigma\, l_c^{-1/D}$$

$$\text{or} \quad l_c = \left[\frac{(D-1)\sigma}{2D|H|}\right]^D \qquad \text{... (24.4.9)}$$

Droplets with $l > l_c$ will be energetically favoured and consequently grow, thus providing the nucleating mechanism by which the metastable state will decay.

To investigate the decay of the metastable state, we need to have an equation for the dynamics. This will be the usual conservation law which we encountered in Chapter 15.

$$\frac{\partial n_l}{\partial t} = J_{l-1} - J_l, \; (l \geq 2) \qquad \text{... (24.4.10)}$$

the change in the number n_l is brought about only by the current set up. We next require an expression for the current. This can be obtained by postulating that the current is due to an evaporation-condensation mechanism, whereby the evaporation of a single 'atom' from a cluster or the condensation of a single 'atom' on the cluster is the dominant mechanism for the current. Consequently

$$J_l = + R_l\, n_l - R'_{l+1}\, n_{l+1} \qquad \text{... (24.4.11)}$$

where J_l is the rate at which the size l grows to size $l + 1$. Clearly R_l is the condensation rate and R'_{l+1} is the evaportion rate. Inserting Eq. (24.4.11) into Eq. (24.4.10), we get the simple picture that n_l is changed because of loss due to condensation (forms n_{l+1}) and evaporation (forms n_{l-1}) and gain due to evaporation from n_{l+1} and condensation on n_{l-1}. Detailed balance would imply that in equilibrium every $J_l = 0$ and hence

$$R_l\, e^{-\varepsilon_l \beta} = R'_{l+1}\, e^{-\varepsilon_{l+1}\beta} \qquad \text{... (24.4.12)}$$

It is convenient to treat the l as a continuous variable and write Eq. (24.4.10) as

$$\frac{\partial n_l}{\partial t} = -\frac{\partial}{\partial l} J_l \qquad \text{... (24.4.13)}$$

and

$$J_l = R_l\, n_l - R_l\, e^{\beta(\varepsilon_{l+1} - \varepsilon_l)}\, n_{l+1}$$

$$\cong R_l\, n_l\, R_l\, n_{l+1} - \beta\, R_l\, (\varepsilon_{l+1} - \varepsilon_l)\, n_l$$

$$= -R_l \frac{\partial n_l}{\partial l} - \beta R_l \frac{\partial \varepsilon_l}{\partial l} n_l \qquad \text{... (24.4.14)}$$

leading to

$$\frac{\partial n_l}{\partial t} = \frac{\partial}{\partial l}\left[\beta\, R_l \frac{\partial \varepsilon_l}{\partial l} + R_l \frac{\partial n_l}{\partial l}\right] \qquad \text{... (24.1.15)}$$

The equilibrium solution corresponds to $J_l = 0$. In the non-equilibrium steady state solution which corresponds to the production of larger than critical sized droplets, we have a constant current I, i.e. $J_l = I$ for all l. The solution for n_l is some time-independent $n_l^{(s)}$ leading to a steady value of I. The boundary conditions on this steady $n_l^{(s)}$ *are*

$$n_l^{(s)} \underset{l \to 0}{\to} \bar{n}_l = N\, e^{-\beta\varepsilon_l}$$

$$n_l^{(s)} \underset{l \to \infty}{\to} 0 \qquad \text{... (24.4.16)}$$

Thus, this steady state describes the formation of droplets at $l = 0$ (i.e. a source at $l = 0$) and removal of large droplets ($l > l_c$) which causes the decay of the metastable state. Using the constant value I of J_l in Eq. (24.4.14), we have

$$\frac{\partial n_l^{(s)}}{\partial l} + \frac{\partial \varepsilon_l}{\partial l}\, n_l^{(s)} = -\, I/R_l \qquad \text{... (24.4.17)}$$

leading to

$$n_l^{(s)} = e^{-\beta\varepsilon_l'}\left\{-I \int_0^l \frac{e^{\beta\varepsilon_l'}}{R_l'}\, dl' + C\right\} \qquad \text{... (24.4.18)}$$

The boundary condition that $n_l \to 0$ for $l \to \infty$ yields

$$C = I \int_0^{\infty} \frac{e^l}{R_l}\, dl$$

and thus $n_l^{(s)} = I\, e^{-\beta\varepsilon_l}\left\{\int_0^\infty \frac{e^{\beta\varepsilon_l}}{R_l}\, dl - \int_0^l \frac{e^{\beta\varepsilon_l'}}{R_{l'}}\, dl'\right\}$

$$= I\, e^{-\beta\varepsilon_l} \int_l^\infty \frac{e^{\beta\varepsilon_l'}}{R_l'\, dl'} \qquad \text{... (24.4.19)}$$

As $l \to 0$, we have the other boundary condition whereby $n_l^{(s)}$ tends to the equilibrium value $n_l = N e^{-\beta\varepsilon_l}$ and hence

$$1 = I \int_0^\infty \frac{1}{R_l\, \bar{n}_l}\, dl \qquad \text{... (24.4.20)}$$

An expression for the current can be obtained by evaluating the integral in a saddle point approximation. This implies evaluating $\bar{n}_l$ about the point where $\bar{n}_l$ has a minimum (ε_l has a maximum). Writing,

$$\varepsilon_l \cong \varepsilon_c - l^2 \left|\frac{\partial^2\varepsilon}{\partial l^2}\right|_{l_c} + \ldots \qquad (24.4.21)$$

$$R_l \cong R_c$$

we find from Eq. (24.4.20)

$$1 = I\,(N/R_c)\, e^{\beta\varepsilon_c} \int_0^\infty e^{-\beta\left|\partial^2\varepsilon/\partial l^2\right|_{l_c} l^2}\, dl$$

$$= I(N/R_c)\, e^{\beta\varepsilon_c}\left[\frac{\pi}{\beta|\partial^2\varepsilon/\partial l^2|}\right]$$

$$= (I/I_0)\, e^{-\beta\varepsilon_c}$$

and thus

$$I = I_0\, e^{-\beta\varepsilon_c} \qquad \text{... (24.4.22)}$$

This is precisely the form of the current in all metastable states (see Chapter 15). The potential barrier is ε_c and the β comes because the decay is thermally activated. The above theory of nucleation is due to Becker and Doring.

Chapter 25
Turbulence

25.1 Introduction

In this chapter we will use methods of statistical mechanics on a problem of fluid dynamics—turbulence. Turbulence deals with a wide class of flows which seem to possess complex and seemingly random structure at some macroscopic scale of dynamical importance. The chief physical characteristic is a tremendous enhancement in transport properties—transport of momentum, energy and particles is orders of magnitude higher than possible by molecular processes. Equally important is the sensitive dependence on initial conditions of a turbulent flow. The turbulent flows which are nearly identical at a given time do not remain so on time scales of dynamical interest. In our dealings with dynamical systems in Chapter 2, we found this sensitivity to initial conditions to be the characteristic of chaotic systems (reflected in positive Lyapunov exponents) and the basis of statistical mechanics. Here again, it will be this sensitivity to initial conditions, which will make it possible for methods of statistical mechanics to be applicable in the study of turbulence.The key to this is the fact that while details of fully developed turbulence are extremely sensitive to triggering disturbances, average properties are not. Hence statistically averaged quantities will be the dynamical variables of interest. We will restrict ourselves to homogeneous isotropic turbulence—which means that all points in space are statistically equivalent (homogeneous) and all directions in space are statistically equivalent (isotropic). In addition, we will consider incompressible fluids only.

The enhanced transport property of turbulence can be exhibited by demonstrating that infinitesimal line elements are stretched by homogeneous isotropic turbulence. Consider the infinitesimal line element $\delta\mathbf{x}(t)$ between two points A and B which were separated by a non-random distance $\delta x(0)$ at $t = 0$. Since $\delta\mathbf{x}$ is infinitesimal, $\delta\mathbf{x}(t)$ must be linearly related to $\delta\mathbf{x}(0)$ and hence there exists a matrix u_{ij} such that (throughout this chapter, we will use the repeated index summation convention)

$$\delta x_j\ (t) = u_{ij}\ \delta x_i\ (0) \qquad \text{... (25.1.1)}$$

The matrix u has to be a random function of time

$$\delta x\ (t)^2 = \delta x_j\ (t)\ \delta x_j\ (t) = u_{ij}\ \delta x_i\ (0)\ u_{kj}\ \delta x_k\ (0) = W_{ik}\ \delta x_i\ (0)\ \delta x_k\ (0) \qquad \text{... (25.1.2)}$$

where $W_{ik} = u_{ij}\, u_{kj}$ is a real symmetric matrix.

$$< \delta x\ (t)^2 > = < W_{ik} > \delta x_i\ (0)\ \delta x_k\ (0) \qquad \text{... (25.1.3)}$$

since $\delta\mathbf{x}\ (0)$ is not random. Isotropy and homogeneity imply that $< W_{ik} > = \lambda\ (t)\ \delta_{ik}$ with $\lambda(b)$ an analytic function of time since no point $\mathbf{x}$ or direction can be preferred. Hence

$$<\delta x\ (t)^2 > = \lambda(t)\ \delta x\ (0)^2 \qquad \text{... (25.1.4)}$$

For a fixed t, let the eigenvalues of W be w_1, w_2 and w_3 and the orthogonal eigenvectors $\mathbf{a}_1$, $\mathbf{a}_2$, and $\mathbf{a}_3$. A small volume bounded by $\varepsilon\mathbf{a}_1$, $\varepsilon\mathbf{a}_2$, $\varepsilon\mathbf{a}_3$ around the point A, becomes a volume bounded by $\varepsilon w_1\, \mathbf{a}_1$ $\varepsilon w_2\, \mathbf{a}_2$ and $\varepsilon w_3\, \mathbf{a}_3$ after a time t, and since the fluid is incompressible, we must have $w_1\, w_2\, w_3 = 1$. The arithmetic-geometric mean inequality now yields

$$\lambda = \frac{1}{3} < \text{Tr}\ W > = \frac{1}{3}\ (w_1 + w_2 + w_3) \geq (w_1\ w_2\ w_3)^{1/3} = 1 \qquad \text{... (25.1.5)}$$

proving that line elements are indeed stretched. It should be noted that vortex lines are not stretched by the turbulence although vortex lines move with the fluid in the inviscid limit. This is because $\vec{\omega} = \vec{\nabla} \times \mathbf{v}$ bears a special relation with the velocity field and the statistical independence required for Eq. (25.1.3) is not there.

We now need to discuss the method of taking averages. The available averages are space averages, time averages and ensemble averages. Spatial averages make sense only if scales over which the turbulence is homogeneous or approximately so are much larger than the scale of turbulent fluctuations. Similarly the time averaging is useful if the time over which the turbulence is statistically stationary is longer than the time for turbulence fluctuations. Experiments usually deal with time averages obtained by using a local probe. The third average is the ensemble average where the average is taken over different realizations of the turbulent field. This ensemble averaging makes sense and can be equal to the time averaging because of the sensitive dependence on the initial conditions which makes the turbulent flow strongly mixing and ergodic. However, this would imply the existence of a stable asymptotic statistical state. It is quite likely that such a state exists but it has not been proved. The advantage of the ensemble averaging is that it applies to inhomogeneous nonstationary turbulence.

The governing dynamical equation for the study of turbulence is the Navier-Stokes equation for the velocity field $\mathbf{v}\,(\mathbf{r}, t)$.

$$\frac{\partial \mathbf{v}}{dt} + (\mathbf{v} \cdot \vec{\nabla})\, \mathbf{v} = -\frac{\vec{\nabla} P}{\rho} + \nu\, \nabla^2\, \mathbf{v} \qquad \text{... (25.1.6)}$$

with the incompressibility condition

$$\vec{\nabla} \cdot \mathbf{v} = 0 \qquad \text{... (25.1.7)}$$

In Eq. (25.1.6), P is the pressure, ρ the density and ν the kinematic viscosity. We have assumed body forces (like gravity) to be irrelevant for our purpose. The pressure is not obtained from any constitutive relation. It serves to maintain the incompressibility condition through the relation

$$\nabla^2 P = -\rho\, \vec{\nabla}\, [(\mathbf{v} \cdot \vec{\nabla})\, \mathbf{v}] \qquad \text{... (25.1.8)}$$

obtained by taking the divergence of Eq. (25.1.6). If we now take a scalar product with $\mathbf{v}$ in Eq. (25.1.6) and integrate over a region confined between rigid stationary boundaries, then

$$\frac{\partial}{\partial t} \frac{1}{2} \int |\mathbf{v}|^2\, d^3x = -\int \mathbf{v} \cdot (\mathbf{v} \cdot \vec{\nabla})\, \mathbf{v}\, d^3x - \frac{1}{\rho} \int (\mathbf{v} \cdot \vec{\nabla} P)\, d^3x$$

$$+ \nu \int \mathbf{v} \cdot \nabla^2 \mathbf{v} \, d^3x \qquad \text{... (25.1.9)}$$

We note that

$$\int \mathrm{v}.\nabla^2 \mathrm{v} d^3x = -\int \frac{\partial v_\alpha}{\partial x_\beta} \frac{\partial v_\alpha}{\partial x_\beta} d^3x \qquad \text{... (25.1.10)}$$

due to the vanishing of surface terms and that

$$\int (\mathbf{v} \cdot \vec{\nabla}) P \, d^3x = \int \vec{\nabla} \cdot (P\mathbf{v}) \, d^3x - \int P \, (\vec{\nabla} \cdot \mathbf{v}) \, d^3x$$

$$= \int \vec{\nabla} \cdot (P\mathbf{v}) \, d^3x$$

is a purely surface term by Gauss's theorem and so is $\int \mathbf{v} \cdot (\mathbf{v} \cdot \vec{\nabla}) \mathbf{v} \, d^3x$ due to the identity

$$\frac{\partial}{\partial x_i} (v^2 v_i) = v^2 \frac{\partial v_i}{\partial x_i} + v_i \frac{\partial}{\partial x_i} (v_i \, v_j)$$

$$= 2v_i \cdot v_j \frac{\partial v_j}{\partial x_i} = 2 \, \mathbf{v} \cdot (\mathbf{v} \cdot \vec{\nabla}) \, \mathbf{v} \qquad \text{... (25.1.11)}$$

Hence over a bounded volume

$$\frac{\partial}{\partial t} \frac{1}{2} \int |\mathbf{v}|^2 \, d^3x = -\nu \int \frac{\partial v_\alpha}{\partial x_\beta} \frac{\partial v_\alpha}{\partial x_\beta} \, d^3x \qquad \text{... (25.1.12)}$$

In the absence of external forces or moving boundaries, all motion in a finite region ceases as $t \to \infty$. To maintain the turbulence we must have external forces—the external forces can be produced by moving boundaries.

Hence the Navier-Stokes equation has to be written as (in component notation)

$$\frac{\partial v_i}{\partial t} + (\mathbf{v} \cdot \vec{\nabla}) \, v_i = -\frac{1}{\rho} \frac{\partial P}{\partial x_i} + \nu \, \nabla^2 v_i + f_i \qquad \text{... (25.1.13)}$$

where f_i is the ith component of the external forces.

We end this section by noting that although the turbulence can be generated by a deterministic equation, it is convenient, given that the solution v_i $(\mathbf{x}, t)$ is going to be a random function, to take the external force f_i to be random. The averaging that we have been discussing becomes straightforward in that case—the averaging of functions of **v** will have to be done over the distribution that produces the random external force f_i. Systems following Eq. (25.1.13) with the f_i random are called randomly stirred fluids. We will return to it later since it is a promising candidate for the development of a perturbation theory.

25.2 Correlation Functions, Energy Balance, etc.

From now on, when we talk about averages, it will be ensemble averaging that we will be concerned with. We begin by noting that spatial homogeneity would make $< \mathbf{v}(\mathbf{x}, t) >$ independent of **x** and hence in Eq. (25.1.6) the spatial derivatives become zero once the averaging has been performed and consequently $\partial < \mathbf{v} >/\partial t$ would be zero. Then $< \mathbf{v}\,(x, t) >$ would be independent of space and time and can without loss of generality be chosen to be zero.

The important quantities are the correlation functions

$$C_{\alpha\beta\ldots\nu} = < v_\alpha\,(\mathbf{x}_1,\ t_1)\ v_\beta\,(\mathbf{x}_2,\ t_2)\ \ldots\ v_\nu\,(\mathbf{x}_n,\ r_n) > \qquad \ldots\ \textbf{(25.2.1)}$$

If all the time arguments are equal, we have an equal time correlation function and for different values of the time argument, it is a time-dependent correlation function. The single most important correlation function is the two-point correlation

$$C_{\alpha\beta}\,(\mathbf{x}_{12},\ t_{12}) = < v_\alpha\,(\mathbf{x}_1,\ t_1)\ v_\beta\,(\mathbf{x}_2,\ t_2) > \qquad \ldots\ \textbf{(25.2.2)}$$

where $\mathbf{x}_{12} = \mathbf{x}_2 - \mathbf{x}_1$ and $t_{12} = t_2 - t_1$.
The equal time version is

$$R_{\alpha\beta}\,(\mathbf{x}_{12}) = < v_\alpha\,(\mathbf{x}_1,\ t)\ v_\beta\,(\mathbf{x}_2,\ t) > \qquad \ldots\ \textbf{(25.2.3)}$$

If we call $\mathbf{x}_{12} = \mathbf{r}$, then from homogeneity

$$R_{\alpha\beta}\,(\mathbf{r}) = R_{\beta\alpha}\,(-\mathbf{r}), \qquad \ldots\ \textbf{(25.2.4)}$$

from incompressibility

$$\frac{\partial R_{\alpha\beta}}{\partial r_\alpha} = \frac{\partial R_{\alpha\beta}}{\partial r_\beta} \qquad \ldots\ \textbf{(25.2.5)}$$

and the Schwartz inequality implies

$$\sum_{\alpha} R_{\alpha\alpha}(0) \sum_{\beta} R_{\beta\beta}(0) \geq |R_{\alpha\beta}(\mathbf{r})|^2. \qquad \text{... (25.2.6)}$$

The vorticity correlation function is given by

$$< \omega_\alpha(\mathbf{x}+\mathbf{r})\, \omega_\beta(\mathbf{x}) > = \nabla^2 R_{\alpha\beta}(\mathbf{r}) - \delta_{\alpha\beta}\, \nabla^2 R_{\gamma\gamma} + \frac{\partial^2}{\partial r_\alpha\, \partial r_\beta} R_{\gamma\gamma} \qquad \text{... (25.2.7)}$$

For isotropic turbulence $R_{\alpha\beta}(\mathbf{r})$ can be further simplified. Consider two arbitrary vectors **a** and **b** and form the scalar

$$F = a_\alpha\, b_\beta\, R_{\alpha\beta} = < \mathbf{a}\cdot\mathbf{v}(\mathbf{x}+\mathbf{r})\ \mathbf{b}\cdot\mathbf{v}(\mathbf{x}) >$$

which has to be invariant under rotations if the turbulence is isotropic and hence can only depend on $a^2, b^2, r^2, \mathbf{a}\cdot\mathbf{b}, \mathbf{a}\cdot\mathbf{r}, \mathbf{b}\cdot\mathbf{r}$ and $\mathbf{a}\cdot\mathbf{b}\times\mathbf{r}$. Since by definition F is linear in **a** and **b**, we must have

$$F = (\mathbf{a}\cdot\mathbf{r})\,(\mathbf{b}\cdot\mathbf{r})\,A(r) + \mathbf{a}\cdot\mathbf{b}\;B(r) + (\mathbf{a}\cdot\mathbf{b}\times\mathbf{r})\;C(r)$$

$$= a_\alpha\, b_\beta\, r_\alpha\, r_\beta\, A(r) + a_\alpha\, b_\beta\, \delta_{\alpha\beta}\, B(r) + \varepsilon_{\alpha\beta\gamma}\, a_\alpha\, b_\beta\, r_\gamma\, C(r)$$

leading to

$$R_{\alpha\beta}(r) = r_\alpha\, r_\beta\, A(r) + \delta_{\alpha\beta}\, B(r) + \varepsilon_{\alpha\beta\gamma}\, r_r\, C(r)$$

If the turbulence is invariant under reflections in space in addition to rotations, then the pseudoscalar $\mathbf{a}\cdot\mathbf{b}\times\mathbf{r}$ is not invariant and we have

$$R_{\alpha\beta}(r) = r_\alpha\, r_\beta\, A(r) + \delta_{\alpha\beta}\, B(r)$$

which is more commonly written as

$$R_{\alpha\beta}(r) = \overline{v^2}\left[\frac{f(r) - g(r)}{r^2}\, r_\alpha\, r_\beta + g(r)\, \delta_{\alpha\beta}\right]. \qquad \text{... (25.2.8)}$$

In the above $\overline{v^2}$ is the average value of each component of velocity and the incompressibility condition of Eq. (25.2.5), leads to the constraint

$$g(r) = f(r) + \frac{1}{2}\, r\, \frac{df}{dr} \qquad \text{... (25.2.9)}$$

The functions $f(r)$ and $g(r)$ are called the longitudinal and transverse velocity correlations since

$$\nabla^2 f(r) = \langle v_P (\mathbf{x} + \mathbf{r})\, v_P (\mathbf{x}) \rangle \qquad \text{... (25.2.10)}$$

$$\nabla^2 g(r) = \langle v_n (\mathbf{x} + \mathbf{r})\, v_n (\mathbf{x}) \rangle$$

where v_p is the velocity component parallel to $\mathbf{r}$ and v_n is one of the components perpendicular to $\mathbf{r}$. From homogeneity $f'(0) = g'(0)$ and hence an expansion around $r = 0$, yields

$$f(r) = 1 - \frac{r^2}{2\lambda^2} + \ldots$$

$$g(r) = 1 - \frac{r^2}{\lambda^2} + \ldots \qquad \text{... (25.2.11)}$$

with

$$\frac{1}{\lambda^2} = -f''(0) = \left\langle \left(\frac{\partial v_1}{\partial x_1}\right)^2 \right\rangle / \overline{v^2} \qquad \text{... (25.2.12)}$$

The length scale is called the Taylor microscale. From Eq. (25.2.7) we find

$$\langle \omega^2 \rangle = -\nabla^2 R_{\alpha\alpha}(\mathbf{r})\,|_{\mathbf{r}=0} = -\overline{v^2}\,\frac{1}{2}\,\frac{d^2}{dr^2}\, r\,(f + 2g)|_{r=0}$$

$$= 15\,\overline{v^2} / \lambda^2 \qquad \text{... (25.2.13)}$$

For homogeneous isotropic turbulence, $\langle \mathbf{v} \cdot \overline{v^2}\, \mathbf{v} \rangle = -\langle \omega^2 \rangle$ and the rate of energy dissipation ε becomes

$$\varepsilon = \nu \langle \omega^2 \rangle = 15\nu\, \overline{v^2}/\lambda^2 \qquad \text{... (25.2.14)}$$

For practical calculations it is more convenient to work in the momentum space. To do so we introduce the Fourier transform

$$\mathbf{v}(\mathbf{x}, t) = \int \mathbf{u}(\mathbf{k}, t)\, e^{i\mathbf{k}\cdot\mathbf{x}}\, d^D k \qquad \text{... (25.2.15)}$$

where

$$\mathbf{u}\,(\mathbf{k}, t) = \frac{1}{(2\pi)^D} \int \mathbf{v}\,(\mathbf{x}, t)\, e^{-i\mathbf{k}\cdot\mathbf{x}}\, d^D x \qquad \text{... (25.2.16)}$$

For the two-point function, we immediately find

$$< u_\alpha(\mathbf{k}, t)\; u_\beta\,(\mathbf{p}, t) > = \frac{1}{(2\pi)^{2D}} \int d^D x \int d^D y < v_\alpha\,(\mathbf{x}, t)\; v_\beta\,(\mathbf{y},\, t) > \mathbf{e}^{-ik\cdot\mathbf{x} - ip\cdot\mathbf{y}}$$

$$= \frac{1}{(2\pi)^{2D}} \int d^D x\; d^D y\; R_{\alpha\beta}\,(\mathbf{x} - \mathbf{y})\; e^{-i\mathbf{k}\cdot\mathbf{x} - i\mathbf{p}\cdot\mathbf{y}}$$

$$= \frac{1}{(2\pi)^{2D}} \int d^D y\; e^{i(\mathbf{k}-\mathbf{p})\cdot\mathbf{y}} \int d^D z\; R_{\alpha\beta}\,(\mathbf{z})\; e^{-i\mathbf{k}\cdot\mathbf{z}}$$

$$= \delta^D\,(\mathbf{k} - \mathbf{p})\; S_{\alpha\beta}\,(\mathbf{k}) \qquad \text{... (25.2.17)}$$

where

$$S_{\alpha\beta}\,(\mathbf{k}) = \frac{1}{(2\pi)^D} \int R_{\alpha\beta}\,(\mathbf{r})\; e^{-i\mathbf{k}\cdot\mathbf{r}}\; d^D r$$

Similarly for the three-point function,

$$< u_\alpha\,(\mathbf{k})\; u_\beta\,(\mathbf{p})\; u_\gamma\,(\mathbf{q}) > = T_{\alpha\beta\gamma}\,(\mathbf{k},\, \mathbf{p})\; \delta\,(\mathbf{p} + \mathbf{k} + \mathbf{q}) \qquad \text{... (25.2.18)}$$

where

$$T_{\alpha\beta\gamma}\;\; (\mathbf{k},\, \mathbf{p}) = \frac{1}{(2\pi)^{2D}} \int d^D y \int d^D z < v_\alpha\;(\mathbf{x}+\mathbf{y})\; v_\beta\;(\mathbf{x} + \mathbf{z})\; v_r\;(\mathbf{x}) >$$

$$e^{-i\mathbf{k}.\mathbf{y}\, -\, i\mathbf{p}\cdot\mathbf{z}} \qquad \text{... (25.2.19)}$$

Isotropy puts constraints on $S_{\alpha\beta}\,(\mathbf{k})$. Once again $a_\alpha\, b_\beta\, S_{\alpha\beta}\,(\mathbf{k})$ can only be functions of k^2, $\mathbf{a}\cdot\mathbf{b}$, $\mathbf{a}\cdot\mathbf{k}$ and $\mathbf{b}\cdot\mathbf{k}$ only and being linear in $\mathbf{a}$ and $\mathbf{b}$

$$S_{\alpha\beta} = A\,(k)\; k_\alpha\; k_\beta + B\,(k)\; \delta_{\alpha\beta} \qquad \text{... (25.2.20)}$$

The incompressibility condition implies that $S_{\alpha\beta}\,(\mathbf{k})$ can be expressed (we consider the special case $D = 3$ now)

$$S_{\alpha\beta}(\mathbf{k}) = \frac{E(k)}{4\pi k^2} P_{\alpha\beta}(k) \qquad \text{... (25.2.21)}$$

where

$$P_{\alpha\beta}(k) = \delta_{\alpha\beta} - k_\alpha k_\beta / k^2 \qquad \text{... (25.2.22)}$$

The scalar function $E(k)$ is to be interpreted as the kinetic energy density in modes of wavenumber k, so that

$$\frac{3}{2}\overline{v^2} = \int_0^\infty E(k)\, dk \qquad \text{... (25.2.23)}$$

We are now in a position to write down the Navier-Stokes equation in momentum space and point out the central difficulty in the problem of turbulence. Writing Eq. (25.1.6) in momentum space leads to

$$\left(\frac{\partial}{\partial t} + \nu k^2\right) u_\alpha(\mathbf{k}, t) = -i \int p_\beta\, u_\beta(\mathbf{k}-\mathbf{p}, t)\, u_\alpha(\mathbf{p}, t)\, d^D p$$

$$- ik_\alpha P(\mathbf{k}, t) \qquad \text{... (25.2.24)}$$

The pressure $P(\mathbf{k}, t)$ can be eliminated using the incompressibility condition and we find

$$\left(\frac{\partial}{\partial t} + \nu k^2\right) u_\alpha(\mathbf{k}, t) = M_{\alpha\beta\gamma}(\mathbf{k}) \int u_\beta(\mathbf{p}, t)\, u_\gamma(\mathbf{k}-\mathbf{p}, t)\, d^D p$$

... (25.2.25)

where the coupling constant $M_{\alpha\beta\gamma}(\mathbf{k})$ is

$$M_{\alpha\beta\gamma}(\mathbf{k}) = -\frac{i}{2}\left[k_\beta P_{\alpha\gamma}(\mathbf{k}) + k_\gamma P_{\alpha\beta}(\mathbf{k})\right] \qquad \text{... (25.2.26)}$$

with $P_{\alpha\beta}(\mathbf{k})$ given by Eq. (25.2.22). From Eq. (25.2.25), the equation for the two-point function $S_{\alpha\beta}(\mathbf{k}, t)$ is found from straightforward algebra to be

$$\left(\frac{\partial}{\partial t} + 2\nu k^2\right) S_{\alpha\beta}(\mathbf{k}, t) = M_{\alpha\rho\sigma}(\mathbf{k}) \int T_{\beta\rho\sigma}(-\mathbf{k}, \mathbf{p})\, d^D p$$

$$+M_{\beta\rho\sigma}(-\mathbf{k})\int T_{\alpha\rho\sigma}(\mathbf{k},\mathbf{p})\,d^Dp \qquad \text{... (25.2.27)}$$

The two-point function $S_{\alpha\beta}$ is determined by the three-point function $T_{\alpha\beta\gamma}$. If we now use Eq. (25.2.25), to find the dynamics of $T_{\alpha\beta\gamma}$, we find

$$\left[\frac{\partial}{\partial t} + \nu\,(k^2 + p^2 + |\mathbf{k} + \mathbf{p}|^2)\right] T_{\alpha\beta\gamma}(\mathbf{k},\ \mathbf{p}) = M_{\alpha\rho\sigma}(\mathbf{k}) \times$$

$$\int U_{\beta\gamma\rho\sigma}(\mathbf{p},\ \mathbf{k} - \mathbf{p},\ \mathbf{q})\,d^Dq \qquad \text{... (25.2.28)}$$

where $U_{\beta\gamma\rho\sigma}$ involves correlations of four velocities and is a four-point function. Thus we see that an n-point function requires the knowledge of a $(n + 1)$-point function. This is the closure problem which we have encountered before in Chapter 4. There are no small parameters in the problem to order a perturbation theory and thus closure problem is what makes turbulence an impossibly difficult proposition.

The energy conservation, discussed before, when cast in Fourier space, acquires the following form for isotropic turbulence and Eq. (25.2.27) becomes

$$\left(\frac{\partial}{\partial t} + 2\nu k^2\right) E\,(k,\ t) = T(k,\ t) \qquad \text{... (25.2.29)}$$

where

$$T(k,\ t) = -\ 4\pi k^2\ k_\rho\ \text{Im}\int T_{\sigma\rho\sigma}(\mathbf{k},\ \mathbf{p})\ d^3p \qquad \text{... (25.2.30)}$$

The nonlinear modes conserve energy and consequently

$$\int_0^\infty T\,(k,\ t)\ dk = 0 \qquad \text{... (25.2.31)}$$

and Eq. (25.2.29) becomes

$$\frac{d}{\partial t}\left(\frac{3}{2}\,\overline{v}^2\right) = -\ \varepsilon \qquad \text{... (25.2.32)}$$

where

$$\varepsilon = 2\nu \int_0^\infty k^2 \, E\,(k,\, t)\; dk \qquad \text{... (25.2.33)}$$

and Eq. (25.2.23) has been used. The quantity ε is the rate of kinetic energy dissipation per unit volume. The high wavenumber modes are dissipated more effectively by the viscosity as can be seen from Eq. (25.2.33).

The advective term $(\mathbf{v} \cdot \vec{\nabla})\, \mathbf{v}$ which involves what are usually called inertial forces and the pressure forces $\vec{\nabla} P$ conserve the energy. In fact, $\int v_i^2 \, d^3 r$ is conserved by the inertia force for each i separately and the pressure force conserves kinetic energy in each Fourier mode. In the absence of other constraints it is plausible then that the inertial forces simply spread the energy over all the wavevectors. In particular, the tendency should be to transfer energy where it is lowest. Since viscosity depletes more effectively the energy in the high wavenumber modes, the typical effect of inertia would be to transfer energy from low wavenumber to high wavenumber modes (large eddies to small eddies).

We now define a length scale L which is essentially the length over which the two-point function $R_{\alpha\beta}$ $(\mathbf{r})$ decays to zero. A useful measure of it is given by the definition

$$L = \int_0^\infty f(r)\; dr \qquad \text{... (25.2.34)}$$

The wavenumber range can then be divided into three parts according to the following criteria:

$$kL << 1 \qquad \text{... (25.3.34a)}$$

referred to as large eddies,

$$kL \sim 1 \qquad \text{... (25.3.4b)}$$

referred to as the energy-containing eddy and finally

$$kL > 1 \qquad \text{... (25.2.34c)}$$

a range referred to as the small eddies. The energy-spectrum tensor $S_{\alpha\beta}$ $(\mathbf{k})$ is a quantity of fundamental interest and the study of this spectrum depends on which of the above momentum ranges one is discussing. The large eddy region $kL << 1$ contains negligible energy. The energy-containing range of eddies are of dynamical importance. The characteristic velocity of these eddies is $\bar{v}$ and the characteristic length is L. The time scale for decay of the turbulence is $L/\bar{v}$ which is also the circulation time of an eddy in the energy-containing range. The decay time and the circulating time being of the same order, establishment of a statistical equilibrium is difficult. For the small eddies, the characteristic time for evolution is shorter and the possibility of setting up an equilibrium state is high.

We now classify the regions in momentum space. With the length scale L that we have discussed so far we will identify the external length scale in any given situation and define a characteristic small momentum scale $k_L \sim 2\pi/L$. The energy-containing eddies are then charaterized by momenta with $k \sim k_L$. The small eddy region corresponds to $k >> k_L$ and is in statistical equilibrium at least approximately. The small scales are generated by the nonlinear terms to dissipate the energy at the rate ε. As the wavenumber increases, there is a maximum wavenumber k_D, beyond which the molecular processes are active dissipating the energy. The scale k_D is determined by the rate of energy dissipation ε and the molecular viscosity ν. Since the dimensions of ε and ν are given by

$$\varepsilon : \text{cm}^2/\text{sec}^3$$

$$\nu : \text{cm}^2/\text{sec}$$

the dimension of k_D is cm^{-1} and hence

$$k_D = (\varepsilon/\nu^3)^{1/4} \qquad \dots (25.2.35)$$

The energy spectral function $E(k)$ Eq. (25.2.23) in the range $k_L << k < k_D$ is then expected to be determined solely by k and ε and since $E(k)$ has the dimension

$$E(k) : \text{cm}^3/\text{sec}^2$$

straightforward dimensional analysis leads to

$$E(k) = C\,\varepsilon^{2/3}\,k^{-5/3} \qquad \dots (25.2.36)$$

The constant C is a universal number which is measured to be around 1.5. The energy spectrum given by Eq. (25.2.36) is called the Kolmogorov spectrum.

The main assumptions behind the spectrum are:

i) The high wavenumber components are not directly influenced by the external conditions setting up the turbulence.

ii) The process in which energy is transferred in many steps of size k_L ($k_L << k$) to maintain the excitation at wavenumber k is dynamically insignificant compared to the direct transfer of excitation from wavenumbers of order $k/2$ (say) to k. The assumption is that small eddies are convected from point to point by the large scale flow but it is a purely kinematic effect but not dynamically important.

To have the scaling form of Eq. (25.2.36) properly established, the scales k_L and k_D must be well separated. To find the condition for this, we introduce a dimensionless parameter, the Reynolds number R, defined as

$$R = vL/\nu \qquad \text{... (25.2.37)}$$

where v is a typical velocity and L the characteristic length. To find the ratio k_D/k_L, we note that in the energy-containing eddies, the rate of energy supply ε (equal to the energy dissipation rate in a statistically steady state) is

$$\varepsilon \sim v^2/T \sim v^2/L/v = v^3/L \qquad \text{... (25.2.38)}$$

and hence

$$k_D/k_L \sim k_D\, L = \frac{\varepsilon^{1/4}}{\nu^{3/4}}\, L = \left(\frac{vL}{\nu}\right)^{3/4} = R^{3/4} \qquad \text{... (25.2.39)}$$

where we have made use of Eq. (25.2.38). Thus $k_D/k_L >> 1$, can be realized for $R >> 1$.

The picture of the high Reynolds number flow in k-space is that of an energy source (low k, i.e. $k \sim k_L$) and an energy sink (high k) with the properties of the sink statistically independent of the source. The dissipation is principally confined to $k > k_D >> k_L$, k_D being the wavenumber at which the molecular viscosity first becomes effective. The principal contribution to $\int_0^\infty E(k)\, dk$ comes from the small k region, i.e. $k \sim k_L^0$ and the principal contribution to

$\int_0^\infty k^2 E(k)\, dk$ comes from the high $-k$, i.e. $k \geq k_D$ region. the Kolmogorov spectrum of Eq. (25.2.36) can easily be seen to be consistent with this picture since $\int_0^\infty k^{-5/3\, dk}$ is indeed divergent from the low-k contributions and $\int_0^\infty k^2 E(k)\, dk$ is divergent from the high-k ones. The source and the sink can be thought of as connected by a lossless transmission line. At any k in this transmission line, we can determine a local rate of energy transfer $\varepsilon(k)$—the rate of energy transfer across the sphere $|\mathbf{k}| = k$. But the lossless line implies that $\varepsilon(k)$ is constant throughout the line and equals the rate of energy input at one end and the rate of dissipation at the other and thus $\varepsilon(k) = \varepsilon$. It is instructive to construct the appropriate expression for $\varepsilon(k)$ from the Navier-Stokes equation and demonstrate that the Kolmogorov spectrum is indeed consistent with $\varepsilon(k) = \varepsilon$ independent of k. The reader is referred to a text on turbulence for this. The range of wavenumbers $k >> k_L$ is the equilibrium range, of which $k_L << k << k_D$ is called the inertial range, $k >> k_D$ the dissipation range and $k >> k_D$ the far dissipation range.

We can now determine the characteristic time scale of the inertial range eddies by a dimensional analysis. As for the energy spectrum, assuming that the time scales determined by ε and k alone at a given wavenumber k, we find

$$\tau(k) = A\, \varepsilon^{1/3}\, k^{-2/3} \qquad \text{... (25.2.40)}$$

The total energy is inertial range eddies of size of the order $2\pi/k$ may be taken to be (somewhat arbitrarily)

$$\frac{1}{2}\, [\Delta v(k)]^2 \approx \int_{1/2k}^{2k} E(p)\, dp \sim \varepsilon^{2/3}\, k^{-1/3} \qquad \text{... (25.2.41)}$$

The quantity $\Delta v(k)$ is a measure of the velocity difference across distances of order $2\pi/k$. The time taken by the eddy of wavenumber k to circulate relative to itself or turn over is of the order $1/k(\Delta v(k)) \sim \varepsilon^{1/2}\, k^{-2/3}$ which is the characteristic time of Eq. (25.2.37). It is also possible to interpret $\tau(k)$ as the time in which the energy of Eq. (25.2.41) is transferred if the input from lower-k modes is turned off. It should be noted that $\tau(k) >> \tau_s(k) = (k\bar{v})^{-1}$, for

$k >> k_L$ where, τ_s is the time an energy-containing eddy takes to sweep an inertial range eddy past an observation point. It should be pointed out that $\tau(k)$ is the time to decorrelate the relative phase of the simultaneous Fourier amplitudes of distinct wavevectors of magnitude of order k. The time scale $\tau(k)$, however, is not the time over which nonsimultaneous phases are decorrelated. This decorrelation occurs over the scale $\tau_s(k)$. Sweeping is effective in this decorrelating because at different times, the positions in space of the eddies corresponding to the mode are changing randomly.

In coordinate space, the structure function tensor is defined as

$$B_{\alpha\beta}(\mathbf{r}, t) = \langle [v_\alpha(\mathbf{x}+\mathbf{r}, t) - v_\alpha(\mathbf{x}, t)]\,[v_\beta(\mathbf{x}+\mathbf{r}, t) - v_\beta(\mathbf{x}, t)]\rangle \quad \text{... (25.2.42)}$$

Carrying out the Fourier transform of $S(k)$, we have for isotropic turbulence

$$B_{\alpha\alpha}(\mathbf{r}, t) = 4\int_0^\infty E(k, t)\left[1 - \frac{\sin kr}{kr}\right] dk \quad \text{... (25.2.43)}$$

where $r = |\mathbf{r}|$. For $k_L r << 1$, the contribution from wavenumbers where $E(k)$ is largest (namely $k \sim k_L$) is suppressed by the factor $(1 - \sin kr/kr) \sim (kr/6!)^2 + \ldots$. The dominant contribution to $B_{\alpha\alpha}$ comes from the equilibrium range and if we restrict ourselves to the region $k_L r << 1 << k_D r$, then $E(k, t)$ is given by Eq. (25.2.36) (independent of time since we are in the equilibrium range) and

$$B_{\alpha\alpha}(r) \sim B\,\varepsilon^{2/3}\,r^{2/3} \quad \text{... (25.2.44)}$$

where

$$B = \frac{9}{5}\,\Gamma\left(\frac{1}{2}\right) C \cong 4.82\ C$$

If we remove the restriction $k_D r >> 1$, then the corresponding $E(k)$ must have the functional form $E(k) = C\,\varepsilon^{2/3}\,k^{-5/3}\,F(k/k_D)$, where $F(0) = 1$ and the corresponding $B_{\alpha\alpha}(r)$ will, from Eq. (25.2.43), acquire the form

$$\begin{aligned} B_{\alpha\alpha}(\mathbf{r}, t) &= \varepsilon^{2/3}\,k_D^{-2/3}\,G(k_0 r) \\ &= (\varepsilon\nu)^{1/2}\,G(k_0 r) \end{aligned} \quad \text{... (25.2.45)}$$

where in the last step, we have used Eq. (25.2.35).

The above results would be the ones obtained from dimensional analysis with the assumption that the probability distribution of $\mathbf{v}(\mathbf{x}+\mathbf{r}, t) - \mathbf{v}(\mathbf{x}, \mathbf{t})$ depends only on ε, r and ν for $k_L r << 1$ and on ε and r if $k_L r << 1 << k_D r$.

An inconsistency with the above equilibrium range theory of Kolmogorov was pointed out by Landau, who noted that the dissipation rate ε is the space average of a fluctuating quantity (since the velocity is random)

$$\varepsilon_x = \frac{\nu}{2}\left(\frac{\partial v_\alpha}{\partial x_\beta} - \frac{\partial v_\beta}{\partial x_\alpha}\right)^2 \qquad \text{... (25.2.46)}$$

and is unlikely to be independent of the Reynolds number as assumed by Kolmogorov. This fact was taken into account by Kolmogorov in the early sixties when he assumed that the distribution of ε_r is log normal which is to say that the distribution of $\ln \varepsilon_r$ is Gaussian. Kolmogorov assumed for the variance σ the form

$$\sigma^2 = <[\ln \varepsilon_r - m]^2> = \begin{cases} A(\mathbf{x}, t) + 9h \ln \dfrac{L}{r}, & k_D r >> 1 \\ A'(\mathbf{x}, t) + 9h \ln (k_D L), & k_D r << 1 \end{cases} \qquad \text{... (25.2.47)}$$

where $m = <\ln \varepsilon_r>$, h is a universal constant and $A(\mathbf{x}, t)$ and $A'(\mathbf{x}, t)$ depend only on the large scale structure of the flow. The log-normal distribution of ε_r requires that the probability density $P_r(a)$ for $\varepsilon_r = a$ is given by

$$P_r(a) = \frac{1}{(2\pi\sigma^2 a^2)^{1/2}} e^{-(\ln a - m)^2/2\sigma^2} \qquad \text{... (25.2.48)}$$

This distribution leads to

$$<\varepsilon_r^p> = \mathbf{e}^{pm + p^2\sigma^2/2} \qquad \text{... (25.2.49)}$$

for all p. It follows that

$$<\varepsilon_r> = e^{m + \frac{1}{2}\sigma^2}$$

and

$$<\varepsilon_r^{2/3}> = e^{2m/3 + 2\sigma^2/9} \propto \varepsilon^{2/3}\left(\frac{r}{L}\right)^h \qquad \text{... (25.2.50)}$$

for $k_D\ r >> 1$. Consequently

$$B_{\alpha\alpha}(r) \sim \varepsilon^{2/3} r^{2/3} (r/L)^h \quad (k_L r << 1 << k_D r)$$

and

$$E(k) \sim \varepsilon^{2/3} k^{-5/3} (kL)^{-h} \quad (k_L << k << k_D) \quad \text{... (25.2.51)}$$

The experimental value of h is of the order of 0.05 and hence the correction to $E(k)$ is expected to be very small. In the higher order structure functions, the contribution of Eq. (25.2.49) is more pronounced (larger p) and that is the experimental technique for measuring h. We should point out that the correction to the Kolmogorov spectrum shown in Eq. (25.2.51) can also be derived from the picture that the turbulent dissipation occurs not over the entire three-dimensional space but over a very small fraction of it (in fact a region of measure zero in $D = 3$). This is called intermittency and leads to the statement that the turbulence is restricted to a space of fractal dimension $D_r < D = 3$. We shall not pursue this any further here.

To end this section, we will provide an expression for $\varepsilon(k)$—the energy transferred across the spherical surface $k = |\mathbf{k}|$ in momentum space. Turning to Eq. (25.2.29), the rate of change of energy at wavenumber k is $T(k, t)$ and hence the rate of total energy flow from all momenta $k' < k$ to value greater than k is obtained from the integral $\int_k^\infty T(k', t)\, dk'$ and thus

$$\varepsilon(k) = \int_k^\infty T(k', t)\, dk' = -\int_0^k T(k', t)\, dk' \quad \text{... (25.2.52)}$$

where

$$T(k, t) = \int d^D p\, M_{\alpha\rho\sigma}(\mathbf{k})\, T_{\alpha\rho\sigma}(-\mathbf{k}, p, t)$$

since energy conservation by the nonlinear term implies

$$\int_0^\infty T(k, t)\, dk = 0 \quad \text{... (25.2.53)}$$

The third order correlation function $T(\mathbf{p}_1, \mathbf{p}_2, t)$ is given in Eq. (25.2.19) and is to be analysed using the randomly stirred model of Eq. (25.1.13)

25.3 The Renormalization Group

As outlined in the previous Section, the problem of turbulence is mathematically intractable since a lower order moment requires a higher order one for its specification. There is no obvious small parameter which can be used to produce a perturbation theory. There is yet another problem which we have not exhibited so far. This is the difficulty of separating out the effects of the sweeping time and the characteristic time τ. One of the more successful methods of calculation is the renormalization group, although there are certain reservations in this approach as well. The model on which we will apply the renormalization group is the Navier-Stokes equation with a random external force, specifically [see Eq. (25.2.25)]

$$\left(\frac{\partial}{\partial t} + \nu k^2\right) u_\alpha(\mathbf{k}, t) = M_{\alpha\beta\gamma} \int d^D p \; u_\beta(\mathbf{p}, t)\, u_\gamma(\mathbf{k} - \mathbf{p}, t) + f_\alpha(k, t) \quad \text{... (25.3.1)}$$

where f_α is a random force. It is specified by a correlation function

$$< f_\alpha(\mathbf{k}_1, \omega_1)\, f_\beta(\mathbf{k}_2, \omega_2) > \; = 2\, F(k)\, P_{\alpha\beta}\, \delta(\mathbf{k}_1 + \mathbf{k}_2)\, \delta(\omega_1 + \omega_2) \quad \text{... (25.3.2)}$$

The function $F(k)$ is specified as

$$F(k) = F_0\, k^{4-D-y} \quad \text{... (25.3.3)}$$

For $y = 2 - D$, $F(k) \sim k^2$ and we have the case of thermal equilibrium where the fluctuation dissipation theorem holds. This was the case first studied by Forster, Nelson and Stephen. As we shall see below, the case $y = 4$, corresponds to the Kolmogorov spectrum. The method is identical to the technique explained in Sec. 24.2 and accordingly the procedure will be to spilt up $u_\alpha(k, \omega)$ and $f_\alpha(k, \omega)$ into two parts, one corresponding to the high momentum part, i.e. between Λ and Λ/b $(b > 1)$, where Λ is the upper cutoff on the wavenumbers and the other corresponding to the part between momenta of zero and Λ/b. As in Sec. 24.2, we write

$$u_\alpha(k, \omega) = \begin{cases} u_\alpha^{>}(k, \omega) \; \Lambda < k \; \Lambda/b \\ u_\alpha^{<}(k, \omega) 0 < k < \Lambda/b \end{cases}$$

$$f_\alpha(k,\ \omega) = \begin{cases} f_\alpha^{>}\ (k,\ \omega)\ \Lambda < k < \Lambda/b \\ f_\alpha^{<}\ (k,\ \omega)\ 0 < k < \Lambda/b \end{cases} \qquad \ldots \ \mathbf{(25.3.4)}$$

Inserting

$$u_\alpha(\mathbf{x},\ t) = \int^{\Lambda/b} e^{i\mathbf{k}\cdot\mathbf{r}}\ u_\alpha^{<}\ (k,\ t)\ \frac{d^Dk}{(2\pi)^D} + \int_{\Lambda/b}^{\Lambda} e^{i\mathbf{k}\cdot\mathbf{r}}\ u_\alpha^{>}\ (k,\ t)\ \frac{d^Dk}{(2\pi)^D}$$

and a similar expression for f_α $(\mathbf{x},\ t)$ in the Navier-Stokes equation in coordinate space and following a procedure identical to that in Sec. 24.2, we arrive at (in Fourier space in both momentum and frequency)

$$(-i\omega + \nu k^2)\ u_\alpha^{<}\ (\mathbf{k},\ \omega) = f_\alpha^{<}\ (\mathbf{k},\ \omega) + M_{\alpha\beta\gamma}^{<}\ (\mathbf{k}) \int \frac{d^Dp}{(2\pi)^D}\ \frac{d\omega_1}{2\pi}$$

$$\times\ (u_\beta^{<}\ (\mathbf{p},\ \omega_1)\ u_\gamma^{<}\ (\mathbf{k}-\mathbf{p},\ \omega - w_1) + 2u_\beta^{<}\ (\mathbf{p},\ \omega_1)\ u_\gamma^{>}\ (\mathbf{k}-\mathbf{p},\ \omega - w_1)$$

$$+\ u_\beta^{>}\ (\mathbf{p},\ \omega^1)\ u_\gamma^{>}\ (\mathbf{k}-\mathbf{p},\ \omega - \omega_1)] \qquad \ldots \ \mathbf{(25.3.5)}$$

$$(-i\omega + \nu k^2)\ u_\alpha^{>}\ (\mathbf{k},\ \omega) = f_\alpha^{>}\ (\mathbf{k},\ \omega) + M_{\alpha\beta\gamma}^{>}\ (\mathbf{k}) \int \frac{d^Dp}{(2\pi)^D}\ \frac{d\omega_1}{2\pi} \times$$

$$[u_\beta^{>}\ (\mathbf{p},\ \omega_1)\ u_\gamma^{>}\ (\mathbf{k}-\mathbf{p},\ \omega - w_1) + 2u_\beta^{<}\ (\mathbf{p},\ \omega_1)\ u_\gamma^{>}\ (\mathbf{k}-\mathbf{p},\ \omega - w_1)$$

$$+\ u_\beta^{<}\ (\mathbf{p},\ \omega_1)\ u_\gamma^{<}\ (\mathbf{k}-\mathbf{p},\ \omega - \omega_1)] \qquad \ldots \ \mathbf{(25.3.6)}$$

The point now is to solve for $u_\alpha^{>}(k,\ \omega)$ from Eq. (25.3.6) and insert the result in Eq. (25.3.5), so that one has an equation of motion for $u_\alpha^{<}(k,\ \omega)$. That equation of motion will have a different dissipation rate $\nu'(k)$ and thus we will be setting up the flow equation.

We now solve for $u_\alpha^{>}$ from Eq. (25.3.6) perturbatively. To do so we imagine that $u_\alpha^{>}$ is expanded in powers of the coupling constant $\overline{M}$ as

$$u_\alpha^{>} = u_\alpha^{>(0)} + u_\alpha^{>(1)} + \ldots \qquad \ldots \ \mathbf{(25.3.7)}$$

and equating identical powers of the two sides of Eq. (25.3.6), we have

$$(-i\omega + \nu k^2)\ u_\alpha^{>(0)} = f_\alpha^{>} \qquad \ldots \ \mathbf{(25.3.8a)}$$

$$(-i\omega + \nu k^2)\ u_\alpha^{>(1)} = M_{\alpha\beta\gamma}^{>}\ (k) \int \frac{d^D p}{(2\pi)^D} \frac{d\omega_1}{2\pi}\ [u_\beta^{>(0)}\ (\mathbf{p},\ \omega_1)$$

$$u_\gamma^{>(\omega)}\ (\mathbf{k} - \mathbf{p},\ \omega - \omega_1) + 2u_\beta^{<}\ (\mathbf{p},\ \omega_1)\ u_\gamma^{>(0)}$$

$$(\mathbf{k} - \mathbf{p},\ \omega - \omega_1) + u_\beta^{<}\ (\mathbf{p},\ \omega_1)\ u_\gamma^{<}\ (\mathbf{k} - \mathbf{p},\ \omega - \omega_1) \quad \text{... (25.3.8b)}$$

and so on.

The solutions are

$$u_\alpha^{>(0)}\ (\mathbf{k},\ \omega) = G\,(\mathbf{k},\ \omega)\ f_\beta^{>}\ (\mathbf{k},\ \omega)\ P_{\alpha\beta} \quad \text{... (25.3.9a)}$$

$$u_\alpha^{>(1)}\ (k,\ \omega) = P_{\alpha\delta}\ G\,(\mathbf{k},\ \omega)\ M_{\delta\beta\gamma}^{>}\ (\mathbf{k}) \int \frac{d^D p}{(2\pi)^D} \frac{d\omega_1}{2\pi}$$

$$[u_\beta^{>(0)}\ (\mathbf{p},\ \omega_1)\ u_\gamma^{>(0)}\ (\mathbf{k}-\mathbf{p},\ \omega-\omega_1) + 2u_\beta^{<}\ (\mathbf{p},\ \omega_1)\ u_\gamma^{>(0)}\ (\mathbf{k}-\mathbf{p},\ \omega-\omega_1)$$

$$+\ u_\beta^{<}\ (\mathbf{p},\ \omega_1)\ u_\gamma^{<}\ (\mathbf{k} - \mathbf{p},\ \omega - \omega_1)] \quad \text{... (25.3.9b)}$$

where

$$G\,(k,\ \omega) = (-i\omega + \nu k^2)^{-1} \quad \text{... (25.3.10)}$$

The explicit use of $P_{\alpha\beta}$ in Eq. (25.3.9a) ensures $\vec{\nabla} \cdot \mathbf{u} = 0$.

The next step is to insert $u_\alpha^{>}$ from Eqs. (25.3.9a) and (25.3.9b) into Eq. (25.3.5), average over the noise $f^{>}$ according to the distribution specified and find the effective equation for $u_\alpha^{<}$. Our interest will centre on picking up the term linear in $u_\alpha^{<}$ after the averaging over the $f^{>}$ has been carried out.

The result of averaging over the $f^{>}$ produces the correlation function

$$\delta^D\,(\mathbf{k} + \mathbf{k}')\ \delta\,(\omega + \omega_1)\ C_{\alpha\beta}\ (k,\ \omega) = \langle\ u_\alpha^{(0)}\ (\mathbf{k},\ \omega)\ u_\beta^{(0)}\ (\mathbf{k},\ \omega')\rangle \quad \text{... (25.3.11)}$$

and if we write

$(-i\omega + \nu k^2)\ u_\alpha^{<} = f_\alpha^{<} - (\Delta\nu)\ k^2 u_\alpha^{<} +$ non-linear terms, then the above process shows

$$\Delta\nu k^2 = -4 \int \frac{d^D p}{(2\pi)^D} \frac{d\omega_1}{2\pi}\ M_{\alpha\beta\gamma}\,(\mathbf{k})\ M_{\delta\sigma\rho}\,(\mathbf{k}-\mathbf{p})\ P_{\beta\delta}\,(\mathbf{k}-\mathbf{p})\ G(\mathbf{k}-\mathbf{p})\ C_{\sigma\gamma}\,(\mathbf{p},\ \omega_1) \quad \text{... (25.3.12)}$$

With

$$C_{\alpha\beta}\ (k,\ \omega) = |G\ (k,\ \omega)|^2\ 2F\ (k)\ P_{\alpha\beta} \qquad \text{... (25.3.13)}$$

we find

$$\Delta\nu\ k^2 = -\ 8 \int \frac{d^D p}{(2\pi)^D} \frac{d\omega_1}{2\pi}\ M_{\rho\beta\gamma}\ (\mathbf{k})\ M_{\delta\sigma\rho}\ (\mathbf{k} - \mathbf{p})\ P_{\sigma\gamma}\ (\mathbf{p})\ P_{\beta\delta}\ (\mathbf{k} - \mathbf{p})$$

$$G\ (\mathbf{k} - \mathbf{p},\ \omega - \omega_1)\ |G(\mathbf{p},\ \omega_1)|^2\ F(p) \qquad \text{... (25.3.14)}$$

This provides the renormalization of ν, with a recursion relation of the form $\nu' = \nu + \Delta\nu$, but before writing down the relation, we have to carry out the rescaling of space and time. We rescale lengths by the factor b, i.e. $x \to x'/b$, which implies that $k \to k' = bk$ (restoring the cutoff to the original Λ) and the velocity rescales as $u \to u' = b^{-a}u$ and the time variable will rescale such that the new frequency ω′ is related to the old by $\omega' = b^z\omega$. The exponents z and a are as yet unknown. These scalings will alter ν according to its naive dimension and also change the expression on the right hand side of Eq. (25.3.14) by a scale factor. We introduce rescaling of the coefficient F_0 in Eq. (25.3.3) and assume that $F_0 \to F'_0 = b^f F_0$ under the spatial rescalings. Thus the dimensional rescalings will lead to:

$$\nu \to \nu' = \nu\ b^{z-2} \qquad \text{... (25.3.15a)}$$

and

$$\Delta\nu \to \Delta\nu' = \Delta\nu\ b^{f+4-y-2z} \qquad \text{... (25.32.15b)}$$

We now note that rescalings cannot change the rate at which the force f supplies energy and hence the correlation function $\langle f_\alpha\ (k_1,\ \omega_1)\ u_\alpha\ (k_2,\ \omega_2)\rangle$ cannot change under rescaling except by its naive dimension and accordingly we can choose a in such a manner that $f = 0$. This choice corresponds to

$$2a = 3z + 2D - 4 + y \qquad \text{... (25.3.16)}$$

The recursion relation under elimination of scales and subsequent rescaling now becomes

$$\nu' = \nu\, b^{z-2} - 8b^{4-y-2z} F_0\, k^{-2} \int \frac{d^D p}{(2\pi)^D} \frac{d\omega_1}{2\pi} M_{\rho\beta\gamma}(\mathbf{k}) \times$$

$$M_{\sigma\rho\rho}(\mathbf{k}-\mathbf{p})\, P_{\beta\delta}(\mathbf{k}-\mathbf{p})\, P_{\sigma\rho}(\mathbf{p})\, G(\mathbf{k}-\mathbf{p}, \omega-\omega_1)$$

$$|G(\mathbf{p}, \omega_1)|^2\, p^{4-D-y} \qquad \text{... (25.3.17)}$$

The integration on p runs over the range Λ/b to Λ and hence we need to work out the integral in Eq. (25.3.17) in the limit $k \to 0$ and retain the first non-vanishing term. But before proceeding any further, we need to ensure that the two terms on the right hand side of Eq. (25.3.17) have the same dimensions and this ensures

$$z = 2 - y/3 \qquad \text{... (25.3.18)}$$

Evaluation of the integral in the limit $k \to 0$, now leads to ($S_D = 2\pi^{D/2}/\Gamma\,(D/2)$)

$$\nu' = \nu\, b^{-y/3}\left[1 + \frac{F_0}{2\nu^3} \frac{D^2 - D}{D(D+2)} \cdot \frac{S_D}{(2\pi)^D} \int_{\Lambda/b}^{\Lambda} \frac{d_p}{p^{1+y}}\right] \qquad \text{... (25.3.19)}$$

The fixed point of this flow can be found perturbatively by expanding about $y = 0$. This leads to

$$\nu' = \nu\left[1 - \frac{y}{3}\ln b + \ldots\right]\left[1 + \frac{F_0}{2\nu^3} \frac{D^2 - D}{D\,(D+2)} \cdot \frac{S_D}{(2\pi)^D} \ln b + \ldots\right]$$

$$= \nu\left[1 + \ln b\left(\frac{F_0}{2\nu^3} \frac{D^2 - D}{D\,(D+2)} \frac{S_D}{(2\pi)^D} - \frac{y}{3}\right) + \ldots\right] \qquad \text{... (25.3.20)}$$

Clearly, the fixed point of ν requires it to acquire the fixed point value given by (to the lowest order in y)

$$\frac{F_0}{\nu^3} = \frac{2y}{3} \frac{(2\pi)^D}{S_D} \frac{D+2}{D-1} \qquad \text{... (25.3.21)}$$

We thus arrive at the result that for small values of k, the behaviour of ν is given by

$$\nu(k) = \nu k^{-y/3} \qquad \text{... (25.3.22)}$$

with the amplitude ν given by Eq. (25.3.21). As for the energy spectrum, we note that

$$k^{D-1}\ E(k) = \int < u\,(k,\ \omega)\ u(-k,\ -\ \omega) > \frac{d\omega}{2\pi}$$

$$= \int |G\,(k,\ \omega)|^2\ < f\,(k,\ \omega)\ f\,(-k, -\omega) > \frac{d\omega}{2\pi}$$

$$= \frac{F_0}{\pi}\int \frac{d\omega}{\omega^2 + \nu\,(k)^2\ k^4}\ k^{D-y+4}$$

$$= F_0\ k^{D+2-2y/3}$$

leading to

$$E(k) = F_0\ k^{1-2y/3} \qquad \text{... (25.3.23)}$$

For $y = 4$, the Kolmogorov spectrum is obtained from Eq. (25.3.23), with F_0 identified as $C\,\varepsilon^{2/3}$ from Eq. (25.2.36).

Using Eq. (25.3.21) in conjunction with Eq. (25.2.52) it is possible to obtain a value for the coefficient C. It should be noted that this uses two different kinds of calculations, a perturbation expansion in powers of y for Eq. (25.3.21) and a direct evaluation at $y = 4$ in Eq. (25.2.52). However, it is interesting to see that this procedure gives the reasonably accurate value of 1.61 for C.

Finally we observe that the flow equations are valid as $k \to 0$. The elimination procedure that we have used generates for $u^<$ terms which are linear in $u^<$, quadratic in $u^<$ and also cubic in $u^<$. It is only for $k \to 0$, that the cubic terms generated by the procedure tend to zero and the renormalization procedure is well defined. On the other hand, as we have discussed earlier, the Kolmogorov scaling refers to comparatively larger wavenumbers of the inertial range and thus the suitability of the renormalization group calculation described here is debatable.

Chapter 26
Disordered Systems

26.1 Introduction

In our study of critical phenomena so far we have been interested only in the inherent thermal fluctuations in a system. The ordering or disordering seen in Ising models or $X-Y$ or Heisenberg models has been due to the competition between aligning forces of atomic origin and the thermal fluctuations. In this chapter, we explore the presence of other fluctuations. In terms of an Ising model these fluctuations could manifest themselves in some sites on the underlying lattice being empty at random or the interaction between all neighbouring sites having a random variation or in the existence of a random external field. The essential question will be whether the transition to an ordered state survives in the presence of these random perturbations and if it does what are the new critical exponents and scaling laws. The two terms to be explained before we start our discussion are the ones commonly used in dealing with disorders—quenched disorder and annealed disorder.

Quenched disorder is one where the disordering variables are fixed rigidly in space over the period of observation time. As an example if we consider a "dilute" magnet—a magnet where some of the magnetic atoms are replaced by non-magnetic ones at randomly chosen sites—then at low temperatures the mobility of the atoms is small and the set of sites at which the non-magnetic atoms sit is fixed in time. Now consider the Hamiltonian for such a system. It can be written as

$$H = -J \sum_{<ij>} p_i\, p_j\, \mathbf{S}_i \cdot \mathbf{S}_j \qquad \text{... (26.1.1)}$$

where $\mathbf{S}_i$ is the n-component spin vector of unit magnitude ($n = 1$, Ising; $n = 2$, X - Y ; $n = 3$, Heisenberg), J is the exchange constant, $< ij >$ denotes sum over nearest neighbours and p_i is the random variable which takes the value unity if the site i is occupied by a magnetic atom and zero if it is occupied by a non-magnetic one. The partition function is

$$Z(\{p_i\}) = \underset{\{S_i\}}{\mathrm{Tr}}\, e^{-\beta H} \qquad \text{... (26.1.2)}$$

and the free energy is given by

$$F\{P_i\}) = -\ln Z(\{p_i\})/\beta L^D \qquad \text{... (26.1.3)}$$

The free energy is dependent on the set $\{p_i\}$ and the correlation functions likewise would be dependent on the set $\{p_i\}$. Thus, there would be no translational invariance and calculations would be extremely difficult.

At this point one simplifies things by trying to imagine what a real free energy would depend upon. For a sufficiently large sample it seems reasonable that the free energy would be determined by the overall impurity concentration p rather than by the set $\{p_i\}$. This implies that the physical free energy will be the one averaged over an ensemble where each member of the ensemble has a different set of sites $\{i\}$ occupied by the non-magnetic impurities. If the distribution $P(\{P_i\})$ for the set $\{P_i\}$ is known, then the observed free energy would be

$$F(P) = \sum_{\{p_i\}} P(\{p_i\})\; F(\{p_i\}) \qquad \text{... (26.1.4)}$$

(this ensemble averaging can be visualized by breaking up the physical dimension L into smaller but still macroscopic sizes, L_j. In each Lj a set of sites are occupied and the number of subsystems of size L_j is L/L_j so that if we take the limits $L_j \to \infty$ and $L/L_j \to \infty$, the above averaging will be accomplished). Thus for the quenched system the relevant free energy is the impurity averaged free energy. Hence, the prescription for dealing with quenched systems is to average ln Z over the impurity distribution. For the above system the $P(\{p_i\})$ takes the simple form

$$P(\{p_i\}) = \prod_i \left(p\, \delta_{p_i,1} + (1-p)\, \delta_{p_i,0}\right) \qquad \text{... (26.1.5)}$$

To understand what annealed systems are we need to return to the above system and consider the diffusivity of the atoms to be high so that the non-magnetic ions could diffuse from one site to another over time scales shorter than that of the experiment. Now, stretching one's imagination to say that the Hamiltonian is still given by Eq. (26.1.1), we would obtain for the partition function

$$Z = \underset{\{p_i\}}{Tr}\; P(\{p_i\})\; Z\left(\{p_i\}\right) = \underset{(\{p_i\},\{S_i\})}{\mathrm{Tr}}\; e^{-\beta H}\, P\left(\{p_i\}\right) \qquad \text{... (26.1.6)}$$

since the averages are measured not only over the possible spin configurations, but also the positions of the non-magnetic ones. The free energy is obtained by simply taking the logarithm. Translational invariance is maintained in the Z of Eq. (26.1.6) and things are relatively straightforward. Henceforth, we will deal with quenched systems only.

26.2 Models with Quenched Disorder

We will consider the model described by the Hamiltonian

$$H = -\sum_{<ij>} J_{ij}\, \vec{S}_i \cdot \vec{S}_j \qquad \text{... (26.2.1)}$$

where the J_{ij} are taken to be random variables. In this manner the above model includes both the random interaction strength as well as the random site occupation case, when we consider $J_{ij} = Jp_ip_j$. To simplify the situation, we consider ferromagnetic models, which is to say that the probability of J_{ij} becoming negative will be taken to be small. The probability distribution of J_{ij} will have to specified and we shall assume that the J_{ij} 's are uncorrelated. Thus the distribution will be of the form

$$P\left(\left\{J_{ij}\right\}\right) = \prod_{<ij>} p\left(J_{ij}\right) \qquad \text{... (26.2.2)}$$

with the mean value of J_{ij} greater than zero.

This sort of disorder is also called homogeneous. We try to retain the ferromagnetic behaviour by requiring the deviation of J_{ij} to be much smaller than its mean.

To see if this model will have a sharp second order transition, we follow an argument due to Harris. Different parts of the sample will, owing to statistical fluctuations, be characterized by different critical temperatures and hence there will be a distribution $p\,(T_c)$ of the critical temperatures across the sample. The number of parts of the system can be taken to be V/ξ^D, where V is the total sample volume and ξ, the correlation length. We can now invoke the central limit theorem to assert that the fluctuation δT_c would be proportional to the square root of the number of individual parts and hence

$$\delta T_c \sim \xi^{-D/2} \qquad \text{... (26.2.3)}$$

For the correlation length to be ξ with a sharp transition temperature T_c, there must be the relation $\xi \propto (T - T_c)^{-\nu} = (\Delta T)^{-\nu}$, or $\Delta T \propto \xi^{-1/\nu}$. If we are to have a sharp transition, then we must have

$$\delta T_c << \Delta T, \qquad \text{... (26.2.4)}$$

which implies

$$\xi^{-D/2} << \xi^{-1/\nu}, \qquad \text{... (26.2.5)}$$

leading to

$$\frac{D}{2} > \frac{1}{\nu}$$

$$D\nu - 2 > 0$$

$$\alpha < 0 \qquad \text{... (26.2.6)}$$

where in the last line we have used the hyperscaling relation $D\nu = 2 - \alpha$. Thus the Harris criterion is that for $\alpha < 0$, there will be a sharp transition to the ferromagnetic phase in the random system. The initial exponents should remain the same as that of the pure system. For $\alpha > 0$ on the other hand, the slightest impurity will change the transition in some way. To find out the quantitative answer, we need to set up a renormalization group argument and that requires a trick special to all problems involving randomness.

We first change from a model on a lattice to a Ginzburg-Landau model. This is done by noting that the characteristic of the randomness is to produce local critical temperatures and hence the coefficient a of the quadratic term which depends on the critical temperature will acquire a spatial dependence and consequently the free energy functional will be ($\varphi(x)$ is a vector field)

$$F = \int d^D x \left[\frac{1}{2} a(x)\, \vec{\varphi}(x)^2 + \frac{1}{2} (\vec{\nabla} \varphi_i)^2 + \frac{u}{4} \left(\vec{\varphi}^2 (x)\right)^2 \right] \qquad \text{... (26.2.7)}$$

The distribution of the $a(x)$ has to be provided and if the disorder is uncorrelated, we must have

$$P\,\{a\,(\mathbf{x})\} = \prod_{\mathbf{x}} p\,(a\,(x)) \qquad \text{... (26.2.8)}$$

and a Gaussian choice of $p\ (a\ \{x\})$ would give

$$p\,(a\,(x)) \propto e^{-(a-a_0)^2/2\Delta} \qquad \text{... (26.2.9)}$$

implying a mean value a_0 for the mass parameter and a fluctuation strength of Δ. From Eq. (26.2.9), it should be clear that

$$<(a\,(\mathbf{x}) - a_0)\ (a\,(\mathbf{o}\,) - a_0)>_{av} = \Delta \delta^D (\mathbf{x}\,), \qquad \text{... (26.2.10)}$$

where $<.....>_{av}$ denotes an averaging over the impurity distribution.

We shall now introduce the trick which will make it possible to address the problem of evaluating the partition function with the Ginzburg-Landau free energy of Eq. (26.2.7), with the supplementary condition that the averaging over the probability distribution has to be done over the free energy and not the partition function. Thus,

$$\beta F = -<\ln Z>_{av} = -\int_{-\infty}^{\infty} D\,a\,(\mathbf{x}\,)\ P\,(\{a\,(\mathbf{x}\,)\})\ \ln\,Z\,(\{a\}) \qquad \text{... (26.2.11)}$$

Using the identity $\ln Z = \underset{n \to 0}{\text{Lt}} \dfrac{Z^n - 1}{n}$, we can easily see that

$$\beta F = -\underset{n \to 0}{\text{Lt}}\ [<Z^n>_{av} - 1]/n \qquad \text{... (26.2.12)}$$

Now

$$Z^n = \int D\varphi^{(\alpha)}(\mathbf{x})\, e^{-\beta F_n} \quad \text{... (26.2.13)}$$

where

$$F_n = \int d^D x \sum_{\alpha=1}^{n} \left[\frac{1}{2} a(x) \left(\vec{\varphi}^{(\alpha)}(\mathbf{x}) \right)^2 + \frac{1}{2} \sum_i (\vec{\nabla} \varphi_i^{(\alpha)})^2 + \frac{u}{4} \left(\vec{\varphi}^{(\alpha)}(x)^2 \right)^2 \right] \quad \text{... (26.2.14)}$$

$$\begin{aligned} \langle Z^n \rangle_{av} &= \int D\vec{\varphi}^{(\alpha)}(\mathbf{x}) \int \mathrm{Da}(\mathbf{x})\, P[\{a(\mathbf{x})\}]\, e^{-\beta F_n} \\ &= \int D\vec{\varphi}^{(\alpha)}(\mathbf{x})\, e^{-\beta F_n^{\text{eff}}} \end{aligned} \quad \text{... (26.2.15)}$$

where

$$F_n^{eff} = \int d^D x \left[\sum_{\alpha=1}^{n} \left\{ \frac{a_0}{2} (\vec{\varphi}^{(\alpha)}(\mathbf{x}))^2 + \frac{1}{2} \sum_i (\vec{\nabla} \varphi_i^{(\alpha)})^2 + \frac{u}{4} (\vec{\varphi}^{(\alpha)2}(\mathbf{x}))^2 \right\} - \Delta \sum_{\alpha,\beta=1}^{n} \left(\vec{\varphi}^{(\alpha)}(x) \right)^2 \left(\vec{\varphi}^{(\beta)}(x) \right)^2 \right] \quad \text{(26.2.16)}$$

The evaluation of $\langle Z^n \rangle_{av}$ involves the translationally invariant free energy functional of Eq. (26.2.16) and hence the renormalization group techniques of Chapter 23 can be taken over directly. The only difference is that instead of the two parameters that we had there (a and u) , we have three parameters now a_0 , u and Δ. Accordingly one can go ahead to find the recursion relations for these three variables. The recursion relations will involve n and in the end the limit $n \to 0$ will have to be taken. This trick of converting the quenched problem into one where an effective translationally invariant Hamiltonian is involved is called the 'replica trick'.

At this point we will not go through the details of the calculation of the recursion relations which is left as an exercise. We merely discuss the outcome of the calculation. There is a fixed point with $\Delta' = 0$, which corresponds to the pure system. On analyzing the stability of this fixed point, it is found that it is stable if $\alpha < 0$, thus establishing the Harris criterion that if $\alpha < 0$, the behaviour of the system with random impurities is identical to that of the pure system. For $\alpha > 0$, a different fixed point is stable. This implies that there is a sharp transition in this case as well, but the critical exponents are changed from the values corresponding to the pure system. In particular, the new α is negative and thus the Harris criterion for a sharp transition is obeyed. It has to be true that if the deviation Δ is large enough, then the sharp transition to the ferromagnetic state will be lost, but it is not clear from the renormalization group flows as to what qualifies as a sufficiently large Δ. What is true is that for small Δ, there is a stable fixed point and a sharp transition and as long as we have a ferromagnetically ordered state at low temperatures, we can call the randomness weak.

The next disordered magnetic system that we want to look at is one which is amorphous. This implies there is no global "easy" direction along which the spins would like to align. We can have local easy axes whose orientation varies from point to point. We consider the orientation to be a random function of position and assume that this captures the essence of amorphousness. The model Hamiltonian for an amorphous magnet is thus

$$H = -J\sum \mathbf{S}_i \cdot \mathbf{S}_j - A\sum_i (\hat{n}_i \cdot \mathbf{S}_i)^2 \qquad \text{... (26.2.17)}$$

where J and A are positive constants and $\hat{n}_i$ is the unit vector at site i and its orientation varies randomly from site to site. The term $(\hat{n}_i \cdot \mathbf{S}_i)^2$ maintains the global $\mathbf{S} \to -\mathbf{S}$ symmetry of the microscopic interactions. The model of Eq. (26.2.17) has competition built into it. The first term requires the spins to align parallel to each other, while the second term favours the ith spin to align in the direction $\hat{n}_i$ or $-\hat{n}_i$. We note in passing that the above model has no meaning if the spins are Ising-like.

The issue that we want to address is whether the above model can show an ordered state. The argument proceeds by assuming that an ordered state exists and we create a domain in which same spins are pointing in a different direction (say at right angles to the original direction). If this new state has a lower energy, then the ordered state cannot exist. The first term of Eq. (26.2.17), as we have seen before, implies that it costs energy to turn the spins and if the domain size is L, then the energy it costs is

$$\Delta E_j \sim JL^{D-2} \qquad \text{... (26.2.18)}$$

The second term of course favours the turning of spins. This is because if all spins are pointing in the same direction, a large number of terms $\hat{n}_i \cdot \mathbf{S}_i$ will be zero or close to it and by turning some of the spins in a different direction a non-zero contribution to $\hat{n}_i \cdot \mathbf{S}_i$ will be achieved and a consequent lowering of the energy. To estimate how big the effect is, we merely note that the energy lowering will be proportional to the excess number of random axes pointing in this new direction and since this number comes from statistical fluctuations, we expect it to be of the order of the square root of the number of sites and thus

$$\Delta E_D \sim -AL^{D/2} \qquad \text{... (26.2.19)}$$

The total energy change is

$$\Delta E = \Delta E_J + \Delta E_D \sim JL^{D-2} - AL^{D/2} \qquad \text{... (26.2.20)}$$

It should be noted that we do not need to consider thermal fluctuations—we are assuming that there is an ordered state in the presence of thermal fluctuations and are testing its stability against the disorder. Hence these calculations refer to $T = 0$ and the thermodynamic stability can be inferred from an energy argument. From Eq. (26.2.20) it is clear that the energy lowering term dominates if

$$D - 2 < \frac{D}{2} \qquad \text{... (26.2.21)}$$

i.e. $D < 4$

Hence in dimensions below four, there can be no ordered state in the amorphous magnet.

The renormalization group argument proceeds by starting with the free energy functional

$$F = \int d^D x \left[\frac{a}{2} \vec{\varphi}^{\,2} + \frac{1}{2} \sum_i (\vec{\nabla} \varphi_i)^2 + \frac{u}{4} (\vec{\varphi}^{\,2})^2 - A(\hat{n} \cdot \vec{\varphi})^2 \right] \qquad \text{... (26.2.22)}$$

where the unit vector $\hat{n}(\mathbf{x})$ is assumed to have an isotropic distribution. The replica trick is now carried out and implementing the ε-expansion, it is found that there are no stable fixed points for $D < 4$.

The last model to be considered in this Section involves a random external field. We imagine a field $\mathbf{h}_i$ at site i and that it varies randomly from site to site. The Hamiltonian for this random field problem has the form

$$H = -J \sum_{<ij>} \mathbf{S}_i \cdot \mathbf{S}_j - \sum_i \mathbf{h}_i \cdot \mathbf{S}_i \qquad \text{... (26.2.23)}$$

Clearly the model is defined for Ising spins as well. As before, our first consideration will be to see if there can be a long-range order in this model. Assuming that we have an ordered phase at $T = 0$, we consider turning over some of the spins namely those in a domain of linear dimension L. The exchange term in Eq. (26.2.23) will oppose this spin flip and it will cost an energy of an amount JL^{D-2} if we are talking about X-Y or Heisenberg models and JL^{D-1} if we are discussing Ising models. The spin flip is favoured by the random magnetic field because of the number of excess sites where the field points in the new spin direction.

Because this excess is due to statistical fluctuations, the number of excess sites is of $O(L^D/2)$. Hence the energy gained by spin flips is $HL^{D/2}$. For X - Y and Heisenberg models the energy change due to the formation of the spin flipped domain is (H is some constant)

$$\Delta E \sim JL^{D-2} - HL^{D/2} \qquad \text{... (26.2.24)}$$

leading to the favouring of disordering for $D < 4$. For Ising systems, on the other hand, the change in energy is

$$\Delta E \sim JL^{D-1} - HL^{D/2} \qquad \text{... (26.2.25)}$$

and hence the disordering is favoured for $D < 2$. This lower critical dimension for the random field Ising model turns out to be quite interesting. If one uses the renormalization group to confirm the above result for the Ising model in the ε-expansion, one comes up with a lower critical dimension of three. The perturbative renormalization group must clearly be in error since the correctness of the lower critical dimension of two obtained from the above heuristic arguments has been rigorously established since. The reason for this failure of the perturbative expansions has been the subject of a number of investigations.

26.3 Spin Glass

In this section we will let the randomness in the interaction strength J of the Ising (or Heisenberg) model be such that positive and negative values of J are almost equally likely. It is now intuitively clear that an ordered state with all spins lined up in a given direction is going to be quite unlikely. We will confine ourselves to Ising systems. A conventional long-range order not being on the cards, we would ask whether there would be any magnetic phase transittion at all in these materials. The answer from experiments is that there is some sort of "freezing" transition to a new kind of "order" at some temperature—an ordered state called 'a spin glass' in which spins are aligned in random directions. The order parameter that we have talked about so far will certainly not do for this transition. So one immediately needs to know what would an appropriate order parameter be. Further, the establishment of a non-zero expectation value of the order parameter is usually associated with a reduction of symmetry. We need to know what is the symmetry that is broken at this transition.

To begin a quantitative description of the transition, we note that below this "freezing" temperature T_f, the spins are "frozen" at given sites, i.e. they do not change over the observation time. However, they do not point in any definite direction. If S_i is the spin at a lattice point R_i, then this implies that for $T < T_j$

$$< S_j > \neq 0 \qquad \text{... (26.3.1)}$$

$$\sum_i \frac{1}{N} <S_i>_t e^{i\mathbf{k} \cdot \mathbf{R}_i} = 0, (N \to \infty) \qquad \text{... (26.3.2)}$$

where $<...>_t$ denotes averaging over a long interval of time. As T_f is approached from above, the range of correlations also increases. To get a quantitative feeling for that, one considers the correlation function $<S_i S_j>_t$. In general, for the spin glass, the sign of this correlation will be random. Hence it is more appropriate to consider the square of this correlation and it is that function which becomes correlated as $T \to T_f$ from above

$$<S_i S_j>_t^2 \propto e^{-|\mathbf{R}_i - \mathbf{R}_j|/\xi_{SG}} \qquad \text{... (26.3.3)}$$

with the correlation length ξ_{SG} increasing as $T \rightarrow T_f$. The relaxation times above T_f are also extremely long getting bigger as $T \rightarrow T_f$.

The increase in ξ_{SG} and the relaxation time should be indicative of collective effects that are taking place. It is not certain whether ξ_{SG} and the relaxation time actually become infinite at $T = T_f$ —the chances are that they actually become infinitely big only at $T = 0$. However, if there is a phase transition at $T = T_f$, it would be possible to replace the time averages by statistical mechanical averages and have a spin glass order parameter defined as

$$q = [< S_i >_T^2]_{av} \qquad \text{... (26.3.4)}$$

where $< ...>_T$ is the thermal average and $[...]_{av}$ is the averaging over the random interactions. In the simplest approximation, namely mean field theory, the spin glass order parameter will turn out to be much more complicated than this. It has to do with the existence of large number of degenerate thermodynamic states with the same macroscopic properties but different microscopic configurations.

The mean field theory for spin glasses was set up by Sherrington and Kirkpatrick. Since there was no a priori knowledge of what the mean field theory should look like, they decided to use the experience of ordinary Ising models where an infinite-ranged interaction (all spins interact with the same J) leads to the mean field answer as the exact result. Consequently, we begin with the Hamiltonian,

$$H = -\sum_{<ij>} J_{ij} \; S_i S_j - \sum H_i \, S_i \qquad \text{... (26.3.5)}$$

where $<ij>$ now means including each pair once. A local field H_i has also been included. The probability distribution $P(J_{ij})$ is given as

$$P(J_{ij}) = \frac{1}{J}\left(\frac{N}{2\pi}\right)^{1/2} e^{-N(J_{ij} - J_0/N)^2/2J^2} \qquad \text{... (26.3.6)}$$

and is the same for all pairs of spins,

$$[J_{ij}]_{av} = J_0/N \qquad \text{(26.3.7)}$$

$$[J_{ij}^2]_{av} = [J_{ij}]_{av}^2 = J^2/N$$

The factors of N are required to get a sensible thermodynamic limit. As explained in Sec. 26.2 the calculation of the partition function for the quenched system is facilitated by use of the replica trick and that leads to

$$[Z^n]_{av} = \sum_{[S_i^\alpha = \pm 1]} \int_{-\infty}^{\infty} P(J_{ij})dJ_{ij} \exp \beta \sum_{<i,j>} J_{ij} \sum_{\alpha=1}^{n} S_i^\alpha S_j^\beta + \beta \sum H_i \sum_{\alpha=1}^{n} S_i^\alpha \tag{26.3.8}$$

where $\alpha = 1, 2, \ldots n$ denotes the n replicas. The integral over J_{ij} is easily performed for the distribution of Eq. (26.3.6) and we have

$$[Z^n]_{av} = \sum_{[S_i^\alpha = \pm 1]} \exp \left\{ \frac{1}{N} \sum_{i,j} \left\{ \frac{1}{2}(\beta J)^2 \sum S_i^\alpha S_j^\alpha S_i^\beta S_j^\beta \right.\right.$$

$$\left.\left. + \beta J_0 \sum_{\alpha} S_i^\alpha S_j^\alpha \right] + \beta \sum_i H_i \sum_{\alpha=1}^{n} S_i^\alpha \right\} \tag{26.3.9}$$

Using $(S_i^\alpha)^2 = 1$ and retaining only the leading order terms in N, one can rewrite Eq. (26.3.9) as

$$[Z^n]_{av} = e^{\frac{1}{4}(\beta J)^2 nN} \sum_{\{S_i^\alpha = \pm 1\}} \exp \left\{ \frac{(\beta J)^2}{2N} \sum_{\alpha < \beta} \left[\sum_i S_i^\alpha S_i^\beta \right]^2 + \frac{\beta J}{2N} \sum_\alpha [\sum_i S_i^\alpha]^2 \right.$$

$$\left. + \beta \sum_i H_i \sum_\alpha S_i^\alpha \right\} \tag{26.3.10}$$

Using the identity

$$e^{\lambda a^2/2} = \left(\frac{\lambda}{2\pi}\right)^{1/2} \int_{-\infty}^{\infty} dx \; e^{\frac{-\lambda x^2}{2} + a\lambda x} \tag{26.3.11}$$

to simplify the squared terms, we get

$$[Z^n]_{av} = e^{\frac{(\beta J)^2 nN}{4}} \int_{-\infty}^{\infty} \prod_{\alpha<\beta} \left(\frac{N}{2\pi}\right)^{1/2} \beta J \, dq_{\alpha\beta} \prod \left[\frac{N\beta J_0}{2\pi}\right] dm_\alpha$$

$$\times\ e^{\frac{-N(\beta J)^2}{2}\sum_{\alpha<\beta} q_{\alpha\beta}^2 - \frac{N\beta J_0}{2}\sum_{\alpha} m_\alpha^2 + N \ln \text{Tr}\, e^{L(q,m)}} \quad \text{... (26.3.12)}$$

where,

$$L(q, m) = (\beta J)^2 \sum_{\alpha<\beta} q_{\alpha\beta}\, S^\alpha S^\beta + \beta \sum_{\alpha} (J_0\, m_\alpha + H) S^\alpha \quad \text{... (26.3.13)}$$

and the trace is over the n spins S^α, and the field H_i has been taken to be uniform i.e. H. There are $n\ (n-1)/2$ variables for $q_{\alpha\beta}$ with $\alpha < \beta$. The diagonal terms $q_{\alpha\alpha}$ do not appear and the $q_{\alpha\beta}$ with $\alpha < \beta$ are defined by the symmetric matrix $q_{\alpha\beta} = q_{\beta\alpha}$.

If we now assume that the limit $N \to \infty$ can be taken before $n \to 0$, then the integral in Eq. (26.3.12) can be done by the method of steepest descent since the argument of the exponential is proportional to N. the free energy per spin is then found as

$$-\beta F = \underset{n\to 0}{\text{Lt}}\ \frac{\partial}{\partial n}\ [Z^n]_{av}$$

$$= \underset{n\to 0}{\text{Lt}} \left\{ \frac{(\beta J)^2}{4}\left[1 - \frac{1}{n}\sum_{\{\alpha,\beta\}} q_{\alpha\beta}^2\right] + \frac{\beta J_0}{2}\ \frac{1}{n}\sum m_\alpha^2 \right.$$

$$\left. + \frac{1}{n} \ln\ \text{Tr}\ e^L \right\} \quad \text{(26.3.14)}$$

where the $q_{\alpha\beta}$ and m_α are self-consistently given by the saddle point conditions

$$\frac{\partial f}{\partial q_{\alpha\beta}} = \frac{\partial f}{\partial m_\alpha} = 0 \quad \text{... (26.3.15)}$$

$$q_{\alpha\beta} = \langle S^\alpha S^\beta \rangle = \underset{n\to 0}{\text{Lt}}\ \frac{\text{Tr}\ S^\alpha\ S^\beta\ e^{L(q,m)}}{\text{Tr}\ e^{L\ (q,m)}} \quad \text{... (26.3.16)}$$

$$m_\alpha = \langle S^\alpha \rangle = \underset{n\to 0}{\text{Lt}}\ \frac{\text{Tr}\, S^\alpha\ e^{L(q,m)}}{\text{Tr}\, e^{L(g,m)}} \quad \text{... (26.3.17)}$$

From Eq. (26.3.14), Tr e^L must tend to unity as $n \to 0$ so that the fator n^{-1} can be cancelled and hence in Eqs. (26.3.16) and (26.3.17) we need not worry about the denominators. The problem now is to find the solutions of Eqs. (26.3.16) and (26.3.17).

The customary procedure is to take the replica symmetric solution $q_{\alpha\beta} = q$ and $m_\alpha = M$. One can then evaluate the trace by writing

$$q \sum_{\alpha, \beta} S^\alpha S^\beta = q\left[\left(\sum_\alpha S^\alpha\right)^2 - n\right] \quad \text{... (26.3.18)}$$

and using the identity given in (Eq. (26.3.11) once again, we find,

$$-\beta F = \frac{(\beta J)^2}{4}(1-q)^2 - \beta\frac{J_0}{2}M^2 + \frac{1}{\sqrt{2\pi}}\int_{-\infty}^{\infty} dz\, e^{-z^2/2}$$

$$= \ln\left[2 \coth \beta\, \overline{H}(z)\right] dz \quad \text{... (26.3.19)}$$

with q and M self-consistently given by

$$q = \frac{1}{\sqrt{2\pi}} \int_{-\infty}^{\infty} e^{-z^2/2} \tan h^2 \{\beta\, \overline{H}(z)\}\, dz \quad \text{... (26.3.20)}$$

$$M = \frac{1}{\sqrt{\pi}} \int_{-\infty}^{\infty} e^{-z^2/2} \tanh \{\beta \overline{H}(z)\}\, dz \quad \text{... (26.3.21)}$$

and $\overline{H}(z)$ related to q and M by

$$\overline{H}(z) = j_q^{1/2} z + J_0 M + H \quad \text{... (26.3.22)}$$

We note that the magnetization $[< S_i >_T]_{av}$ is the average $< S_i^\alpha > = M$ and further $[< S_i >_T^2]_{av}$ is the average $< S_i^\alpha S_i^\beta > = q$. Clearly, q is non-zero so long as $< S_i >_T$ is non-zero even if $< S_i >_T$ is random and has no preferred sign. Consequently, the spin glass phase is characterized by $M = 0,\ q \neq 0$.

To understand the solutions given by Eqs. (26.3.20) and (26.3.21), we set the external field $H = 0$ and look for the regions in T vs J_0 space, where $m = q = 0$ (paramagnet), $m \neq 0$, $q = 0$ (ferromagnet) and $M = 0$, $q \neq 0$ (spin

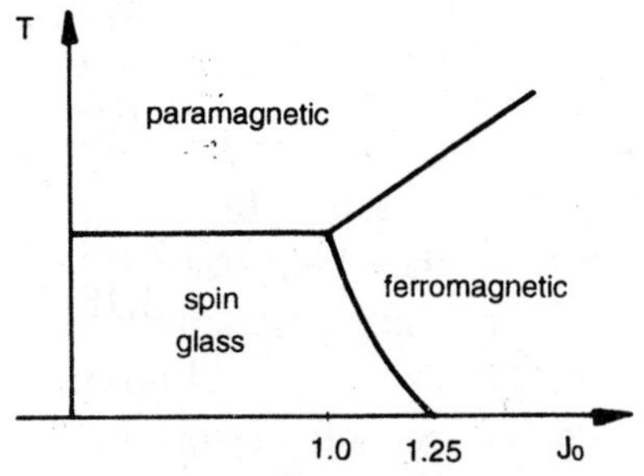

Fig. 26.1 Phase diagram for the Sherrington-Kirkpatric model

glass). The numerical results are shown in Fig. (26.1). The curious fact about the phase diagram is that in the range $1 < J_0/J \leq 1.25$ the system passes from an ordered to ferromagnet to a spin glass state as the temperature is lowered. Given that entropy has been taken into account, this is quite clearly absurd as at lower temperature, the more disordered spin glass state is being shown to be thermodynamically stable while at a higher temperature it is the ferromagnetic state. This is the first indication that our solution is not quite right. Indeed, if the entropy is evaluated, it turns out to be negative at low temperatures which is impossible. For further simplification, we set $j_0 = 0$ and now for small q, an analytic expansion in powers of q can be carried out for the free energy and one gets

$$\beta F = \beta F_0 + \frac{1}{4}\left(1 - \frac{T^2}{J^2}\right) Q^2 - \frac{Q^3}{3} + \frac{17}{24} Q^4 \qquad \text{... (26.3.23)}$$

where $Q = (\beta J)^2 q$

and $\beta F_0 = -\ln 2 - \beta^2 J^2/4$

The transition occurs when the coefficient of the quadratic term changes sign, i.e. at

$$T_f = J \qquad \text{... (26.3.24)}$$

But, it is clear from Eq. (26.3.23) that the for $T > T_f$, the state q = 0 has a maximum in the free energy, whereas we would have expected it to be a minimumn. Further for $T < T_f$ one finds that the state with $q = 0$ has a lower free energy than the one with $q \neq 0$. This is contrary to expectation.

The above paradoxical situation can be understood from the fact that the number of variables $q_{\alpha\beta}$ is $\dfrac{n(n-1)}{2}$ and hence when $n \to 0$, we are dealing with a negative number of variables. The saddle point calculation makes sense only if the chosen solution is a local minimum (i.e. it is locally stable). A criterion for local stability is to form the matrix of second derivatives $\frac{\partial^2 f}{\partial q_\alpha \, \partial q_\beta}$ at the saddle point and check for the eigenvalues. Existence of negative eigenvalues will imply instability. The solution with $q = 0$ happens to be stable for $T > T_f$, but is unstable for $T < T_f$ and hence must be rejected. In fact for $T < T_f$ the solution with $q \neq 0$ is also unstable and thus we have not been able to find the correct solution for $T < T_f$. One is looking for the stable solutions of Eqs. (26.3.16) and (26.3.17) with the lowest free energy. The correct solution has been obtained by Parisi. The interested reader should consult the original literature.

26.4 Electron Localization in Disordered Systems

In this last Section we discuss briefly a problem that has been of considerable interest in the last decade—the motion of electrons in a random medium. A random medium consists of static distribution of scattering or binding centres placed at random in a D-dimensional space. We will ignore the spin of the electron and all interactions among the electrons themselves. The last assumption will allow us to concentrate on the motion of a single electron in a random medium. The classical version of this model is the classical Lorentz model where one considers the motion of a classical particle through a random array of fixed scatterers (hard spheres in $D = 3$, hard discs in $D = 2$). If the scatterers are placed completely at random without regard to the position of the scatterers, we have the overlapping version of the model and if there is interaction among the scatterers and consequently a constraint on the positions of the scatterers, we have the non-overlapping model. Yet another variant is the wind tree model in $D = 2$, where the scatterers (trees) are placed randomly with their diagonals in the x and y directions. The wind is allowed to travel only in the x and y directions, changing directions if it collides with a tree.

In the classical models, the quantity of interest is the diffusion coefficient D, which is related to the velocity correlation function by the Green-Kubo formula

$$D = \int_0^\infty C(t)\, dt = \int_0^\infty \langle v_x(t)\; v_x(0) \rangle\, dt \qquad \text{... (26.4.1)}$$

where v_x is the x-component of the velocity and the average denoted by angular brackets is an average over all possible configurations of the scatterers. For the quantum problem it is more usual to talk about the conductivity σ which measures the amount of current produced in response to a small applied electric field by the relation $J_x = \sigma E_x$. The conductivity σ can be related to the diffusion coefficient D. For the classical Lorentz model, the most interesting result is that the correlation function decays slowly (power law) at large times and the behaviour is $C(t) \sim t^{-(1+D/2)}$.

In the quantum model, the quantum mechanical result is well known if the medium is perfectly ordered—a perfect crystal. The electron eigenstates are Bloch functions and Bloch's theorem implies that the conductivity is infinite for a perfect crystal. The eigenstates themselves are extended over the crystal instead of being localized about any particular point in the solid. The energy levels of the system form bands with characteristic gaps between bands. The fact that the conductivity of a solid is finite is due to the lattice vibrations which disturb the perfect crystal structure and provide the mechanism—phonon-electron scattering—for the finite conductivity. In this Section our concern is with static disorder and we consider circumstances where the atoms of the solid do not form a crystalline solid. The basic results on the electron motion in a disordered system were obtained by Anderson who considered a tight-binding model where atoms are placed at the regular sites of a lattice, but the atoms differ from each other in a random way. For example, the binding energy of an electron at the various sites may be a random variable with some distribution. In addition, an electron can hop from one site to another. The parameters of importance in the Anderson model are W the width of the binding energy distribution and V the measure of the matrix element of the part of the Hamiltonian responsible for hopping. Anderson found that, in D = 3, if W/V is large enough then the eigenstates for an electron are all localized. Localized states fall away exponentially from some central point. To have a non-zero value of the conductivity, the electron's wave packet must have some contribution from extended states since a wave packet constructed entirely from localized states cannot travel across the solid under the influence of a weak electric field. For small values of W/V, however, there are both extended states and localized states separated by a mobility edge. In $D = 1$, however, the smallest non-zero value of W/V causes all states to be localized and it is more than likely that the same situation holds in $D = 2$. Hence the transition from localized to conducting states can occur only in $D = 3$.

The calculational tricks that one employs for such problems are usually some special techniques that generally work in $D = 1$, the scaling theory which is presumably quite accurate for $D > 2$ but whose conclusion for $D = 2$ may just have been built into the theory and finally the evaluation of velocity correlation functions which lead to evaluation of the frequency-dependent diffusivity $D(\omega)$ and conductivity $\sigma(\omega)$. If there is complete localization, then the mean square displacement $< \Delta x(t)^2 >$ must remain bounded and hence $\sigma(\omega)$ must approach zero at least as fast as ω as $\omega \to 0$,

We begin with a disordered one-dimensional solid where it is universally agreed that all states are localized and that the conductivity vanishes. We will demonstrate first that the electrical conductance $G(L) = \sigma/L$, where σ is the electrical conductivity of a material of length L, is given by the formula

$$G(L) = e^2 \, T/2\pi h \, R \qquad \text{... (26.4.2)}$$

where T and R are the transmission and reflection coefficients of a spinless electron of energy E across the sample. In the above formulae e is the electronic charge and h is Planck constant. Instead of the electrical conductivity, we will consider the diffusion of an electron across the sample in response to a density gradient. We imagine a steady state maintained by having a wave incident from the left with probability P_L and from the right with probability P_R. If the transmission probability is T, then the probability of the wave incident on the left ending up at the right is TP_L and that for the one incident from the right ending up at the left is TP_R. The reflection probability R means there will be a fraction P_LR on the left and a fraction $P_R \cdot R$ on the right due to reflection. The total densities of particles on the left and right are then

$$n_L = P_L + T P_R + R P_L \qquad \text{... (26.4.3)}$$

$$N_R = P_R + T P_L + R P_R$$

If v is the velocity of the particles in free space then the current moving into the sample from the left is

$$j_L = v(P_L - T P_R - R P_L) \qquad \text{... (26.4.4)}$$

and that moving in from the right is

$$j_R = v(P_R - T P_L - R P_R) \qquad \text{... (26.4.5)}$$

Using $R + T = 1$, we find from Eq. (26.4.3)

$$n_L - n_R = (P_L - P_R)\ 2R \quad \text{... (26.4.6)}$$

and from Eqs. (26.4.4) and (26.4.5)

$$j_R = v\,T\,(P_R - P_L) = -j_L \quad \text{... (26.4.7)}$$

Thus the current from the left is

$$j_L = \frac{vT}{2R}\,(n_L - n_R) = -\frac{vT}{2R}\,L\,\frac{\partial n}{\partial x} \quad \text{... (26.4.8)}$$

and remembering that $j = -D\,\frac{\partial n}{\partial x}$, where D is the diffusion coefficient, we have

$$D = \frac{vT}{2R}\cdot L \quad \text{... (26.4.9)}$$

The connection between the electrical conductivity and the diffusivity is given by the relation

$$\sigma = e^2\,N(E)\,D(E) \quad \text{... (26.4.10)}$$

where $N(E)$ is the density of states per unit length. For a one-dimensional system

$$N(E) = \frac{1}{\pi h}\,\frac{dp}{dE} = \frac{1}{\pi h v} \quad \text{... (26.4.11)}$$

where P is the momentum of the particles at energy E and $v = dE/dp$ is the group velocity. Using Eq. (26.4.11) in Eq. (26.4.10), we get back Eq. (26.4.2).

We now present arguments about why σ should vanish for a random medium. The medium is a set of scatterers (each scatterer is a group of atomic scatterers) placed at random separations along a line. The phase of the electron's wavefunctions at any scatterer is the random variable. Consider two adjacent scatterers with transmission coefficients t_1 and t_2 and reflection coefficients r_1 and r_2. Using straightforward quantum mechanics, it is easy to show that the transmission coefficient for the two scatterers together is

$$t = \frac{t_1 t_2}{1 - r_1 r_2} \quad \text{... (26.4.12)}$$

The total conductance of the two scatterers is then

$$G = \frac{e^2}{2\pi\hbar} \frac{|t|^2}{1 - [t|^2} = \frac{e^2}{2\pi\hbar} \frac{|t_1 t_2|^2}{1 + |r_1 r_2|^2 - |t_1 t_2|^2 - (r_1 r_2 + r_1^* r_2^*)}$$

... (26.4.13)

To use this formula as a basis for a theory of localization, we consider a dimensionless resistance ρ defined as

$$\rho = e^2 (2\pi\hbar\ G)^{-1} = R/T \quad \text{... (26.4.14)}$$

The total resistance of the two scatterers in series is then given by Eq. (26.4.13) as

$$\rho = \frac{1 + |r_1|^2 |r_2|^2}{t_1|^2 t_2|^2} - 1 - \frac{r_1 r_2 + r_1^* r_2^*}{|t_1|^2 |t_2|^2} \quad \text{... (26.4.15)}$$

At this point , we need to average over the randomness—the random interval between the scatterers and the random interval between the atomic scatterers in each of the two scatterers considered. Since we assume that there is no correlation between the two scatterers, the last term in Eq. (26.4.15) averages to zero and we arrive at

$$\begin{aligned} <\rho> &= \frac{1 + |r_1|^2 |r_2|^2}{|t_1|^2 |t_2|^2} - 1 = \left(\frac{1}{|t_1|^2} - 1\right) + \left(\frac{1}{|t_2|^2} - 1\right) \\ &\quad + 2\left(\frac{1}{|t_2|^2} - 1\right)\left(\frac{1}{|t_2|^2} - 1\right) \\ &= <\rho_1> + <\rho_2> + 2 <\rho_1> <\rho_2> \end{aligned} \quad \text{... (26.4.16)}$$

The above equation is the composition law of two resistances in series and the important point is that it is not Ohm's law. We now use Eq. (26.4.16) by assuming that $<\rho_1>$ corresponds to a region of length L and $<\rho_2>$ corresponds to a macroscopically small length δL at some random distance from L. If we write $<\rho_2> = \alpha\delta L$, where α is some undetermined proportionality constant, then we can cast Eq. (26.4.16)) as

$$\delta\rho = <\rho> - <\rho_1> = <\rho_2> [1 + 2 <\rho_1>]$$

$$= \alpha \delta L \ [1 + 2 <\rho_2>]$$

$$= \alpha \delta L \ (1 + 2 <\rho>) \qquad \text{... (26.4.17)}$$

where, in the last step, we have dropped terms of $O\,(\delta L)^2$). Integration of Eq. (26.4.17) leads to

$$<\rho \ (L) > = (e^{2\alpha L} - 1)/2 \qquad \text{... (26.4.18)}$$

Thus we have an exponentially growing resistance or equivalently an exponentially decaying conductance. This exhibits the localization of all states in a disordered one dimensional solid.

The above manipulation which depended crucially on the one-dimensional nature of the problem cannot be repeated in higher dimensions. Instead we will explain a technique due to Abrahams et al. We follow a demonstration due to Thouless.

The solid is taken to be a D-dimensional cube of length L on a side. Thouless pointed out that the extended and localized states respond differently to perturbations at the boundaries of the cube. The localized states which are confined to the interior of the cube are insensitive to the perturbations while the extended states are sensitive to it. The main idea is that if two cubes with possibly different energy levels are put into contact, then if the perturbations produced by the new interface can shift the energy levels sufficiently, then electrons will be able to move freely from one cube to the other and diffusion will occur. On the other hand, if motion from cube to cube is unlikely, localization will occur. Thus, what we need to characterize is energy shift produced by a perturbation.

Let δE be the energy shift produced by altered boundary conditions. Using the uncertainty principle

$$\delta E \sim \hbar \ / t_D \qquad \text{... (26.4.19)}$$

where t_D is the time taken by the electron to diffuse to the boundary and is estimated as

$$t_D \sim L^2/D(L) \qquad \text{... (26.4.20)}$$

where D (L) is the diffusion coefficient. Thus

$$\delta E \sim \hbar \; D(L)/L^2 \qquad \text{... (26.4.21)}$$

and using the connection between conductivity and diffusivity, we have

$$\delta E \sim (\sigma(L)\hbar/e^2) \; (L^2 N(E))^{-1} \qquad \text{... (26.4.22)}$$

where N (E) is the density of states per unit volume. If ΔE is the average spacing between energy levels, then

$$N(E) \sim \frac{1}{\Delta E} \frac{1}{L^D} \qquad \text{... (26.4.23)}$$

and thus

$$\frac{\delta E}{\Delta E} \sim \frac{\sigma(L)\hbar \; L^{D-2}}{e^2} \qquad \text{... (26.4.24)}$$

If $\delta E/\Delta E \ll 1$, then boundary alterations have little effect on the electron and one would have localization. On the other hand, diffusion occurs for $\delta E/\Delta E \gg 1$. The dimensionless quantity on the right hand side of Eq. (26.4.24) is called the scale dependent conductance $g(L)$. Thus

$$g(L) = \frac{\sigma(L)\hbar \; L^{D-2}}{e^2} \qquad \text{... (26.4.25)}$$

This quantity determines what happens to the electrical conductivity when cubes of volume L^D are put together. We will now concentrate on the L-dependence of $g(L)$ for large L.

The central question is as follows. Suppose we know $g(L)$ for some large but finite value L_0 of L. What would happen if cubes of side $L_0(1+\varepsilon)$ were put together. The scaling assumption is that $g\,[L_0(1+\varepsilon)]$ is determined by $g(L_0)$ and the scale factor $(1+\varepsilon)$ only, i.e.

$$g\{L_0(1+\varepsilon)\} = f\{\,g(L_0),\; (1+\varepsilon)\} \qquad \text{... (26.4.26)}$$

If we expand the two sides of the above equation in powers of ε,

$$g(L_0) + \varepsilon L_0 \frac{\partial g}{\partial L_0} + \ldots = f\{g(L_0), 1\}$$

$$+ \varepsilon \left. \frac{\partial f\{g(L_0), (1 \mp \varepsilon)\}}{\partial \varepsilon} \right|_{\varepsilon=0} + \ldots \qquad \ldots (26.4.27)$$

Noting that $\left.\frac{\partial f}{\partial \varepsilon}\right|_{\varepsilon=0}$ is a function of $g(L_0)$ alone, we can write it as $g\beta(g)$ and obtain from the $O(\varepsilon)$ part of Eq. (26.4.27)

$$\frac{d \ln g}{d \ln L} = \beta(g) \qquad \ldots (26.4.28)$$

The sign of is vital. If $\beta(g)$ is positive, then for large L, g scales up to a non-zero value, while for negative $\beta(g)$, g will approach zero as $L \to \infty$.

For large g, the system would be conducting and hence $\sigma(L)$ should be non-zero. Returning to Eq. (26.4.25), we assume that $\sigma(L)$ is independent of L and hence for large g obtain

$$\beta(g) = D - 2 \qquad (g \text{ large}) \qquad \ldots (26.4.29)$$

For small g, on the other hand, the system will be localized and we can assume a form

$$g = g_0\, e^{-L/\xi} \qquad \ldots (26.4.30)$$

which leads to

$$\beta(g) = \ln \frac{g}{g_0}, \quad g \ll g_0 \qquad \ldots (26.4.31)$$

A further assumption of monotonic passage from one end [Eq. (26.4.31)] to the other [Eq. (26.4.29)], yields the curves shown in Fig. 26.2 for the β-function. For $D = 1, 2$, all states are localized. while for $D = 3$, there is a critical g_0 above which there is diffusive behaviour, but below which the system scales down to $g = 0$ and hence all states are localized.

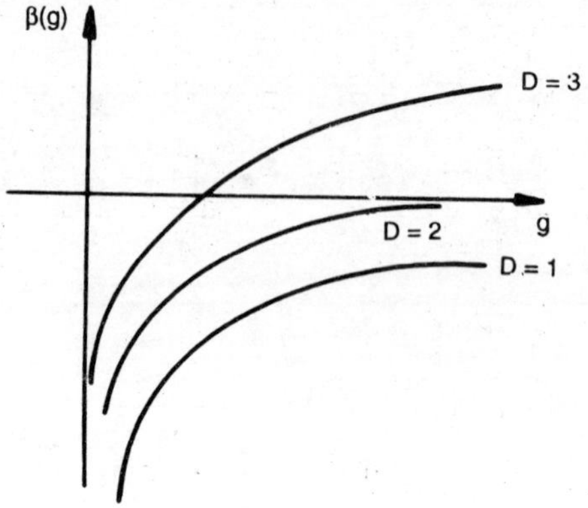

Fig. 26.2 β-Functions in different drmensions for the electron localization problem

It should be noted that the prediction of the above analysis that all states are localized in $D = 2$ is in a way built into the theory due to the assumption that $\sigma(L)$ is L-independent for large values of g and by the assumption of monotonic behaviour of $\beta(g)$.

Chapter 27

Miscellaneous Topics in the Form of Problems

27.1 Kac-ring Model

Consider a series of n points P_1, P_2 ... P_n arranged on a circle. Some of these points do and some do not have a marker. There are m such markers. At the n half way points between any two successive points P_i and P_{i+1} there are n balls, some black and some white. The balls move together in elementary steps. In one step, all balls move one unit counterclockwise along the ring. Those which sweep across a marker change colour, the others do not. If $\Delta(0)$ is the difference between the number of black and white balls at $t = 0$, one is interested in $\Delta(t)$ as $t \to \infty$. To get a definite answer, one would need to make an assumption. Assume that at time t if the number of black and white balls are $B(t)$ and $W(t)$ respectively and the corresponding number with a marker in front of them are $b(t)$ and $w(t)$, then

$$\frac{b(t)}{B(t)} = \frac{w(t)}{W(t)} = \frac{m}{n} = \mu$$

i.e. the colour of the ball is uncorrelated with the property of having a marker in front. With this assumption, show that

$$\Delta(t) = (1 - 2\mu)^t \, \Delta(0)$$

This answer clearly loses the Poincare cycle that exists in the system (the fact that after two complete revolutions, the original state must be reached). Further, while the above answer is satisfying in the statistical sense i.e. one the average there will be equal number of black and white balls as $t \rightarrow \infty$, the answer certainly violates the reversibility which existed in the prescribed dynamics of the system. For a better insight, a more detailed anslysis is worthwhile. Introduce a quantity ε_i at each site such that

$$\varepsilon_i = \begin{matrix} +1 \text{ if there is no marker at } P_i \\ -1 \text{ if there is a marker at } P_i. \end{matrix}$$

and a quantity $\eta_i(t)$ such that

$$\eta_i(t) = \begin{matrix} +1 \text{ if the ball at the position preceding } P_i \text{ is black at time } t \\ -1 \text{ if the ball at the position preceding } P_i \text{ is white at time } t \end{matrix}$$

Now show that

$$\Delta(t) = \sum_{m=1}^{n} \varepsilon_{m-1}\, \varepsilon_{m-2}\, \varepsilon_{m-3} \ldots \varepsilon_{m-t}\, \eta_{m-t}(0)$$

To make any further progress, one needs to know the distribution of the markers. Since the exact position of markers is not of interest, one can consider an ensemble of Kac rings each having n balls and each starting from the same configuration at $t = 0$. The quantity of interest is the ensemble average and hence instead of $\Delta(t)$, one considers $< \Delta(t) >$, where

$$< \Delta(t) > = \sum_{m=1}^{n} < \varepsilon_{m-1}\, \varepsilon_{m-2} \ldots \varepsilon_{m-t} > \eta_{m-t}\,(0)$$

If the distribution is random, then it does not matter in the correlation function, where one starts from and hence show

$$< \Delta(t) > = < \varepsilon_1\, \varepsilon_2 \ldots \varepsilon_t > \Delta(0)$$

Let μ be the probability of having a marker at a given point P_i, then probability of having s markers in t-steps is going to be binomial and one has

$$p\,(s, t) = \frac{t!}{s!(t-s)!}\, \mu^s\, (1-\mu)^{t-s}, \quad 0 \le t \le n$$

while

$$p\,(s, t) \;=\; \frac{(2n-t)!}{s!(2n-t-s)!}\;\mu^{s}\,(1-\mu)^{2n-t-s},\;\; n \le t \le 2n$$

Now, show that

$$<\Delta\,(t)> \;=\; (1-2\mu)^{n-|n-t|}\;\Delta(0)$$

Clearly the Poincare cycle which is of length $2n$ exists in the system. However, if n is very large $<\Delta\,(t)>$ is very nearly given by $<\Delta(t)> \sim (1-2\mu)^{t}\,\Delta(0)$ which is the result previously obtained, showing that for the infinitely large system size, Poincare recurrence time is extremely large and hence irrelevant.

27.2 Transport Coefficients from Boltzmann Equation

We will focus on the coefficient of shear viscosity. If there exists a mean flow in the x-direction with velocity $\bar{u}$, and if $\bar{u}$ happens to depend upon z, then there will be a resistive force between two successive layers trying to reduce the relative motion. This drag force per unit area is proportional to the gradient $d\bar{u}/dz$ and the constant of proportionality is the shear viscosity η. Turning to the Eoltzmann transport equation, we note that if the velocity $\bar{u}$ is a constant, then $f_0(v_x-\bar{u}, v_y, v_z)$ is the equilibrium distribution, where

$$f_0\,(v_x, v_y, v_z) \;=\; n\left(\frac{m}{2\pi kT}\right)^{3/2} e^{-\frac{m}{2kT}(v_x^2+v_y^2+v_z^2)}$$

If, $\bar{u}$ depends on z, then we will be picking up a correction δf to the distribution $f_0(v_x-\bar{u}, v_y, v_z)$ since f_0 does not satisfy the transport equation. Approximating the collision term as $-\,\delta f/\tau$ where τ is the characteristic time involving collisions, we can write

$$v_z\,\frac{\partial f_0}{\partial z} \;=\; -\,\frac{\delta f}{\tau}$$

or

$$\delta f = \tau v_z \frac{\partial f_0}{\delta V_x} \frac{\partial \bar{u}}{\partial z}$$

where $V_x = v_x - \bar{u}$

The $x - z$ component of the stress tensor is

$$T_{xz} = m \int dv^3 \, f \, V_x V_z,$$

where $f = f_0 + \delta f$. Averaging with f_0 gives zero and hence

$$T_{xz} = m \int d^3V \frac{\partial f_0}{\partial V_x} V_x \, V_z^2 \, \tau \frac{\partial \bar{u}}{\partial z}$$

where $V_y = v_y$ and $V_z = v_z$. The shear viscosity obtains from$T_{xz} = -\eta \dfrac{\partial \bar{u}}{\partial z}$.

Show that

$$\eta = -\frac{m^2}{15} \int dV^3 \frac{df_0}{d\varepsilon} \tau \, V^4,$$

where $f_0 = f_0(\varepsilon) = f_0\left(\frac{1}{2} mV^2\right)$,

which with the specific form of f_0 given above yields

$$\eta = nkT <\tau>.$$

where

$$<\tau> = \frac{16}{15\sqrt{\pi}} \int_0^{\infty} dse^{-s^2} s^6 \, \tau(\tilde{v}s),$$

$\tilde{v}$ being the most probable speed $(2kT/m)^{1/2}$. A constant value of τ, given by $1_{mfp}/\bar{v} = \dfrac{1}{\sqrt{2}\,n} \sigma_0 \bar{v}$, where σ_0 is the scattering cross section and $\bar{v}$ the root mean square velocity, leads to

$$\eta = \frac{1}{6} \frac{\sqrt{mkT}}{\sigma_0},$$

about 20% lower than the fairly precise value obtained on the basis of an accurate evaluation of the collision term. (See F. Reif, "Fundamentals of statistical and thermal physics")

27.3 Zero Sound in a Neutrino Gas

Neutrinos are massless fermions with a short ranged weak interaction mediated by the Z—boson. Find the velocity of the zero sound mode of dense neutrino gas.

27.4 Models of Kinetic Roughening

The free energy associated with the surface tension at an interface can be written as $F = \int d^D x\, (\vec{\nabla} h)^2$, where D is dimension of the surface and h is the height fluctuation on the surface. The dynamics of the height fluctuations can be described by the Langevin equation $\dot{h} = -\delta F/\delta h + \eta$, where the noise η has the correlation

$$< \eta\, (\mathbf{x}, t)\, \eta\, (\mathbf{x}', t') > \; = \; 2D_0\, \delta\, (\mathbf{x} - \mathbf{x}')\, \delta\, (t - t').$$

This is known as the Edwards Wilkinson (EW) model. Show by explicit solution of the above Langevin equation that the correlation function $<h\, (\mathbf{x}, t)\, h\, (\mathbf{x} + \mathbf{r}, t)>$ can be written as $r^{2\alpha} f\, (t/r^z)$. Find α and z.

For $\alpha > 0$, the interface is described as being rough.

If the surface tension vanishes, then the free energy is curvature—controlled and hence $F = \int d^D x\, (\nabla^2 h)^2$. The dynamics of the height fluctuations is once again governed by

$$\dot{h} = -\frac{\delta F}{\delta h} + \eta,$$

with the η-correlation as given above. Once again find α and z.

Repeat the above calculation of α and z if the noise is conserved i.e. the correlation is

$$< \eta(\mathbf{x}, t)\, \eta\, (\mathbf{x}', t')> \; = \; -\; 2D_0\, \nabla^2\, \delta\, (\mathbf{x} - \mathbf{x}')\, \delta\, (t - t')$$

27.5 Stochastic Resonance

Consider the Langevin equation

$$\frac{dx}{dt} = -\frac{\partial U}{\partial x} + a \cos \omega t + N,$$

where the potential $U(x) = -\frac{ax^2}{2} + \frac{bx^4}{4}$, with $a, b > 0$ and N is the usual white noise.

In the absence of the periodic drive, the noise induces transition from one stable minimum to the other of the bistable position, the time for the transition being the Kramer's time, Show that there is a resonance in this system i.e. large periodic jumps in the power spectrum when the oscillation time of the imposed force is half the Kramer's time for the potential. (R. Benzi et al.)

27.6 KPZ Equation

The first non linear term that needs to be added to the EW model of the problem 27.4 is $\lambda (\vec{\nabla} h)^2$. The resulting growth dynamics is

$$\frac{\partial h}{\partial t} = \Gamma \nabla^2 h + \lambda (\nabla h)^2 + \eta$$

This is known as the Kardar—Parisi—Zhang equation. Use the dynamic renormalization group technique (explained in Chapters 24 and 25) to find α and z as functions of the substrate dimension. Also study

$$\frac{\partial h}{\partial t} = -\nabla^2 (\Gamma \nabla^2 h + \lambda (\nabla h)^2) + \eta$$

and

$$\frac{\partial h}{\partial t} = -\nabla^2 [\Gamma \nabla^2 h + \lambda (\nabla h)^2) + \eta]$$

27.7 Roughening and Crumpling

Let the position of an interface (a crystal in equilibrium with its own vapour e.g.) be described by the single-valued function $h\,(x_1,\, x_2)$ which rules out the possibility of overhangs. Microscopically h is quantized in terms of the spacing between Bragg planes perpendicular to the h-axis. At high temperatures this discreteness is washed out by thermal fluctuation and the free energy of the interface can be described in terms of the surface tension σ. For an arbitrary parametrization $\mathbf{r}\,(y_1, y_2)$ of the surface, the free energy is

$$F \doteq \sigma \int \sqrt{g}\; dy_1\, dy_2,$$

where

$$g = \det g_{ij},\ g_{ij} \text{ being the metric tensor,}$$

$$g_{ij} = \frac{\partial \mathbf{r}}{\partial y_i} \cdot \frac{\partial \mathbf{r}}{\partial y_j}.$$

For the particular representation $\mathbf{r} = (x_1, x_2, h\,(x_1, x_2))$ show that

$$g_{ij} = \begin{pmatrix} 1 + \left(\dfrac{\partial h}{\partial x_1}\right)^2 & \dfrac{\partial h}{\partial x_1}\dfrac{\partial h}{\partial x_2} \\[2ex] \dfrac{\partial h}{\partial x_1}\dfrac{\partial h}{\partial x_2} & 1 + \left(\dfrac{\partial h}{\partial x_2}\right)^2 \end{pmatrix}$$

This is called the Monge representation. With this coordinate system, show that

$$F = \sigma \int d^2x\; \sqrt{1 + (\vec{\nabla} h)^2}$$

For small fluctuations,

$$F = \text{const} + \frac{\sigma}{2} \int d^2x (\vec{\nabla} h)^2$$

Show that

$$< [h(\mathbf{y}) - h(\mathbf{O})]^2 > \approx \frac{kT}{\pi\sigma} \ln (y/a)$$

as $y \to \infty$, where 'a' is a microscopic length. The large y behaviour is a signature of roughening.

Now consider a membrane with vanishing surface tension. The fluctuation of the surface will now be constrained by a bending rigidity K and the free energy can be written in terms of the curvature. With the curvature proportional to $\nabla^2 h$, we have the free energy $F = K \int (\nabla^2 h)^2 d^2x$.

Show that the normal-normal correlation, $<\mathbf{n}(\mathbf{r}).\mathbf{n}(\mathbf{O})>$ decays exponentially, showing that there is no long range correlation between orientations of different parts of the membrane. Such a membrane is called crumpled.

27.8 Dangerously Irrelevent Variables

In the mean field theory of critical phenomena, the specific heat exponent $\alpha = 0$ and hence the hyperscaling relation $D\nu = 2 - \alpha$ seems to be violated. To explore this situation further, imagine the existence of one irrelevent field u_1, so that Eq. (23.3.12) reads

$$F = t^{D/y_t} f\left(\frac{h}{t^{y_h/y_t}}, \frac{u_1}{t^{y_1/y_t}}\right)$$

with $y_1 < 0$. As $t \to 0$, we expect

$$F = t^{D/y_t} f\left(\frac{h}{t^{y_h/y_t}} \quad 0\right)$$

This implies the existence of the limit of $f(x, y)$ as $y \to 0$. However, this limit may not be well defined, i.e.

$$f(x, y) \sim y^{-\mu} f(x), \text{ with } \mu > 0 \text{ as } y \to 0.$$

This requires a modification of the hyperscaling relation. Now, the coupling constant λ of the four point coupling is irrelevent for $D > 4$ with $y_1/y_t = -(D-4)/2$. By considering the free energy for the model, show that $\mu = 1$ and hence show that the modified hyperscaling relation is satisfied. The irrelevent variable u_1 is called a dangerous irrelevent variable.

References

Thermodynamics

M.W. Zemansky, 'Heat and Thermodynamics' 1957 McGraw Hill, New York.

J. Wilks 'Third Law of Thermodynamics' 1961 Oxford University Press (London).

E. Fermi, 'Thermodynamics,' 1956 Dover (New York).

Statistical Mechanics

L.D. Landau and E.M. Lifshitz, 'Statistical Physics' 1980 Pergamon Press (London)

K. Huang, 'Statistical Mechanics' 1987 John Wiley (New York).

S.K. Ma, 'Statistical Mechanics' 1985 World Scientific (Singapore).

R.P. Feynman, 'Statistical Mechanics' 1972 Adision Wesley.

F. Reif 'Fundamental of Statistical and Thermal Physics' McGraw-Hill 1965.

G. Wannier 'Statistical Physics' 1966 John Wiley (New York).

Foundations and Chaotic Dynamics

Y. Sinai, Introduction to Ergodic Theory 1977 Princeton University Press (Princeton).

N.S. Krylov 'Works on the Foundations of Statistical Physics' 1979 Princeton University Press (Princeton).

E. Ott, 'Chaos in Dynamical Systems' 1992 Cambridge University Press (Cambridge).

Solid State Physics

C. Kittel 'Introduction to Solid State Physics' 1974 John Wiley (New York).

N. Ashcroft and N.D. Mermin, 'Solid State Physics' 1976, Holt, Reinhart and Winston (New York).

Superfluidity

I.M. Khalatnikov 'Introduction to the Theory of Superfluidity' 1965 W.A. Benjamin (New York).

F. London 'Superfluids' 1961, Dover (New York).

Superconductivity

P.G. DeGennes 'Superconductivity of Metals and Alloys' 1966, W.A. Benjamin (New York)

J.R. Schrieffer, 'Superconductivity' 1968, W.A. Benjamin (New York)

Quantum Hall Effect

R.E. Prange and S.M. Girvin, ed. 'Quantum Hall Effect' 1987 Springer-Verlag (New York).

Non-Equilibrium Statistical Mechanics

H. Risken, "Fokker Planck Equations" 1985 Springer Verlag, New York.
S.A. Rice, K.F. Freed and J.C. Light, ed. "Statistical Mechanics; New Concepts, New Problems, New Applications" 1972 University of Chicago Press, Chicago.
S.R. DeGroot and P. Mazur, "Non-Equilibrium Thermodynamics" 1984 Dover, New York.
P. Meijer ed. "Quantum Statistical Mechanics" 1962.
L. Van Hove, "Quantum-Mechanical Perturbations giving rise to a statistical transport theory" 1995 Physica **21** 517.
H.B. Callen and T.A. Welton, "Irreversibility and Generalized Noise" 1951 Phys. Rev. **83** 34.

Phase Transitions

H.E. Stanley, "Introduction to Phase Transitions and Critical Phenomena, 1971 Oxford University Press.
D.J. Amit, "Field Theory, Renormalization group and critical phenomena" 2nd edition, 1978 World Scientific (Singapore).
S.K. Ma, "Modern Theory of Critical Phenomena" 1975 Benjamin.
J. Kogut, "Introduction to lattice gauge theory and spin systems" 1979 Rev. Mod. Phys. **51** 659.
N. Goldenfeld "Lectures on phase transitions and the renormalization group" 1992 Addison Wesley.
P. DeGennes, "Scaling concepts in polymer physics" 1979 Cornell University Press.
P. DeGennes, "The physics of liquid crystals", 1975 Oxford University Press.
D.R. Nelson, T. Piran and S. Weinberg, "Statistical Mechanics of membranes and surfaces" 1989 World Scientific, Singapore.
D. Forster. D.R. Nelson and M.J. Stephen, 1976 Phys. Rev. **A16** 732.

Dynamic Critical Phenomena

P.C. Hohenberg and B.I. Halperin, "Dynamic Critical Phenomena" 1977 Rev. Mod. Phys. **49** 435.

J.D. Gunton and M. Droz, "Introduction to the theory of metastable and unstable states" 1983 Lecture Notes in Physics Vol. 183 Springer-Verlag.

A.J. Bray, "Exact renormalization group results for domain-growth scaling in spinudal decomposition" 1989 Phys. Rev. Lett. **62** 2841.

Turbulence

E.M. Tennekes and J.M. Lumley, "Introduction to turbulence" 1971 Cornell Univ. Press.

A.S. Monin and A.M. Yaglom, "Statistical Fluid Mechanics" Vol. I and II, 1975 MII Press.

W. McComb, "The Physics of Fluid Turbulence" 1990 Oxford University Press.

D.C. Leslie "Developments in the Theory of Turbulence" 1973 Oxford University Press.

Disordered Systems

D. Chowdhury, "Spin Glasses and Other Frustrated Systems" 1986 World Scientific (Singapore).

Index